Distilled Spirits Sensory Science: A Systematic Approach to Flavor Evaluation and Analysis

By: George F Manska

Chief of Research & Development

Arsilica, Inc., Dedicated to Spirits Sensory Research

ISBN and Publisher Information

Distilled Spirits Sensory Science: A Systematic Approach to Flavor Evaluation and Analysis
© 2025 by George F Manska
All rights reserved.

For permissions or inquiries, contact:

George F Manska, Sensorium Publishing Co., LLC

Las Vegas, Nevada, USA

Email: george.manska@sensoriumllc.com

Website: www.sensoriumllc.com

Published by Sensorium Publishing Co., LLC

Las Vegas, Nevada, USA

Disclosure: The author contributed to the design of a sensory-optimized spirits vessel discussed herein. Discussion and analysis are presented solely for scientific and educational purposes and carry no commercial intent or endorsement.

First Edition — 2025

Paperback ISBN: 979-8-9935658-0-4

Library of Congress Control Number 2026902672

eBook ISBN: 979-8-9935658-1-1

Selected editorial and linguistic tasks were assisted by **ChatGPT-5 (OpenAI, 2025)**, a large-language model used under human guidance for structuring complex research material and serving as a digital aid to improve organization and clarity, while all analytical reasoning, factual accuracy, and interpretive judgment remain exclusively those of the author.

10 9 8 7 6 5 4 3 2 1

Body text set in Minion Pro, tables in Myriad Pro. Composed in Microsoft Word 365
Printed on demand

Dedications

My spouse, soulmate, and best friend, Lynda Grace Schultz, my personal guide and love of my life, has always provided a pragmatic and insightful perspective, personal encouragement, and patience. Her sensory abilities have been an essential contribution to our research, product development, and the content of this book.

My business partner, close friend, and confidant, Christine Renea Crnek, whose sensory abilities, organizational talents, and personal insights guided product design and development of our patents, and provided positive impetus in the achievement of our sensory research objectives

Acknowledgements

To the distillers, artisans, and craftspersons who transform raw materials into sensory art; to the researchers and scientists who seek to understand, measure, and quantify that art; and to every mind working to define and perfect the emerging discipline of spirits sensory science. Your persistence, curiosity, and refusal to accept easy, superficial answers are shaping a new era of truth in flavor and understanding. This book stands on the shoulders of those who build, measure, and question—those who believe that mastery begins with evidence, not tradition. Special acknowledgement, Dr. Spencer Steinberg (ret.) Professor of chemistry at the University of Nevada, Las Vegas, whose GC-MS studies provided the much-needed validation to continue with our first sensory project.

Distilled Spirits Sensory Science:
A Systematic Approach to Flavor Evaluation and Analysis
Table of Contents

List of Figures, Case Studies, Evaluation Notes

Section V: Presenting the Flavors—The Role of the Evaluation Vessel

Appendices

Case Studies

Evaluation Notes

Preface

As Maître and a member of the Commanderie de Bordeaux (CdB), Las Vegas Chapitre, I spent more than fifteen years among some of the world's most respected wine professionals. A wine aficionado and collector of fine wines for over six decades, I learned through extensive worldwide travel and formal tastings with the International CdB Board of Governors members. I discovered that even seasoned palates often lacked a consistent, objective, and scientific framework for sensory evaluation. Traditional methods were insufficient for accurately assessing flavor, aroma, and quality. We needed a deeper description foundation—one grounded in science and fact.

My engineering education (BSME, North Carolina State University, 1967) and thirteen years of experience in automotive engine design, product development, luxury car product planning, and project management instilled in me a disciplined approach to organizing my endeavors using scientific, technical, and analytical methods. That background gave me the tools to look beyond opinion and description toward measurable cause-and-effect relationships and consumer perceptions. This approach is readily applied to sensory evaluation and analysis.

As a decades-long collector and aficionado of fine wines and spirits, I have long been interested in the magic that gives rise to the flavors of well-crafted alcoholic beverages. A pivotal moment occurred in March 2000 at a Commanderie de Bordeaux Parliament in Las Vegas, where I met the renowned glass artist Dale Chihuly, who encouraged me to enroll in a glass-blowing class. In one session, an unintended wide rim flare produced an oddly shaped tumbler. I set it aside with other projects, unaware that it would one day redirect my professional life.

Weeks later, while sampling a newly purchased bottle of cask-strength Scotch, I bypassed my usual tulip-shaped glass and reached for the flared tumbler. The result was startling. The harsh ethanol pungency was gone, and hidden aromas became clear and distinct. Retesting in the traditional tulip glass confirmed that glass shape alone had altered my perception, subjectively proving that concentrating ethanol distorts accurate evaluation. Graham's Law of gaseous diffusion through an orifice, as well as other gas behavior physics, factually explained what I had observed—the whisky's true character emerged through the dispersion of ethanol.

"The real voyage of discovery consists not in seeking new landscapes, but in having new eyes."— Marcel Proust

Ten years of focused research followed. We tested more than 50 glass shapes, manufacturing them to precise specifications using wooden lathe-turned cores to build slip molds, producing uniform porcelain clay prototypes, and firing and glazing several pieces to support a full-panel evaluation. These were followed by a thorough investigation of the physics and sensory dynamics associated with each. The result was a sensory-designed vessel, co-invented and patented with my business partner, Christine Crnek. Quickly adopted by many spirits competitions worldwide as their official judging glass, our company, Arsilica, Inc., became devoted to sensory research and the design of tools to improve the evaluation of alcoholic beverages.

My practical experience of over 15 years as a professional consultant to spirits competitions, combined with decades of study in judging methods, attitudes, and olfactory fatigue, culminated in a peer-reviewed publication: "Applying Physics and Sensory Sciences to Spirits Nosing Vessel Design to Improve Evaluation Diagnostics and Drinking Enjoyment" (*MDPI Beverages*, 2018).

This book is an in-depth continuation of that research. It challenges long-held misconceptions and offers a clear, evidence-based framework for analyzing spirits and liqueurs. It is not a guide for casual drinkers or a celebration of brands, but rather a scientific text for students, judges, distillers, and professionals who seek measurability, precision, reproducibility, and a deeper understanding of the sources of sensory characteristics.

The methods presented herein are the product of well over twenty years of research in physics, chemistry, and sensory psychophysics of spirits—and of the monumental contributions of researchers who built today's scientific foundation for sensory evaluation. That framework moves far beyond the amateur alchemy of popular critics and bloggers.

The goals of this text are threefold: (1) to provide readers with the tools for understanding what they truly smell and taste in spirits, (2) to build sensory literacy that replaces folklore and myth with verifiable science, and (3) to connect students and professionals through a bridge of practical application between the creative art and formal science of spirits.

This book is a pathway to precision evaluation for students, judges, educators, collectors, and serious enthusiasts who wish to move beyond tradition and emotional marketing toward measurable truth. It is a hands-on bridge from passionate interest to professionalism in distilled spirits sensory science.

— George F Manska

Introduction, Overview, and Mission

State of the Art

Within the spirits industry, scientific principles are most often secondary to marketing and profit. Sensory education receives minimal attention, and evaluation practices rely more on inherited ritual than on verifiable method. As a result, misplaced priorities reward tradition and myth over fact and scientific discipline.

On the consumer side, tasting is frequently compromised by poor technique, reliance on ineffective aroma vessels, and misplaced value on ethanol's pharmacodynamic "burn." Pungency serves as the head-nodding validation of quality among casual drinkers, while ethanol simultaneously masks faults, suppresses nuance, and dominates flavor perception.

Meanwhile, much of the scientific literature remains inaccessible—locked behind paywalls or written in a language that few outside academia and scientific research can understand. Even where research is sound, its message and conclusions rarely reach product development, judging standards, or consumer education, allowing commercial interests to distort science for profit.

What This Book Changes

This book bridges those gaps by presenting a simple, evidence-based methodology for spirits evaluation grounded in physics, chemistry, and sensory science. It challenges unsupported traditions and replaces them with testable, repeatable practices that improve aroma and taste perception. The goal is to clarify perception—to help serious enthusiasts, judges, and professionals understand, interpret, and communicate their experiences accurately.

Who the Book Is For

- Foremost, this book is a text for sensory science students, professors, and educators seeking structured methodology and a standardized teaching framework
- Spirits judges, evaluators, enthusiasts who value disciplined analysis over opinion or branding
- Distillers, blenders, and producers who link sensory outcomes to process control and integrity
- Collectors, connoisseurs pursuing evidence-based evaluation rather than anecdotal narrative
- Industry executives, ambassadors who wish to align marketing objectives with scientific facts

What the Reader Will Learn

Readers will learn to detect and describe aroma and flavor using a scientifically valid process. They will explore fault-detection techniques, lexicons anchored to chemical families, and reproducible approaches to sensory evaluation, including olfaction (nosing) and gustation (tasting). Most importantly, they will learn to distinguish between persuasive marketing rituals and methodical sensory observation.

Mission

The mission is to elevate both consumer and industry understanding of spirits evaluation by integrating principles of physics, chemistry, and sensory science into a practical, evidence-based framework and to replace myths and misplaced traditions with reproducible methods that empower serious enthusiasts and professionals to evaluate with clarity, accuracy, and intellectual integrity.

How to Use This Book

The book supports both linear reading and reference use. Each chapter presents core concepts, supported by diagrams or tables. While the sequence builds progressively, each chapter also stands alone for independent study. References provide technical citations; an index aids topical navigation. This text serves as both a guide and a training tool, applying, testing, and adapting each concept to the reader's own sensory objectives, whether professional or personal.

Tone and Intellectual Contract

This work does not appeal to sentiment, branding loyalty, or nostalgia. It avoids marketing language, untestable claims, and emotional positioning. Its content is empirical; its voice direct. It seeks to teach and inform, not to impress. Terms are defined. Claims are supported by measurable or published evidence to provide a precise, verifiable path to understanding spirits.

Evaluation Notes, Observations, and Case Studies

Throughout the text, selected notes, case studies, and observations illustrate how scientific principles apply directly to sensory evaluation. These items are consecutively numbered and serve as practical tools for sharpening sensory skills and critical thinking.

Reference Appendices

Specific chapters include technical appendices that cover advanced material relevant to in-depth study or literature review. While a detailed discussion of these topics lies beyond the book's primary scope, familiarity will enhance comprehension and expand expertise in sensory evaluation.

Section I:

Flavor Detection—The Anatomy of Sensory Science

The Spirits' Sensory Journey in Perspective

Across all alcoholic beverages, more than 99% of the liquid is water and ethanol. The compounds that define character—the congeners—make up much less than one-half percent. Nevertheless, they are responsible for nearly the entire spectrum of sensory perception: aroma, flavor, and texture. This course of study concerns that tiny fraction as it applies to distilled spirits. Within this chemical band, sensory identity is created, styles diverge, and evaluation acquires meaning. By examining spirits through this lens, we focus on the subtle compounds that shape complexity, determine quality, and provide sensory enjoyment.

Alcohol Beverage Composition by % Volume			
Beverage	Ethanol	Water	Congeners
Spirits	38-40	59-61	0.1-0.5
Wine	11-13	87-89	0.2-0.4
Beer	4-5	95	0.1-0.3
Alcohol Beverage Composition by % Weight			
Spirits	32-33	67-68	0.1-0.5
Wine	10-11	89-90	0.2-0.4
Beer	4	96	0.1-0.3

Understanding Spirits Through a Sensory Lens

Distilled spirits occupy a unique position in sensory science because the very compound that carries aroma—ethanol—is also the primary source of sensory interference. Fermentation creates it, distillation concentrates it, and maturation reshapes it, making ethanol both the vehicle of flavor and the greatest obstacle to perceiving it. Unlike food, beer, or wine, where water and low alcohol levels allow flavors to emerge easily, spirits push the senses to their limits. Meaningful evaluation requires learning to distinguish ethanol's physical and sensorial effects from those of the congeners it contains. Mastery in spirits analysis begins with recognizing this duality: ethanol opens the door to aroma, but it can just as quickly drown the message—if the evaluator is not prepared to consider its silent but significant effects.

Chapter 1: What Is Sensory Science?

Sensory science in spirits is the systematic study of how humans perceive and interpret the chemical complexity of distilled alcohol beverages. It connects measurable physical stimuli—volatile compounds, textures, and temperatures—to the brain's subjective experience of aroma, taste, and mouthfeel. In this context, sensory analysis bridges perception and chemistry, revealing how distilled spirits communicate their identity through the senses. The principles of sensory science form the foundation for precise and methodical evaluation.

Understanding how perception becomes sensory knowledge begins with recognizing both the precision and the inherent variability in human data processing and response. Each individual builds their own sensory library to quantify perceptions, store sensory experiences, and act as a structured model for future comparisons and orderly classification. Fig. 1.1 summarizes the diverse major factors that both consciously and subconsciously contribute to the building and expansion of one's personal sensory library.

Sensory science is not an exact discipline

Sensory science examines how humans perceive stimuli—particularly smell and taste—using methods from physics, chemistry, and statistics. Unlike purely quantitative fields, sensory science involves subjective human responses and relies on trained observers applying relative scales anchored to consistent reference standards. Dependable results demand methodical panel training, practice, and calibration. This approach does not eliminate subjectivity but manages it, creating structured links between measurable stimuli and internal perception (Lawless & Heymann, 2010).

Macro-influences on sensory personalization

Sensory preferences are shaped early on and continue to evolve. Genetics and fetal exposure to maternal diets form baseline sensitivities. Childhood experiences—such as regional ingredients, cultural cooking practices, and family traditions—further define familiarity, acceptability, and preferences. Religious rituals, social habits, and early emotional memories establish long-lasting predispositions (Prescott, 2015). These foundational biases evolve and are refined throughout life, influencing how individuals perceive and experience flavors, often with limited conscious control.

Individual differences in perception

No two people experience taste and aroma in the same way. Variability arises from differences in sensory receptors, neural processing pathways, and cognitive associations. Expectations, mood, and linguistic precision also influence perception. Even knowing a product is "premium" or "budget" can skew evaluation. Without rigorous calibration and controlled vocabulary, individual variability can derail analysis. Sensory science acknowledges these differences and seeks methodologies to overcome them (Delwiche, 2012).

Fig. 1.1 Major factors that influence the sensory library

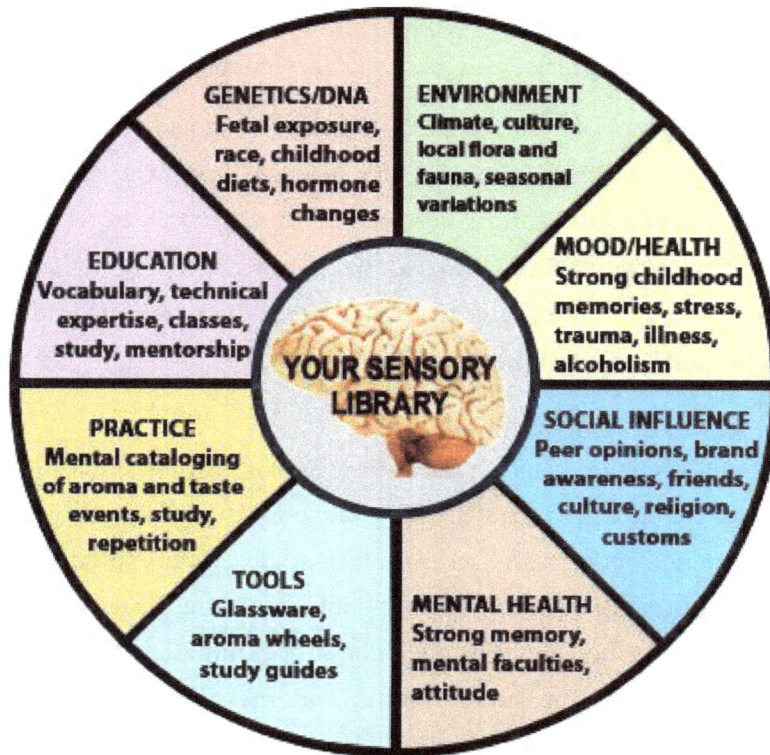

Tools for reducing subjectivity

Scientific methods enhance the reliability of sensory evaluation. Aroma wheels and lexicons attempt to standardize descriptors using principal component analysis (PCA), which reduces complex sensory datasets into principal components, enabling visualization of relationships between samples and descriptors. Reference compounds create reproducible aroma anchors.

Neuroimaging reveals which brain regions respond to smells. Psychometric models help visualize patterns in sensory data. Cross-cultural studies identify overlapping descriptors across cultures, while also documenting culturally specific differences (Meilgaard et al., 2016). Research into the psychology of flavor perception reveals how memory and context influence recognition and preference, thereby complicating interpretation (Shepherd, 2015).

Aroma wheels are often presented as tools to guide sensory evaluation, but their circular design and predefined categories can mislead rather than instruct. They imply symmetry and equivalence among aroma groups but can misrepresent perceptual relationships and tend to promote recognition by suggestion rather than by accurate detection. This can foster bias, leading evaluators to focus on listed terms rather than accurately describing what is genuinely perceived. In addition, wheels rarely convey intensity, temporal development, or the complex interactions among volatiles that shape the actual sensory experience. For cautionary reasons, aroma wheels are better viewed as teaching aids than as critical analytical instruments. The most reliable path to consistent and accurate aroma description lies in developing and using a structured sensory lexicon.

Developing sensory skills

Sensory proficiency is developed rather than innate. Improvement requires learning standardized sensory terms, concepts, and methods; conducting structured, blind tastings; calibrating against reference standards; and recording observations methodically. Such training dismantles entrenched myths and marketing-driven misconceptions, redirecting decision-making toward genuine personal sensory preferences. Like any disciplined practice, whether learning a musical instrument or driving, a combination of study, feedback, and repetition yields lasting improvement (Lawless & Heymann, 2010).

Summary

Sensory science is a discipline that connects human perception with measurable stimuli through structured methods and techniques. It emphasizes that while subjectivity cannot be eliminated, it can be managed through training, calibration, and the use of standardized tools and procedures. Genetics, culture, and experience shape individual sensory preferences; no two people perceive flavor in exactly the same way, and scientific methods help reduce variability. Ultimately, sensory proficiency is a learned skill that requires systematic practice, anchoring, and feedback, forming the foundation for reliable and reproducible evaluation of spirits.

Beyond: The strength of sensory science emerges from how each sense contributes distinct perceptual information, integrated and weighted by the brain to create the taster's total impression. Weighting the importance of each sense in the overall perception is key to developing a consistent approach to sensory outcomes. Weighting enables discourse between individuals at a pre-agreed, somewhat uniform, basic level of importance.

"Every aroma is a construct of the brain, not the nose; perception is memory in motion."
— Gordon M. Shepherd, Yale School of Medicine

"The senses do not operate in isolation; awareness emerges from their interaction."
— Charles Spence, Crossmodal Research Laboratory, University of Oxford

"We are not chasing flavors; we are chasing consistency. Flavor is the outcome of discipline in process and perception."— Dr. Bill Lumsden, Director of Whisky Creation, Glenmorangie

"To taste with understanding is to train the mind to connect chemistry with sensation."
— Harold McGee, Author of On Food and Cooking (2004)

References

Delwiche, J. (2012). You eat with your eyes first. *Physiology & Behavior, 107*(4), 502–504. https://doi.org/10.1016/j.physbeh.2012.07.011

Lawless, H. T., & Heymann, H. (2010). *Sensory evaluation of food: Principles and practices* (2nd ed.). Springer. https://doi.org/10.1007/978-1-4419-6488-5

Meilgaard, M. C., Civille, G. V., & Carr, B. T. (2015). *Sensory evaluation techniques* (5th ed.). CRC Press. https://doi.org/10.1201/b19493

Prescott, J. (2015). Multisensory processes in flavour perception and their influence on food choice. *Current Opinion in Food Science, 3*, 47–52. https://doi.org/10.1016/j.cofs.2015.02.007

Shepherd, G. M. (2015). Neuroenology: How the brain creates the taste of wine. *Flavour, 4*, Article 19. https://doi.org/10.1186/s13411-014-0030-9

Observations on Sensory Science

A Mirror of Perception: Sensory science reminds us that perception is not a passive reception, but an active construction. What we sense is filtered through memory, language, and expectation; to understand flavor is to study how the mind assembles reality from fragments of chemistry.

The Discipline of Awareness: The practice of sensory analysis cultivates awareness—the deliberate act of noticing. It teaches that refinement of the senses is refinement of thought, and that careful observation, whether of a spirit or of life, depends on disciplined attention.

The Integrity of Method: Scientific rigor in sensory work depends on honesty in observation and consistency in method. A single unexamined assumption can distort a conclusion; therefore, discipline lies not only in seeing but in questioning how we see. Reliability begins with methodical practice and ends with reproducible truth.

Chapter 2: The Flavor Equation – Sensory Weighting

Flavor is often confused with taste. This misunderstanding is so widespread that even experienced spirits drinkers often describe flavor notes that were never tasted, only smelled—a classic example of "taste–smell confusions" described by Rozin (1982). Scientifically, flavor is distinct from taste. Flavor is a perceptual construction that the brain builds by integrating signals from multiple sensory channels, the most important of which is smell; the others are taste and mouthfeel.

Smell contributes the most significant portion of the information the brain uses to form what we call "flavor." While taste is limited to a few broad categories—sweet, salty, sour, bitter, and umami—smell includes thousands of distinct aromatic compounds, and with that, thousands of possible perceptual combinations. In spirits, where non-volatile flavor components are minimal, and ethanol impairs taste, tactile, and chemosensory stimuli, the brain relies more on olfactory input. In distilled spirits, smell stimuli may account for approximately 90% of flavor perception (Spence, 2015).

This concept—the relative contribution of each sensory system to flavor—is the Flavor Equation. It is not a literal mathematical formula but a conceptual model that emphasizes sensory weighting in flavor perception and serves as a valuable tool for spirits evaluators. For distilled spirits, the general approximation is:

$$\text{Flavor} \approx 90\% \text{ aroma} + 5\% \text{ taste} + 5\% \text{ mouthfeel}$$

This ratio is not exact, nor can it be applied universally to all foods or beverages. However, in the context of spirits, it is both functionally accurate and necessary to maintain an evaluative perspective. Charles Spence, a leading researcher in multisensory perception, has stated that anywhere between 75% and 95% of what we experience as flavor originates in the sense of smell (Spence, 2017; Shepherd, 2012). The remaining input comes from the tongue (taste), the trigeminal nerve, and tactile receptors in the oral cavity (chemosensory mouthfeel). (See Fig. 2.1)

The flavor equation also explains why spirits drinkers often respond so strongly to nose feel—such as pungency, burn, or numbing—as if those sensations were part of the flavor. While these effects are real, they are not flavor in the true sense. They are chemosensory reactions, mediated mainly by the trigeminal nerve, that modulate perception rather than define it (Green, 1996).

Ethanol's trigeminal masking effects are central to understanding this modulation. Ethanol vapors stimulate trigeminal nociceptors in the nasal epithelium, producing sensations of burn or irritation that can dominate the sensory field. This stimulation activates a defensive neural pathway that prioritizes nociceptive input over olfactory signals, thereby "masking" or attenuating the perceived intensity and complexity of aromas (Cometto-Muñiz et al., 2005; Doty & Cometto-Muñiz, 2003). Classified as an anesthetic, ethanol reduces receptor sensitivity and narrows the range of detectable volatiles. Ethanol's pungency is a trigeminal irritant that completely distracts the evaluator from the balanced relationship in the flavor equation. Practical methods for controlling ethanol delivery and maintaining balance will be key to accurate evaluation.

Fig. 2.1   The Flavor Equation

Flavor = 90% Aroma + 5% Taste + 5% Mouth-Feel

- 5,000+ aromas, only 5 tastes, 25 mouth-feel sensation groups
- We don't taste raspberries - we smell raspberries, taste sweet, and mouth-feel fuzzy, juicy, tiny, popping raspberry flavor beads
- The olfactory, palate, and mouth-feel sensors send a single signal packet to brain triggering the Aha! Moment, **"It's raspberries!"**

Aroma and mouthfeel are often mistaken for taste. Only 5 tastes

The implications for spirits evaluation are profound. Olfaction is responsible for the overwhelming majority of what we perceive as flavor. Any weakness in olfactory perception—whether due to poor technique, impaired receptors, olfactory congestion, inappropriate aroma delivery vessels, or excessive ethanol exposure—can significantly reduce the accuracy of sensory judgment. Worse, it can lead to confident but incorrect conclusions, as the brain fills in missing data with assumptions or past learned expectations.

For this reason, olfactory training must be prioritized first in any structured evaluation framework to be effective. Developing accurate aroma recognition skills, the ability to differentiate among aromas, and effective use of learned vocabulary are essential for achieving reliable, reproducible results. Tasting technique alone is insufficient. Without proper control of ethanol concentration and optimal aroma delivery, the olfactory system becomes saturated, irritated, or numbed, resulting in degraded signals.

It is not enough to taste a spirit and describe what comes to mind. The process must begin with conscious control of aroma inputs, especially those that could compromise accurate evaluation, and end with structured interpretation based on unencumbered olfactory stimuli. Taste and mouthfeel play supporting roles, but the nose carries the weight.

Psychophysics in Spirits Sensory: Thresholds, JNDs, and Scaling

Every claim about "what is there" and "how much" is constrained by the way sensation maps onto numbers. Psychophysics, the study of the human reaction to stimuli, describes that mapping. It explains why two evaluators can be honest and careful, yet report different values, and why minor changes can be noticeable in one context and invisible in another.

Weber's law is a fundamental principle in psychology that quantifies the perceived change in a stimulus. It states that the smallest detectable difference in stimulus intensity is always a constant proportion of the original intensity. This means that as the intensity of a stimulus increases, the amount of change needed for a person to notice a difference also increases. However, at a consistent ratio (Fig. 2.2), to make the change in intensity equal 1 + 1, 12 candles would have to be added to 12 candles.

- **Detection threshold, DT,** is the lowest level at which an attribute becomes *just perceptible* against the background. Below this level, the attribute behaves as noise.
- **Recognition threshold, RT**, is the level at which the attribute is not only noticed but identified (e.g., "lemon oil," not merely "something"). Recognition usually sits above detection because it requires a pattern match to memory and language.
- **Difference threshold, JND** (the just-noticeable difference), is the smallest change that can be reliably discriminated from a reference. It answers a different question: "Is Sample B different from Sample A?" rather than "Is lemon present?"

In mixtures, thresholds are not fixed properties of a compound; they are context-dependent. Ethanol elevates many detection thresholds (trigeminal load and volatility effects). Some notes suppress others (mixture suppression), whereas others are enhanced by contrast or congruence—such as temperature, dilution, and time after pouring—thereby shifting thresholds again. Treat published thresholds as orders of magnitude, not absolutes, unless conditions match them closely.

Weber's law: Why background matters.

For many perceptual dimensions, the JND is proportional to the baseline intensity: the higher the background intensity, the larger the increment required to detect a change. Practically: a 2-point increase on a 0–10 intensity scale may be evident at "2 → 4" but barely visible at "7 → 9." This is one reason late-stage barrel effects are more complex to confirm than early extraction phases—the background is already high.

Fig. 2.2

Add one candle to one, change
in brightness is noticeable

Weber's Law

Add one candle to twelve change in
brightness is barely noticeable

Weber's Law: The smallest perceptible difference in intensity (ΔI)between two stimuli scales with the magnitude of the baseline intensity (I). The ratio $\Delta I/I$ is constant for a given sense modality. Also known as the Weber-Fechner Law

Examples:

- The weight of a single coin placed in the hand is easily detected. When the same coin is added to a handful of coins, the change is nearly imperceptible.
- A quiet violin note introduced into silence is immediately distinct. The same, introduced during a loud concert, vanishes into the background.
- The glow of a streetlamp is obvious at dusk. The same lamp appears invisible in midday.
- A trace of smoky phenols is easily detected in a light grain spirit. The same trace amount is imperceptible if added to a heavily peated Scotch.

From a slightly different perspective, the Weber-Fechner Law suggests that our perception of a stimulus grows more slowly than its physical intensity.

Psychometric Function – Floor and Ceiling Effects

When stimulus intensity is plotted against the probability of detection or perceived strength, the result follows a psychometric function. This S-shaped curve describes how sensation grows with increasing stimulus concentration. Near the lower end, responses cluster at the floor, where stimuli are too weak to rise above noise; many trials yield no detection, even when the compound is present. At the upper end, judgments compress at the ceiling, where increases in concentration no longer change the perceived intensity because receptors or cognitive scale limits are saturated. Between these extremes lies the sensitive mid-region where small concentration changes produce measurable perceptual differences. Reliable sensory data come from this central band; readings near the floor or ceiling are unstable because perception has reached its functional limits. Fig. 2.3 illustrates this relationship, showing that valid sensory judgments occur within the central band of the curve, where perception changes predictably with concentration.

Fig. 2.3 PSYCHOMETRIC FUNCTION: FLOOR AND CEILING EFFECTS

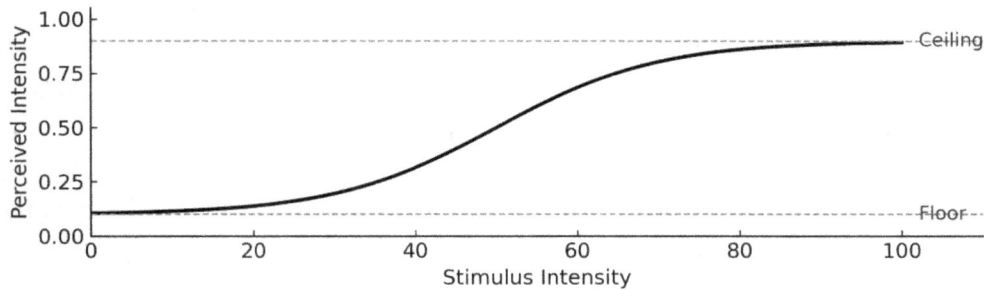

Scaling: what numbers mean (and do not).

Not all numbers are equal.

- **Nominal,** names only (e.g., lexicon categories).
- **Ordinal**, ordered but uneven steps (typical category intensity ratings).
- **Interval:** equal steps but no true zero.
- **Ratio,** equal steps with a true zero ("twice as intense" is meaningful).

Most classroom and panel intensity ratings are at best ordinal unless they are based on carefully trained anchors. Means and standard deviations for ordinal data can be misleading; they remain useful summaries, but interpretation should be done cautiously. Methods such as magnitude estimation or cross-modal matching can approximate ratio properties, but they introduce their own instabilities (context, range, expectation). The point is to treat numbers as structured impressions, not instruments.

Mixtures, masking, and time.

Spirits are dynamic systems. High ethanol concentrations can mask early-arising volatiles and reveal others after dilution or with time. Some attributes crest on the finish rather than the onset. A single "intensity" can collapse these temporal profiles; a time-aware reading avoids that error.

Summary

Flavor is a perceptual construction, dominated by olfactory input. In distilled spirits, smell accounts for approximately 90% of the experience, whereas taste and mouthfeel play supporting roles. Ethanol complicates this balance by triggering strong trigeminal responses that can mask nuance. For accurate evaluation, olfactory training and control of ethanol delivery are essential. Technique, vocabulary, and calibration—not casual tasting—determine reliability. Thresholds and JNDs are properties of a context, not just compounds. Most intensity numbers are ordinal summaries—useful, but not absolute. Interpretation improves when you respect floor/ceiling effects, Weber-like dependence on background, and temporal unfolding.

These principles underlie the Flavor Equation and explain why careful language, controlled context, and time awareness matter throughout the text. Together, these mechanisms explain why precise control of sensory conditions—not subjective impression—defines reliable flavor evaluation.

Beyond: Grasping the concept of sensory weighting is important. Recognizing the advantages and shortcomings of human senses compared to those of other animals, as well as the role of human evolution, places the importance of sensory perception into a realistic perspective.

"Flavor begins in the mouth but is perfected in the brain."— Linda Bartoshuk, professor of food science and nutrition

"Smell is 80 percent of taste."— Julia Child, chef and author

"Flavor is not just taste. It's taste, smell, texture, temperature, and the way it makes us feel."— Harold McGee, American food writer

References

Cometto-Muñiz, J. E., Cain, W. S., & Abraham, M. H. (2003). Quantification of chemical vapors in chemosensory research. *Chemical Senses, 28*(6), 467–477. https://doi.org/10.1093/chemse/28.6.467

Doty, R. L., & Cometto-Muñiz, J. E. (2003). Trigeminal chemoreception: Perception of odorless chemicals. In R. L. Doty (Ed.), *Handbook of Olfaction and Gustation* (2nd ed., pp. 981–1000). CRC Press.

Green, B. G. (1996). Chemesthesis: Pungency as a component of flavor. *Trends in Food Science & Technology, 7*(12), 415–420. https://doi.org/10.1016/S0924-2244(96)10004-5

Rozin, P. (1982). "Taste–smell confusions" and the duality of the olfactory sense. *Perception & Psychophysics, 31*(4), 397–401. https://doi.org/10.3758/BF03202667

Shepherd, G. M. (2012). *Neurogastronomy: How the brain creates flavor and why it matters.* Columbia University Press.

Spence, C. (2015). Just how much of what we taste derives from the sense of smell? *Flavour, 4,* Article 30. https://doi.org/10.1186/s13411-015-0044-2

Spence, C. (2017). Professor Charles Spence on the role of smell in taste. *CLASS Magazine.* https://classbarmag.com/news/fullstory.php/aid/689/Professor_Charles_Spence_on_the_role_of_smell_in_taste.html

Observations on Psychophysics in Sensory

Perception emerges from contrast, not absolute intensity. A fruity ester detected at 5 µg/L in water might require 25 µg/L to be perceived in 40% ethanol, demonstrating how the sensory matrix alters perceptual thresholds through suppression, masking, and mixture interaction effects.

Chapter 3: Human Olfactory Perspectives and Evolution

One must examine the human position within the animal kingdom to develop a realistic perspective of human abilities. Comparing the diversity of functions and adaptations of the senses in other animals and reflecting on human sensory evolution will underscore the importance of a solid perspective in sensory analysis.

Perspectives on human olfactory sensitivity

Our sense of smell is often misunderstood, undervalued, and dismissed as a primitive sense. For centuries, it has been overshadowed by vision and hearing, both of which dominate modern life through reading, speech, and media. However, in the evaluation of spirits—and the construction of flavor—smell is the primary input channel. While our olfactory system is far from the most powerful, it remains surprisingly adaptable, trainable, and scientifically fascinating.

While humans have fewer OR genes and a smaller epithelium than many olfaction-specialist mammals, modern work (e.g., McGann, 2017) shows our smell is **not** categorically 'poor'; it is adaptable and highly trainable. For spirits evaluation, the key is less raw acuity and more disciplined training. Figure 3.1 provides a general overview of human olfactory sensitivity.

Fig. 3.1   **Perspectives on Human Olfactory Sensitivity**

Female Human:	1.4 X > male	nurturing instincts, pheromones
Bloodhound:	40 X human	crucial to law enforcement
Grizzly Bear:	2,100 X human	food detection range 20 miles
Elephant:	100,000+ X human	water detection range 12 miles
Salmon:	1,000,000 X human	acute chemical sensitivity

Sensitivity in bloodhounds is often reported but not empirically confirmed to exceed human thresholds by an order of magnitude, and said to be forty times that of humans (Lazarowski et al., 2020). Bloodhounds have approximately three hundred million scent receptors (not the same as olfactory receptor genes). This remarkable ability (likely overstated by the media and not scientifically verifiable) supposedly enables them to detect scents from up to 130 miles away and to follow a subject's trail for over 300 hours (American Kennel Club, 2021; PBS Nature, 2008). These reports suggest extraordinary sensitivity, though some controlled studies indicate thresholds that are more modest yet still remarkable. Their unique physical features interact with their olfactory abilities: loose skin and long, drooping ears enhance scent detection, allowing them to trap and collect odors (American Kennel Club, 2021). Robust, peer-reviewed quantitative data are lacking to support many of the commonly cited extreme figures about animal olfactory sensitivity. These loosely assembled non-scientific claims are therefore included solely to suggest orders of magnitude for species whose survival depends heavily on olfaction—not as authoritative evidence. Use them only as speculative contextual markers, not as scientific facts.

- Pacific salmon rely heavily on their sense of smell to navigate back to their natal spawning streams. Studies show that juvenile salmon imprint on the unique chemical signature of their birth waters, and adult salmon use olfactory cues to guide migration (Dittman & Quinn, 1996).
- Kiwi birds (*Apteryx* species) olfactory bulbs are among the largest of any bird examined, indicating a heightened reliance on smell, supporting their ability to forage in dense vegetation (Bang & Cobb, 1968).
- Sharks—especially great whites and lemon sharks (also not scientifically verifiable) are said to detect blood at extremely low concentrations, debatably between 1 part in 25 million to 1 part in 10 billion in water (unverified by controlled studies) (BBC Science Focus, 2016).

Evolution of Human Sensory Abilities

Human social evolution likely reduced the need for an acute sense of smell, which was once critical for hunting and survival. Nonetheless, developing and valuing olfaction remains possible—sensitivity can be improved with effort and training (Li et al., 2008). Over millennia, as human evolution favored other systems, comparative genomics shows accelerated pseudogenization in olfactory receptor (OR) genes relative to some primates (Gilad et al., 2004).

Concurrently, anatomical studies indicate that the surface area of the human olfactory epithelium—approximately 5–10 cm²—is significantly smaller than in many mammals, placing humans at the lower end of the spectrum of olfactory sensitivity (McGann, 2017). Despite these lower levels, modern neuroscience demonstrates that olfactory perception remains trainable, and structured exposure and identification practice can improve detection thresholds and alter cortical responsiveness (Al Aïn et al., 2019). See Fig. 3.2 for comparisons.

Paleontological and comparative anatomical evidence suggests that Neanderthals possessed craniofacial features and nasal passages consistent with adaptation to cold, low-light climates, while overall brain allometry differed between Neanderthals and modern humans (Bruner et al., 2003; Trinkaus & Holliday, 2008). Airflow modeling also indicates convergent nasal adaptations in Neanderthals and *Homo sapiens* (Azevedo et al., 2017). Some primates (e.g., baboons) likely had larger olfactory epithelia than Neanderthals, underscoring variability even among close relatives. Today's humans prioritize vision, hearing, and social cognition over raw sensory acuity, a shift shaped by agriculture, culture, and group living.

Fig. 3.2			Animal Species Olfactory Sensitivity		
Species	**OR Genes (count)**	**Epithelium Area (cm^2)**	**Olfactory bulb/brain size ratio**	**Detection thresholds (parts/trillion)**	**Ecological/ Evolutionary Adaptation**
Human*	**400**	**5**	**0.01**	**1000**	**Broad, generalist**
Bloodhound	1000	150	0.31	1.0	Tracking, scavenging
Grizzly Bear	1000	200	0.35	0.5	Foraging, scavenging
Mouse	1100	60	2.0	10.0	Pheromone detection
Elephant	2000	200	0.10	5.0	Foraging, social cues
Scientific data sparse, somewhat unreliable, values illustrative, as interspecies comparability varies by method. *Reference: Early Neanderthal epithelium = 8-10 cm^2, Baboon = 10-20cm^2 Inference: Social development + evolution decreases olfactory sensitivity					

Despite this evolutionary shift away from olfactory prioritization, it is important to remember that our systems are atrophied, not destroyed. With conscious attention—purposeful scent engagement, training protocols, and environmental awareness—we can recover a meaningful portion of our olfactory potential (Al Aïn et al., 2019).

Genetic and structural constraints

The human brain is highly adaptable, yet our olfactory capacity is modest. Several measurable factors contribute to this reality. The human genome contains approximately a few hundred functional OR genes alongside many pseudogenes, reflecting an evolutionary de-emphasis on olfaction (Gilad et al., 2004; Trimmer et al., 2019). The human olfactory epithelium is small—about 5–10 cm^2—compared with many mammals (McGann, 2017).

The human olfactory bulb, which relays signals from the nose to the brain, is proportionally smaller than in many olfactory-dependent mammals.

These factors place us near the lower end of the spectrum of vertebrate olfactory sensitivity. Age-related decline, anosmia, and hyposmia are practical concerns; however, olfactory training, education, and calibration can help mitigate these limitations. For the sensory evaluator, reliable assessment derives from training and calibration rather than from reliance on raw receptor counts (Li et al., 2008; Al Aïn et al., 2019).

Gender-based differences in olfactory sensitivity

There is a notable exception within our species: women often outperform men across multiple olfactory tasks. This advantage has been documented in quantitative histology of the olfactory bulb (Oliveira-Pinto et al., 2014) and in a meta-analysis of detection, identification, and discrimination thresholds (Sorokowski et al., 2019). Females are also more likely to detect certain chemosignals that may be imperceptible to males.

Training Implications

For the sensory evaluator, the evolutionary decline of human olfactory acuity underscores the need for deliberate training. Unlike animals with specialized adaptations, humans must rely on structured calibration, reference standards, and repeated exposure to extend natural limits. Recognition of aromas in spirits is therefore not a matter of raw ability but of cultivated perception. Sensory science provides the frameworks to manage subjectivity and compensate for biological constraints, transforming an atrophied sense into a disciplined tool for evaluation and assessment.

Summary

Human olfactory capacity is modest by evolutionary standards but remains adaptable and trainable. While animals such as bloodhounds and salmon far exceed humans in raw sensitivity, structured training enables humans to regain significant function. Evolution and social development have diminished the primacy of smell, yet sex differences and neuroscience confirm its potential. For sensory evaluators, disciplined training transforms a weakened sense into a reliable tool.

The human sensory pathway is a progression from detecting the existence of stimuli, converting these signals to neural impulses, fitting them into a pattern, quantifying the experiences and characteristics, and finally, translating them into language, sharing the experiences, evaluating, classifying, and cognitively comparing the results, while building a memory bank of the characteristics and finally determining preferences and dislikes. Fig. 3.3 visualizes the path.

Beyond: A brief synopsis of each human sensory, anatomical, and physiological system will elucidate how stimuli become perceptual information, beginning with the olfactory system.

Fig. 3.3 The Human Sensory Pathway

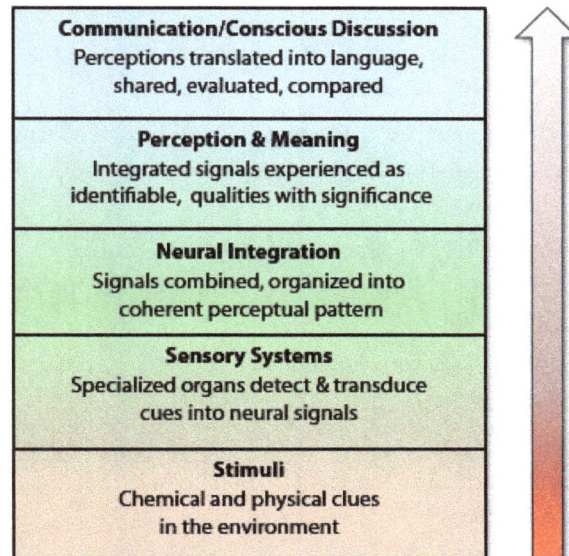

Communication/Conscious Discussion
Perceptions translated into language,
shared, evaluated, compared

Perception & Meaning
Integrated signals experienced as
identifiable, qualities with significance

Neural Integration
Signals combined, organized into
coherent perceptual pattern

Sensory Systems
Specialized organs detect & transduce
cues into neural signals

Stimuli
Chemical and physical clues
in the environment

"To recognize flavor, you must first recognize your own limits. Sensory training is the distiller's true still."— attributed to Dr. Jim Swan, chemist

"Smell is the most direct of all the senses—it connects the brain to the world in just two synapses."— Richard Axel, Nobel Laureate in physiology

References

Al Aïn, S., Poupon, D., Hétu, S., Mercier, N., Steffener, J., & Frasnelli, J. (2019). Smell training improves olfactory function and alters brain structure. *NeuroImage, 189*, 45–54. https://doi.org/10.1016/j.neuroimage.2019.01.008

American Kennel Club. (2021, August 27). Radar the Bloodhound helps solve dozens of murder cases. https://www.akc.org/expert-advice/news/radar-the-bloodhound-helps-solve-dozens-of-murder-cases/

Azevedo, F. C., González-Mateo, I., Cintas, C., Ramallo, V., & Quinto-Sánchez, M. (2017). Nasal airflow simulations suggest convergent adaptation in Neanderthals and modern humans. *Proceedings of the National Academy of Sciences, 114*(47), 12442–12447. https://doi.org/10.1073/pnas.1703790114

Bang, B. G., & Cobb, S. (1968). The size of the olfactory bulb in 108 species of birds. *The Auk, 85*(1), 55–61. https://doi.org/10.2307/4083624

BBC Science Focus. (2016, December 14). How do sharks smell blood underwater? https://www.sciencefocus.com/the-human-body/how-do-sharks-smell-blood-underwater

Bruner, E., Manzi, G., & Arsuaga, J.-L. (2003). Encephalization and allometric trajectories in the genus Homo: Evidence from the Neandertal and modern lineages. *Proceedings of the National Academy of Sciences of the*

United States of America, 100(26), 15335–15340. https://doi.org/10.1073/pnas.2536671100

Delwiche, J. (2012). You eat with your eyes first. *Physiology & Behavior, 107*(4), 502–504. https://doi.org/10.1016/j.physbeh.2012.07.011

Dittman, A. H., & Quinn, T. P. (1996). Homing in Pacific salmon: Mechanisms and ecological basis. *Journal of Experimental Biology, 199*(1), 83–91. https://doi.org/10.1242/jeb.199.1.83

Gilad, Y., Wiebe, V., Przeworski, M., Lancet, D., & Pääbo, S. (2004). Loss of olfactory receptor genes coincides with the acquisition of full trichromatic vision in primates. *PLoS Biology, 2*(1), e5. https://doi.org/10.1371/journal.pbio.0020005

Lazarowski, L., Krichbaum, S., DeGreeff, L. E., Simon, A., Singletary, M., Angle, C., & Waggoner, L. P. (2020). Methodological considerations in canine olfactory detection research. *Frontiers in Veterinary Science, 7*, 408. https://doi.org/10.3389/fvets.2020.00408

Lawless, H. T., & Heymann, H. (2010). *Sensory evaluation of food: Principles and practices* (2nd ed.). Springer. https://doi.org/10.1007/978-1-4419-6488-5

Li, W., Howard, J. D., Parrish, T. B., & Gottfried, J. A. (2008). Aversive learning enhances perceptual and cortical discrimination of indiscriminable odor cues. *Science, 319*(5871), 1842–1845. https://doi.org/10.1126/science.1152837

McGann, J. P. (2017). Poor human olfaction is a nineteenth-century myth. *Science, 356*(6338), eaam7263. https://doi.org/10.1126/science.aam7263

Meilgaard, M. C., Civille, G. V., & Carr, B. T. (2015). *Sensory evaluation techniques* (5th ed.). CRC Press. https://doi.org/10.1201/b19493

Oliveira-Pinto, A. V., Santos, R. M., Coutinho, R. A., Oliveira, L. M., Santos, G. B., Alho, A. T., … Lent, R. (2014). Sexual dimorphism in the human olfactory bulb: Females have more neurons and glial cells than males. *PLoS ONE, 9*(11), e111733. https://doi.org/10.1371/journal.pone.0111733

PBS Nature. (2008, June 9). The Bloodhound's amazing sense of smell. https://www.pbs.org/wnet/nature/the-bloodhounds-amazing-sense-of-smell/2982/

Prescott, J. (2015). Multisensory processes in flavour perception and their influence on food choice. *Current Opinion in Food Science, 3*, 47–52. https://doi.org/10.1016/j.cofs.2015.02.007

Shepherd, G. M. (2015). *Neuroenology: How the brain creates the taste of wine. Flavour, 4*, Article 19. https://doi.org/10.1186/s13411-014-0030-9

Sorokowski, P., Karwowski, M., Misiak, M., Marczak, M., Dziekan, M., Hummel, T., & Sorokowska, A. (2019). Sex differences in human olfaction: A meta-analysis. *Frontiers in Psychology, 10*, 242. https://doi.org/10.3389/fpsyg.2019.00242

Trimmer, C., Keller, A., Murphy, N. R., Snyder, L. L., Willer, J. R., Nagai, M. H., Katsanis, N., Mainland, J. D., & Matsunami, H. (2019). Genetic variation across the human olfactory receptor repertoire alters odor perception. *Proceedings of the National Academy of Sciences, 116*(19), 9475–9480. https://doi.org/10.1073/pnas.1804106115

Trinkaus, E., & Holliday, T. W. (2008). Cold adaptation and Neanderthals? *American Journal of Physical Anthropology, 136*(3), 361–371. https://doi.org/10.1002/ajpa.20857

Chapter 4: The Olfactory System

Accepted Olfactory Theory and Mechanism of Olfaction

The foundation of modern olfactory science rests on the Lock-and-Key model of receptor–ligand interaction: specific odorant molecules bind to olfactory receptor neurons (ORNs) based on molecular shape, size, and chemical affinity—a body of work that culminated in the Nobel Prize in Physiology or Medicine in 2004 (Buck & Axel, 2004). While vibration-based mechanisms have been proposed, they remain unsubstantiated by the weight of current evidence (Block, 2015). The lock-and-key model is incomplete but remains a widely accepted framework for receptor binding and signal transduction. See Fig. 4.1.

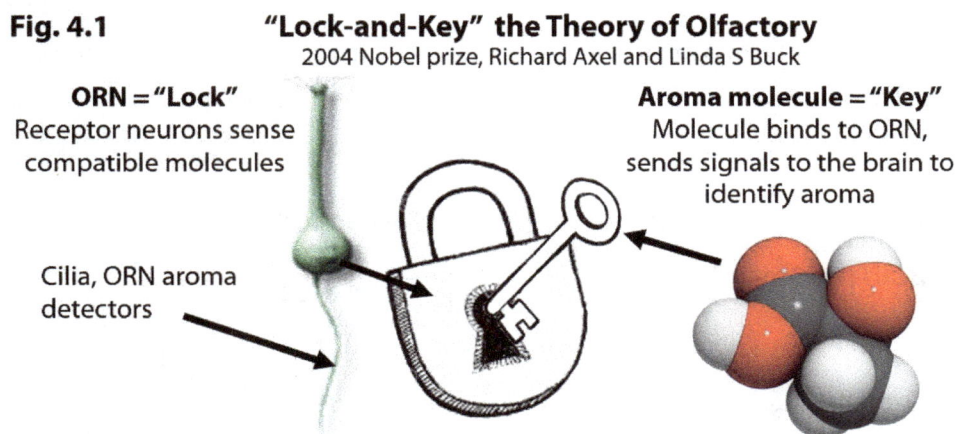

Fig. 4.1   **"Lock-and-Key" the Theory of Olfactory**
2004 Nobel prize, Richard Axel and Linda S Buck

ORN = "Lock"
Receptor neurons sense compatible molecules

Aroma molecule = "Key"
Molecule binds to ORN, sends signals to the brain to identify aroma

Cilia, ORN aroma detectors

Humans possess 6-10 million ORNs located within the olfactory epithelium at the roof of the nasal cavity (Standring & Lobo, 2024). Each ORN expresses a single receptor type, yet coding is highly combinatorial—one odorant can activate multiple receptors, and one receptor can respond to multiple odorants (Malnic et al., 1999). The concept is sometimes extended as "odotope theory," which classifies molecular features into structural "families" recognized across different receptors, even when absolute shapes differ (Rinaldi, 2007). While much of the sensory-level processing is understood, the mechanism by which ORNs group complex molecules into perceptual categories remains an active frontier for future research.

The Process of Odorant-Receptor Binding

The primary step in olfactory perception is the binding of odorants to neuron receptors. Volatile aroma compounds entering the nasal cavity must dissolve into the mucous layer and contact ORN cilia. When an odorant's shape and chemical features complement a receptor site, ORN activation triggers an electrical signal to the olfactory bulb (Sanz et al., 2005). Binding is selective, not absolute—some compounds elicit irritation via the trigeminal nerve rather than olfactory receptors (Green, 1996).

Ethanol—abundant in spirits—illustrates this distinction. While some weak receptor interactions may occur, ethanol's dominant sensory impact is trigeminal (pungency, anesthesia) rather than olfactory (Hummel et al., 2017; Green, 1996). Other examples include acetic acid, carbon dioxide, menthol, and capsaicin, which can produce burning, cooling, or tingling with limited ORN binding (Green, 1996).

Anatomy of the Human Olfactory System

The olfactory bulbs are paired structures situated just above the nasal cavity, separated from the cranial vault by the cribriform plate of the ethmoid bone. Each bulb primarily processes input from one nostril, although cross-hemispheric connections exist via the anterior commissure (Purves et al., 2012). See Fig. 4.2.

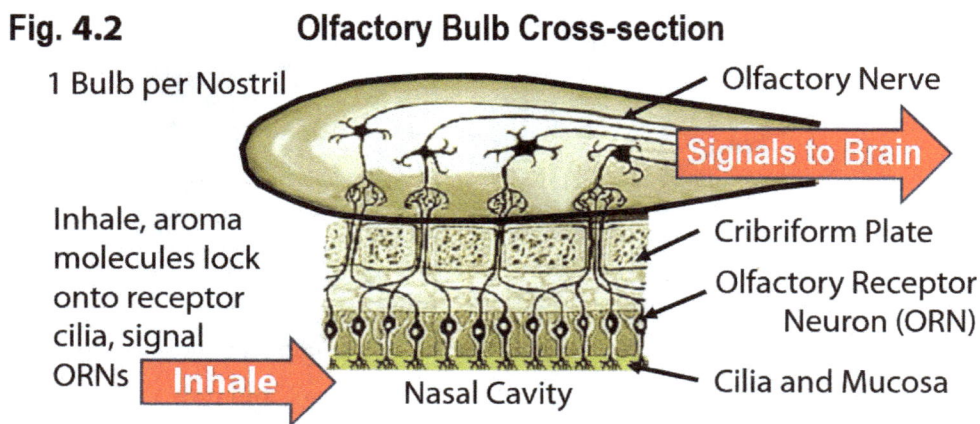

Fig. 4.2   **Olfactory Bulb Cross-section**

1 Bulb per Nostril — Olfactory Nerve — Signals to Brain

Inhale, aroma molecules lock onto receptor cilia, signal ORNs — Inhale — Nasal Cavity — Cribriform Plate — Olfactory Receptor Neuron (ORN) — Cilia and Mucosa

Evaluation Note 4.1: Using both nostrils is essential for broad detection, as airflow, mucous coating, and blockage can vary side to side. It is advisable to gently blow the nose before any evaluation session to remove contaminants and refresh the mucous layer, and test for a favored nostril (the side with stronger detection). Nostril restrictions naturally cycle in 25-200 minutes.

Inhaling aromas draws volatile compounds upward to the olfactory epithelium, where they encounter the cilia of ORNs embedded in the mucus. The first molecule to reach and bind to a cilium initiates a signal that is relayed via the olfactory nerve to the olfactory bulb and ultimately to the cerebral cortex for identification and memory tagging (Purves et al., 2001).

Orthonasal and Retronasal Olfaction

There are two distinct pathways by which we perceive aroma: the orthonasal and retronasal pathways. See Fig. 4.3. Orthonasal olfaction occurs when we inhale through the nose and encounter environmental volatiles. This form of smelling serves as an anticipatory system, allowing us to assess whether a substance is safe to consume. It also prompts cognitive associations, drawing on memory to recognize familiar or novel stimuli. Orthonasal signals are sent directly to the brain via the olfactory nerve and processed by the olfactory cortex.

Fig. 4.3   **Two Paths: Palate - Nose Connection**

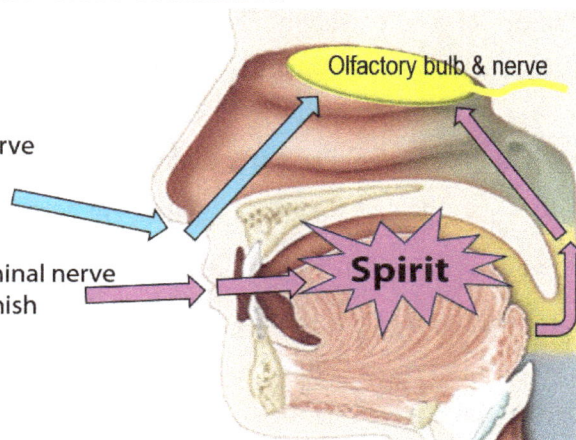

"Finish" is not entirely on the palate

Smell prior to tasting = Orthonasal Olfactory nerve
only sends signal to brain
Ask: *"Is it safe to drink? What is it?"*

Oral tasting (saliva dilution) = Retronasal Trigeminal nerve
to brain includes Taste + smell + mouthfeel = finish
Ask: *Do I like it? Why? or why not?"*

Olfactory bulb & nerve

Spirit

Retronasal olfaction begins when a spirit or food is placed into the mouth. As we exhale, volatiles are released from the sample and travel through the nasopharynx (pharyngeal opening) to the back of the nasal cavity, where they again interact with the olfactory epithelium—but now from behind. Unlike orthonasal olfaction, retronasal perception integrates taste and chemosensory inputs. All these signals—olfactory, gustatory, and trigeminal—are integrated in the brain and conveyed as a unified "flavor experience" (Auvray & Spence, 2008).

Evaluation Note 4.2: Retronasal tasting of spirits introduces body-temperature heat into the oral cavity, increasing vapor volatility by elevating the vapor pressure of aroma compounds. Saliva immediately dilutes the spirit and alters its physicochemical environment—changing viscosity, pH, and volatile partitioning—while taste receptors, trigeminal input, and mechanical mouthfeel cues (oily, grainy, chewy, etc.) integrate with retronasal olfaction as volatiles move through the pharyngeal opening to the olfactory epithelium (Ployon et al., 2017). Saliva also introduces enzymatic activity, dominated by α-amylase with minor and variable esterase contributions, which modulates the matrix and subtly shifts aroma-release dynamics.

Many experienced evaluators mistakenly treat "finish" as the ease of swallowing and reduce it to simplistic terms such as "smooth," overlooking the combined sensory processes that actually produce it. Finish arises from the evolving interaction of dilution, temperature, enzymatic action, matrix chemistry, and texture, all shaped by the composition and flow of saliva. Misunderstanding these factors leads to vague conclusions; in reality, saliva is a primary driver of retronasal perception, materially influencing the final expression of aroma and flavor (Ployon et al., 2017).

Evaluation Note 4.3: Critically, excessive orthonasal exposure to ethanol-rich vapors can numb the olfactory epithelium through its anesthetic action, thereby diminishing retronasal sensitivity. Evaluators who take deep inhalations of high-proof spirits risk muting the very receptors needed for accurate finish assessment. Therefore, minimizing ethanol exposure during the orthonasal stage is crucial. We know from the flavor equation (Chapter 2, page 7) that olfaction contributes the overwhelming share to perceived flavor; ethanol, being a pungent and numbing irritant, must be controlled during orthonasal nosing (Spence, 2015).

The Faces of Ethanol: Volatile Carrier, Flavor Fault, Fragrance Engine

Ethanol's sensory role extends well beyond spirits. In cooking, it dissolves and volatilizes aromatic compounds in sauces and reductions, carrying flavors upward in steam and deepening complexity in extracts such as vanilla. In perfumery, it serves as a neutral, high-volatility carrier that both stabilizes and stages the release of fragrance notes. In premium olive oils, trace ethanol produced by fermentative activity can be a fault marker, imparting a fermenty or wine-like off-note. As a food additive, ethanol functions as a solvent for flavors and colorants, enabling the uniform dispersion of otherwise insoluble compounds. In each context, ethanol's volatility and solvency dictate how aromas reach the nose — sometimes desirable, sometimes not.

Olfactory Neural Processing

Interpretation of olfactory and flavor signals is not localized to a single brain region. Instead, separate neural circuits process different sensory modalities. Taste signals are routed to the gustatory cortex, while aromas are processed in the primary olfactory cortex. The somatosensory cortex is responsible for processing chemesthesis sensations, including burning, tingling, or cooling. Emotional and visceral responses—those tied to memory and personal meaning—are generated in the amygdala and hippocampus. All of these processed outputs are relayed to the frontal cortex, where they are integrated into a single conscious perception (Gottfried, 2010). See Fig. 4.4.

Emotional Context of Smell

This multisensory convergence helps explain how complex spirits tasting becomes a profound personal experience. The same whiskey, consumed at different life moments, may evoke profoundly different emotional interpretations—not because the liquid changes, but because the brain reconstructs flavor as a composite memory. Building on the multisensory tasting experience, we examine how multisensory interaction can alter initial impressions.

Scenario 4.1: Bob celebrates his graduation with a rare bottle of whiskey gifted by his father. Tasting and appreciating it together in an auspicious moment, he remembers it not just for the flavor but also for the context—the accomplishment, the setting, the emotions, and his father's continued support. Years later, tasting the same bottle brings back those vivid moments, perhaps even more intensely with nostalgia; it has become Bob's favorite whiskey for life.

Scenario 4.2: Sue associates the aroma of cinnamon and nutmeg with her grandmother's kitchen and a moment alone with her as the first slice of a freshly baked apple pie was presented to her to taste. That olfactory memory, tied to warmth, love, and security, can return powerfully with just a single sniff of pie spices, or perhaps unexpectedly, along with a vision of her grandmother's smile and a recollection of their special relationship.

Fig. 4.4   **"Finish" Signal Flow Diagram (whiskey example)**

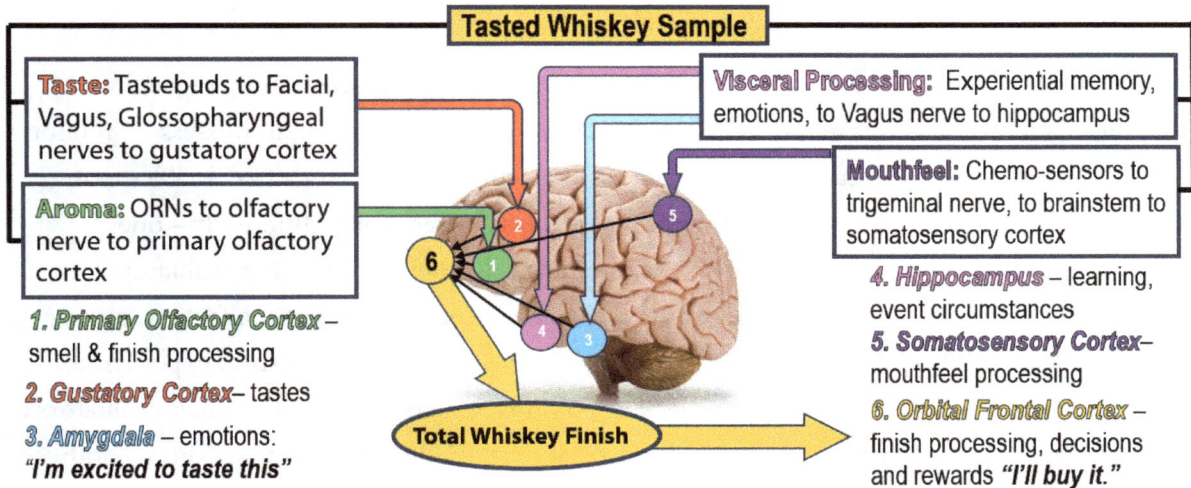

Tasted Whiskey Sample

Taste: Tastebuds to Facial, Vagus, Glossopharyngeal nerves to gustatory cortex

Aroma: ORNs to olfactory nerve to primary olfactory cortex

Visceral Processing: Experiential memory, emotions, to Vagus nerve to hippocampus

Mouthfeel: Chemo-sensors to trigeminal nerve, to brainstem to somatosensory cortex

1. Primary Olfactory Cortex – smell & finish processing

2. Gustatory Cortex – tastes

3. Amygdala – emotions: *"I'm excited to taste this"*

Total Whiskey Finish

4. Hippocampus – learning, event circumstances

5. Somatosensory Cortex – mouthfeel processing

6. Orbital Frontal Cortex – finish processing, decisions and rewards *"I'll buy it."*

Proust Phenomenon

Scenario 4.2 illustrates the Proust Phenomenon, in which odor-triggered memories surface unexpectedly and vividly, often rooted in early childhood or formative life events. Unlike other sensory inputs, olfactory signals can project directly to limbic and cortical targets, which helps explain why aroma-driven memories are so emotionally potent and long-lasting (Chu & Downes, 2002; Herz & Engen, 1996).

The Olfactory Sensory Pathway

The olfactory pathway (Fig. 4.5) bears a striking similarity in signal routing to the human sensory pathway previously illustrated in Fig. 3.3.

Fig. 4.5   The Human Olfactory Pathway

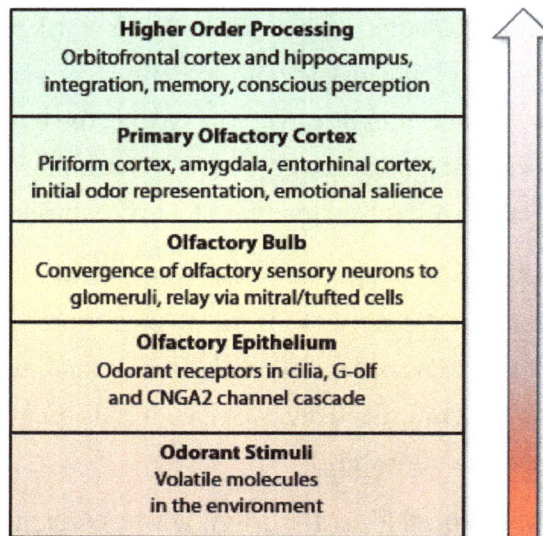

Higher Order Processing
Orbitofrontal cortex and hippocampus,
integration, memory, conscious perception

Primary Olfactory Cortex
Piriform cortex, amygdala, entorhinal cortex,
initial odor representation, emotional salience

Olfactory Bulb
Convergence of olfactory sensory neurons to
glomeruli, relay via mitral/tufted cells

Olfactory Epithelium
Odorant receptors in cilia, G-olf
and CNGA2 channel cascade

Odorant Stimuli
Volatile molecules
in the environment

Summary

The olfactory system is the dominant channel for flavor perception, functioning through receptor–ligand interactions described by the Lock-and-Key theory. Humans possess ~6 million olfactory receptor neurons, each tuned to odorants in a combinatorial code that enables vast perceptual diversity. Orthonasal and retronasal pathways deliver aroma differently—one anticipatory, the other integrated with taste and mouthfeel—thereby requiring evaluators to control ethanol exposure to preserve sensitivity. Neural processing integrates olfactory, gustatory, and trigeminal signals with memory and emotion, explaining why aroma-driven experiences are vivid and deeply personal. Despite anatomical limitations, structured training allows evaluators to transform biological constraints into reliable sensory tools. Calibration exercises should address both orthonasal and retronasal perception.

Beyond: Olfactory sensations combine with gustatory to build identifiable, recallable perception.

"Smell is a potent wizard that transports you across thousands of miles and all the years you have lived."— Helen Keller, deaf and blind author

"The sense of smell can be extraordinarily evocative, bringing back pictures as sharp as photographs of scenes we had forgotten."— Oliver Sacks, writer

"Nothing revives the past so completely as a smell that was once associated with it."— Vladimir Nabokov, novelist, poet

"Olfaction is our most ancient sense, and still our most emotionally immediate."
— Rachel Herz, neuroscientist and author

Receptor Reference Appendix for Further Study

Although a more detailed discussion is beyond the scope of this text, familiarity with receptor function enriches comprehension and aids the development of evaluation expertise.

Appendix 4.1		Olfactory Receptors and Transduction Mechanisms
Sensor	**Definition**	**Description/Location**
ORs	Olfactory Receptors	G-protein-coupled receptors (GPCRs); detect volatile odorants; located on ORNs in olfactory epithelium
G-olf	Olfactory G Protein subtype	Activates adenylate cyclase upon binding to an odorant; initiates a signal cascade. Olfactory sensory neurons are located in the ciliary membranes, as well as in the striatum and the bulb.
ACIII	Adenylate Cyclase III	Converts ATP to cAMP; essential for signal transduction in olfaction.
CNG	Cyclic Nucleotide-Gated Channels	Allow Ca^{2+}/Na^+ influx after cAMP binding; depolarize ORNs. Primarily cilia in epithelium, and retina (visual), limited in some non-sensory tissues.
CNGA2	Cyclic Nucleotide-Gated Channel Alpha 2	Ion channel component with olfactory signal transduction via cAMP cascade. Cilia of olfactory receptor neurons in the epithelium.
ANO2	Anoctamin 2	Calcium-activated chloride channel; amplifies depolarization in ORNs.
TAARs	Trace Amine-Associated Receptors	Detect volatile amines; contribute to social, predator-prey chemical signaling. Axons in specific glomeruli in the olfactory bulb.

References

Auvray, M., & Spence, C. (2008). The multisensory perception of flavor. *Consciousness and Cognition, 17*(3), 1016–1031. https://doi.org/10.1016/j.concog.2007.06.005

Block, E. (2015). Implausibility of the vibrational theory of olfaction. *Proceedings of the National Academy of Sciences, 112*(21), E2766–E2774. https://doi.org/10.1073/pnas.1503054112

Buck, L., & Axel, R. (2004). The Nobel Prize in Physiology or Medicine 2004. NobelPrize.org. https://www.nobelprize.org/prizes/medicine/2004/summary/

Chu, S., & Downes, J. J. (2002). Proust nose best: Odors are better cues of autobiographical memory. *Memory & Cognition, 30*(4), 511–518. https://doi.org/10.3758/BF03194952

Gottfried, J. A. (2010). Central mechanisms of odour object perception. *Nature Reviews Neuroscience, 11*(9), 628–641. https://doi.org/10.1038/nrn2883

Green, B. G. (1996). Chemesthesis: Pungency as a component of flavor. *Trends in Food Science & Technology, 7*(12), 415–420. https://doi.org/10.1016/S0924-2244(96)10043-1

Herz, R. S., & Engen, T. (1996). Odor memory: Review and analysis. *Psychonomic Bulletin & Review, 3*(3), 300–313. https://doi.org/10.3758/BF03210754

Hummel, T., Whitcroft, K. L., Andrews, P., Altundag, A., Cinghi, C., Costanzo, R. M., … Welge-Lüssen, A. (2017). Position paper on olfactory dysfunction. *Rhinology Supplement, 54*(26), S1–S30. https://doi.org/10.4193/Rhino16.248

Malnic, B., Hirono, J., Sato, T., & Buck, L. B. (1999). Combinatorial receptor codes for odors. *Cell, 96*(5), 713–723. https://doi.org/10.1016/S0092-8674(00)80581-4

Ployon, S., Morzel, M., & Canon, F. (2017). The role of saliva in aroma release and perception. *Food Chemistry, 226*, 212–220. https://doi.org/10.1016/j.foodchem.2017.01.055

Purves, D., Augustine, G. J., & Fitzpatrick, D. (Eds.). (2001). The transduction of olfactory signals. In *Neuroscience* (2nd ed.). Sinauer Associates. https://www.ncbi.nlm.nih.gov/books/NBK11039/

Purves, D., Augustine, G. J., & Fitzpatrick, D. (Eds.). (2012). The organization of the olfactory system. In *Neuroscience* (5th ed.). Sinauer Associates. https://www.ncbi.nlm.nih.gov/books/NBK493175/

Rinaldi, A. (2007). The scent of life: The exquisite complexity of the sense of smell in animals and humans. *EMBO Reports, 8*(7), 629–633. https://doi.org/10.1038/sj.embor.7401029

Sanz, G., Schlegel, C., Pernollet, J. C., & Briand, L. (2005). Comparison of odorant specificity of two human olfactory receptors from different phylogenetic classes and evidence for antagonism. *Chemical Senses, 30*(1), 69–80. https://doi.org/10.1093/chemse/bji002

Spence, C. (2015). Just how much of what we taste derives from the sense of smell? *Flavour, 4*, Article 30. https://doi.org/10.1186/s13411-015-0044-2

Standring, S., & Lobo, M. (2024). Cranial nerve I (Olfactory). In *StatPearls*. StatPearls Publishing. https://www.ncbi.nlm.nih.gov/books/NBK538260/

Observations on Building Olfactory Memory

Olfactory memory is built through repeated, mindful exposure—each scent is linked to its context, emotion, and experience, thereby strengthening neural encoding.

The hippocampus and olfactory cortex interact to encode and retrieve odors, making smell one of the most durable and emotionally salient forms of memory.

Expanding olfactory memory requires deliberate training—systematic smelling, note-taking, and cross-referencing scents to their molecular families and real-world sources.

Olfactory memory strengthens when scents are revisited under varying temperatures, concentrations, and mixtures, forcing the brain to refine and stabilize recognition patterns.

Associating aromas with verbal descriptors, visual cues, and tactile experiences enhances cross-modal reinforcement, deepening neural connectivity and recall accuracy.

Retrieval practice—actively recalling scents without re-exposure—consolidates long-term olfactory memory by strengthening hippocampal–orbitofrontal pathways.

Chapter 5: The Gustatory System

Classification and Standards for Taste Designation

Gustatory is the next step in defining the perception of finish. The Western classification of taste is based on clear, objective criteria. Academic research and consensus define a basic taste stimulus as one that: activates a specific receptor on taste buds; elicits a perceptible, unique sensory experience distinguishable from the five accepted basic tastes; and is relevant to human nutrition or survival.

Standards and procedures for sensory evaluation are established by the International Organization for Standardization (ISO), ASTM International, and the Institute of Food Science and Technology (IFST).

The five accepted tastes are sweet, salty, sour, bitter, and umami, mediated by distinct receptor pathways. Fig. 5.1 (Chandrashekar, Hoon, Ryba, & Zuker, 2006; Roper & Chaudhari, 2017).

Fig. 5.1	Taste Description and Detection	
Taste	**Description**	**Taste-bud Detection**
Sweet	Pleasant, sugars, amino acids	T1R2/T1R3 heterodimer receptor
Sour	Sharp, acidic (H^+ protons), lemon, vinegar	OTOP1, the current principal transducer, ion channels PKD2L1 secondary
Salty	Sodium and other minerals	ENaC sodium channel, others
Bitter	Sharp, often unpleasant, possibly toxic	T2R receptors (25 known)
Umami	Savory, meaty, glutamates, nucleotides	T1R1/T1R3 receptors

Some stimuli do not fully meet the criteria for acceptance as a basic taste but require further consideration. See Fig. 5.2. While some cultures refer to spicy or pungent sensations as 'tastes,' these are not classified as tastes in the scientific sense. International scientific and regulatory communities are increasingly adopting these definitions to ensure clarity and consistency (Chandrashekar et al., 2006).

Fig. 5.2	Candidates for Basic Taste Acceptance	
Taste	**Description**	**Tastebud Detection**
Fat	Pleasant, from long-chain fatty acids	CD36, GPR120, GPR40
Kokumi	Mouth fullness	CaSR calcium receptor
Starch	Complex carbohydrates, amylase	Incomplete, evidence unknown

The Mechanism of Taste

The tongue's surface is covered with small projections called papillae: fungiform, foliate, circumvallate, and filiform (Fig. 5.3). The first three house taste buds. Filiform papillae lack taste buds and instead provide tactile function (Chandrashekar et al., 2006; Roper & Chaudhari, 2017). Fungiform papillae host many sweet and umami-responsive cells; foliate papillae include sour- and salt-responsive fields; and circumvallate papillae are enriched in bitter-responsive taste buds. However, modern data show all basic taste qualities can be detected across regions where taste buds are present (Roper & Chaudhari, 2017; Collings, 1974).

Fig. 5.3   **Tastebuds in the Tongue Papillae**

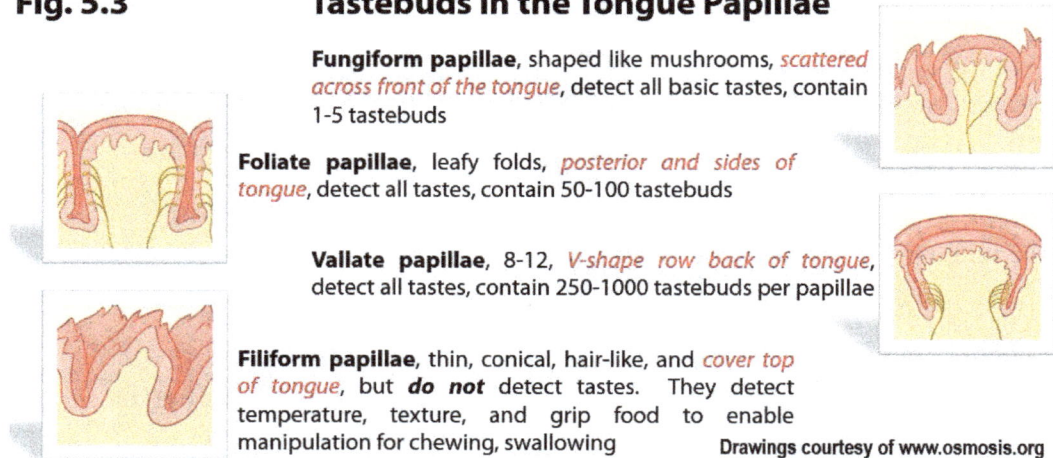

Fungiform papillae, shaped like mushrooms, *scattered across front of the tongue*, detect all basic tastes, contain 1-5 tastebuds

Foliate papillae, leafy folds, *posterior and sides of tongue*, detect all tastes, contain 50-100 tastebuds

Vallate papillae, 8-12, *V-shape row back of tongue*, detect all tastes, contain 250-1000 tastebuds per papilla

Filiform papillae, thin, conical, hair-like, and *cover top of tongue*, but **do not** detect tastes. They detect temperature, texture, and grip food to enable manipulation for chewing, swallowing

Drawings courtesy of www.osmosis.org

Are You a Supertaster?

Supertasters experience taste with far greater intensity, especially for compounds in coffee, dark chocolate, grapefruit, and cruciferous vegetables. This variation is characterized by a higher density of fungiform papillae and increased responsiveness to bitter ligands, such as PROP (6-*n*-propylthiouracil). Estimates suggest roughly 25% of the population qualify as supertasters, with sex and genetic variation influencing prevalence (Bartoshuk, Duffy, & Miller, 1994; Duffy, Hayes, Sullivan, & Faghri, 2004). Simple field test: place a drop of blue food coloring on the mid-tip of the tongue and view through a 6-mm hole in white card stock; counting >30 pink papillae suggests supertaster[G] status (Bartoshuk et al., 1994; Duffy et al., 2004).

Gustatory Neural Processing

Taste buds on the tongue, soft palate, epiglottis, and pharynx transduce stimuli via cranial nerves VII (facial), IX (glossopharyngeal), and X (vagus) to the nucleus of the solitary tract in the medulla, relay to the ventral posteromedial nucleus of the thalamus, and project to primary gustatory cortex in the anterior insula and frontal operculum (Fig. 5.4; Purves, Augustine, & Fitzpatrick, 2012; Small, 2010).

Higher-order integration occurs with olfactory and somatosensory inputs in the orbitofrontal cortex, amygdala, and related limbic structures, supporting reward valuation and decision-making choice (Small, 2010; Rolls, 2015).

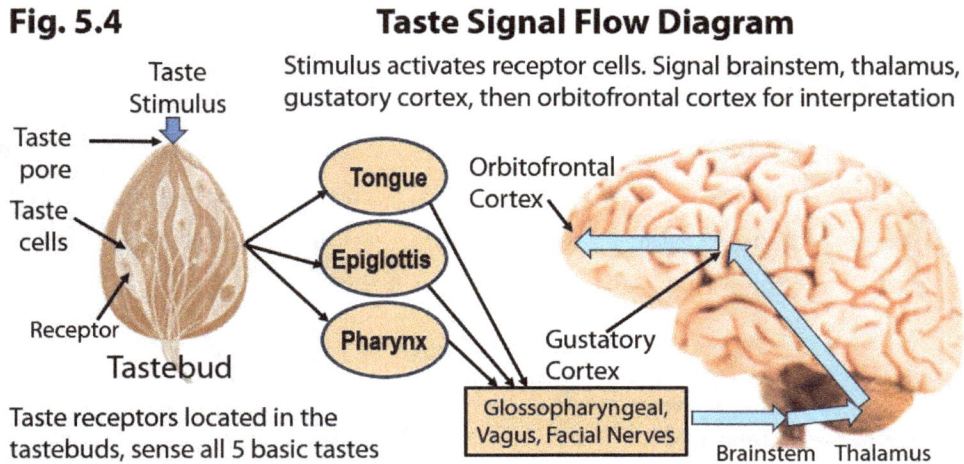

Fig. 5.4   Taste Signal Flow Diagram

Stimulus activates receptor cells. Signal brainstem, thalamus, gustatory cortex, then orbitofrontal cortex for interpretation

Taste Stimulus

Taste pore

Taste cells

Receptor

Tastebud

Tongue

Epiglottis

Pharynx

Orbitofrontal Cortex

Gustatory Cortex

Glossopharyngeal, Vagus, Facial Nerves

Brainstem   Thalamus

Taste receptors located in the tastebuds, sense all 5 basic tastes

Case Study 5.1: The 'tongue map' myth claims that different basic tastes are confined to distinct tongue regions. The notion stems from a misinterpretation of Hanig's 1901 thresholds and was amplified by Boring's 1942 historical account (Hanig, 1901; Boring, 1942). Virginia Collings (1974) directly tested regional sensitivity on the tongue and soft palate and found only minor threshold differences, not exclusive regions; modern reviews reaffirm the broad distribution of taste sensitivity wherever taste buds exist (including soft palate and epiglottis) (Collings, 1974; Peng & Peng, 2022). See Fig. 5.5), related to Evaluation Note 5.1.

Fig. 5.5   The Tongue Map Myth

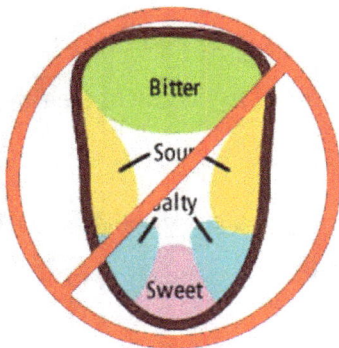

Bitter

Sour

Salty

Sweet

Evaluation Note 5.1: Occasionally, marketing narratives exploit the tongue-map trope, suggesting that glass shape or rim diameter changes taste. Evaluators should be mindful of suggestibility. Marketing demonstrations that claim the shape of glass alters taste exploit expectation bias. Studies show that such expectations arise from context and psychological priming, rather than gustatory biology (Rolls, 2015; Plassmann, O'Doherty, Shiv, & Rangel, 2008). Attendees often respond affirmatively to suggestive tasting prompts out of social obligation or expectation. Such feedback loops can reinforce myths without proper sensory validation. Subliminal suggestion is an integral and effective marketing tool for product promotion. In controlled settings, the shape of glass primarily affects aroma delivery and expectation, rather than tongue receptor biology; reported taste differences are primarily due to contextual or priming effects (Rolls, 2015; Plassmann et al., 2008).

Summary

Taste is mediated by five well-established receptor pathways, with candidates such as fat and kokumi under investigation. Unlike olfaction, gustation provides limited diversity but plays a crucial role in balance, modulation, and signaling potential toxicity. Myths, such as the tongue map, mislead evaluators, whereas differences, such as supertaster status, highlight biological variation. For spirits evaluation, compared to olfaction, taste plays a supporting role. Similar to the olfactory pathway of Fig. 4.5, the gustatory pathway is defined in Fig. 5.6.

Fig. 5.6 The Human Gustatory Pathway

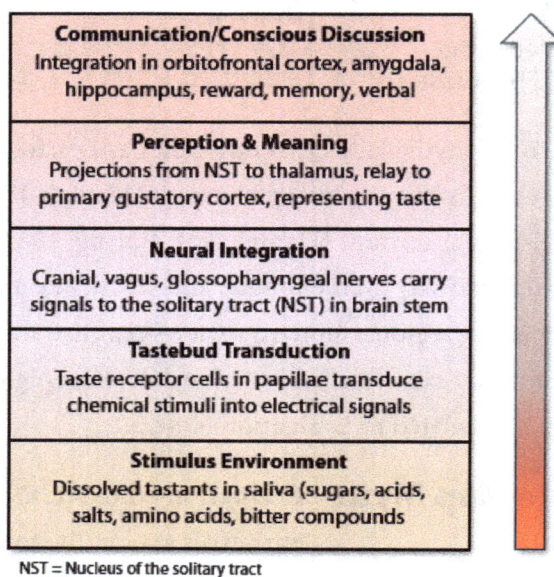

Communication/Conscious Discussion Integration in orbitofrontal cortex, amygdala, hippocampus, reward, memory, verbal
Perception & Meaning Projections from NST to thalamus, relay to primary gustatory cortex, representing taste
Neural Integration Cranial, vagus, glossopharyngeal nerves carry signals to the solitary tract (NST) in brain stem
Tastebud Transduction Taste receptor cells in papillae transduce chemical stimuli into electrical signals
Stimulus Environment Dissolved tastants in saliva (sugars, acids, salts, amino acids, bitter compounds

NST = Nucleus of the solitary tract

Beyond: Olfactory and gustatory sensations combine with mouthfeel to complete the final, identifiable, and recallable perception aptly called "finish."

"Great is the power of steady misrepresentation; but the history of science shows that fortunately this power does not long endure." — Charles Darwin, naturalist

Receptor Reference Appendix

This information may be found in advanced studies or technical literature. Although a deeper discussion is beyond the scope of this text, familiarity will enrich comprehension and aid in developing evaluation expertise (Appendix 5.1).

Appendix 5.1		Gustatory Receptors
Sensor	Definition	Description / Location
T1R1	Taste Receptor Type 1 Member 1	Part of the umami receptor forms a heterodimer with T1R3, which is located on Type II taste cells.
T1R2	Taste Receptor Type 1 Member 2	Combines with T1R3 to detect sweet compounds and is expressed in Type II taste receptor cells.
T1R3	Taste Receptor Type 1 Member 3	Common subunit in sweet and umami receptors; dimerizes with T1R1 or T1R2 depending on the modality.
T2Rs	Taste Receptor Type 2 Family	A family of ~25 receptors mediating bitter taste; expressed in Type II cells.
ENaC	Epithelial Sodium Channel	Mediates detection of salty taste through direct Na^+ entry; found in Type I cells.
TRPM5	Transient Receptor Potential Cation Channel Subfamily M Member 5	Intracellular Ca^{2+}-activated channel involved in signal transduction for sweet, umami, and bitter tastes.
PKD2L1	Polycystic Kidney Disease 2-Like 1	Previously thought to mediate sour detection; now associated with broader chemosensory roles.
PKD1L3	Polycystic Kidney Disease 1-Like 3	Forms complex with PKD2L1; contributes to general chemosensory functions.
CALHM1	Calcium Homeostasis Modulator 1	Forms an ion channel complex that facilitates ATP release from Type II taste cells for neurotransmission.
CALHM3	Calcium Homeostasis Modulator 3	Works with CALHM1 in ATP release; required for proper neurotransmitter release in Type II taste cells.
OTOP1	Otopetrin 1	Proton-selective ion channel enabling sour taste detection; expressed in Type III taste receptor cells.
GPR120	G-Protein Coupled Receptor 120	Candidate receptor for detecting dietary fats; proposed to respond to long-chain fatty acids.
CD36	Cluster of Differentiation 36	Fatty acid translocase is involved in lipid detection and contributes to the oral perception of fat.

References

Bartoshuk, L. M., Duffy, V. B., & Miller, I. J. (1994). PTC/PROP tasting: Anatomy, psychophysics, and sex effects. *Physiology & Behavior, 56*(6), 1165–1171. https://doi.org/10.1016/0031-9384(94)90361-1

Boring, E. G. (1942). *Sensation and perception in the history of experimental psychology.* Appleton-Century-Crofts.

Chandrashekar, J., Hoon, M. A., Ryba, N. J. P., & Zuker, C. S. (2006). The receptors and cells for mammalian taste. *Nature, 444*(7117), 288–294. https://doi.org/10.1038/nature05401

Collings, V. B. (1974). Human taste response as a function of locus of stimulation on the tongue and soft palate. *Perception & Psychophysics, 16*(2), 169–174. https://doi.org/10.3758/BF03203282

Darwin, C. (1871). *The descent of man, and selection in relation to sex.* John Murray. https://darwin-online.org.uk/converted/pdf/Descent_F937.pdf

Duffy, V. B., Hayes, J. E., Sullivan, B. S., & Faghri, P. (2004). Surveying food and beverage liking: A tool for epidemiological studies to connect chemosensation with health outcomes. *Chemical Senses, 29*(7), 531–543. https://doi.org/10.1093/chemse/bjh054

Hanig, D. P. (1901). Zur Psychophysik des Geschmackssinnes [*On the psychophysics of the sense of taste*]. *Philosophische Studien, 17*, 576–623. https://archive.org/details/philosophischest17wund/page/576

Peng, Y., & Peng, Y. (2022). The tongue map and spatial modulation of taste perception: A review. *Chemosensory

Perception, 15(3), 73–83. https://doi.org/10.1007/s12078-022-09320-y

Plassmann, H., O'Doherty, J., Shiv, B., & Rangel, A. (2008). Marketing actions can modulate neural representations of experienced pleasantness. *Proceedings of the National Academy of Sciences, 105*(3), 1050–1054. https://doi.org/10.1073/pnas.0706929105

Purves, D., Augustine, G. J., & Fitzpatrick, D. (Eds.). (2012). *Neuroscience* (5th ed.). Sinauer. https://www.ncbi.nlm.nih.gov/books/NBK10804/

Rolls, E. T. (2015). Taste, olfactory, and food texture processing in the brain, and the control of food intake. *Physiology & Behavior, 152*, 431–443. https://doi.org/10.1016/j.physbeh.2015.05.018

Roper, S. D., & Chaudhari, N. (2017). Taste buds: Cells, signals and synapses. *Nature Reviews Neuroscience, 18*(8), 485–497. https://doi.org/10.1038/nrn.2017.68

Small, D. M. (2010). Taste representation in the human brain. *Behavioral and Brain Sciences, 33*(2–3), 189–190. https://doi.org/10.1017/S0140525X10000463

Observations on the Sense of Taste

Taste is foundational but narrow: Compared with olfaction, the gustatory system provides limited qualitative information—sweetness, sourness, saltiness, bitterness, and umami—but it provides critical signals of balance, structure, and safety that define the framework of flavor.

Mouthfeel completes what taste begins: Pure taste sensations are transient and two-dimensional until joined by tactile and chemesthetic feedback. The senses of viscosity, temperature, and astringency transform basic tastes into perceived textures, roundness, or attack (initial onset impression).

Olfaction provides identity: taste without smell is anonymous; it conveys intensity and valence but not character. Flavor identity emerges only when retronasal olfaction integrates with gustatory and tactile signals in the orbitofrontal cortex.

Taste calibrates the evaluator: Recognizing the limits of gustation refines professional judgment: accurate sensory evaluation depends less on distinguishing "taste" than on understanding its modulation by ethanol, mouthfeel, and aroma synergy.

Taste functions as an early warning system: bitterness, sourness, and saltiness evolved as protective cues, guiding ingestion decisions long before conscious evaluation.

Adaptation reshapes perception: Prolonged exposure to a single taste quality suppresses sensitivity to other qualities, altering the balance and emphasizing contrast in sequential tasting.

Training the palate refines discrimination: A systematic comparison of calibrated standards builds neural precision, enabling evaluators to detect subtler gradients of sweetness, acidity, or bitterness within complex matrices.

Chapter 6: Chemesthetic Mouthfeel, and Cross-modal Effects

Chemesthesis (mouthfeel) is the final component of the flavor equation, defining "finish."

Contemporary Classification of Mouthfeel

The contemporary model comprises about 25 mouthfeel sensations (Guinard & Mazzucchelli, 1996), about 10 chemesthetic (trigeminal), 10 tactile/mechanical, and five cross-modal (not standardized). There are three categories of mouthfeel. The grouping appears arbitrary, yet it reflects the sensory system's division of labor and evolutionary purpose. It also offers practical utility in evaluating spirits by shaping both immediate perception and the lingering "finish."

Category 1: Trigeminal Chemesthesis

Trigeminal chemesthesis encompasses sensations such as burning, cooling, tingling, stinging, and numbing. These sensations originate from receptors such as TRPV1, TRPM8, and TRPA1 (Caterina et al., 1997; McKemy et al., 2002; Story et al., 2003; Green, 1996), which signal the brain about potential harm or novel stimuli—signals essential for survival. In spirits tasting, these sensations define how ethanol may mask aroma, numb the palate, or alter drinkers' focus. Without accounting for them, we risk conflating chemesthetic intensity with flavor quality. Irritation, pungency, burn, itchiness, and pain are not flavors and interfere with accurate evaluation. Fig. 6.1 describes mouthfeel, often confused with taste. (Compare to Fig. 5.1). Note that trigeminal sensors are also located in the nasal passages. See p. 316 (oral vs nasal cavities).

Fig. 6.1	Trigeminal Irritant Mouthfeel Sensations*	
Mouthfeel	**Stimuli**	**Primary Mechanism**
Ethanol heat	High ABV, cask strengths	TRPV1/TRPA1, also anesthetic
Burning	Capsaicin, chili, piperine	TRPV1 ion channels
Cooling	Menthol, eucalyptus	TRPM8 channels
Tingling	CO_2, Szechuan pepper	TRPA1/acid-sensitive ion channels
Stinging	Acetic acid, carbonyls	TRPV1/acid-sensitive receptors
Numbing	Eugenol, clove, benzocaine	Voltage-gated Na^+ channel inhibition
Pungent	Allyl isocyanate, wasabi	TRPA1 activation
Scratchy	Sulfur, onion, garlic	TRPV1/TRPA1 activation
Cool/hot shift	High-concentrate menthol	Overlap of TRPM8/TRPV1
Metallic	Metal salts (Fe/Cu), blood, CO_2	TRPA1/acid receptors/trigeminal electrochemical
*Chemical irritants that activate trigeminal nerve fibers. NOTE: metallic sensations may also arise from oral galvanic interactions or lipid-oxidation volatiles; not all 'metallic' is chemesthetic.		

Evaluation Note 6.1: When assessing spirits, pause to note where trigeminal burn dominates. Irritation and burn intensity overlap with aroma or taste, potentially obscuring detection and identification and misleading scoring and evaluation, separating trigeminal irritation from flavor.

Category 2: Tactile/Mechanical Sensations

By contrast, tactile/mechanical sensations such as viscosity, smoothness, roughness, and dryness are mediated by mechanoreceptors (Haggard & de Boer, 2014) in the mouth's mucosa and skin layers. These signals are critical for perceiving texture and consistency (Fig. 6.2). In high-fat or highly viscous spirits, this classification helps explain why some liquids feel "full-bodied" or "texturally rich," shaping mouthfeel after aromas fade.

Evaluation Note 6.2: During evaluation, separate tactile impressions from trigeminal burn. A viscous texture can enhance the sense of balance, even if the ethanol burn is high. Familiarity with various tactile and mechanical sensations, as well as the ability to distinguish them from trigeminal irritation, enhances evaluation accuracy and fairness.

Fig. 6.2	Tactile/Mechanical Mouthfeel Sensations*	
Mouthfeel	**Stimuli**	**Primary Mechanism**
Viscosity	Sugar, dense syrups, glycerol	Merkel cells, Meissner corpuscles
Smoothness	Cream, custard	Slowly adapting mechanoreceptors
Rough/grainy	Gritty foods, sandy texture	Pacinian/Meissner activation
Crunchiness	Chips, nuts, brittle solids	Rapid mechanoreceptor bursts
Hardness	Crusty bread	Mechanoreceptors (periodontal)
Slipperiness	Oils, fats	Lubrication + mechanoreceptors
Tenderness	Cooked meat, soft gels	Deformation cues + mechanoreceptors
Wetness	Juicy fruits, soups, and free water	Thermoreceptors + mechanoreceptors
Coating	Fats, oils, dairy residues	Persistent lubricating film, tribology
Drying	Crackers, high tannins	Lubricant loss, saliva-polyphenol
*Somatosensory mechanoreceptors, physical properties, texture, surface, temperature		

Category 3: Cross-Modal/Integral Sensations

The final group is cross-modal or integral sensations, which emerge where chemesthetic and mechanical channels overlap. Creaminess, oiliness, fullness, and prickle combine multiple receptor types simultaneously. They are not discrete in isolation, but they arise predictably from specific compound families. For example, carbonation produces a "prickle" via simultaneous activation of TRPA1 and mechanoreceptive neurons (Dessirier et al., 2000; Dessirier et al., 2001; Silver & Finger, 2009; Wang et al., 2011). See Fig. 6.3.

Evaluation Note 6.3: Recognize and record cross-modal sensations. They often predict consumer liking more accurately than isolated descriptors. Consumer acceptance evaluators should become familiar with cross-modal sensations that appeal to specific consumer demographics.

Completing the "Finish" Experience

These mouthfeel sensations are essential to understanding a spirit's "finish" which, as established in Chapter 4, is a composite of retronasal aroma, taste, and tactile/chemical sensory. A long, warm whiskey finish may include heat from ethanol (trigeminal), weight from viscosity, and a lingering oily tube of flavor. All three classification groups converge to produce a coherent mouth exit, revealing why a spirit's finish is as much felt as smelled or tasted.

Fig. 6.3	Cross-Modal/Integrating Mouthfeel Sensations*	
Mouthfeel	**Stimuli**	**Primary Mechanism**
Creaminess	Emulsified fats, dairy	Lube/viscosity + stable emulsion droplets
Fullness	High viscosity/fat	Viscosity + mechano ± fat (CD36, GPR120)
Oiliness	Oils, fatty emulsions	Mechano + fat-sensing (CD36, GPR120); TRPV1 only if irritants present (chili oil)
Prickle	CO_2 commonly comes from carbonated mixers	Carbonic anhydrase → acidification → TRPA1 (± ASICs) + mechanosensory
Astringency	Polyphenols, tannins	Salivary protein binding, lubricant loss, oral tribology, ± TRP contribution (Pires, et al., 2020)
*Combined input from trigeminal, mechanoreceptive, thermal, and gustatory paths		

Evaluation Note 6.4: The flavor equation (Chapter 2, page 7) provides a rule of thumb for finish evaluation, yet the impact of certain mouthfeels can upset the balance, which the evaluator must consider when rating and scoring.

Why This Classification Structure Endures

Sensory scientists value this tripartite framework for consistency with receptor biology, ease of measurement, and practical relevance. Older classifications loosely group sensations, and this grouping reflects neural pathways and supports objective evaluation (e.g., Calibrating samples to isolate trigeminal from mechanical sensations improves reproducibility in sensory panels).

The current receptor-based model of mouthfeel builds upon earlier frameworks developed by sensory scientists such as Guinard & Mazzucchelli (1996), who pioneered structured terminology and experiential mapping of oral texture, irritation, and flavor persistence. Although their model was not explicitly receptor-driven, it highlighted consumer-relevant categories that remain consistent with neurobiological insights.

Future Directions in Chemosensory Research

As interest grows in how mouthfeel influences consumer preferences, particularly for low-alcohol or novel spirits, the classification provides a roadmap for future research. Genetic studies of TRP receptor variants may explain sensitivity to heat or cold. Advances in haptic instrumentation could enable more accurate modeling of tongue sensation. Moreover, cross-modal scent/tactile synergy—research into why carbonation feels brighter when paired with aroma—may unlock new avenues for flavor design. The future of chemosensory science may redefine both spirits evaluation and culinary art/food formulation.

Evaluation Note 6.5: The tripartite framework for chemosensory mouthfeel is a solid, helpful foundation for determining which spirits' characteristics are not taste or aroma. Know and understand the difference; remember that, regardless of taste or aroma, mouthfeel makes or breaks the overall spirit rating.

Summary

Spirits' mouthfeel stems from trigeminal chemesthesis (burn/cool/tingle), tactile texture (viscosity/smoothness), and cross-modal blends (creaminess, prickle). Ethanol masks the aroma; the carbonation prickle arises from carbonic anhydrase-mediated acidification. Robust "finish" scoring distinguishes irritation from flavor and weights texture toward retronasal aroma. The chemesthetic pathway is similar to the olfactory and gustatory pathways. See Fig. 6.4.

Beyond: The creation and origin of flavors and aromas, and how they are organized and classified.

Fig. 6.4 The Human Chemesthetic Pathway

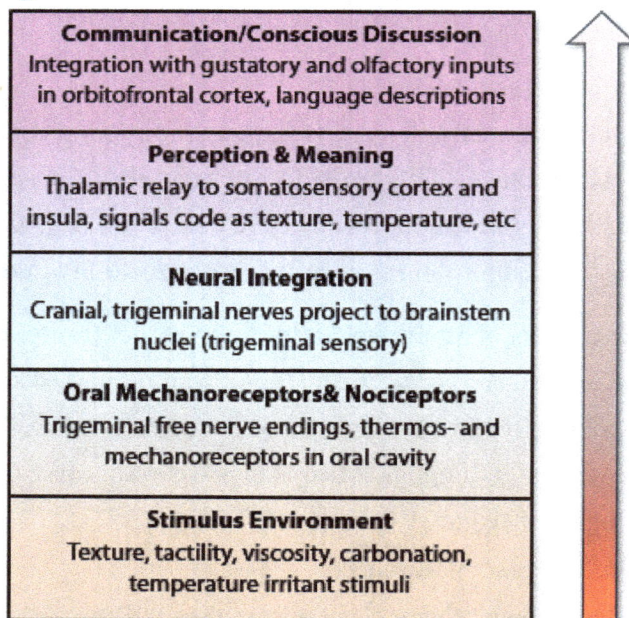

Communication/Conscious Discussion
Integration with gustatory and olfactory inputs in orbitofrontal cortex, language descriptions
Perception & Meaning
Thalamic relay to somatosensory cortex and insula, signals code as texture, temperature, etc
Neural Integration
Cranial, trigeminal nerves project to brainstem nuclei (trigeminal sensory)
Oral Mechanoreceptors& Nociceptors
Trigeminal free nerve endings, thermos- and mechanoreceptors in oral cavity
Stimulus Environment
Texture, tactility, viscosity, carbonation, temperature irritant stimuli

"What you feel in your mouth can override what you taste on your tongue."— Dr. Charles Spence, experimental psychologist

Mouthfeel Receptor Appendix

Appendix 6.1	Mouthfeel Receptor Appendix	
Sensor	**Definition**	**Description / Location**
TRPV1	Transient Receptor Potential Vanilloid 1	Detects heat, capsaicin, and acidity; located in the mucosa, these fibers are part of the trigeminal system.
TRPA1	Transient Receptor Potential Ankyrin 1	Responds to pungent irritants like mustard oil and wasabi, found in trigeminal neurons.
TRPM8	Transient Receptor Potential Melastatin 8	Detects cooling sensations (menthol); found in trigeminal afferents.
TRPV3	Transient Receptor Potential Vanilloid 3	Activated by warm temperatures; expressed in keratinocytes.
TRPV4	Transient Receptor Potential Vanilloid 4	Sensitive to osmotic pressure and warmth; expressed in mucosal tissues.
ASICs	Acid-Sensing Ion Channels	Activated by protons (low pH); found in trigeminal nerve endings.
Piezo1/2	Mechanosensitive Ion Channels	Detect pressure and stretch; located in Merkel cells and other mechanoreceptors (inferred, not confirmed)
Merkel cells	Tactile cells	Detect sustained pressure; communicate with afferent nerves.
Meissner corpuscles	Rapidly adapting mechanoreceptors	Detect light touch and flutter in the dermal papillae of the tongue and lips.
Pacinian corpuscles	Deep pressure and vibration sensors	Located in the deep submucosal and subcutaneous layers.
Ruffini endings	Stretch receptors	Detect sustained tension, present in oral tissues and skin.

References

Caterina, M. J., Schumacher, M. A., Tominaga, M., Rosen, T. A., Levine, J. D., & Julius, D. (1997). The capsaicin receptor: A heat-activated ion channel in the pain pathway. *Nature, 389*(6653), 816–824. https://doi.org/10.1038/39807

Dessirier, J.-M., O'Mahony, M., & Carstens, E. (2000). Psychophysical and neurobiological evidence that the oral sensation elicited by carbonated water is of chemogenic origin. *Chemical Senses, 25*(3), 277–284. https://doi.org/10.1093/chemse/25.3.277

Dessirier, J.-M., Simons, C. T., Sudo, M., Sudo, S., & Carstens, E. (2001). The perception of carbonation: Oral trigeminal and taste interactions. *Chemical Senses, 26*(5), 601–608. https://doi.org/10.1093/chemse/26.5.601

Green, B. G. (1996). Chemesthesis: Pungency as a component of flavor. *Trends in Food Science & Technology, 7*(12), 415–420. https://doi.org/10.1016/S0924-2244(96)10043-1

Guinard, J.-X., & Mazzucchelli, R. (1996). The sensory perception of texture and mouthfeel. *Trends in Food Science & Technology, 7*(7), 213–219. https://doi.org/10.1016/0924-2244(96)10025-X

Haggard, P., & de Boer, L. (2014). Oral somatosensory awareness. *Neuroscience & Biobehavioral Reviews, 47*, 469–484. https://doi.org/10.1016/j.neubiorev.2014.09.015

McKemy, D. D., Neuhausser, W. M., & Julius, D. (2002). Identification of a cold receptor reveals a general role for TRP channels in thermosensation. *Nature, 416*(6876), 52–58. https://doi.org/10.1038/nature719

Pires, M. A., Gonçalves, B., & Rocha, S. (2020). Sensorial perception of astringency: Oral mechanisms and consequences for food texture. *Foods, 9*(8), 1127. https://doi.org/10.3390/foods9081127

Silver, W. L., & Finger, T. E. (2009). Cellular and molecular basis for CO_2 detection in the nose. *Proceedings of the National Academy of Sciences, 106*(28), 10644–10649. https://doi.org/10.1073/pnas.0903146106

Story, G. M., Peier, A. M., Reeve, A. J., et al. (2003). ANKTM1 (TRPA1), a TRP-like channel expressed in nociceptive neurons, is activated by cold. *Cell, 112*(6), 819–829. https://doi.org/10.1016/S0092-8674(03)00158-2

Wang, Y.-Y., Chang, R. B., & Liman, E. R. (2011). TRPA1 is a component of the nociceptive response to CO_2. *Journal of Neuroscience, 31*(8), 3015–3021. https://doi.org/10.1523/JNEUROSCI.5757-10.2011

Observations on Chemesthesis

The "Cask Burn" Lesson — Glenlivet Cooper's Story: A cooper at Glenlivet once noted that casks left to dry before filling produced whisky with a sharper "nose burn." Later GC–MS analysis confirmed elevated levels of ethyl acetate and acetic acid, produced via micro-oxygenation, in dry staves. This imbalance in volatile compounds increased trigeminal irritation far more than aroma, suggesting that harshness often arises from compositional shifts rather than sole alcohol strength.

Overcut Rum — The Solvent Edge: A Caribbean distiller described a rum batch where head cuts ran too deep. Although the ethanol content was normal, tasters reported sharp nasal prickling and numbness on the palate. Analysis revealed excess acetaldehyde, accompanied by low levels of esters, both of which intensified the trigeminal sting. "It smelled sweet but felt like vapor,"—a case where an overcut shifted sensation toward chemesthetic burn.

The Dilution Misstep — Vodka's Vanishing Texture: A craft vodka producer found that direct dilution from high-proof to bottling strength, instead of gradual proofing over days, produced a "thin" mouthfeel. Transient ethanol–water re-equilibration created short-lived microheterogeneity, shifting volatility and trigeminal response—an invisible structural lag that altered perceived smoothness despite identical final chemistry.

The Peated Whisky Paradox: A master blender noted that heavily peated whiskies often feel softer than expected despite their intense smoke. Phenolics, such as guaiacol and cresols, dominate the aroma and can mask the trigeminal bite, making the ethanol seem gentler, even at high strength—an example of sensory competition shaping perceived heat.

The Rehydration Shock — Mezcal's First-Pour Sting: A mezcalero noted that the first pour from clay pot stills delivered a sharp nasal bite. Elevated dissolved CO_2 and micro-bubbles from transfer released carbonic acid and light volatiles, temporarily amplifying trigeminal burn. After a brief rest, the harshness faded, showing micro-aeration alters chemesthesis independently of ethanol.

Section II:

Flavor and Aroma Classification

Quick Reference to Link Chemical Names and Families to Aromas

Gaining familiarity with the chemical names of aromas can be daunting without a handy reference guide. These tables improve familiarity by linking aromas to the chemistry of fermentation, distillation, and barrel aging, a key to becoming a knowledgeable and proficient evaluator.

Identifying Chemical Families in Spirits

Chemical Family	Typical Odor/Flavor	Source in Spirit	Common Suffix	Example Compound
Esters	Fruity, floral, sweet	Fermentation, esterification	-ate	Ethyl acetate, isoamyl acetate
Aldehydes	Green, nutty, grassy, sometimes pungent	Oxidation of ethanol or acids	-al	Acetaldehyde, vanillin
Ketones	Creamy, buttery, caramel	Maillard reactions, wood aging	-one	Diacetyl (butanedione), cyclopentanone
Alcohols (higher/fusel)	Solvent, floral, oily, sweet	Yeast metabolism (amino acid catabolism)	-ol	Isoamyl alcohol, 2-phenylethanol
Acids (volatile)	Sour, vinegary, rancid (low MW); cheesy (high MW)	Oxidation or bacterial metabolism	-ic acid	Acetic acid, butyric acid
Lactones	Coconut, creamy, woody	Oak wood, oxidation of lipids	-lactone	β-methyl-γ-octalactone (whisky lactone)
Phenols	Smoky, medicinal, tarry	Peat smoke, oak char, degradation of lignin	-phenol	Guaiacol, cresol, eugenol
Terpenes	Citrus, floral, resinous	Botanicals, grapes, aged wood	-ene / -ol / -one	Linalool, limonene, α-pinene
Sulfur Compounds	Meaty, vegetal, onion, rubbery	Yeast metabolism, reduction, copper catalysis	thiol / -sulfide	Methanethiol, dimethyl sulfide
Pyrazines	Nutty, roasted, earthy	Toasted wood, Maillard reactions	-pyrazine	2,3,5-trimethylpyrazine
Furans	Caramel, toasted, nutty	Heat degradation of sugars, wood toasting	-furan	5-hydroxymethylfurfural
Amines	Fishy, ammoniacal	Fault compounds from yeast autolysis	-amine	Putrescine, cadaverine

Structural Clues in Compound Naming

Suffix	Chemical Function	Common Sensory Character	Example
-ol	Alcohol	Sweet, oily, solvent-like	Isoamyl alcohol
-al	Aldehyde	Green, grassy, nutty	Acetaldehyde, vanillin
-one	Ketone	Buttery, creamy, soft	Diacetyl
-ate	Ester	Fruity, sweet	Ethyl acetate
-ic acid	Acid	Sour, pungent	Acetic acid
-ene	Unsaturated hydrocarbon (terpene)	Citrus, pine, floral	Limonene
-lactone	Cyclic ester	Creamy, coconut, oak	Whisky lactone
-phenol	Aromatic hydroxyl compound	Smoky, medicinal, clove	Guaiacol

Chapter 7: The Origin of Flavors in Spirits

Introduction

Spirits are born through a chain of biological and chemical transformations, each contributing distinct flavor compounds and sensory attributes. From the initial selection of raw materials to the complex interactions during fermentation, distillation, and maturation, every step contributes to shaping the spirit's final character. A comprehensive understanding of where and how these flavors originate is essential for any serious spirits evaluator. This chapter traces the full production arc—from raw materials to post-distillation additions—while introducing two new terms to distinguish spirits' flavor origins from those commonly used for wine. Raw Material Provenance (RMP) and Agrochemical Signature.

Case Study 7.1: Setting Sensory Objectives for Raw Materials: Dakota Rader, founder and CEO of Golden Eagle vodka, set specific objectives for their flagship vodka by first defining sensory objectives, including viscosity and body, which are driven by glycerol and protein-derived compounds, resulting in a spirit with texture on the palate rather than the water-like neutrality. Aromatic nuance – subtle but identifiable esters and aldehydes, creating soft notes of vanilla, toasted nut, or gentle sweetness, much like the secondary flavors in wine. Resilience under minimal processing – a grain that could shine with one careful distillation instead of being stripped down by many still passes. Most producers focus primarily on the complex distillation process without deliberately defining the sensory process, thereby failing to tightly control the production factors that can achieve the sensory objective, which frequently results in post-production "tweaking," blending, or additions to mask flaws.

Raw Material Provenance (RMP) and Agrochemical Signature

The wine industry uses the term *terroir* to describe the sum of environmental factors (climate, soil, elevation, and viticultural practices) that influence a wine's flavor profile (Jackson, 2008). However, in the case of distilled spirits, this term becomes problematic. Most distillers do not grow their own grains or sugarcane, and raw materials are often transported long distances from their place of origin (Piggott, 2012). Among wine critics, the term terroir has become a "dumping ground" for traits that are difficult to quantify, thereby confusing the original intent of identifying and classifying aromas. Regarding spirits, terroir cannot account for production variables that dominate post-harvest flavor, since raw material agriculture and spirit manufacturing are rarely co-located. To resolve inconsistencies, two new, precise terms must be defined:

- **Raw Material Provenance (RMP)** includes the distinct environmental, geographical, and biological origins of the fermentable substrate, including its species, variety, harvest time, and post-harvest handling. RMP specifically excludes production and processing effects.

- **Agrochemical Signature** refers to the unique composition of primary and secondary plant metabolites, nutrients, and trace compounds found in raw materials as a result of environmental exposure, soil chemistry, agricultural inputs, and climatic stress (Ebeler & Thorngate, 2009). This signature is independent of fermentation, distillation, or maturation.

These concepts provide a more accurate framework for sensory science in spirits, allowing evaluators to recognize environmental origins without misapplying erroneous wine-centric terminology, yet keeping raw-material origins analytically separate from production effects (fermentation, distillation, maturation). Also see Glossary.

Case Study 7.2 Raw Material Provenance (RMP) in Grain: Careful material selection led to the creation of a new spirits category, Character Vodka. Having set sensory objectives (Case Study 7.1), Golden Eagle vodka evaluated wheat species to identify those that could provide a more rounded, pleasant mouthfeel. Soft, white wheats produced thin spirits, hard red winter wheats improved body, but were light in flavor. Specifically, a less common strain of hard red spring wheat, grown in the northern Midwest has a geographical provenance and extremely high protein, and thus greater free amino nitrogen (FAN) after mashing (higher FAN → more Ehrlich pathway flux to fusel alcohols/esters) (Hazelwood et al., 2008), while redox and osmotic stress response produce more glycerol for a higher viscosity and rich, full, mouthfeel. Hot days, cold nights, and mineral-rich soils in this location induce stress cycles in the grain, which, in turn, elevate protein content and secondary metabolite levels. Processing and distillation are not located on the grow site, providing a prime example of the newly coined terms "Raw Material Provenance" (RMP) and "Agrochemical Signature," both of which are important to define key differences from the commonly misused and misapplied term "terroir" when discussing spirits.

Fermentation: Phases, Conditions, and Flavor Origins

Fermentation is the primary generator of volatile aroma compounds in all spirits. Fermentation far exceeds the contributions of raw materials in both magnitude and diversity, particularly in the number of volatile compounds (Rojas & Dellaglio, 2001). The flavor profile established in fermentation is carried through distillation and often defines spirit identity.

Phases of Fermentation

Fermentation progresses through predictable microbial stages, though exact timing depends on temperature, substrate, and yeast strain (Fleet, 2003):

Lag Phase (0–6 hours): Yeast acclimates to its environment. No significant alcohol or flavor production has occurred at this time. During the lag phase, yeast cells synthesize and incorporate sterols, primarily ergosterol, into their membranes, thereby maintaining membrane fluidity, resisting ethanol toxicity, and preparing for budding and nutrient uptake (Walker, 2011). All of these processes are essential for a successful fermentation.

Exponential (Log) Phase (6–36 hours): Yeast begins rapid sugar metabolism. Primary flavor production happens here. Lactic acid bacteria (LAB), such as *Lactobacillus* and *Pediococcus*, typically become active during the early Exponential Phase of fermentation, especially in warm, nutrient-rich, low-oxygen mashes around 30–38°C (86–100°F) (Bokulich & Bamforth, 2013). These microbes convert residual sugars into lactic acid, esters, and other volatile compounds that can contribute either desirable fruity and sour complexity or, if left unchecked, undesirable spoilage. As fermentation progresses, the environment becomes anaerobic, acidic, and alcoholic. Strong yeast membranes play a crucial role in conferring ethanol tolerance and acid stress resistance. As budding and cell division occur, ergosterol and fatty acids build new membranes.

Stationary Phase (36–72 hours): Nutrient depletion slows growth; alcohol and fusel oil concentrations increase (Pretorius, 2000).

Decline Phase (72+ hours): Yeast activity drops; esters and off-compounds may accumulate under stress (Bisson & Karpel, 2010).

Conditions During Fermentation: Temperature, pH, and oxygen exposure influence the dominance of each phase, affecting the yield and sensory balance of fermentation-derived compounds.

Origin of Major Flavor Compounds from Fermentation
Fermentation produces hundreds of esters, alcohols, acids, and aldehydes through yeast metabolism (Swiegers et al., 2005). Primary contributors include:
- Esters (e.g., isoamyl acetate, ethyl hexanoate): Fruity, floral, banana, and solvent-like aromas.
- Higher Alcohols (fusel oils): Waxy, spicy, or solventy; some enhance complexity, others overwhelm balance (Hazelwood et al., 2008).
- Volatile Fatty Acids (e.g., butyric, caproic): May yield cheesy, rancid, or tropical notes depending on concentration.
- Sulfur Compounds (e.g., hydrogen sulfide, mercaptans): Arise under stress or poor nutrition; often undesirable.
- Carbonyls and Aldehydes (e.g., acetaldehyde): Contribute sharp, green apple, or metallic character (Moreira et al., 2002).

Managing yeast strain, temperature (e.g., commonly 25–33°C for rum and whiskey), nutrient supplementation, and fermentation length allows producers to emphasize or suppress different compound families. Although there are many more complex steps in fermentation, those chemical stages are not pertinent to this text. The resultant, simplified end reaction is represented as:

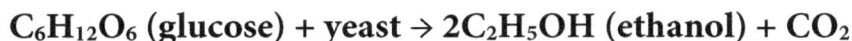

$$C_6H_{12}O_6 \text{ (glucose)} + \text{yeast} \to 2C_2H_5OH \text{ (ethanol)} + CO_2$$

Case Study 7.3: Fermentation Difficulties to Protect RMP: The ongoing case study of Golden Eagle vodka illustrates the trade-offs a distiller must make during fermentation to achieve the final sensory and flavor objectives. High-protein wheat is notoriously challenging to work with in the fermentation process, as high foam, stalled fermentation, and vegetal aromas are key reasons why most vodka distillers avoid high-protein grains. The vast majority concentrate on high production at the lowest cost, achieving flavor neutrality and high ethanol with short fermentation times. Golden Eagle preferred to tackle the fermentation issues as a value proposition with flavor objectives, rather than following the usual financial-value course, and this chosen path required more patience, technical expertise, and a slight increase in production costs. This integrated approach to production, aimed at achieving a well-defined sensory objective, continued from the agrochemical signature to fermentation process control.

Distillation: Separation and Selective Flavor Retention

Distillation does not create flavor—it selects which fermentation-derived and thermal reaction compounds to retain based on their boiling points, volatilities, and solubilities (Piggott et al., 1993). The still's shape and design, reflux ratio, distillation rate, and cut points determine which volatiles reach the final spirit.

Heads (Foreshots): These are the earliest volatiles collected. They contain:
- Low boiling point compounds: methanol, acetaldehyde, ethyl acetate, and light esters.
- Sensory notes: harsh, solvent-like aromas, green, sharp, or chemical (Zhang et al., 2021).
- Methanol and high concentrations of acetaldehyde are toxic.

Head cuts are made based on temperature (65–72°C), with an illustrative approximation depending on ABV, atmospheric pressure, vapor composition, and alcoholmeter/hydrometer/proof or sensory sniff tests. Most spirits discard this fraction or recycle it for redistillation.

Hearts (Spirit Run): This middle fraction contains:
- Desirable ethanol fraction and mid-range volatiles, including ethyl esters, isoamyl alcohol, and sweet fusels.
- Sensory notes: fruity, creamy, floral, grainy, balanced (Russell & Stewart, 2014).

Cutting into hearts is based on stable temperature plateaus (illustrative 78–83°C), consistent spirit strength (65–72% ABV), and continuous sensory monitoring for off-odors.

Tails (Feints): The final portion of distillation includes:

- High boiling point compounds (higher than ethanol): long-chain fatty acids, higher level alcohols (fusels), phenols, sulfur, and waxes.
- Sensory notes: soapy, vegetal, cheesy, bitter, sulfuric, and occasionally smoky or nutty (Barbosa et al., 2020).
- Use: Often recycled for oil extraction or blended back for flavor layering.

Tails are cut when the vapor temperature is approximately 90°C (illustrative) or the spirit's ABV falls below 50%, depending on still size and load. Prolonged collection can introduce off-flavors if not carefully controlled.

Azeotropic behavior and still design are more important determinants of vapor composition than is temperature. A deeper discussion of the art of making cuts and their effects on spirit flavors is presented in Chapter 22.

Flavor from Barrel Aging

While a more comprehensive discussion of barrel aging is presented in Chapter 23, it is worth noting that maturation in oak barrels contributes vanillin, lactones (such as those found in coconut), furfurals (like those found in caramel), and phenolic smoke notes. These flavors arise from thermal degradation of hemicellulose, lignin, and cellulose during toasting and charring (Pérez-Coello & Díaz-Maroto, 2009). Oxidative processes also convert aldehydes to acids and esters, increasing complexity and smoothness over time.

Flavor Additions and Botanical Infusions

Some spirits derive flavor from external additions rather than fermentation or maturation:

- Gin: Juniper and other botanicals are infused via maceration, vapor infusion, or basket distillation. Compounds include terpenes, aldehydes, and phenolics.
- Flavored Vodkas, Rums, and Liqueurs: Artificial or natural flavorings such as citrus oils, vanilla extracts, or syrups are added post-distillation, as well as various spices.

Additions and infusions are typically disclosed but may complicate sensory evaluation by masking base spirit flaws or altering perception (Gómez-Cortés & Moreno-Rojas, 2019). Trained panels must distinguish between natural and artificial volatiles by building familiarity with the compound families.

Case Study 7.4: Results of Defining RMP Sensory Objectives Prior to Production. In the final case study on Golden Eagle vodka, comparing the sensory results of a disciplined approach with those of conventional neutral vodkas, the tasting panel's findings demonstrated quantifiable improvements in body, aroma, and mouthfeel. Extended fermentation produced a creamy mouthfeel. Slowing distillation and careful heart-cut preservation retained trace levels of ethyl acetate and ethyl lactate, adding a subtle sweetness and complexity. High-protein raw material (wheat) produced secondary metabolites that were carried into the distillate, yielding a unique, recognizable agrochemical signature distinct from that of other thin, neutral, tasteless vodkas.

Sensory Training Notes

To recognize origin-specific flavors, the student should focus on compound families by process phase:

- Fermentation esters: Identify fruit-forward esters like isoamyl acetate (banana) or ethyl hexanoate (apple).
- Heads indicators: Note ethyl acetate (nail polish), acetaldehyde (green apple), and solvent notes.
- Tails indicators: Learn to identify caprylic acid (goat), phenols (smoke), or vegetal off-notes.
- Barrel indicators: Calibrate to vanillin, cis- and trans-oak lactone, and caramelized sugars.

Evaluators should use reference standards and repeated smell-memory training to distinguish between overlapping profiles, such as charred oak vs. peat smoke, or fruit esters vs. added flavors.

Summary

Every spirit carries the sensory fingerprint of its manufacture. Fermentation defines the base profile, distillation shapes and refines it, maturation adds complexity, and infusion may extend or distort it. By learning to recognize the origin of each aroma or flavor, the sensory evaluator becomes fluent in the chemical language of spirits.

The Distiller's Mission: Create a spirit that balances purity, complexity, and character by managing fermentation to shape flavor, distillation to refine it, and, when appropriate, aging to enhance it while eliminating flaws and capturing only what elevates the final experience.

The Sensory Evaluator's Mission: Objectively assess the spirit's aroma, taste, mouthfeel, and finish to document strengths, flaws, and character. Provide clear, accurate insight into how fermentation, distillation, and aging have shaped the final product, enabling others to make informed decisions based on an honest, unbiased evaluation.

Beyond: Fermentation is the engine of chemical changes: yeast is the fuel that drives the engine.

"Fermentation is life without air." — Louis Pasteur, chemist, microbiologist

"Good whiskey is made in the still. Great whiskey is made in the fermenter." — Dave Pickerell, master distiller

"Distillation is not invention. It is refinement. It is a way of revealing what has always been there, but hidden." — Aeneas Coffey, Coffey still inventor

Appendix 7.1: Recognizing production mistakes

Appendix 7.1	Recognizing Spirit Production Mistakes		
Mistake	**Sensory Impact**	**Recognizable Aromas**	**Description**
Poor head cuts	Sharp solvent, harsh burn	Nail polish remover, paint thinner, pear	Methanol, ethyl acetate, may contain acetone
Incomplete tail separation	Wet, moldy, vegetal, bitter	Wet cardboard, old cheese, damp cellar	Tails rich in fusel oils, fatty acids, heavy volatiles
Still running too hot	Burnt, heavy, harsh, oily	Burnt plastic, rubber, heavy oil	High temperatures force nasty congeners into the distillate
Contaminated wash/mash	Sour, rancid, putrid	Vomit, baby diapers, rotten butter	Bacteria produce butyric and isovaleric acids
Stressed or bad yeast	Rotten eggs, rubber, sulfur	Rotten egg, boiled cabbage, burnt rubber	Hydrogen Sulfide (H_2S) and mercaptans from stressed yeast. Insufficient copper contact
Wrong still materials	Chemical, metallic, toxic	Plastic, diesel, sharp, metallic, twangy taste	Plastics or reactive metals (zinc) contaminate the spirit
Over-distilling (stripping)	Thin, harsh, bland	Faint chemical water, faint alcohol, loss of body	Over-rectifying strips desired flavors
Bad dilution water	Chlorine, earth, musty	Pool water, wet stone, mildew	Impurities in dilution water
Lack of proper aging, airing	Raw, harsh, green, sharp	Raw grain, green wood, biting alcohol	No oxidation, esterification, or mellowing has occurred
Dirty still	Rotten eggs, rubber, sulfur	Rotten egg, boiled cabbage, burnt rubber	Poor cleaning leaves tainting residue for the next run.

References

Barbosa, A., Nascimento, D., & da Silva, E. (2020). Volatile composition of tail cuts in artisanal cachaça: Analytical and sensory implications. *Journal of Food Science, 85*(10), 3120–3129. https://doi.org/10.1111/1750-3841.15362

Bisson, L. F., & Karpel, J. E. (2010). Genetics of yeast impacting wine quality. *Annual Review of Food Science and Technology, 1*, 139–162. https://doi.org/10.1146/annurev.food.080708.100734

Bokulich, N. A., & Bamforth, C. W. (2013). The microbiology of malting and brewing. *Microbiology and Molecular Biology Reviews, 77*(2), 157–172. https://doi.org/10.1128/MMBR.00060-12

Ebeler, S. E., & Thorngate, J. H. (2009). Wine chemistry and flavor: Looking into the crystal glass. *Proceedings of the National Academy of Sciences, 106*(34), 14191–14192. https://doi.org/10.1073/pnas.0906938109

Fleet, G. H. (2003). Yeast interactions and wine flavour. *International Journal of Food Microbiology, 86*(1–2), 11–22. https://doi.org/10.1016/S0168-1605(03)00245-9

Gómez-Cortés, P., & Moreno-Rojas, J. M. (2019). Flavored spirits: A sensory and compositional review. *Beverages, 5*(2), 35. https://doi.org/10.3390/beverages5020035

Hazelwood, L. A., Daran, J.-M., van Maris, A. J. A., Pronk, J. T., & Dickinson, J. R. (2008). The Ehrlich pathway for fusel alcohol production: A century of research on *Saccharomyces cerevisiae* metabolism. *Applied and Environmental Microbiology, 74*(8), 2259–2266. https://doi.org/10.1128/AEM.02625-07

Jackson, R. S. (2008). *Wine science: Principles and applications* (3rd ed.). Academic Press.

https://doi.org/10.1016/C2009-0-00417-5

Moreira, N., Mendes, F., Guedes de Pinho, P., Hogg, T., & Vasconcelos, I. (2002). Volatile compounds contribution of *Saccharomyces cerevisiae* strains to the aroma of fermented apple juice. *Journal of Agricultural and Food Chemistry, 50*(10), 2879–2886. https://doi.org/10.1021/jf011503f

Pérez-Coello, M. S., & Díaz-Maroto, M. C. (2009). Chemical and sensory effects of wood aging on wine and spirits. *Food Science and Technology International, 15*(6), 579–590. https://doi.org/10.1177/1082013209350140

Piggott, J. R. (2012). *Alcoholic beverages: Sensory evaluation and consumer research.* Woodhead Publishing. https://doi.org/10.1533/9780857095176.1.3

Piggott, J. R., Conner, J. M., & Paterson, A. (1993). Flavour development in whisky. In *Whisky: Technology, production and marketing* (pp. 215–237). Academic Press.

Pretorius, I. S. (2000). Tailoring wine yeast for the new millennium: Novel approaches to the ancient art of winemaking. *Yeast, 16*(8), 675–729. https://doi.org/10.1002/1097-0061(20000615)16:8<675::AID-YEA585>3.0.CO;2-B

Rojas, V., & Dellaglio, F. (2001). Yeasts and their production of volatile compounds in alcoholic fermentations. *Food Microbiology, 18*(1), 45–65. https://doi.org/10.1006/fmic.2000.0364

Russell, I., & Stewart, G. G. (2014). *Whisky: Technology, production and marketing* (2nd ed.). Academic Press.

Swiegers, J. H., Bartowsky, E. J., Henschke, P. A., & Pretorius, I. S. (2005). Yeast and bacterial modulation of wine aroma and flavour. *Australian Journal of Grape and Wine Research, 11*(2), 139–173. https://doi.org/10.1111/j.1755-0238.2005.tb00285.x

Walker, G. M. (2011). *Pichia anomala: Cell physiology and biotechnology. Yeast.* https://doi.org/10.1002/yea.1854

Zhang, H., Liu, X., Zhang, L., & Guo, X. (2021). Chemical profiling of head and tail fractions in Baijiu distillation. *Journal of the Institute of Brewing, 127*(2), 197–208. https://doi.org/10.1002/jib.649

Observations on Distilling Mistakes

Low Wines Overcharge: At a Scotch distillery, a spirit still was accidentally charged with low wines (the first distillate 20–25% ABV) at a higher fill than specified. The extra liquid reduced copper contact, allowing more sulfur compounds to survive. The whisky retained faint sulfur notes even after aging. Caught later during nosing, the batch could not be blended into core stock.

Boiler Carryover: In a Kentucky distillery, a steam boiler with insufficient anti-foam treatment pushed boiler water into the still, introducing sodium salts and giving the spirit a "soapy edge." It took several runs before operators realized that an upstream mechanical fault had carried over into the spirit character.

Condenser Flooding: At a rum distillery, a blocked condenser drain trapped refluxing vapor, causing pressure fluctuations that smeared the heads–hearts cut. The resulting spirit exhibited harsh solvent notes and an inconsistent ester balance throughout the run.

Chapter 8: Yeast Strain Contribution to Spirits' Aromas

Yeast selection is one of the most consequential decisions in distilled spirits production, not only for ethanol yield but, critically, for the generation of volatile aroma compounds that survive into the new-make and, after maturation, shape the identity of the final spirit. The strain's genotype influences flux through pathways that yield higher alcohols, esters, organic acids, carbonyls, phenolics, and sulfur compounds; each with characteristic descriptors and detection thresholds (Saerens et al., 2010). Understanding these biochemical origins enables the evaluator to distinguish between production choices, raw-material signatures, and cask effects, to identify process faults, and to make informed inferences about provenance and technique (Hazelwood et al., 2008).

Yeast Biology and Strain Diversity

The workhorse for most beverage fermentations remains *Saccharomyces cerevisiae*, which has been domesticated into numerous lineages that differ in stress tolerance, nutrient uptake, esterification propensity, and fusel alcohol formation (González et al., 2017). Related or co-occurring species—*S. bayanus*, *Torulaspora delbrueckii*, *Hanseniaspora* spp., *Pichia* spp., and, in some artisanal contexts, *Brettanomyces/Dekkera*—can contribute either desirable complexity or faults, depending on process control and intended style (Dzialo et al., 2017). Hybridization and selection programs (and, more recently, rational genetic engineering) have produced strains tuned for particular aroma outcomes, including elevated acetate-ester production or modified higher-alcohol profiles (Celińska et al., 2019). In several well-characterized cases, differences map to changes in alcohol acetyltransferase genes (notably *ATF1/ATF2*) or to changes in amino acid catabolism that feed the Ehrlich pathway (Verstrepen et al., 2003).

Metabolic Pathways Relevant to Aroma Formation

Primary carbohydrate metabolism (glycolysis → pyruvate → acetaldehyde → ethanol) provides the carbon and redox background for aroma formation (Boulton & Quain, 2001). Secondary metabolism branches determine which volatiles accumulate above threshold. Two pathways are critical. First, the Ehrlich pathway converts amino acids to higher (fusel) alcohols via transamination, decarboxylation, and reduction; leucine → isoamyl alcohol and phenylalanine → 2-phenylethanol are canonical examples (Hazelwood et al., 2008). Second, ester biosynthesis couples acyl-CoA moieties (e.g., acetyl-CoA, hexanoyl-CoA) with alcohols via alcohol acetyltransferases (AAT) encoded by *ATF1/ATF2*, yielding acetate esters (e.g., isoamyl acetate, phenethyl acetate) and ethyl esters (e.g., ethyl hexanoate, ethyl octanoate) (Saerens et al., 2010).

Regulation is multifactorial, involving gene expression, lipid composition, oxygenation status (for sterol/unsaturated fatty acid synthesis), nitrogen availability, and temperature, all of which shift the balance between fusel alcohols and esters (Christiaens et al., 2014).

Major Aroma Compound Classes Influenced by Yeast

Methodology note on thresholds. Odor detection thresholds reported in the literature are highly sensitive to the matrix and method used. At a minimum, three variables must be considered:

- **ethanol concentration** (raising ABV elevates thresholds for most odorants, sometimes by orders of magnitude)
- **temperature** (warmer matrices increase headspace partitioning and can lower practical thresholds)
- **matrix composition and testing protocol** (salts, acids, sugars, and panel design alter detectability).

The values in Fig. 8.1 are indicative ranges for hydroalcoholic matrices relevant to new-make spirits; in-bottle perception at different ABVs and temperatures, or with complex congeners, may differ substantially. For ethanol–odorant mixture effects relevant to spirits (e.g., isoamyl acetate and whisky lactone), see Le Berre et al. (2007).

Environmental and Process Factors Modulating Yeast Aroma

Fermentation temperature controls enzyme kinetics and membrane composition. Generally, cooler fermentations suppress the production of fusel alcohols and favor the retention of esters, while warmer regimes increase the production of higher alcohols and shift the ester-to-fusel ratio (Pires et al., 2014). An oxygenation strategy implemented early in fermentation supports sterol/unsaturated fatty acid synthesis, indirectly influencing ester formation (Saerens et al., 2010). Free amino nitrogen (FAN) nutrition and micronutrients are crucial, as amino acid availability drives the Ehrlich pathway substrates and modulates sulfur metabolism. Deficiency risks lead to stuck fermentations and elevated H_2S (Swiegers et al., 2005). Pitching rate and viability influence growth-associated ester formation; under-pitching elevates fusel and H_2S levels, and over-pitching can reduce ester intensity (Verbelen et al., 2009). Stressors (high gravity, osmotic stress, high ethanol) trigger protective responses (Gibson et al., 2007). Pitching rate refers to the number of yeast cells inoculated into a fermenting medium (wort, must, or mash). It is usually expressed as cells/mL/^{0}Plato in brewing, or as cells/unit volume in winemaking and distilling.

Common Yeast Selection in Spirits

The practical question for production, and the diagnostic question for evaluators, is not "does yeast matter?" but "which yeast under which conditions, aiming for which aroma balance?" The table (Fig. 8.1) below summarizes representative use cases. Values for "key compounds" are qualitative (present above threshold, not absolute concentration), because process settings frequently dominate absolute numbers.

Fig. 8.1		Major Aromas Influenced by Yeast			
Compound	Class	Primary Pathway	Descriptor	Threshold (Approx.)	Typical Spirit
Isoamyl alcohol	Higher alcohol	Ehrlich (leucine)	Solvent, fusel, banana oil (at low conc.)	40 mg/L	Pot still rum, malt whisky
Isoamyl acetate	Acetate ester	Alcohol acetyltransferase	Banana, pear drop	1.2 mg/L	Light rum, unaged brandy
Ethyl hexanoate	Ethyl ester	Fatty acid metabolism	Apple, pineapple	0.21 mg/L	Cognac, Calvados
Phenethyl alcohol	Aromatic alcohol	Shikimate (phenylalanine)	Rose, floral	14 mg/L	Brandy, gin (base spirit)
Phenethyl acetate	Acetate ester	Acetyltransferase	Honey, rose	0.25 mg/L	Cognac, malt whisky
Ethyl lactate	Ethyl ester	Esterification of lactic acid	Buttery, creamy	150 mg/L	Rum, grain whisky
Acetaldehyde	Carbonyl	Pyruvate decarboxylation	Green apple, cut grass	25 mg/L	New make whisky
4-Vinyl guaiacol	Phenolic	Phenolic acid decarboxylase	Clove, spicy	0.3 mg/L	Some rums, agave spirits
Dimethyl sulfide (DMS)	Sulfur	Amino acid catabolism	Sweet corn, vegetal	25 µg/L	Molasses rum, immature whisky
Hydrogen sulfide (H_2S)	Sulfur	Sulfate reduction	Rotten egg	1–3 µg/L	Fermentation fault in any spirit

Note: Threshold ranges reflect multi-study variability: isoamyl acetate and ethyl hexanoate thresholds drop at lower ABV; phenethyl acetate and 2-phenylethanol show panel/matrix dependence; DMS and H_2S thresholds are highly protocol-sensitive. Selected sources include Saerens et al. (2010), Verstrepen et al. (2003), and Le Berre et al. (2007). Thresholds rise with ABV; values are in hydroalcoholic matrices, not water.

Faults and Off-Flavors Linked to Yeast Activity

Three recurring yeast-related problems are central to the evaluation of spirits.

- **Sulfur** defects (H_2S, mercaptans, DMS) arise from sulfur assimilation stress and amino-acid catabolism; even µg/L concentrations are conspicuous (Siebert et al., 2010).

- Excess **fusel alcohols** elevate harshness/solvent character and mask esters; typically linked to high temperature or nitrogen imbalance (Pires et al., 2014).

- **Phenolic** taint (4-ethylphenol, 4-vinylphenol) from wild yeasts produces medicinal/smoky notes. Artisanal context; generally faults in premium spirits (Chatonnet et al., 1995).

Note: Primary pathways are of interest to those who need to know the connection between sensory and chemistry and the neurological aspects. Not pertinent to developing evaluation acumen, they are important for integration with other sciences.

Analytical Methods for Studying Yeast-Derived Aroma

GC–FID/GC–MS identifies and quantifies volatiles, while high-resolution MS can resolve isomers and attribute pathways (Kataoka et al., 2000). Descriptive sensory analysis links peaks to perception, and metabolomics/genomics connect gene expression (ATF1/ATF2, amino-acid turnover) to aroma output (Holt et al., 2018). Stable-isotope labeling with ^{13}C-glucose or ^{15}N-amino acids confirms precursor-product relationships (Hazelwood et al., 2008).

Sensory Implications for the Spirits Evaluator

In perspective, yeast leaves a fingerprint distinguishable from raw materials and cask effects. Fruity esters (banana, apple, pear) are yeast-derived; solvent-like fusel edges trace to higher alcohols; clove/spice phenolics (4-vinylguaiacol) and sulfur notes indicate fermentation origin (not aging) (Wang et al., 2024). Evaluators should calibrate expectations by category: rum and agave spirits often show overt fermentation esters; vodka and gin should be neutral; brandies balance grape precursors with yeast esters (Li et al., 2023). Fig. 8.2 details yeast use by spirit.

Fig. 8.2	Yeast Selection by Spirit Type			
Spirit Category	**Common Yeast Choice**	**Compounds >Threshold**	**Sensory Emphasis**	**Rationale**
Scotch malt whisky	Distilling *S. cerevisiae* with moderate–high ester potential	Ethyl hexanoate, isoamyl acetate	fresh apple/pear, soft banana over malt	Fruity complexity without masking cereal notes
American bourbon	Robust *S. cerevisiae* (high ethanol tolerance)	Isoamyl alcohol, ethyl acetate	fusel warmth with light fruit	Performs in high-gravity, high-corn mashes
Rum (molasses)	Mixed *S. cerevisiae/S. bayanus*; NS* yeasts possible	Ethyl butyrate, ethyl acetate	pineapple/ fruity solvent	Supports esterification in high-acid molasses matrices
Brandy/ Cognac	Wine-derived *S. cerevisiae*	Phenethyl acetate, ethyl hexanoate/ lactate	floral/honey, creamy	Preserves grape-derived fruit delicacy
Gin (neutral base)	Low-congener *S. cerevisiae*	Minimal esters	neutral/clean	Avoids interference with botanical distillate
Agave spirits	Wild mixed fermentations (*S. cerevisiae, Torulaspora,* others)	4-vinylguaiacol + ethyl esters	spicy phenolics + tropical fruit	Reflects artisanal substrates and open fermentation
Vodka	High-purity ethanol strains	Trace esters only	neutral	Facilitates charcoal polishing/rectification
Novel	*S. cerevisiae* with enhanced ester/thiol release	isoamyl acetate, 3MH/3MHA	banana/ tropical	Designer signatures for new spirit styles
Variance notes for Fig. 8.2: The ">threshold" designation depends on ABV at evaluation (thresholds rise with ABV) and on distillation cut strategy (carryover varies). Documented cases of strain-driven ester shifts (via *ATF1/ATF2* modulation) and category-specific ester importance (Verstrepen et al., 2003; Saerens et al., 2010). NOTE: Thiols depend on the cut/rectification process and are more likely in pot-still new-make. *NS = non-Saccharomyces yeasts.				

Summary

Yeast is an active architect of spirits' aroma. Strain genetics and fermentation management (temperature, oxygenation, nitrogen status, pitching rate, and stressors) direct flux through the Ehrlich and ester pathways, producing a portfolio of volatiles. Their detectability depends on ABV, temperature, and matrix. Mastering relationships enables accurate attribution, improved fault diagnosis, and more persuasive sensory arguments regarding origin and process.

Beyond: The fermentation engine performs with the production of many different alcohols

"We don't make alcohol — yeast does. We just try not to get in the way." — Dave Pickerell, master distiller

"Yeast is probably your most important flavor component in whisky." — Don Livermore, master blender

References

Abbott, N., Puech, J. L., Bayonove, C., & Baumes, R. (1995). Determination and sensory evaluation of cis- and trans- oak lactones in wines. *Food Chemistry, 51*(2), 135–141. https://doi.org/10.1016/0308-8146(94)P4179-F

Boulton, C., & Quain, D. (2001). *Brewing yeast and fermentation.* Blackwell Science.

Celińska, E., Kubiak, P., Borkowska, M., et al. (2019). Genetic engineering of the Ehrlich pathway modulates production of higher alcohols. *FEMS Yeast Research, 19*(2), foy122. https://doi.org/10.1093/femsyr/foy122

Chatonnet, P., Dubourdieu, D., Boidron, J.-N., & Pons, M. (1995). *Brettanomyces* spoilage compounds in beverages. *Journal of the Science of Food and Agriculture, 67*(3), 309–315. https://doi.org/10.1002/jsfa.2740670306

Christiaens, J. F., et al. (2014). The fungal aroma gene *ATF1* promotes dispersal of yeast cells through insect vectors. *Cell Reports, 9*(2), 425–432. https://doi.org/10.1016/j.celrep.2014.09.009

Dzialo, M. C., Park, R., Steensels, J., Lievens, B., & Verstrepen, K. J. (2017). Non-*Saccharomyces* yeasts in fermentation: Impact on aroma and quality. *Trends in Food Science & Technology, 71*, 39–51. https://doi.org/10.1016/j.tifs.2017.10.006

Gibson, B. R., Lawrence, S. J., Leclaire, J. P. R., Powell, C. D., & Smart, K. A. (2007). Yeast responses to stress in high-gravity fermentations. *Journal of Applied Microbiology, 102*(2), 461–471. https://doi.org/10.1111/j.1365-2672.2006.03122.x

González, R., Quirós, M., & Morales, P. (2017). Wine secondary aroma: Understanding yeast production of higher alcohols. *Microbial Biotechnology, 11*(1), 1–13. https://doi.org/10.1111/1751-7915.13010

Hazelwood, L. A., Daran, J.-M., van Maris, A. J. A., Pronk, J. T., & Dickinson, J. R. (2008). The Ehrlich pathway for fusel alcohol production: A century of research on *Saccharomyces cerevisiae* metabolism. *Applied and Environmental Microbiology, 74*(8), 2259–2266. https://doi.org/10.1128/AEM.02625-07

Holt, S., Mukherjee, V., Lievens, B., Verstrepen, K. J., & Thevelein, J. M. (2018). Polygenic analysis of aroma production in yeast in the absence of major effector *ATF1*. *G3: Genes, Genomes, Genetics, 8*(9), 2909–2923. https://doi.org/10.1534/g3.118.200383

Kataoka, H., Lord, H. L., & Pawliszyn, J. (2000). Applications of solid-phase microextraction in food analysis. *Journal of Chromatography A, 880*(1–2), 35–62. https://doi.org/10.1016/S0021-9673(00)00309-5

Le Berre, E., Atanasova, B., Langlois, D., Nicklaus, S., & Etievant, P. (2007). Impact of ethanol on perception of odorant mixtures. *Food Quality and Preference, 18*(7), 1013–1019. https://doi.org/10.1016/j.foodqual.2007.04.006

Li, H., Zhang, J., Wang, W., et al. (2023). Aroma-active compounds and sensory contributions in Fenjiu. *Foods, 12*(6), 1245. https://doi.org/10.3390/foods12061245

Pires, E. J., Teixeira, J. A., Brányik, T., & Almeida, C. (2014). The influence of fermentation temperature on yeast aroma compounds: A review. *Food and Bioprocess Technology, 7*(1), 145–158. https://doi.org/10.1007/s11947-013-1133-3

Saerens, S. M. G., Verstrepen, K. J., Van Laere, S. D. M., et al. (2010). Production and biological function of volatile esters in *Saccharomyces cerevisiae*. *Microbial Biotechnology, 3*(2), 165–177. https://doi.org/10.1111/j.1751-7915.2009.00106.x

Siebert, T. E., Solomon, M. R., Pollnitz, A. P., & Jeffery, D. W. (2010). Hydrogen sulfide and other sulfur compounds in fermentation: Causes and control. *American Journal of Enology and Viticulture, 61*(3), 221–227. https://doi.org/10.5344/ajev.2010.61.3.221

Swiegers, J. H., Bartowsky, E. J., Henschke, P. A., & Pretorius, I. S. (2005). Yeast nitrogen and aroma formation in wine fermentation. *Australian Journal of Grape and Wine Research, 11*(2), 139–173. https://doi.org/10.1111/j.1755-0238.2005.tb00286.x

Verbelen, P. J., De Schutter, D. P., Delvaux, F., Verstrepen, K. J., & Delvaux, F. R. (2009). The influence of pitching rate on yeast fermentation and ester formation. *Journal of the Institute of Brewing, 115*(2), 134–142. https://doi.org/10.1002/j.2050-0416.2009.tb00356.x

Verstrepen, K. J., Van Laere, S. D. M., Vanderhaegen, B. M., Derdelinckx, G., Dufour, J.-P., Pretorius, I. S., Winderickx, J., Thevelein, J. M., & Delvaux, F. R. (2003). Expression levels of *ATF1* and *ATF2* control acetate esters. *Applied and Environmental Microbiology, 69*(9), 5228–5237. https://doi.org/10.1128/AEM.69.9.5228-5237.2003

Wang, X., Li, P., Zhou, C., et al. (2024). Molecular insights into aroma differences between beer and wine. *Foods, 13*(10), 1587. https://doi.org/10.3390/foods13101587

Observations on Yeast

Scotch whisky: Some distillers recount that when a brewer's yeast was accidentally used in place of the standard distiller's yeast, the ferment finished sluggishly and produced excess diacetyl, giving the new make spirit a heavy, buttery fault that carried even after distillation.

Rum: Distillers have reported that switching from a wild, molasses-tolerant yeast to a high-ethanol "neutral" strain stripped the wash of its fruity esters, leaving a flat and uninspiring spirit.

Brandy: Producers experimenting with mixed-culture fermentations found that introducing a vigorous Saccharomyces bayanus strain—common in sparkling wine—shortened fermentation time but muted the aromatic complexity derived from native yeasts, resulting in a spirit that is technically clean yet lacking the floral nuance expected of fine brandy.

Chapter 9: Alcohols in Spirits - A Broader View

The Role of Alcohols in Spirits

Alcohols are a foundational class of compounds in spirits, contributing both directly and indirectly to aroma, flavor, texture, and the perception of complexity. While ethanol is the dominant alcohol by volume and will be addressed separately due to its unique psychophysical effects, a diverse group of other alcohols, collectively known as higher alcohols or fusel alcohols, plays a significant role in the sensory signature of distilled beverages.

Alcohols are a diverse group of organic compounds sharing the hydroxyl (-OH) functional group. These alcohols are produced as secondary metabolites during yeast fermentation, primarily via the Ehrlich pathway (from amino acid catabolism) or from glycolytic intermediates. Survival through distillation depends on volatility, solubility, and the specific distillation method employed.

They typically appear in concentrations ranging from tens to hundreds of milligrams per liter, depending on fermentation conditions, yeast strain, and nutrient availability. While some fusel alcohols contribute pleasant, faintly fruity or floral notes at low levels, they are also capable of generating harsh, solvent-like, or oily aromas and mouthfeel when present in excess. Their volatility, solubility, and structural diversity influence perceptual thresholds and palate persistence.

From an evaluation standpoint, these alcohols influence more than just aroma. Their impact on mouthfeel, particularly perceived heat, oiliness, and viscosity, can enhance or degrade the sensory impression depending on concentration and balance with esters, acids, and other volatiles. Interactions between fusel alcohols and esters can also alter the volatility and headspace distribution of key aroma molecules, thereby shifting the evaluator's perception and potentially masking subtle notes. Therefore, understanding the identity, formation, and sensory behavior of those alcohols beyond ethanol is essential for accurate assessment.

Trained evaluators must be able to differentiate between the desirable complexity contributed by moderate levels of higher alcohols and the sensory defects associated with excessive fusel content. This awareness is critical not only for judging product quality but also for diagnosing fermentation or distillation faults and for understanding house-style differences among producers. The following sections will examine the primary alcohols found in spirits, their sensory profiles, and the analytical and perceptual implications of these compounds.

The primary alcohols found in spirits are grouped by carbon chain length, and their interactions with central nervous system (CNS) receptors, sensory contributions, and toxicological implications are highlighted. Understanding these alcohols deepens the evaluator's appreciation for the subtle complexity of spirits and the physiological responses they can induce (Fig. 9.1).

Fig. 9.1	Classification of Alcohols (-OH)	
Category	Examples	Sensory
Simple Aliphatic	Ethanol, Methanol, 1-Propanol, 1-Butanol	Pungent, solvent, warming effect, slightly bitter
Higher, Fusel	Isoamyl, Isobutanol, 2-Methyl-1-Butanol	Fruity, fusel notes, harsh at high levels
Aromatic	Phenylethanol, Benzyl, Tyrosol	Floral, sweet, rose-like, almond
Polyols (Sugar)	Glycerol, Sorbitol, Xylitol, Mannitol	Sweetness, viscous, full-body
Terpenoids	Linalool, Geraniol, Nerol, Terpineol	Citrus, floral, herbal, piney
Miscellaneous	1-Hexanol, Octanol, Cetyl	Green, grassy, waxy, fatty mouthfeel
(-OH) = a single hydroxyl radical bonded to a saturated carbon atom (alkyl). *Over 40 alcohols exist in spirits* produced through yeast fermentation; whiskey, rum, and brandy have the highest number of alcohols.		

Delving into neuroreceptor function is beyond the scope of this text, but the association is critical to understanding the physiological effects of alcohols. Three receptors are affected. These are:

- **GABA$_A$ receptors** inhibit neurotransmission in the central nervous system (CNS), producing sedative, muscle-relaxant effects when irritated by alcohols.
- **NMDA receptors**, located throughout the central nervous system, are responsible for excitatory neurotransmission, synaptic plasticity, and memory formation. Alcohols suppress cognitive function and motor coordination.
- **Glycine receptors** are predominantly in the spinal cord and brainstem. Alcohols contribute to sedation and loss of motor control.

Methanol
- Simplest alcohol with a one-carbon structure.
- Does not bind to GABA$_A$, NMDA, or glycine receptors (Barceloux, 2003).
- Toxicity arises from metabolism to formaldehyde and formic acid, which affect mitochondrial function (Jacobsen & McMartin, 1986). ~10× more toxic than ethanol
- No psychotropic or sensory value in spirits; presence is strictly regulated.
- Ingestion may cause blindness, organ failure, or death at low concentrations.

1-Propanol (n-propanol)
- Three-carbon linear alcohol.
- Functions as a GABA$_A$ receptor agonist (Arora & Ohlrich, 2020).
- Interacts with NMDA and glycine receptors, contributing to sedative effects (Zakhari, 2006).
- Low concentrations, contributing mild solvent and fruity notes.
- CNS depressant with greater toxicity than ethanol (Kalant, 1996).

- Higher in pot stills than column stills (higher boiling point).

2-Propanol (Isopropanol)

- Short, three-carbon structure increases volatility and CNS penetration.
- Acts as a $GABA_A$ receptor agonist, enhancing inhibitory signaling and sedation (Arora & Ohlrich, 2020).
- Also shows interaction with NMDA and glycine receptors, contributing to anesthetic effects (Arora & Ohlrich, 2020).
- Rapid CNS depressant; more potent than ethanol on a molar basis but more toxic in overdose (Kalant, 1996).
- Common in industrial and medical applications (e.g., rubbing alcohol); not safe for consumption but may appear in trace amounts during poor distillation practices.
- Volatility and polarity limit its olfactory contribution, but higher concentrations can induce solvent-like aromas.

1-Butanol

- Four-carbon chain increases hydrophobicity and receptor affinity.
- $GABA_A$ potentiator and glycine receptor modulator (Lovinger, 1997).
- Produces stronger motor impairment and respiratory depression than ethanol (Wallner et al., 2003).
- Less volatile than ethanol; contributes more to body and mouthfeel than aroma.

Isobutanol (2-Methyl-1-Propanol)

- Branched-chain isomer of 1-butanol.
- Functions as a $GABA_A$ agonist and a weak NMDA inhibitor (Woodward, 2000).
- Exhibits mild alcoholic and fusel oil aromas; contributes to heavier, oily sensory notes in some spirits (Lachenmeier & Sohnius, 2008).
- Present in higher concentrations in poorly controlled fermentations and may enhance perceived harshness.

2-Butanol

- Secondary butanol with limited volatility.
- Activates $GABA_A$ and glycine receptors (Lovinger, 1997).
- Perhaps mild CNS depressant; less studied than primary butanols.
- Contributes solvent-like aromas and burning sensations.

Isoamyl Alcohol (3-methyl-1-butanol)

- Five-carbon, branched alcohol
- Potentiates $GABA_A$ and inhibits **NMDA** receptor activity (Woodward, 2000).
- Common fusel alcohol with a pungent, banana-like aroma at low concentrations.
- In higher levels, it adds to the spirit's harshness and solvent-like notes.

- Derived from leucine metabolism during fermentation (Hazelwood et al., 2008)

Phenylethanol
- Aromatic alcohol with an eight-carbon structure.
- Possible weak CNS receptor interaction; minimal $GABA_A$ activity (Hoffman & Tabakoff, 1994).
- Known for rose-like floral aroma, contributes to the perceived elegance of some spirits.
- Acts more through olfactory receptors than CNS pathways.
- Higher boiling point results in lower volatility and late emergence during distillation.

Ethanol
- Primary psychoactive component of spirits with a two-carbon structure.
- Modulates CNS function allosterically, rather than binding directly as a classical ligand.
- Enhances $GABA_A$ and glycine receptor activity.
- Inhibits NMDA receptor function, contributing to intoxication and memory impairment (Peoples & Stewart, 2000).
- Alters membrane fluidity and ion channel dynamics across multiple receptor systems (Hoffman & Tabakoff, 1994).
- Provides structure, volatility, and sensory lift to distilled spirits. Its dose-response curve is nonlinear, and its effects are highly context-dependent.

In essence, ethanol is an atypical CNS agent—not a "key" for a receptor "lock," but rather a modulator of the brain's electrical and chemical operations.

Evaluation Note 9.1: Understanding the chemical behavior of all alcohols provides essential background for grasping ethanol's unique place in the sensory and neurological landscape of spirits. These comparisons also help explain why hangovers, harsh mouthfeel, and disagreeable odors are more common in impure distillates, due to the presence of receptor-binding fusel alcohols at higher concentrations. Evaluators' complacent acceptance of ethanol, without understanding how it affects aroma evaluation through its anesthetic effects alone, leads to erroneous evaluations of spirits, particularly cask strengths, when evaluated in vessels that concentrate ethanol on the olfactory. Order of magnitude (size relationship) of alcohols in a 1 oz. serving to place the relationship in clearer perspective (Fig. 9.2), do not over-interpret as a rule.

"Above 40 percent alcohol, ethanol begins to dominate the nose, masking many volatile compounds and dulling the taster's sensitivity."— Andreas Buettner, Flavour Science

"Fusel alcohols are derived from amino-acid catabolism via a pathway that was first proposed a century ago by Ehrlich." — Lucie A. Hazelwood

Fig. 9.2		Illustrative Example: Diverse Alcohols in Whiskey		
Alcohol	Carbon Chain	Typical Flavor/ Mouthfeel/Impact	Dose= 1 oz Whiskey	DT (mg/L)
Methanol	1	Subtle, nearly undetectable, ~10× more toxic than ethanol, causes CNSD,* organ failure, blindness, fatal 30mg/100mL	TTB Legal Limit 300mg/ 100mL anhydrous	2
Ethanol	2	Alcoholic heat, sweet, body, pungency, intoxication. High doses cause CNSD.	Abundant 40-95% ABV	10,000
1-Propanol	3	Mildly alcoholic, solventy, 4X>toxic ethanol, severe CNSD	11mg	800
Isobutanol	4	Fruity, slightly solventy, >toxicity than ethanol, possible CNSD, nausea, headache	19mg	75
1-Butanol	4	Alcoholic, slightly sweet	0.6mg	800
2-Butanol	4	Bitter, solvent-like	negligible	800
1-Pentanol	5	Bready, slightly fruity	0.3mg	800
3-methyl-1-butanol	5	Solventy, banana when esterified, CNSD, dizzy (isoamyl)	30mg	70
2-methyl-1-butanol	5	Alcoholic, harsh, marzipan, CNSD, nausea (active amyl)	4mg	0.020
1-Hexanol	6	Grassy, woody, dizzy, nausea, irritates skin /eyes	<0.3mg	8
1-Octanol	8	Fatty, oily, slightly floral CNSD, weak muscles	negligible	0.7
1-Nonanol	9	Waxy, citrusy, rose-like, liver/kidney damage	negligible	0.5
1-Decanol	10	Waxy, floral, heavy, **CNSD**, skin/eye irritant	negligible	0.2
Phenethyl	2+ phenyl	Rose-like, floral	0.44mg	10

DT= detection threshold in milligrams or parts per million. *CNSD= central nervous system depression. 1oz = 29.57mL. **NOTE:** The USA & Europe have legal limits on methanol content. Homemade spirits cause most instances of methanol poisoning. CNSD: central nervous system depression. Carbon chain, 3+ denotes high alcohols, 6+ denotes fusel alcohols. Regulatory methanol limits vary by category and jurisdiction; fruit spirits often permit higher methanol levels. Verify current TTB/region-specific limits for the spirit in question.

Summary

Alcohols beyond ethanol are active players in the sensory and physiological landscape of spirits. Their diversity shapes aroma, mouthfeel, and balance. At low concentrations, they can enrich complexity; in excess, they drive harshness and off-notes and can create a physiological burden. For the evaluator, mastery lies in recognizing both the desirable contributions and the liabilities of higher alcohols, understanding their origins, and distinguishing their effects from those of ethanol's dominant role. This perspective supports sharper fault diagnosis, clearer attribution of house style, and greater confidence in sensory judgment.

Beyond: The ethanol-water matrix exhibits many less-obvious behaviors that affect evaluation.

"Understanding fusel alcohols and heads and tails isn't optional. It's how you protect flavor — and people." — Nancy Fraley, master blender

"The effects of alcohol on the brain are not uniform; they vary depending on dose, timing, and

neural circuitry. GABA, NMDA, dopamine, all are in play." — Dr. Nora D. Volkow, psychiatrist

Reference Appendix: Alcohols in Spirits Receptor Binding (Fig. App. 9.1)

App. 9.1	Receptor Binding Profiles for Alcohols in Spirits				
Alcohol	Chain	GABA$_A$	NMDA	Glycine	Other Effects
Methanol	1	-	-	-	Primarily toxic via metabolism to formic acid. Does not bind to CNS receptors
Ethanol	2	*	*	*	Allosteric modulation: alters membrane fluidity
1-Propanol	3	✓	✓	✓	Direct GABA$_A$ and Glycine agonist; CNS depressant
2-Propanol	3	✓	✓	✓	Direct GABA$_A$ agonist; anesthetic effects
1-Butanol	4	✓	✓	✓	Direct GABA$_A$ and NMDA interaction; intoxicating and sedative effects
2-Butanol	4	✓	✓	✓	Similar to 1-butanol; CNS depressant via GABA$_A$ agonism and NMDA antagonism; mild anesthetic properties.
Isobu-tanol	4	✓	✓	✓	Moderate GABA$_A$ affinity; contributes to CNS depression.
Isoamyl	5	✓	✓	✓	Potent CNS depressant; enhances GABAergic signaling; contributes to hangover and off-flavors at high levels.
2-Phenyl ethanol	8	✓	✓	NA	Primarily aromatic; may have olfactory and mild sedative effects; no strong CNS receptor binding.

* Ethanol does not bind as a classical ligand but modulates receptor function indirectly (allosterically). It enhances GABA$_A$ and glycine inhibition and inhibits NMDA excitation, but does not bind; NA data inconclusive. Chain refers to Carbon atoms. Methanol = high toxicity, Ethanol = toxic at high levels. Some receptors have limited data

Reference Appendix: Proof vs ABV: Definition and Origins

ABV (Alcohol by Volume): Alcohol by volume is the modern, internationally recognized standard for expressing alcoholic strength. It measures the percentage of pure ethanol in a beverage relative to the total liquid volume at 20 °C. For example, a whiskey labeled 40% ABV contains 40 mL of ethanol per 100 mL of solution.

Proof: The term "proof" predates ABV and originated in England in the 16th–17th centuries. Excise officers tested spirits by soaking gunpowder in the liquid and attempting to ignite it. If the mixture burned, the spirit was deemed "above proof" (sufficiently strong to be taxed at a higher rate). If it failed to ignite, it was "under proof." Later, the British government codified 100 proof as equal to 57.15% ABV, which is the point at which gunpowder would still ignite after being soaked in spirit (British Customs and Excise Act, 1816). This was the British sailors' method for testing their rum ration.

U.S. Proof System: The U.S. adopted a simpler system in 1848: proof is defined as exactly twice the ABV. Thus, a 40% ABV spirit = 80 proof. This remains the U.S. labeling standard today, even though most of the world uses ABV exclusively. (Fig. App. 9.2)

Proof Summary:

- ABV: scientific, volumetric, universal.
- Proof (UK historical): 100 proof = 57.15% ABV.
- Proof (U.S.): proof = 2 × ABV.
- Proof arose from a taxation test, not from sensory or scientific necessity, but the term persists in marketing and cultural identity.

App. 9.2	Ethanol Concentration/Proof Equivalency	
ABV (%) Alcohol by Volume	**U.S. Proof (2×ABV)**	**UK Historic Proof Equivalent**
40	80	70
45	90	78.7
50	100	87.5
55	110	96.2
57.15	114.3	100
60	120	105
65	130	113.7
70	140	122.5
NOTE: Proof definitions vary historically; the table shows UK historic equivalencies		

Reference Appendix: Formation of Alcohols During Distillation

Rising distillation temperatures provide the thermal energy required to vaporize alcohols with higher boiling points, which generally correspond to molecules with longer carbon chains, driving them from the mash into the distillate. (Fig. App. 9.3)

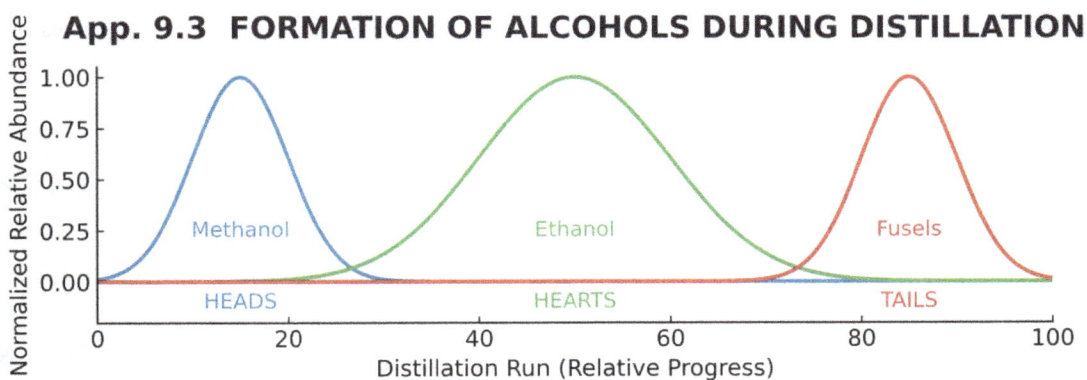

App. 9.3 FORMATION OF ALCOHOLS DURING DISTILLATION

"The concept of 'proof' embodies the intersection of commerce, law, and culture: it is at once a measure of taxation, authenticity, and identity."— Nicholas Faith, wine and spirits author

"Proof was less about science and more about trust—a way to reassure buyer and seller that the spirit had strength enough to ignite."— David Wondrich, writer

"Proof means strength, not quality, yet some still confuse the two."-Unattributed, common

References

Arora, S., & Ohlrich, R. (2020). Mechanisms of alcohol-induced neurotoxicity. *Alcohol Research: Current Reviews, 40*(1), 13. https://doi.org/10.35946/arcr.v40.1.13

Barceloux, D. G. (2003). Methanol. *Clinical Toxicology, 41*(2), 155–164. https://doi.org/10.1081/CLT-120018840

Chatonnet, P., Dubourdieu, D., Boidron, J.-N., & Pons, M. (1995). *Brettanomyces* spoilage compounds in beverages. *Journal of the Science of Food and Agriculture, 67*(3), 309–315. https://doi.org/10.1002/jsfa.2740670306

Hoffman, P. L., & Tabakoff, B. (1994). The role of cell membrane structure in ethanol action. *Journal of Molecular Neuroscience, 5*(2), 101–107. https://doi.org/10.1007/BF02736771

Jacobsen, D., & McMartin, K. E. (1986). Methanol and ethylene glycol poisonings. *Medical Toxicology, 1*(5), 309–334. https://doi.org/10.1007/BF03259816

Kalant, H. (1996). Problems in the use of the concept of CNS depression. *Addiction Biology, 1*(1), 3–13. https://doi.org/10.1080/13556219610000001

Koob, G. F., & Le Moal, M. (2001). Drug addiction, dysregulation of reward, and allostasis. *Neuropsychopharmacology, 24*(2), 97–129. https://doi.org/10.1016/S0893-133X(00)00195-0

Kyzar, E. J., Collier, A. D., Garner, J. P., & Kalueff, A. V. (2012). Effects of alcohol on behavior in C57BL/6J and DBA/2J inbred mouse strains. *Behavioural Brain Research, 230*(1), 39–51. https://doi.org/10.1016/j.bbr.2012.01.046

Lachenmeier, D. W., & Sohnius, E. M. (2008). The role of alcohols other than ethanol in the safety of beverages. *Food and Chemical Toxicology, 46*(11), 2903–2911. https://doi.org/10.1016/j.fct.2008.06.072

Lovinger, D. M. (1997). Alcohols and neurotransmitter-gated ion channels: Past, present and future. *Naunyn-Schmiedeberg's Archives of Pharmacology, 356*(3), 267–282. https://doi.org/10.1007/PL00005050

Peoples, R. W., & Stewart, R. R. (2000). Alcohols modulate NMDA receptor-mediated transmission. *Neuropharmacology, 39*(1), 89–98. https://doi.org/10.1016/S0028-3908(99)00083-9

Wallner, M., Hanchar, H. J., & Olsen, R. W. (2003). Ethanol enhances $\alpha 4\beta 3\delta$ and $\alpha 6\beta 3\delta$ receptors. *Proceedings of the National Academy of Sciences, 100*(25), 15218–15223. https://doi.org/10.1073/pnas.2435171100

Woodward, J. J. (2000). Ethanol and NMDA receptor signaling. *Critical Reviews in Neurobiology, 14*(1), 69–89. https://doi.org/10.1615/CritRevNeurobiol.v14.i1.50

Zakhari, S. (2006). Overview: How is alcohol metabolized by the body? *Alcohol Research & Health, 29*(4), 245–254. (Archived NIAAA resource; URL retired 2024)

Chapter 10: Ethanol Properties - Effect on Evaluation

Ethanol occupies a central position in the study of spirits not only because it is the primary psychoactive component but also because of its multifaceted roles in shaping aroma, flavor, and human perception. To understand the full impact, this chapter divides ethanol into three critical domains: physico-dynamics, pharmacodynamics, and psychodynamics. Each section explores a distinct perspective on how ethanol functions and influences both the material properties of spirits and the evaluator's perceptual capabilities.

Source of ethanol in spirits

Ethanol is the byproduct of the fermentation of sugars and starches in grains, fruits, and sugarcane by yeast. As previously discussed in Chapter 7, chemical fermentation has many complex steps, but in its final form, the basic equation (Legras et al., 2007) stands as:

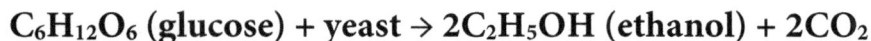

$$C_6H_{12}O_6 \text{ (glucose)} + \text{yeast} \rightarrow 2C_2H_5OH \text{ (ethanol)} + 2CO_2$$

Physicodynamics: How Ethanol Molecules Behave

At the molecular level, ethanol (C_2H_5OH) is a small, polar compound with both hydrophilic (hydroxyl) and hydrophobic (ethyl) characteristics. This amphipathic nature allows ethanol to dissolve in both aqueous and lipid environments, making it uniquely suited to traverse biological membranes and mix with a wide range of flavor compounds. Comparing to water sets the perspective (Lide, 2004). See Fig. 10.1.

Fig. 10.1	Physical Properties of Ethanol and Water (20 °C, 1 atm.)	
Property	**Ethanol**	**Water**
Density (g/cm³)	0.789	.998
Dynamic Viscosity (mPa·s ≈ cP)	~1.074 (25 °C)	~1.002 (20 °C)
Surface Tension (mN/m)	22.27	72.75
Boiling Point (°C)	78.2	100
Freezing Point (°C)	-114.1	0
Heat Capacity C_p (J/(mol·K)	~112.4	~75.3
Vapor Pressure (kPa)	~5.95	2.33

*40% ABV (alcohol by volume) spirit freezes at -23 °C. ~ = Approximate
NOTE: Ethanol dissolves esters, aldehydes, oils, and acids, key to flavor extraction, and forms flavor compounds like ethyl acetate and esters during fermentation and aging. Values at 20 °C unless noted. Vapor pressure values from NIST (Antoine fits); water surface tension from NIST tables; ethanol surface tension from CRC; viscosities: water from IAPWS; ethanol from CRC/NIST-compiled datasets. Source: Linstrom & Mallard (2001).

When considering the constituents of a 40% ABV (alcohol by volume) spirit, we see that the remaining spirit, after alcohol is removed, is nearly 60% water, with traces of other compounds that contribute to its flavor. Leaving a glass of whiskey on the counter overnight, it is not surprising that the ethanol is gone by morning, and some water and residue remain in the glass. Sniffing the residue after ethanol evaporation provides clues to the spirit's character.

Ethanol's vapor pressure is about two-and-a-half times that of water, which accounts for its higher evaporation rate and volatility. Water's surface tension is roughly three-and-a-half times higher than ethanol's, but surface tension does not control evaporation; volatility is governed primarily by vapor pressure (Lide, 2004; Tuckermann, 2007).**Density:** Ethanol's lower density than water affects weight and mouthfeel of spirits (Lide, 2004).

Boiling Point and Volatility: Ethanol's lower boiling point makes it highly volatile under normal conditions. Its volatility plays a crucial role in aroma delivery, serving as a carrier for more hydrophobic aroma molecules while adding its own pungent, solvent-like note (Reid et al., 1987; Poole, 2003).

Solvent Behavior: Ethanol serves as a cosolvent, enabling nonpolar congeners (e.g., esters, terpenes) to remain in solution. This affects both mouthfeel and flavor intensity, particularly in high-ABV spirits (Poole, 2003).

Surface Tension and Evaporation: Ethanol lowers the surface tension of aqueous solutions and evaporates more rapidly than water, thereby contributing to the concentration gradient of aromas that rise from the glass. The rate and profile of evaporation determine how different aroma compounds are perceived over time (Tuckermann, 2007).

Molecular Interactions: Ethanol forms hydrogen bonds with water and interacts with other volatiles, altering their vapor pressure. This dynamic affects retronasal and orthonasal perception, influencing both initial nosing and flavor persistence (Walrafen, 1966; Poole, 2003).

Heat Capacity: Ethanol's lower heat capacity means it warms and cools more quickly. Hand heat through vessel walls releases more aromas, intensifying aroma scent and ethanol pungency with small temperature changes. The higher the ethanol content, the more sensitive it is to temperature (De Groot & Mazur, 2013). Slight temperature rises increase vapor pressures exponentially (Clausius–Clapeyron), enriching headspace VOCs; dynamic APCI-MS (Atmospheric-Pressure Chemical Ionization Mass Spectrometry) headspace studies of ethanolic systems show higher ethanol fractions suppressing other volatiles in air, whereas dilution lowers ethanol in headspace and improves odorant detectability (Tsachaki, Linforth, & Taylor, 2005; Wang & Cadwallader, 2024). Regarding Fig. 10.1, per mass, ethanol's specific heat (~2.44 J/g·K) is lower than water's (~4.18 J/g·K), so small handling temperature changes warm/cool spirit faster as ABV rises. (Molar heat capacities listed in Fig. 10.1 are higher for ethanol due to its molar mass).

Dynamic Viscosity: Ethanol flows more easily than water and does not cling to oral surfaces, contributing to a thinner, sharper mouthfeel. At 20–25 °C, ethanol's viscosity is only slightly higher than water's, and mixtures near 40% ABV behave similarly; perceived "thinness" is driven more by reduced molecular cohesion and rapid volatilization than by viscosity. When water freezes, ethanol remains liquid because of its much lower freezing point (Watanabe et al., 1996).

Azeotrope Formation: At 1 atmosphere, ethanol and water form a minimum-boiling azeotrope at 95.6 % by weight ethanol (78.2 ^{0}C), beyond which further distillation cannot increase purity. An azeotrope is a mixture of two or more liquids that boils at constant temperature and composition, behaving like a single substance during distillation. This physical limit underpins the practical boundaries of spirit concentration and purification (Gmehling et al., 1998; Reid et al., 1987).

Summarizing the sensory effects of the physical properties of ethanol, one can easily deduce that each physical property of ethanol has a decided influence on the sensory perception of the drinker and evaluator. The key factor is that ethanol rapidly diffuses across the oral and nasal surfaces, intensifying trigeminal sensations such as burning or tingling and reducing olfactory acuity (Green, 1992).

Ethanol facilitates clear expression of volatile notes at moderate concentrations but also obscures complexity when it dominates the aromatic landscape. We love ethanol—it is why we drink—but it presents significant problems during aroma assessment (Kobal & Hummel, 1991).

Pharmacodynamics: How the Human Body Responds to Ethanol

The effects of ethanol on human physiology are complex, as it acts on multiple receptor systems and organ systems. Its influence extends from acute intoxication to long-term neuroadaptation and ethanol dependency (Koob & Le Moal, 2006). Included in each category are specific examples of the possible effects on evaluators (in italics).

Absorption and Distribution: Ethanol is rapidly absorbed through the stomach and small intestine, with peak blood concentrations occurring within 30–90 minutes post-ingestion, depending on matrix and gastric emptying (Merck Manual Professional Version, n.d.). Due to its solubility in both water and lipids, it distributes widely throughout the body, including the brain (Zakhari, 2006; Mitchell, 2008). **Effect:** *The accumulation of small sips during evaluation leads to rapid cerebral ethanol presence, causing sensory dampening before the evaluator is consciously aware, degrading accuracy and panel consistency.*

Receptor Activity: Ethanol enhances the activity of GABA$_A$ and glycine receptors (inhibitory) while inhibiting NMDA glutamate receptors (excitatory), leading to central nervous system (CNS) depression. This receptor-level modulation accounts for ethanol's sedative, anxiolytic, and amnestic effects. **Effect:** *These subtle changes alter attention, memory formation, and olfactory signal processing, making judgments less precise and often irreproducible under blind or time-separated conditions* (Lovinger, 1997; Oscar-Berman & Marinković, 2007).

Metabolism: Ethanol is primarily metabolized in the liver via alcohol dehydrogenase (ADH) to acetaldehyde, which is then metabolized by aldehyde dehydrogenase (ALDH) to acetic acid. Acetaldehyde is toxic and contributes to hangover symptoms and long-term tissue damage (Zakhari, 2006). **Example:** *During long tasting sessions, continuous low-level ingestion outpaces metabolic clearance, producing fatigue and early signs of sensory burnout that reduce evaluative accuracy.*

Dose-Dependent Effects: Low doses produce disinhibition and mild euphoria; moderate doses impair motor control and memory; high doses can result in respiratory depression, unconsciousness, or death. Chronic exposure alters receptor expression and brain structure (Oscar-Berman & Marinković, 2007). **Example:** *Even within a single flight, disinhibition and impaired sensory memory subtly shift preferences and heighten subjectivity, particularly under time pressure or fatigue.*

Physiological Consequences: Acute effects include vasodilation, hypothermia, and increased urine production (diuresis). Chronic consumption is linked to liver disease, neurodegeneration, cardiovascular dysfunction, and addiction through complex neurobiological pathways (Rehm et al., 2009; World Health Organization, 2018). **Example:** *These physiological effects compound evaluator variability, leading to inconsistent results across sessions and reduced reliability of scoring systems when ethanol intake is unregulated.*

Judging panels, competition organizers, evaluators, collectors, connoisseurs, and consumers must become acutely aware of the biological effects of ethanol. Even minimal ingestion during evaluation alters perception, attention, and memory, compromising accuracy and consistency. Recognizing these pharmacodynamic impacts is essential for credible ratings, informed purchasing decisions, and maintaining focus and effectiveness throughout the sensory process (Spence, 2015). Avoiding olfactory ethanol is a top priority for practical evaluation.

Psychodynamics: How Ethanol Alters and Misleads Evaluation

While ethanol facilitates relaxation and sociability, it also impairs sensory perception, cognitive function, and judgment. For the sensory analyst, this presents a critical paradox: the very substance under evaluation interferes with the evaluator's ability to remain objective and accurate during analysis (Sayette, 2017).

Disinhibition and Bias: Ethanol lowers inhibitions and enhances perceived pleasure, which can artificially inflate ratings of aroma or flavor and conflate emotional response with sensory quality (Bujarski et al., 2019).

Sensory Adaptation: Continued exposure to ethanol vapor desensitizes the olfactory epithelium, reducing sensitivity to other compounds. This adaptation leads to underestimation of certain aromas and overemphasis on ethanol (Dalton, 2000).

Attentional Narrowing: Ethanol impairs prefrontal cortex function and working memory, thereby narrowing attention and reducing the evaluator's ability to detect subtle flavor transitions or identify faults (Fillmore & Vogel-Sprott, 2000).

Expectation Effects: Under the influence of ethanol, suggestibility increases. The power of labeling, brand reputation, and verbal suggestion may disproportionately influence judgment (McClure et al., 2004).

Temporal Distortion: Ethanol alters time perception and processing speed, interfering with accurate timing during nosing, tasting, and scoring protocols (Weafer & Fillmore, 2008). Understanding psychodynamic interference is essential for designing objective evaluation protocols. Techniques such as blind tasting, spitting, controlled pacing, and low-ABV dilution are necessary safeguards.

Ethanol is not merely an ingredient; it is an active participant in the perceptual and biological dynamics of spirits evaluation. Its molecular behavior, pharmacological effects, and psychological influences demand deliberate awareness. Mastery of ethanol's properties is thus foundational to any rigorous and credible approach to spirits sensory science.

Social and Industry Awareness

Despite its central role in the enjoyment of spirits, ethanol remains deeply misunderstood by the public. Most drinkers view it solely as the agent of pleasure—of warmth, laughter, and social ease—without recognizing its full physiological impact (Room et al., 2005). Its effects on perception, memory, mood, and behavior are often invisible to the drinker, particularly during tastings or social events. The popular ethos of "drink to relax" blinds many to the fact that ethanol, even in small quantities, impairs olfactory clarity, dulls evaluation accuracy, and promotes fatigue. This ignorance is not due to a lack of intelligence, but rather a lack of science-based education.

The spirits industry, including producers, marketers, judges, and influencers alike, has historically benefited from this blind spot. However, as consumer sophistication grows and demand for quality intensifies, so too must the standards of transparency and responsibility. A shift in attitude is overdue. Acknowledging ethanol's pharmacological effects does not threaten sales or culture; it is a catalyst for trust, credibility, and long-term growth (Mosher & Johnsson, 2005; World Health Organization, 2018). The industry has a chance to lead not just in flavor or craftsmanship, but in elevating how we think about drinking. Teaching proper sensory evaluation—how to taste without distortion—is a service to both connoisseurs and casual drinkers, ultimately benefiting industry.

This reframing is not about temperance or moral posturing. It is about helping people engage with spirits more intelligently, safely, and meaningfully. Panels, judges, educators, and even collectors must understand that ethanol can mislead perception, degrade focus, and alter judgment without conscious awareness. By designing tasting protocols and utilizing well-engineered evaluation vessels that minimize ethanol interference, the industry can enhance the integrity of competitions and panel ratings, improve the consistency of consumer experiences, and ultimately promote public health. Ethanol is why we drink—but it should not be on the evaluator's nose during diagnostics. Only when that distinction is respected will true sensory excellence in spirits be possible. These sensory and pharmacological realities align with public-health guidance to reduce alcohol-related harm while improving evaluation integrity (Wang & Cadwallader, 2024).

The Misleading Role of Pungency in Spirit Evaluation

Pungency is not a flavor, yet it is often mistaken for one. In the context of spirit evaluation, pungency arises from chemosensory irritation rather than from olfactory or gustatory perception. Ethanol, due to its physicochemical properties, activates the trigeminal nerve endings in the nasal and oral mucosa, producing sensations of burning, stinging, or tingling. These responses are governed by transient receptor potential (TRP) channels, particularly TRPV1 and TRPA1, which are not involved in odor recognition but in nociception, the detection of harmful stimuli (Green, 1996). This response is dose-dependent: the higher the ethanol concentration in the vapor phase, the more intense the irritation.

To the trained evaluator, this interference masks the subtleties of aroma compounds by dominating the perceptual field with a nonspecific burning sensation. Ethanol's low molecular weight and high volatility enable it to rapidly saturate the headspace, displacing less volatile aroma molecules from the olfactory pathway (Buettner, 2007). Consequently, when pungency is misinterpreted as an aroma quality, it degrades evaluation accuracy by conflating tactile pain with olfactory complexity.

For many consumers, pungency is paradoxically interpreted as a hallmark of strength and authenticity. The burning sensation is perceived as a proxy for alcohol content, reinforcing the illusion of quality or value. This cultural bias is not accidental; it arises from conditioned associations between stronger ethanol sensations and potency, age, or exclusivity. Social rituals surrounding the consumption of spirits often valorize the ability to endure high-alcohol beverages without flinching, fostering an atmosphere in which tolerance becomes synonymous with connoisseurship (McCabe & West, 2017).

This expectation fuels demand for high-proof or "cask strength" spirit offerings, of which many contain ethanol concentrations exceeding 60% ABV. Such products deliver intense trigeminal impact but compromise evaluative clarity, both by anesthetizing the sensory field and by elevating blood ethanol levels during tasting panels. The spirits industry, recognizing the marketability of strength, continues to respond with increasingly bold expressions, an ironic reversal of the evaluator's need for olfactory precision. In many cases, panels may mistake ethanol sharpness for complexity, awarding higher scores to spirits with greater pungency rather than greater nuance (Prescott, 2012).

Unless the physiological role of pungency is clearly distinguished from aroma and flavor, both evaluators and consumers remain susceptible to this fundamental perceptual error. Education in the chemical, neurological, and behavioral aspects of ethanol irritation is necessary to reframe pungency not as a flavor or virtue, but as a barrier to truthful sensory evaluation.

Ethanol rapidly stimulates trigeminal nociceptors by activating/potentiating TRPV1 and engaging the TRPA1 pathway, producing burning, stinging, and tingling ('chemesthetic') sensations and triggering neurogenic inflammation (CGRP/SP release, vasodilation, mucus secretion). This chemesthetic drive precedes and can mask odor coding at the olfactory epithelium. Volatile agents, including ethanol, also modulate transduction at the sensory cilia and can produce olfactory masking at the receptor/channel level (Wang & Cadwallader, 2024; Nicoletti et al., 2008; Bessac & Jordt, 2008).

Regarding odorant binding proteins (OBP), Human OBP isoforms (OBP2A/OBP2B) are expressed in nasal tissue. At the same time, ethanol-specific competitive displacement data are sparse, and solvent composition can alter OBP–ligand equilibria (Gonçalves, Ribeiro, Silva, & Cavaco-Paulo, 2021).

Evaluation Note 10.1: Awareness of ethanol during spirits evaluation is key to maintaining good performance. According to some claims, highly volatile ethanol can be present in the headspace of tulip glassware at a concentration of 65-75% of all aromas, excluding air molecules, due to its extremely high volatility. Although not experimentally quantified to a high degree of accuracy, the fact remains that an extremely high percentage of anesthetic ethanol is found in vessels with tiny rim diameters compared to larger bowl diameters, and in vessels with rim heights too high to allow many heavier aroma compounds to reach the rim for detection. Under typical nosing conditions, ethanol constitutes the major fraction of the volatile mass in the headspace and increases sharply with ABV.

The data in Fig. 10.2 have been extrapolated from recent experiments using Henry's Law constant (K_h) to assess the effect of ethanol on aromas (Wang & Cadwallader, 2024). Henry's law states that the amount of dissolved gas in a liquid is directly proportional to its partial pressure above the liquid (at equilibrium).

Fig. 10.2	Ethanol Effect on Key Aromas (K_h)*		
Key Aroma	**Description**	**20% ABV**	**40% ABV**
Ethyl isobutyrate	Tropical fruit, pineapple	3.6 x	3.8 x
Ethyl butyrate	Orange, citrus, fruit	1.9 x	2.0 x
β-Damascenone	Rose, apple, plum	2.3 x	3.8 x
Guaiacol	Smoky, medicinal	1.2 x	1.5 x
Isoeugenol	Spicy, sweet cloves	1.5 x	2.3 x
*Baseline = 0%ABV Henry's Law Constant (K_H) increases as ethanol concentration increases, decreasing aroma detection, identification by x times over 0% ABV baseline. Data derived from (Wang & Cadwallader, 2024)			

Evaluation Note 10.2: Regarding Fig. 10.2, the evaluator should be aware of the distortion in aroma detection and identification by ethanol, especially given the elevated headspace ethanol concentrations in tulip-shaped vessels.

Evaluation Note 10.3: Three separate clues can indicate that the evaluator is experiencing olfactory fatigue: Evaluator realizes (1) "I can't identify this," (2) "I don't smell anything," or (3) "These samples all smell the same." Each of these is a sign of olfactory fatigue, accelerated by exposure to olfactory ethanol, and is particularly common among spirits competition judges, sniffing several high-ethanol spirits in a single 5–20-minute sitting. Olfactory "fatigue" (adaptation) from brief, repeated sniffs of high-ABV headspace resolves on the order of minutes rather than seconds, and recovery depends systematically on how intense and how long the prior stimulation was. Using two-part psychophysical methods, Steinmetz, Pryor, and Stone demonstrated that sensitivity returns along a negatively accelerated (approximately exponential) time course. A slower approach toward baseline follows a rapid early rebound. Stronger or longer exposures require longer clean-air intervals, and repeated exposures further slow recovery. Practically, this means evaluators should interleave sniffing with several minutes of odor-free air, extend breaks after heavy ethanol exposure, expect substantial—but not necessarily complete— recovery within a short rest window, and allow longer intervals when judgment precision is critical. (Steinmetz, Pryor, & Stone, 1970). Coffee bean "resets" show no consistent physiological benefit versus odor-free air; perceived benefit is likely cognitive (Grosofsky, Haupert, & Versteeg, 2011). Smelling coffee beans, the crook of the elbow, or the armpit are most probably psychological crutches. Effective recovery happens only by refreshing the epithelium with fresh mucous. Olfactory fatigue, sometimes called "lockout," is a temporary anosmia, and became evident decades ago in wine tastings, during which samples contain much lower concentrations of ethanol than do spirits samples (Dalton, 2000; Green, 1992), even considering wine sample sizes commonly run larger than spirits. The effects are similar when evaluating perfumes and alcohol based fragrances.

Evaluation Note 10.4: Olfactory adaptation is the sensory "safety net" that moves persistent aromas to the background and is the reason some aromas seem to disappear from a flight of similar spirit samples. Adaptation works on two levels: (1) ORNs decrease continuous stimulus signals over time to increase the possibility of detecting new aromas, and (2) the brain reduces neural activity, diminishing aroma awareness. A reset is necessary when this occurs, as described in Evaluation Note 10.3. This is why pungency disappears as the evaluator becomes accustomed to the stimulus, yet ethanol still blocks sensory channels. The evaluator is usually unaware of the sensation or presence of ethanol, and/or a particular flavor prevalent in a flight of samples with similar characteristics. Since the evaluator misses signs of ethanol presence, he is also unaware that his analysis and evaluation are now flawed (Green, 1992; Dalton, 2000).

Summary

Ethanol is both the medium and the disruptor in spirits evaluation. Its physicochemical traits carry aromas yet overwhelm them, its pharmacological effects dampen memory and attention, and its psychodynamic influence biases judgment and inflates evaluation scores. For the professional sensory analyst, mastery lies in recognizing ethanol not as a neutral background, but as an active force that must be managed through protocol: spitting, dilution, pacing, and rest, if evaluations are to remain credible and reproducible. Placing the pharmacological aspects into perspective alerts the evaluator that, even at low doses, accuracy in detection and identification can be severely compromised if due caution is not taken regarding olfactory exposure.

Beyond: The sensory effects of the ethanol-water matrix are complex, and the quality of dilution water is often overlooked; its quality can significantly impact the flavors and aromas of spirits.

"Alcohol is the anesthesia by which we endure the operation of life."— George Bernard Shaw, playwright, critic

"Here's to alcohol, the cause of — and solution to — all of life's problems."— Homer Simpson, the animated poster child of parental dysfunction and ignorance

Reference Appendix 10.1:

Detection thresholds are significantly altered (raised) in key aromatic compounds as ABV increases (suppressing aromas). Data extrapolated from Wang & Cadwallader (2024).

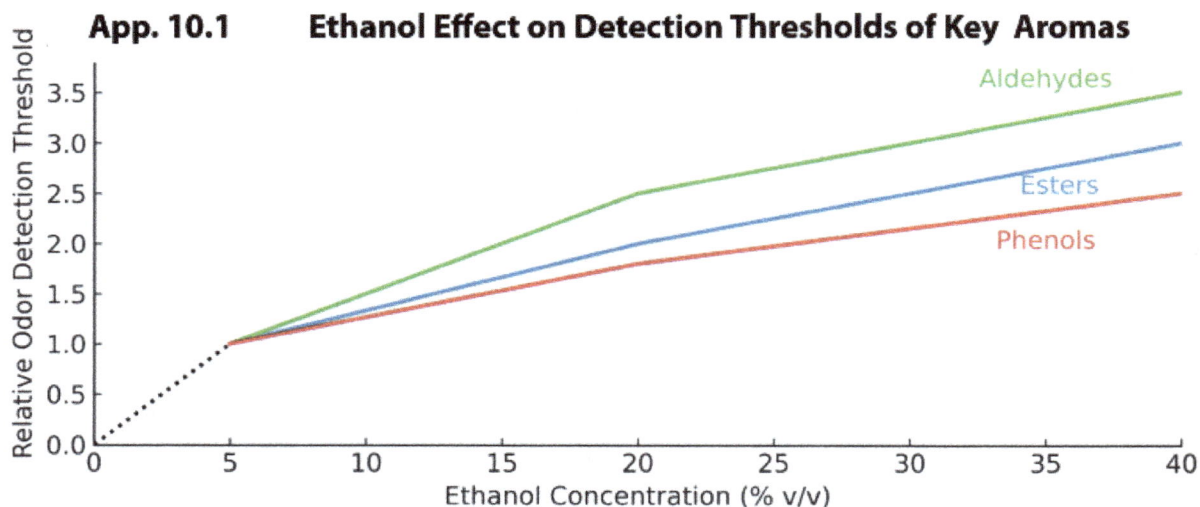

App. 10.1   Ethanol Effect on Detection Thresholds of Key Aromas

References

Archunan, G. (2018). Odorant binding proteins: A key player in the sense of smell. *Bioinformation, 14*(1), 36–39. https://doi.org/10.6026/97320630014036

Bessac, B. F., & Jordt, S.-E. (2008). Breathtaking TRP channels: TRPA1 and TRPV1 in airway chemosensation and inflammation. *Pflügers Archiv – European Journal of Physiology, 455*(4), 623–636. https://doi.org/10.1007/s00424-007-0313-2

Bujarski, S., Ray, L. A., & Roche, D. J. O. (2019). Alcohol effects on attention and response inhibition. *Drug and Alcohol Dependence, 197*, 81–87. https://doi.org/10.1016/j.drugalcdep.2018.12.018

Buettner, A. (2007). Influence of human salivary enzymes on odorant concentrations. *Journal of Agricultural and Food Chemistry, 55*(18), 7427–7433. https://doi.org/10.1021/jf070664x

Dalton, P. (2000). Psychophysical and behavioral characteristics of olfactory adaptation. *Chemical Senses, 25*(4), 487–492. https://doi.org/10.1093/chemse/25.4.487

De Groot, S. R., & Mazur, P. (2013). *Non-equilibrium thermodynamics.* Courier.

Fillmore, M. T., & Vogel-Sprott, M. (2000). Response inhibition under alcohol. *Journal of Studies on Alcohol, 61*(2), 239–247. https://doi.org/10.15288/jsa.2000.61.239

Gmehling, J., Wittig, R., Lohmann, J., & Joh, R. (1998). Azeotropic data—A survey. *Journal of Chemical & Engineering Data, 43*(6), 1049–1062. https://doi.org/10.1021/je980145x

Gonçalves, F., Ribeiro, A., Silva, C., & Cavaco-Paulo, A. (2021). Biotechnological applications of mammalian odorant-binding proteins. *Critical Reviews in Biotechnology, 41*(4), 441–455. https://doi.org/10.1080/07388551.2020.1869690

Green, B. G. (1992). The sensory effects of ethanol. *Physiology & Behavior, 52*(3), 479–486. https://doi.org/10.1016/0031-9384(92)90333-A

Green, B. G. (1996). Chemesthesis: Pungency as a component of flavor. *Trends in Food Science & Technology, 7*(12), 415–420. https://doi.org/10.1016/S0924-2244(96)10042-6

Grosofsky, A., Haupert, M. L., & Versteeg, S. W. (2011). A test of the myth that smelling coffee aroma restores the ability to smell. *Perceptual and Motor Skills, 113*(2), 529–537. https://doi.org/10.2466/24.22.PMS.113.5.529-537

International Association for the Properties of Water and Steam. (2008). *Release on the IAPWS formulation 2008 for the viscosity of ordinary water substance.* https://www.iapws.org/relguide/viscosity.html

Kobal, G., & Hummel, T. (1991). Olfactory evoked potentials. *Behavioural Brain Research, 48*(1), 85–94. https://doi.org/10.1016/S0166-4328(05)80141-7

Koob, G. F., & Le Moal, M. (2006). *Neurobiology of addiction.* Academic Press.

Legras, J.-L., Merdinoglu, D., Cornuet, J.-M., & Karst, F. (2007). Bread, beer and wine: *Saccharomyces cerevisiae* diversity reflects human history. *Molecular Ecology, 16*(10), 2091–2102. https://doi.org/10.1111/j.1365-294X.2007.03266.x

Lide, D. R. (Ed.). (2004). *CRC handbook of chemistry and physics* (85th ed.). CRC Press.

Linstrom, P. J., & Mallard, W. G. (2001). The NIST Chemistry WebBook: A chemical data resource on the internet. *Journal of Chemical & Engineering Data, 46*(5), 1059–1063. https://doi.org/10.1021/je000236i

McCabe, S. E., & West, B. T. (2017). Social drinking context and alcohol misuse. *Journal of Studies on Alcohol and Drugs, 78*(3), 394–403. https://doi.org/10.15288/jsad.2017.78.394

McClure, S. M., Li, J., Tomlin, D., Cypert, K. S., Montague, L. M., & Montague, P. R. (2004). Neural correlates of behavioral preference for culturally familiar drinks. *Neuron, 44*(2), 379–387. https://doi.org/10.1016/j.neuron.2004.09.019

Merck Manual Professional Version. (n.d.). *Alcohol toxicity and withdrawal.* https://www.merckmanuals.com/

Mitchell, M. C. (2008). Alcohol-induced alterations in hepatic lipid metabolism. *Journal of Gastroenterology and Hepatology, 23*(s1), S38–S41. https://doi.org/10.1111/j.1440-1746.2007.05291.x

Mosher, J. F., & Johnsson, D. (2005). Flavored alcoholic beverages: An international marketing campaign that targets youth. *Journal of Public Health Policy, 26*(3), 326–342. https://doi.org/10.1057/palgrave.jphp.3200021

National Institute of Standards and Technology. (n.d.). *NIST Chemistry WebBook.* https://webbook.nist.gov/

Nicoletti, P., Trevisani, M., Manconi, M., Gatti, R., De Siena, G., & Geppetti, P. (2008). Ethanol causes neurogenic vasodilation by TRPV1 activation and CGRP release in the trigeminovascular system of the guinea pig. *Cephalalgia, 28*(1), 9–17. https://pubmed.ncbi.nlm.nih.gov/17888011/

Oscar-Berman, M., & Marinković, K. (2007). Alcohol: Effects on neurobehavioral functions and the brain. *Neuropsychology Review, 17*(3), 239–257. https://doi.org/10.1007/s11065-007-9038-6

Poole, C. F. (2003). *The essence of chromatography.* Elsevier.

Prescott, J. (2012). Chemosensory learning and flavor. In R. J. Shepherd (Ed.), *Handbook of the senses: Chemical senses* (pp. 253–274). Springer. https://doi.org/10.1007/978-1-4614-6435-8_12

Rehm, J., Mathers, C., Popova, S., Thavorncharoensap, M., Teerawattananon, Y., & Patra, J. (2009). Global burden of disease attributable to alcohol use. *The Lancet, 373*(9682), 2223–2233. https://doi.org/10.1016/S0140-6736(09)60746-7

Reid, R. C., Prausnitz, J. M., & Poling, B. E. (1987). *The properties of gases and liquids* (4th ed.). McGraw-Hill.

Room, R., Babor, T., & Rehm, J. (2005). Alcohol and public health. *The Lancet, 365*(9458), 519–530. https://doi.org/10.1016/S0140-6736(05)17870-2

Sayette, M. A. (2017). The effects of alcohol on emotion in social drinkers. *Behaviour Research and Therapy, 88*, 76–89. https://doi.org/10.1016/j.brat.2016.07.012

Spence, C. (2015). Multisensory flavor perception. *Cell, 161*(1), 24–35. https://doi.org/10.1016/j.cell.2015.03.007

Steinmetz, G., Pryor, G. T., & Stone, H. (1970). Olfactory adaptation and recovery in man as measured by two psychophysical techniques. *Perception & Psychophysics, 8*, 327–330. https://doi.org/10.3758/BF0321260

Tsachaki, M., Linforth, R. S. T., & Taylor, A. J. (2005). Dynamic headspace analysis of VOC release from ethanolic systems by direct APCI-MS. *Journal of Agricultural and Food Chemistry, 53*(21), 8328–8333. https://doi.org/10.1021/jf051120x

Tuckermann, R. (2007). Surface tension and evaporation rate of liquids. *Journal of Physical Chemistry B, 111*(40), 11670–11676. https://doi.org/10.1021/jp074088w

Walrafen, G. E. (1966). Raman spectral studies of hydrogen bonding in liquids. *Journal of Chemical Physics, 44*(10), 3726–3729. https://doi.org/10.1063/1.1727158

Wang, Z., & Cadwallader, K. R. (2024). Ethanol's pharmacodynamic effect on odorant detection in distilled spirits models. *Beverages, 10*(4), 116. https://doi.org/10.3390/beverages10040116

Watanabe, A., Nishimura, K., & Arai, S. (1996). Viscosity and flavor perception. *Journal of Texture Studies, 27*(3), 257–271. https://doi.org/10.1111/j.1745-4603.1996.tb00024.x

Weafer, J., & Fillmore, M. T. (2008). Acute alcohol effects on attentional bias. *Addiction, 103*(4), 689–697. https://doi.org/10.1111/j.1360-0443.2008.02143.x

World Health Organization. (2018). *Global status report on alcohol and health 2018.* https://www.who.int/publications/i/item/9789241565639

Zakhari, S. (2006). Overview: How is alcohol metabolized by the body? *Alcohol Research & Health, 29*(4), 245–254.

Observations on Ethanol and Sensory Evaluation

- **Moderate ethanol vapor** suppresses olfactory receptor signaling within minutes, reducing sensitivity to delicate volatiles and narrowing aroma range.
- **Ethanol's trigeminal stimulation**—burn, warmth, or numbing—creates false intensity cues, often mistaken for body or flavor strength.
- **Rising ethanol partial pressure** near the rim shifts vapor composition toward lighter compounds, masking heavier congeners critical to a spirit's identity.
- **Ethanol's effects:** slower reaction time, elevated confidence, and distorted judgment. Panels and evaluators who do not expectorate produce inflated scoring and reduced discrimination.

Chapter 11: Sensory Effects of Ethanol-Water Solutions

Ethanol and water are the twin pillars of all spirituous beverages. Their physical and chemical interactions determine not only the concentration of psychoactive ethanol but also the volatility, solubility, and perception of hundreds of congeners dissolved within. Spirits are not inert mixtures; they are dynamic, thermodynamically active solutions. A proper evaluation requires understanding the properties of this binary mixture, how dilution alters volatile behavior, and why the addition of water is a critical tool in sensory analysis.

The term "spirit" originates from the Latin "spiritus," meaning "breath," "breeze," or "soul." Initially used in alchemy, the term "spirit" entered Middle English in the mid-1300s, credited to Ramon Llull, the first to document it as a term for the vapors produced during distillation for medicinal purposes (Forbes, 1970). A synonym for liquor, the term "spirit" has been used to define alcoholic beverages with an alcohol content as low as 20% ABV (alcohol by volume).

The TTB classifies distilled spirits as products made via distillation. Many core spirits (e.g., whisky, vodka, gin, brandy) are bottled at a minimum of 40% ABV; others (e.g., liqueurs) are bottled at lower percentages. Proof is defined as twice the alcohol by volume (ABV) at 60 °F (15.56 °C) in the U.S. system; other countries use ABV only (Alcohol and Tobacco Tax and Trade Bureau [TTB], 2021). For this discussion, ethanol is the alcohol in question. Water makes up the remaining percentage of the mix, reaching nearly 100%, with the remainder comprising small contributions of character aroma molecules that define the spirit's aroma, taste, and mouthfeel.

The Importance of Polarity

Alcoholic beverages are ethanol-water mixtures, and factors such as polarity, hydrophilicity (water-loving), hydrophobicity (water-avoiding), temperature, molecular size, and chain length determine ethanol's behavior in these mixtures. Polarity is crucial to sensory science because it governs how aroma and flavor molecules interact with solvents, receptors, and delivery systems, ultimately determining their volatility, solubility, and perceptual impact. The polarity of water molecules is crucial for the sensory detection of flavor molecules. Understanding common terminology associated with polarity is essential.

- **Polar:** Molecule has positive (+) charge on one end, (-) negative on other. Water molecules are polar and form hydrophilic hydrogen bonds (Atkins & de Paula, 2010).
- **Hydrogen bond:** Formed by attraction of water molecules or other polar molecules (positive & negative attracted to each other) (Atkins & de Paula, 2010).

- **Hydrophilic:** Water-attracting flavor molecules that readily engage with water, often dissolving or wetting in water. Polar (Atkins & de Paula, 2010).
- **Hydrophobic:** Water-repelling flavor molecules that do not dissolve, interact well, or wet with water. Non-polar (Atkins & de Paula, 2010).
- **Amphipathic:** Possesses both hydrophilic and hydrophobic properties (Israelachvili, 2011).

The Existence of Flavors in Ethanol-Water Solutions (Fig. 11.1)

There are seven basic states of flavor molecules in ethanol-water solutions; the physical and chemical properties of the compounds determine which state they occupy in a particular solution (Poole & Poole, 1991).

1. **True Molecular Solutions:** Aroma/flavor molecules individually dissolved at the molecular level without forming aggregates or clusters, fully solvated by water-ethanol solvent molecules. *Highly soluble alcohols, aldehydes, ketones, short-chain esters, acids, and phenolics.*

2. **Micelles:** Small aggregates (5–100 nanometers) formed by amphiphilic aroma molecules, clustering hydrophobic tails inward, hydrophilic heads outward. Driven by hydrophobic interactions, stabilized by ethanol-water hydrogen bonding (Attwood & Florence, 2012), *such as longer-chain esters, terpenoids, fatty acid derivatives, fusel oils, larger phenolics, and amphiphilic aroma compounds.*

3. **Microemulsions:** Thermodynamically stable, transparent/translucent nanometer-sized droplets (5–100 nm) of hydrophobic aroma compounds suspended in water-ethanol mixtures. Stabilized by ultra-low interfacial tension due to ethanol and natural surfactants (Lawrence & Rees, 2012). *Oil-rich ingredients, strongly hydrophobic essential oils, heavy terpenes, lipid-rich botanicals.*

4. **Macroemulsions:** Larger droplets (>100 nm to several microns) that are thermodynamically unstable but kinetically stable (opaque/cloudy appearance) (Lawrence & Rees, 2012). Formed by agitation or emulsifiers, *oil-rich ingredients, strongly hydrophobic essential oils, heavy terpenes, and lipid-rich botanicals.*

5. **Colloidal Suspensions:** Finely dispersed insoluble solid particles or precipitates (typically 10 nm – 1μm), suspended by Brownian motion or weak surface charges (Dickinson, 1992). *Polyphenols, tannins, proteins, haze-forming compounds, fatty acid precipitates, insoluble salts.*

6. **Solvent Clusters:** Small clusters of solvent (water-ethanol molecules) surrounding and loosely bound to aroma compounds (Poole & Poole, 1991); intermediate state between solution and micelle. *Moderately hydrophobic esters, alcohols, phenolics, fatty acids, and aldehydes.*

7. **Free Volatile Vapor Phase:** Aroma molecules freely evaporating above the liquid interface, responsible for perceived aromas (Ferreira, 2010). *highly volatile short-chain aldehydes, esters, ketones, and alcohols.*

The dominant flavor states change with increasing ethanol concentration. For spirits, the most prevalent and important factors when diluting a spirit for tasting and evaluation are: proper molecular solution, micelles and microemulsions, solvent clusters, and the free vapor phase (Lawrence & Rees, 2012; Attwood & Florence, 2012; Poole & Poole, 1991; Ferreira, 2010).

Fig. 11.1	Physical States of Flavor Molecules by Beverage/ABV					
ABV range→ ↓State	Beers ≤13%	Wines 8-18%	Fortified Wines 18-25%	Liqueurs 20-39%	Spirits 40-60%	High-Proof Spirits 45-90+%
True Molecular Solution	Moderate, noticeable	Strongly prevalent	Strongly prevalent, influential	Strongly prevalent, influential	Dominant, critical importance	Dominant, critical importance, defining behavior
Micelles	Minimal sporadic	Limited, but meaningful	Moderate, noticeable	Strongly prevalent, influential	Dominant, critical importance, defining behavior	Strongly prevalent
Micro-emulsions	n/a	Minimal, sporadic	Minimal sporadic	Strongly prevalent, influential	Dominant, critical importance, defining behavior	Strongly prevalent
Macro-emulsions	Limited, but meaningful	Minimal, sporadic	n/a	n/a	n/a	n/a
Colloidal Suspensions	Moderate, noticeable	n/a	Limited, but meaningful	n/a	n/a	n/a
Solvent Clusters	Minimal, sporadic	Moderate, noticeable	Strongly prevalent	Strongly prevalent, influential	Dominant, critical importance, defining behavior	Dominant, critical importance, defining behavior
Free Volatile Vapor Phase	Strongly prevalent, influential	Strongly prevalent, influential	Strongly prevalent	Strongly prevalent	Strongly prevalent, influential	Strongly prevalent
NOTE: Flavors exist in various physical forms, which define the solution's structure and affect how aromas and flavors are available for detection and identification. n/a = n/a=negligible/not available						

The Importance of Glassware Shape

For over a century, drinkers and evaluators alike have used tiny, convergent-rim (rim diameter < bowl diameter) tulip glasses under the erroneous assumption that their tiny rims concentrate all aromas directly under the nose, making them easy to access and detect. This assumption was further supported by the widely accepted belief that it would be impossible to separate anesthetic ethanol molecules from character aroma molecules, leaving the only alternatives as: (1) change the evaluation method, and (2) "deal with it." The common tulip glass was introduced to the drinking public when sherry and fortified wines gained popularity, primarily due to the rise of British world trade in the 1700s (Moss, 2015). Almost every household had them, so they were adopted for the spirits trade and became the go-to Scotch glass, particularly in the 1960s, when the Scotch industry launched an intensive, successful worldwide marketing campaign.

The lack of scientific knowledge regarding the physico-dynamics of the ethanol-water solution, and the convenience of using the tulip glass, which was widely available, in use in many households, and accepted as a suitable vessel for alcoholic beverages, dispelled any notion that changing glass shape was ever an option. As a result, the accepted glass for drinking high-ABV whiskeys has been firmly established as the tulip style, with many slight variations. More details are provided in Chapter 40 and Appendix VI.

Both drinkers and evaluators noticed the high concentrations of pungent, distracting, olfactory-numbing, and anesthetic ethanol. The Scotch industry proposed procedural changes for tasting spirits in tulip glasses, which were accepted and incorporated as standard procedure. Drinking spirits from tulips remains unchallenged due to these modifications. They are:

- Do not swirl, swirling releases the ethanol (Likar & Jepsen, 2003). True, but a problematic solution, because swirling also releases flavor aromas.
- Breathe through the mouth and nose simultaneously, reducing nasal inhalation by sharing it with the oral cavity, which reduces the velocity of ethanol inhalation (Doty, 2015). Also true, however, fewer aroma molecules are available for detection and identification.
- Wafting or successive closer positioning to acclimate. Acclimation reduces the strong, initial shock effect of pungent ethanol (Shepherd, 2006), and reducing the ethanol initial dose definitely acclimates, but it does not prevent irritation of the trigeminal nerve endings and numbing olfactory to character aromas
- Add water to "open up" the spirit. Although water dilution lowers ethanol's partial pressure and reduces its trigeminal impact, it also changes the matrix in ways that suppress or re-rank the volatility of many aroma compounds (Meilgaard et al., 2006; Ebeler, 2001). Water does not "shut down" ethanol evaporation through surface tension; instead, dilution alters partition coefficients and reduces the overall release of both ethanol and specific congeners. The relief from burn creates the impression that the spirit "opened up," when in fact some character aromas are subdued or reshaped. The effect is context-dependent: dilution can reveal specific notes while masking others, but it does not universally enhance aroma. The anesthetic effect on the trigeminal system resets with each sniff, whether acknowledged or not.
- None of these methods accomplishes their intended purpose and is mainly ineffective, except for providing slight relief from *perceived* irritancy. Spirits of 40% ABV or higher generally yield an ethanol concentration in the headspace of commonly used tulip glasses of around 60-75% due to high ethanol volatility (see Chapter 10), which contributes to quick olfactory

overload and olfactory fatigue.

Adding Water to Spirits

Adding water (dilution) significantly alters the availability and detectability of aromas. Several physical and chemical factors alter the sensory experience by changing the aroma profile and intensity. These factors are explored in more detail:

Surface Tension and Dilution: Adding water increases surface tension (see Fig. 10.1); water's surface tension is more than three times that of ethanol. Surface tension describes the cohesive forces that minimize a liquid's surface area—hence the spherical shape of droplets. However, static surface tension does not impede aroma release into the headspace; evaporation is primarily governed by vapor pressure, activity coefficients, and gas-phase diffusion, not by resistance at the liquid surface. When water is added to spirits, the key effect is dilution, which lowers ethanol's partial pressure and alters the partitioning of volatiles. These matrix changes suppress or re-rank the volatility of many aroma compounds, especially those that are heavier or more hydrophobic. Hydrogen bonding among water molecules shapes interfacial structure but does not impede molecular escape; it is the thermodynamic shift induced by dilution, rather than increased surface tension, that reduces aroma evaporation (Brochard-Wyart, de Gennes, & Quéré, 2003; Weast, 1985). Reduced Kinetic Energy & Volatility: Kinetic energy describes the motion of molecules. Represented by the equation:

$$K = \frac{1}{2} m v^2$$

where K = kinetic energy, m = mass, and v = velocity of the molecule.

Adding water reduces lower mass molecular velocities and aroma volatility, slowing the transfer of aromas into the headspace. All compounds are affected, with the largest effect on lower-mass, higher-volatility aromas, such as light alcohols, esters, aldehydes, and light phenols, all of which are sensory flavor molecules.

Aromatic compounds become harder to detect as more water is added—not because they must overcome surface tension, but because dilution changes vapor–liquid partitioning. Increasing the water fraction lowers ethanol's partial pressure and alters activity coefficients, reducing the volatility of many congeners, especially heavier or more hydrophobic ones. Hydrogen bonding strengthens water's solvation environment and further suppresses volatility, but it does not physically block molecular escape (Atkins & de Paula, 2010).

Miscibility and Solubility Shifts: Miscibility is the ability of water and ethanol to mix. Solubility is the amount of a substance that can be dissolved in a water-ethanol mixture. Higher water content in the spirit decreases solubility, as pure ethanol has much higher solubility than water. Esters, oils, and fatty acids may separate out of the solution (precipitate), aggregate, or evaporate more readily as water is added. Detectability depends on the polarity of the oil and ester, as well as the chain length and structure of the acids. Water alters the sensory profile of the spirit (Belitz, Grosch, & Schieberle, 2009; Rowe, Sheskey, & Quinn, 2009).

Microemulsions and Micelles: A microemulsion is a stable colloidal dispersion of tiny droplets of immiscible oils or flavor compounds that do not settle out. Micelles are spherical structures formed by the aggregation of surfactant molecules in water, loosely bound, made up of several atoms, ions, or molecules to form a colloidal particle. Oils, such as those in ouzo, absinthe, pastis, and sambuca, readily form micelles upon addition of water (Arndt & Loos, 2014; Lawrence & Rees, 2012). Detectability depends on the polarity of oil and ester, and for acids, on both chain length and structure (Attwood & Florence, 2012).

Hydration Effects: Adding water can convert certain compounds into hydrated forms and describe how water molecules surround and interact with solute ions or molecules (e.g., when salts dissolve in water). Adding water makes the mixture more polar, triggering the release of new aromas. Compounds previously soluble in ethanol begin to separate, aggregate, or evaporate more readily. Oils, esters, and fatty acids become less soluble and separate from the solution. Water becomes dominant. Spirit may smell less harsh, with a reduced palate and nose burn, and less pungent, tending toward aromatic, floral, and fruity notes (Belitz et al., 2009; Ferreira, 2010).

Hydrolysis Effects: Hydrolysis occurs when a water molecule breaks one or more chemical bonds. Adding water to an ethanol-water mix enables ester hydrolysis over time, releasing fewer aromas. Fragile esters and lactones are affected by hydrolysis. Subtle aroma shifts occur, and fewer aromas become detectable.

Louche Effect: The louche or ouzo effect is defined by spontaneous emulsification, forming a milky oil-in-water emulsion when water is added. Water releases insoluble, hydrophobic esters and oils, causing precipitation, cloudiness, and turbidity. Compounds affected include fatty acid esters, terpenes, and anethole (an aromatic compound found in anise seed). Results include visual cloudiness, sensory richness, and a less harsh but oily mouthfeel, especially in ouzo, sambuca, anisette, absinthe, arak, and other liqueurs, spirits, and Amari containing anethole (Belitz et al., 2009; Ferreira, 2010).

Marangoni Effect: Fluid motion generated by gradients in surface tension along an interface, typically at a liquid–gas boundary or within miscible liquid layers. These gradients arise from temperature differences (thermal Marangoni effect) or compositional variations, such as changes in ethanol–water concentration (solutal Marangoni effect). During swirling, Marangoni flows contribute to surface renewal and mixing, which can modestly influence the transport of volatile compounds near the liquid–air interface, though they are one of several mechanisms acting in parallel (Scriven & Sternling, 1960; Brochard-Wyart et al., 2003).

The thermal effect may seem inconsequential, but it becomes significant when considering hand heat conduction through the vessel walls or when dealing with extremely cool or warm room temperatures compared to the spirit's temperature. The very fact that it seems inconsequential indicates that the evaluator is not placing temperature in proper perspective or giving it the respect and awareness it deserves.

Evaluation Note 11.1: Be aware that the vessel transfers hand heat faster than one generally realizes. Glass is not a perfect thermal insulator; a few degrees in temperature will produce an entirely different evaluation.

Surface tension arises from cohesive forces between like molecules in a liquid. Molecules at the surface experience a net inward force because they lack neighboring molecules above them, creating an energy imbalance that pulls surface molecules into a tighter arrangement. When surface tension is nonuniform, due to spatial temperature differences or variations in the ethanol-water composition, molecules move from regions of low surface tension to regions of high surface tension, setting up a lateral flow along the surface. This flow induces bulk liquid motion beneath the surface, leading to convective currents known as Marangoni convection. The Marangoni equation (Ma = Marangoni number) (Scriven & Sternling, 1960) is expressed as:

$$Ma = \frac{(\partial\gamma/\partial x \cdot L)}{\mu D}$$

Simplified: where $\partial\gamma/\partial x$ = gradient of surface tension along the surface, L = gradient length, μ = dynamic viscosity, D = diffusion coefficient for thermal or solutal variations

High Marangoni numbers indicate strong convective motion, and low values suggest that diffusion or bulk viscosity dominates. As a result, visual "legs" or "tears" appear on the vessel sides, which the layperson may mistakenly interpret as an indication of high alcohol, spirit quality, or viscosity, when in truth it is a physical phenomenon governed by:

- Ethanol–water concentration differentials vary by ethanol concentration in the mix.
- Evaporation rate and surface replenishment (Ebeler, 2001).

- Adhesion to the surface of the glass vessel, which is affected by detergent, glassware material, and surface porosity, use of spot removers, and cleanliness (Kawaguchi, 2005).
- Gravitational drainage (vertical shear on vessel sides, due to gravitational pull as liquid returns after a swirl from the sides to the liquid pool) counteracted by surface tension flow. Vertical shear releases aromas in addition to those that were released by circular shear created by swirling. Shear breaks surface tension (Brochard-Wyart et al., 2003; Ebeler, 2001).

There is a lot more to the Marangoni contribution to aroma release than meets the eye, with "tears" or "legs" often misinterpreted as quality markers.

Nothing is simple about the complex relationship between ethanol and water in alcoholic beverages. From low to high concentration (ABV), each of the above factors takes a distinct approach to preventing flavor molecules from entering the headspace for detection, identification, and evaluation. The state of the ethanol-water continuum in any alcohol beverage cannot be reduced to a single, simple explanation of flavor molecule behavior, as each plays a separate but significant part in flavor suppression or release. The final effect of each spirit is the sum of its parts, and the importance of these parts varies with ethanol concentration and other factors.

Water pH Effect on Sensory Evaluation

The pH scale ranges from 1 to 14 and measures the concentration of hydrogen ions in a liquid. pH<7 = Acidic, pH = 7 = Neutral, pH>7 = basic or alkaline. The effect of pH on spirits can be characterized by:

- pH affects aroma sensory in concert with ethanol, volatile compounds, and temperature. To a distiller, the right water can be the difference between a harsh experience and a clean expression (Belitz et al., 2009; Mirlohi, 2022).
- pH exhibits a significant influence on the ionization of organic acids and phenolics; low pH tastes sharp or sour, high pH masks or lowers aroma intensity (Belitz et al., 2009; Sakurai et al., 2009).
- Alkaline water = smooth, round mouthfeel, extremely high alkalinity = chalky mouthfeel. Low acidity = crisp mouthfeel, high acidity = metallic.
- Low pH can shift color when exposed to air, and high pH can cause precipitation or haze in liqueurs (Belitz et al., 2009).
- Low pH<6 enhances acidic notes, clashes with congeners; pH>8.5 introduces alkaline bitterness, suppresses aromatics (Belitz et al., 2009; Mirlohi, 2022).

Adding Same Source Water to Scotch Whisky

Many whiskey enthusiasts and specialty providers now offer bottled spring waters sourced from distilleries or their source streams, marketed explicitly as ideal for adding a few drops to whisky to mellow harsh ethanol or unlock aromatic complexity. For example, Uisge Source (a Scottish company) sells spring water from the Islay, Speyside, and Highland regions, touting its chemical similarity to the original water used in mash and proofing processes. Similarly, Glenfiddich notes that it bottles Robbie Dhu spring water for on-site cask dilution, inviting discerning drinkers to taste the same water used in production (Uisge Source, 2025).

In practice, however, sensory gains are minimal. Most dilution proponents and distillers recommend using neutral, low-mineral water such as deionized or distilled to preserve the whisky's inherent flavors and prevent mineral-induced alterations to mouthfeel or aroma. Adding region-specific spring water may introduce trace minerals or bicarbonate ions. However, these are often imperceptible after dilution levels (e.g., down to 40% ABV) and can even mask subtle esters or congeners. Thus, while marketing narratives emphasize terroir or brand continuity, the scientific value of adding branded distillery spring water is largely symbolic, with genuine sensory improvements measurable only in extreme cases, if at all. There is also always the possibility that the water is not from the specified source (Belitz et al., 2009).

Summary

The ethanol–water solution is the foundation of every spirit, and its behavior governs how aroma and flavor molecules are expressed, suppressed, or transformed. Polarity, solubility, volatility, and molecular clustering determine which compounds are detectable and how dilution alters their release. While traditional practices such as tulip glassware and "opening" with water persist, scientific evidence indicates that these methods often suppress aromas rather than reveal them. Matrix effects—including changes in activity coefficients, micelle formation, hydrolysis, and surface-layer flows such as Marangoni motion—demonstrate that adding water can suppress or reshape aromatic expression. Even pH and trace minerals in dilution water influence mouthfeel and aromatic clarity. For the evaluator, recognizing these dynamics is essential: water is not a neutral additive but an active participant in sensory perception, altering every stage of evaluation. The only functional reason to dilute is to reduce palate burn from high ethanol; in such cases, both an undiluted and a slightly diluted sample should be assessed for the same spirit. Remember that saliva also dilutes the spirit once it enters the mouth, further modifying perception.
Beyond: Fragrances in spirits come from esters, a significant sensory component.

"It is vital to use water that won't degrade the flavor or appearance of the final spirit." — Calum Fraser, Chief Blender at Bowmore

"Water doesn't just make the bourbon; it defines it." — Jimmy Russell, (Wild Turkey)

"It's the most underrated ingredient in bourbon making." — Denny Potter, (Maker's Mark)

"The purity of water directly affects how the vodka interacts with the palate. Water that is too hard or too soft can change the mouth-feel dramatically." — Bomb City Distillery (blog entry)

References

Alcohol and Tobacco Tax and Trade Bureau. (2021). *Distilled spirits.* https://www.ttb.gov/spirits

Arndt, U., & Loos, H. H. (2014). The ouzo effect. *Langmuir, 30*(21), 6206–6211. https://doi.org/10.1021/la501500r

Atkins, P., & de Paula, J. (2010). *Physical chemistry* (9th ed.). Oxford University Press.

Attwood, D., & Florence, A. T. (2012). *Surfactant systems: Their chemistry, pharmacy and biology.* Springer. https://doi.org/10.1007/978-94-011-1282-1

Belitz, H. D., Grosch, W., & Schieberle, P. (2009). *Food chemistry* (4th ed.). Springer. https://doi.org/10.1007/978-3-540-69934-7

Brochard-Wyart, F., de Gennes, P. G., & Quéré, D. (2003). *Capillarity and wetting phenomena: Drops, bubbles, pearls, waves.* Springer. https://doi.org/10.1007/978-1-4757-4249-2

Dickinson, E. (1992). *An introduction to food colloids.* Oxford University Press.

Doty, R. L. (2015). *Handbook of olfaction and gustation* (3rd ed.). Wiley-Blackwell. https://doi.org/10.1002/9781118971757

Ebeler, S. E. (2001). Analytical chemistry of wine flavor. *Analytica Chimica Acta, 428*(1), 73–80. https://doi.org/10.1016/S0003-2670(00)01227-5

Ferreira, V. (2010). Volatile aroma compounds and wine sensory attributes. *Comprehensive Reviews in Food Science and Food Safety, 9*(4), 425–447. https://doi.org/10.1111/j.1541-4337.2010.00118.x

Forbes, R. J. (1970). *Short history of the art of distillation.* Brill.

Israelachvili, J. N. (2011). *Intermolecular and surface forces* (3rd ed.). Academic Press. https://doi.org/10.1016/C2009-0-21560-1

Kawaguchi, M. (2005). Surface chemistry of glass. *Journal of Non-Crystalline Solids, 351*(4), 342–348. https://doi.org/10.1016/j.jnoncrysol.2004.09.046

Lawrence, M. J., & Rees, G. D. (2012). Microemulsion-based media. *Advanced Drug Delivery Reviews, 64*(1), 175–193. https://doi.org/10.1016/j.addr.2012.09.018

Likar, M. D., & Jepsen, T. (2003). Ethanol evaporation and olfactory exposure. *Chemical Senses, 28*(8), 661–666. https://doi.org/10.1093/chemse/bjg061

Meilgaard, M. C., Civille, G. V., & Carr, B. T. (2006). *Sensory evaluation techniques* (4th ed.). CRC Press. https://doi.org/10.1201/9781420006687

Mirlohi, S. (2022). Characterization of metallic off-flavors in drinking water. *Food Quality and Safety, 6*(1), fyac008. https://doi.org/10.1093/fqsafe/fyac008

Moss, S. (2015). *The curious bartender's whiskey road trip.* Ryland Peters & Small.

Piggott, J. R., Conner, J. M., & Paterson, A. (1995). *Fermented beverage production* (2nd ed.). Springer. https://doi.org/10.1007/978-1-4615-2177-3

Poole, C. F., & Poole, S. K. (1991). *Chromatography today.* Elsevier.

Rowe, R. C., Sheskey, P. J., & Quinn, M. E. (Eds.). (2009). *Handbook of pharmaceutical excipients* (6th ed.). Pharmaceutical Press / American Pharmacists Association.

Sakurai, T., Misaka, T., Ueno, Y., Ishiguro, M., Matsuo, S., Ishimaru, Y., Asakura, T., & Abe, K. (2009). The inhibition of bitter taste receptors by acidic pH and peptides. *Biochemical and Biophysical Research Communications, 381*(4), 703–707. https://doi.org/10.1016/j.bbrc.2009.02.107

Scriven, L. E., & Sternling, C. V. (1960). The Marangoni effects. *Nature, 187*(4733), 186–188. https://doi.org/10.1038/187186a0

Observations on Water Quality

Chlorine Surprise: Municipal water supplies are sometimes shock-treated with chlorine after storms. One distillery ran an untreated batch, and chlorophenols formed in the mash. The new make carried antiseptic/band-aid notes that aging could not disguise, and the entire run was lost.

The Iron Bite: A producer relying on well water began to notice metallic, blood-like aroma notes in the spirit. Testing revealed dissolved iron concentrations exceeding 0.3 ppm, sufficient to taint both the wash and the distillate. Iron also encouraged the formation of reddish microbial films in the tanks, prompting the distillery to install a filtration system.

The Sulfur Wash: High-sulfate water went untreated in a Caribbean ferment, and yeast reduced the sulfates to hydrogen sulfide. The wash reeked of rotten eggs, and even copper contact could not scrub the overload. The spirit carried cabbage-like notes that persisted in the barrel.

The Biofilm Breeder: Warm, untreated source water was piped directly into fermenters at a rum operation. Microbial biofilms quickly colonized the tanks and produced recurring sour, acetic notes. The problem was only solved after UV and inline filtration were added to the water system.

Seasonal Swings: A Midwestern whiskey plant found that winter water yielded clean ferments, but summer water carried farm runoff nitrates. Extra nitrogen shifted yeast metabolism, producing more esters and fusels and altering house style. Reverse osmosis was installed to stabilize the profile year-round.

The Alkaline Mash: A bourbon distillery using carbonate-rich well water found that mash pH stayed above 6.0. Competing bacteria flourished, and butyric notes began to creep into the wash. The issue was controlled once the distillers began acidifying with backset and water treatment.

The Acid Crash: In a Caribbean rum ferment, acidic molasses combined with unchecked lactic growth dropped the pH below 3.0. Yeast performance declined markedly, resulting in a sour, under-attenuated wash and a lower yield. Buffering molasses charges with lime prevented further collapses.

The Scaling Problem: A spring-water source high in calcium and magnesium left heavy scale deposits in cookers and heat exchangers. The buildup harbored bacteria, imparting lactic taint to the ferments and reducing heat-transfer efficiency. Regular acid washing and eventual water softening corrected the issue.

The Hollow Ferment: One gin producer attempted to use ultra-pure deionized water for all mashing and dilution. Lacking minerals, the yeast ran poorly, and the wash attenuated weakly. The resulting spirit nosed hollow and lacked texture until mineralized water was restored.

The Copper Stain: A small distillery used untreated water high in dissolved sulfur for condenser cooling. Over time, the sulfides reacted with copper, resulting in black copper sulfide films and metallic off-notes in the spirit. Switching to filtered water eliminated both fouling and flavor taint.

The Nitrate Boost: A rum producer discovered unusually rapid fermentations after heavy rainfall. Nitrates washed into the local water supply, supercharging yeast activity and driving ester levels far above house style. The overly fruity spirit had to be segregated from regular production.

The Silica Haze: A whiskey distiller using well water rich in silica noticed persistent cloudiness in proofed-down spirit. The colloidal haze resisted filtration, creating an unmarketable appearance. Installing a silica filter corrected the fault and restored visual clarity.

The Sodium Slip — Salinity from Dilution Water: A coastal distillery detected a faint savory note in proofed spirits. Municipal water had spiked in sodium following seawater intrusion, and the increased salinity was evident only after dilution. Switching to low-TDS water restored the clean profile.

The Manganese Shadow — Oxidative Taint on Aeration: A mountain distillery saw light browning and stale notes in aerated mashes. Elevated manganese oxidizes upon contact with oxygen, producing off-odors. Filtration and reduced aeration cleared the problem.

Water in mashing and dilution is not passive. Mineral load, pH, and buffering capacity alter enzyme kinetics and Maillard reactivity, thereby influencing the distribution of flavor precursors. High-bicarbonate waters tend to favor darker, maltier notes due to elevated wort alkalinity, whereas soft waters accelerate fermentation and emphasize ester brightness. Each water source imparts a unique pre-fermentation signature to the distillate.

Chapter 12: Esters – The Fragrance Architect of Spirits

The Role of Esters in Spirits

Esters are among the most influential aromatic compounds in distilled spirits, significantly contributing to their perceived fruitiness, sweetness, and overall complexity. They are responsible for many desirable sensory traits in well-crafted spirits, from young whiskies to rums and brandies. Esters are formed as byproducts of fermentation and aging. Understanding the origin, structure, behavior, and sensory contribution is essential to mastering spirit evaluation (Piggott, Conner, & Paterson, 1993; Nykänen & Suomalainen, 1983).

From a chemical standpoint, esters are derived from the reaction between an alcohol and a carboxylic acid. In spirit production, this typically occurs during fermentation, where yeast metabolizes sugars and produces alcohols and organic acids as byproducts. These compounds subsequently react—either enzymatically or spontaneously—to form esters. The balance of esters, their volatility, and their interactions with ethanol and water define much of a spirit's aromatic fingerprint (Ferreira, López, & Cacho, 2000).

Evaluators must learn to recognize the impact of esters on both the orthonasal and retronasal regions. Due to their relatively high volatility and moderate polarity, esters readily vaporize into the headspace of a glass, contributing to the initial aromatic impression (Pino, 2014). Their sensory expression is further influenced by ethanol concentration, temperature, and vessel geometry (Ferreira & López, 2019). At moderate levels, esters enhance a spirit's complexity; in excess, they can obscure subtle characteristics and skew evaluation toward fruity or solvent-like perceptions. In poorly controlled fermentation, off-flavor esters may emerge, such as ethyl acetate, which imparts undesirable aromas reminiscent of nail polish or glue (Poisson & Schieberle, 2008).

In summary, esters are a double-edged sword in the aroma of spirits: they are essential for defining character but can also lead to excess or imbalance. Assessing their presence, type, and interactions with other aroma compounds is a key skill for trained sensory professionals.

Chemical Structure and Nomenclature of Esters

Esters are organic compounds that result from the condensation reaction between an alcohol and a carboxylic acid, eliminating a molecule of water in the process. The general structural formula of an ester is R–COO–R', where R is the hydrocarbon chain from the acid and R' is the hydrocarbon chain from the alcohol. The following general reaction illustrates ester formation:

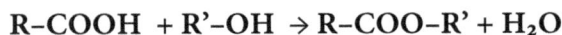

$$\text{R-COOH} + \text{R'-OH} \rightarrow \text{R-COO-R'} + \text{H}_2\text{O}$$

$$\text{Carboxylic acid} + \text{alcohol} \rightarrow \text{ester} + \text{water}$$

This process, commonly known as "Fischer esterification," was first described by German chemist Emil Fischer near the end of the 19th century. In the context of spirits, both primary and secondary alcohols—such as ethanol, isoamyl alcohol, or 1-butanol—can participate, combining with acids such as acetic, butyric, hexanoic, or lactic acid. The resulting esters vary in aroma from pineapple- and green-apple-like to sweaty, cheesy, or solvent-like, depending on the precursor molecules (Jackson, 2014).

The nomenclature of esters reflects their dual origin. The name begins with the alkyl group from the alcohol and ends with the carboxylic acid root, replacing the acid's "-ic acid" suffix with "-ate." For example, ethanol reacts with acetic acid to form ethyl acetate. Structurally, esters may be classified by chain length, branching, and degree of saturation, all of which affect volatility and aroma intensity. Short-chain esters (e.g., ethyl acetate, ethyl formate) tend to be highly volatile and contribute sharp, fruity notes. Medium-chain esters (e.g., ethyl butyrate, isoamyl acetate) produce rounded fruit and candy-like aromas. Long-chain esters (e.g., ethyl decanoate) are less volatile and contribute creamy or waxy notes that typically become perceptible only after dilution or warming (Pozo-Bayón & Moreno-Arribas, 2011).

Ester naming examples:
- **Ethyl acetate** (ethanol + acetic acid): pear, solvent
- **Isoamyl acetate** (isoamyl alcohol + acetic acid): banana, candy
- **Ethyl butyrate** (ethanol + butyric acid): pineapple
- **Ethyl hexanoate** (ethanol + hexanoic acid): apple, green fruit
- **Ethyl lactate** (ethanol + lactic acid): creamy, mild
- **Phenethyl acetate** (phenethyl alcohol + acetic acid): floral, honey

These esters differ not only in aroma type but also in detection threshold and solubility, properties discussed later. Evaluators should understand these names, their structural implications, and their aromatic signatures to accurately identify esters in both simple and complex spirits.

Sensory Characteristics and Evaluation of Esters
The sensory profile of esters in spirits is defined by their volatility, molecular structure, and interaction with ethanol and water. Esters are typically perceived orthonasally because they rapidly diffuse into the headspace of the nosing vessel. Their low molecular mass and moderate polarity enable many esters to evaporate at ambient temperatures, allowing them to reach the olfactory epithelium even at low concentrations. This characteristic contributes to the bright,

inviting aromas often associated with fruit-forward spirits such as rum, cognac, or certain styles of unaged whiskey (Pino, 2014).

Esters exhibit a wide range of odor descriptors, from pleasant fruity, floral, and sweet notes to undesirable pungent, solvent-like, or cheesy characters, depending on the specific precursor molecules and their concentrations. For instance, isoamyl acetate is commonly associated with banana or pear drops, while ethyl butyrate evokes pineapple or tropical fruit. These pleasant aromas are particularly noticeable in freshly distilled spirits and those produced with high-ester yeast strains or in pot stills with longer reflux times (Lehtonen, 2020). In contrast, elevated levels of ethyl acetate or ethyl lactate can lend a sharp, glue-like, or overly creamy quality, respectively, especially when the balance with other compounds is lacking (Ferreira & Lopez, 2019).

One of the critical challenges in ester evaluation is their concentration-dependent sensory impact. Many esters exhibit an inverted-U-shaped relationship between desirability and level: low levels contribute complexity and freshness, moderate levels enhance fruitiness, but excessive levels can overwhelm or unbalance the aroma profile. Compounds such as ethyl hexanoate and phenethyl acetate exhibit this pattern, providing notes of apple or floral honey when properly integrated, but becoming cloying or artificial when overexpressed (Poisson & Schieberle, 2008). Evaluators must develop a calibrated sense of such thresholds to interpret esters not in isolation, but in harmony with the spirit's style and intended character.

Another layer of complexity arises from matrix effects (interactions among esters, ethanol, water, and other volatiles). Ethanol alters the vapor pressure and partition coefficient of esters, sometimes suppressing their volatility until dilution or warming occurs. This phenomenon explains why certain long-chain esters (e.g., ethyl decanoate) are only detected after adding water or warming the spirit in hand (Gutzwiller & Chambers, 2010).

Ester aroma perception is highly variable and influenced by genetics, experience, and olfactory training. Compounds such as isoamyl acetate and ethyl butyrate are detectable by nearly all evaluators due to their low thresholds and familiarity from food exposure. Others, such as branched or high-mass esters, may be less widely recognized. Trained evaluators build reference banks by isolating ester compounds from training kits or model matrices and aligning their chemical identities with verbal descriptors. Developing this vocabulary is foundational to any sensory program that seeks to identify esters in both technical and descriptive analyses (Kemp & Gilbert, 2006).

Esters are central contributors to a spirit's aromatic fingerprint, but evaluation requires more than passive smelling. The trained evaluator considers their structure, concentration, balance, and interaction within the context of the entire matrix and the intended style. Mastering esters enhances the ability to distinguish between production techniques, maturation choices, and even faults, making it an essential component of professional spirits evaluation (Fig. 12.1).

Fig. 12.1	Common Esters in Spirits					
Ester	Aroma	Chain Size	MM (g/mol)	DT (ppm)	Boiling Point (°C)	Source*
Ethyl formate	rum, raspberry	Short	74.08	19.0	54.3	F
Methyl acetate	solvent, fruity	Short	74.08	5.0	56.9	F
Ethyl acetate	pear, solvent	Short	88.1	7.0	77.1	F
Propyl acetate	pear, fruity	Short	102.13	50.0	101.6	F
Isopropyl acetate	fruity, nail polish	Short	102.13	35.0	89.5	F
Ethyl butyrate	pineapple, apple	Medium	116.16	0.4	121.5	F
Ethyl lactate	creamy, buttery	Short	118.13	5.0	154.0	B
Isoamyl acetate	banana, candy	Medium	130.19	1.6	142.5	F
Ethyl valerate	green apple, fruity	Medium	130.19	0.3	146.5	F
Ethyl hexanoate	apple, green fruit	Medium	144.21	0.21	168.5	F
Isoamyl butyrate	pineapple, banana	Medium	158.24	0.4	179.0	F
Phenethyl acetate	floral, honey	Medium	150.17	1.0	221.0	F
Ethyl octanoate	pineapple, floral	Long	172.26	0.2	211.5	A
Ethyl decanoate	waxy, fruity	Long	200.31	0.01	243.5	A
Ethyl dodecanoate	creamy, coconut	Long	228.36	0.005	271.5	A
Less Common Esters in Spirits						
Methyl butyrate	apple, pineapple	Short	102.13	0.01	102	F
Ethyl isobutyrate	fruity, strawberry	Short	116.16	0.02	123	F
Isobutyl acetate	fruity, banana, solvent	Short	116.16	0.03	118	F
Methyl hexanoate	green apple, pineapple	Short	130.19	0.02	151	F
Ethyl 2-methylbutyrate	apple, fruity	Medium	130.18	0.02	151	F
Butyl butyrate	pineapple, sweet	Medium	144.21	0.05	151	F
Hexyl acetate	green apple, floral	Medium	144.21	0.06	171	F
Benzyl acetate	floral, jasmine, sweet	Long	150.17	0.01	213	A
Ethyl phenylacetate	honey, floral	Medium	164.2	0.05	250	A
Ethyl cinnamate	cinnamon, balsamic	Long	176.21	0.02	271	A
Geranyl acetate	floral, rose	Long	196.29	0.03	240	A
Some references conflict on DT, ordinally correct: DT=detection threshold approx, depending on consolidated studies, F=fermentation, A=aging, B=both: Short=2-5 C atoms, Medium= 6-10 C atoms, Long =11+ C atoms; ppm=parts per million, MM=Molar mass						

Note: Fig. 12.1 Readers are encouraged to verify these values against threshold compilations (e.g., Guth, 1997; Czerny et al., 2008) for precise odor threshold data under varying sensory contexts. Molecular mass increases by the number of carbon atoms in the chain. Long chains and higher masses make escape of esters from liquid into headspace at room temperature more difficult, and many esters may be present in the liquid but undetectable to the olfactory as they have not been released in sufficient numbers for detection (see detection threshold column). Temperature, volatility, dilution, and glass shape are key factors in the detection and identification of esters.

Temperature, Volatility, and Dilution Effects

Understanding how these factors influence ester perception is vital to evaluating the aromatic complexity of spirits. These three interrelated variables govern the behavior of esters in glass and their presentation to the olfactory system.

Temperature directly affects the volatility of ester compounds. As the temperature rises, molecular kinetic energy increases, enhancing their tendency to evaporate from the liquid matrix into the headspace. This increased volatilization increases ester concentrations in the vapor phase, thereby intensifying perceived aroma strength, particularly for short- and medium-chain esters with lower boiling points. At cool serving temperatures (18–22 °C), esters such as ethyl acetate and ethyl butyrate contribute noticeably. Thus, esters are a double-edged sword in the aroma of spirits: they are essential for defining character but can also lead to excess or imbalance. Assessing their presence, type, and interactions with other aroma compounds is a key skill for trained sensory professionals. to fruity or solvery notes (Gutzwiller & Chambers, 2010).

Volatility itself is determined by molecular weight, vapor pressure, and polarity. Short-chain esters, being lighter and less polar, exhibit high vapor pressure and are readily released into the atmosphere. Conversely, long-chain esters, such as ethyl decanoate or ethyl laurate, are less volatile and require elevated temperatures or dilution to reach detectable thresholds. This principle helps explain why certain creamy or waxy esters may appear muted until water is added or the spirit is warmed slightly in the hand (Jackson, 2014).

Dilution alters both the ethanol concentration and the solubility behavior of esters. Water reduces ethanol content, lowering its suppressive effect on ester volatility. This allows some esters that may have been previously solubilized in ethanol-rich matrices to become more volatile, enhancing the fruity or floral dimensions of the aroma profile (Ferreira & Lopez, 2019). However, excessive dilution can alter micellar structure or induce phase separation, particularly for long-chain or hydrophobic esters, thereby attenuating or altering their aromatic contribution (Attwood & Florence, 2012). Fig. 12.2 illustrates the headspace volatility of esters.

Speculatively, knowledge that ester release due to water dilution may also be why many tulip users believe water "opens up" the spirit. However, a few drops will not change the concentration to the degree that heavy esters begin to vaporize noticeably. Increased water content can temporarily suppress ethanol aroma when droplets sit on the surface without mixing—not because surface tension limits volatility, but because unmixed water locally dilutes the surface layer. This temporary suppression can create the impression that the spirit has "opened up."

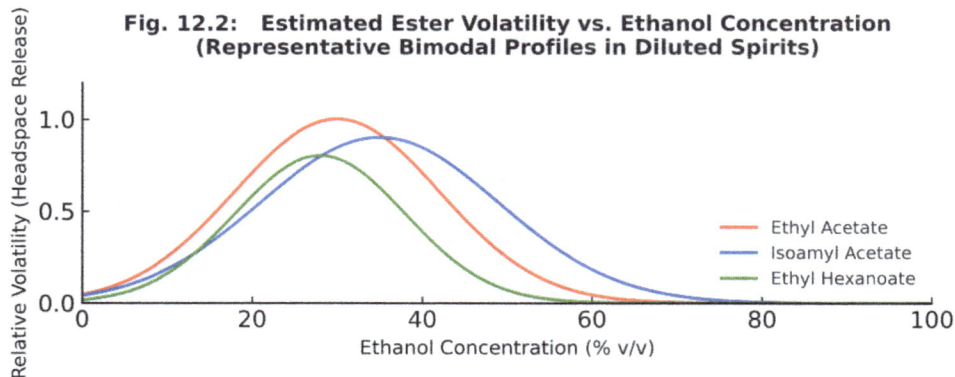

Fig. 12.2:   Estimated Ester Volatility vs. Ethanol Concentration (Representative Bimodal Profiles in Diluted Spirits)

Evaluation Note 12.1: Temperature is a key influencer in changing aroma profiles. Most evaluators are unaware that their hand heat will warm the sample enough to change the aroma profile. In fact, many evaluators, particularly the distiller, will deliberately warm the sample vessel in their hands to detect which aromas will enter the headspace as the temperature rises. As noted, excessive handling of the vessel noticeably alters the aroma profile. Fig. 12.3 describes the volatility of 3 common esters in spirits from a range of cool serving temperatures to average human body temperature. Setting the volatility of ethyl acetate at 1.0 at 20 ^{0}C, one can easily see the volatility of ethyl acetate quadruples if the temperature reaches human body temperature (approx. 37 ^{0}C), stressing the importance of sample temperature control, and pointing to large heat sink bases or stems on the sample vessel as logical barriers to hand heat. Chart generated using the Clausius-Clapeyron equation to estimate vapor pressures.

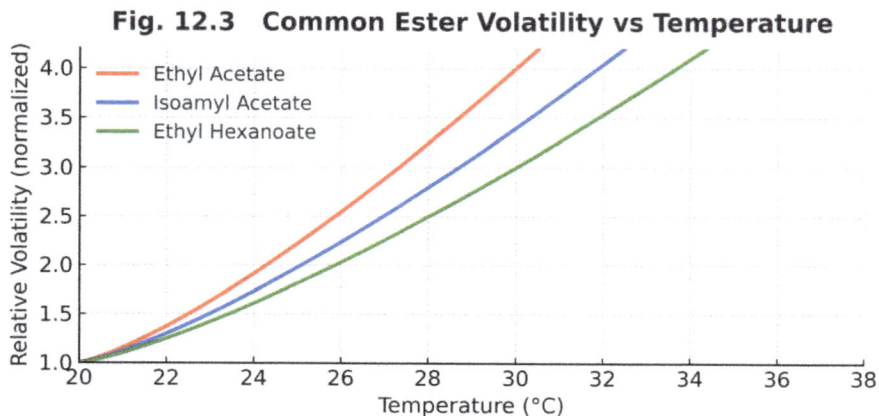

Fig. 12.3   Common Ester Volatility vs Temperature

Research shows headspace chemical composition and sensory characteristics evolve and vary by glass shape—after 5 to 10 minutes, aroma intensity can shift significantly (Hirson, Heymann, & Ebeler, 2012).

Tulip or bowl-shaped glassware concentrates ethanol vapors alongside aromas, which can numb the olfactory system and mask subtler notes. Alternative vessel designs that allow ethanol to diffuse away can mitigate this effect (Hirson et al., 2012; Lee et al., 2001). A simple tumbler can change detection and perception.

Evaluation Note 12.2: For decades, evaluators and drinkers have used tulip-shaped vessels without giving thought to their major drawback, the concentration of distracting, nose-numbing, pungent, anesthetic ethanol. The unconditional acceptance of the tulip as an evaluation vessel has led to the widespread practice of adding water to prevent ethanol from evaporating, in search of character aromas that will define the spirit. This practice disregards the effects of dilution water on the aroma profile and focuses on a problem better addressed by the vessel. Change vessel shape to at least a tumbler or, better still, a flared-rim vessel. Flares give headspace ethanol an escape route, provided the nose is not inserted below the rim or glass "neck," the ethanol will remove itself from detection by the laws of diffusion, with diffusion occurring quicker if the vessel has a "neck" or narrow portion to redistribute kinetic energy to low mass molecules (Graham's law of diffusion).

Overall, mastering the interplay between temperature, volatility, dilution, and vessel shape provides critical insight into how esters are perceived—and how they can dominate or diminish during evaluation. Skilled assessors learn to modulate these variables to reveal specific aromatic layers and avoid misinterpreting aroma balance based on temperature-related volatility shifts or inappropriate dilution protocols.

Masking of Flaws by Esters

While esters are celebrated for imparting fruity and floral aromas to spirits, they can also mask flaws that arise during fermentation, distillation, or maturation. This phenomenon—known as flavor masking—occurs when esters dominate the aromatic profile to such an extent that off-notes are perceptually suppressed or undetectable. From a production standpoint, this may seem advantageous, but for the trained evaluator, it represents a significant challenge and a potential misrepresentation of spirit quality.

Esters, especially short- and medium-chain variants like ethyl acetate, ethyl butyrate, and isoamyl acetate, exhibit high volatility and low detection thresholds. Their potent and often pleasant aromas can saturate olfactory receptors, effectively crowding out detection of compounds that signal faults—such as excessive sulfur (e.g., hydrogen sulfide, dimethyl trisulfide), bacterial byproducts (e.g., butyric or caproic acids), or fusel oils (e.g., higher alcohols like propanol or isobutanol). This sensory occlusion is particularly problematic in spirits with naturally high ester concentrations, such as Jamaican rum, Calvados, or certain Armagnacs (Pino, 2014; Poisson & Schieberle, 2008).

Producers sometimes exploit this phenomenon—intentionally or unintentionally—by crafting ester-rich profiles to mask incomplete fermentation, poor yeast health, or harsh distillation cuts. In some commercial flavored whiskeys or young spirits, flavoring or finishing in sweet wine casks may amplify ester-like aromas, further reducing the evaluative clarity of the spirit's base character (Moss, 2015). This can mislead both consumers and judges into overrating a product's quality based solely on its surface pleasantness. This tactic is less prevalent, though craft distillers use it to some extent. Economics and cash flow play a key role in the decision.

To counteract this masking effect, sensory training must focus on differentiation beneath the surface of complexity. Evaluators should be trained to:

- Realize that sample dilution may reduce ester saturation, but the departure from the aroma profile will make the sample unsuitable for evaluation comments.
- Use dry-glass or low-ABV sniffing techniques to bypass dominant volatiles, especially if using tulip-shaped glasses, which concentrate ethanol. Lower airflow to olfactory receptors (ORs) compromises aroma detection, not to mention obliterating proper identification.
- Compare across control samples of known flaw presence to learn the specific olfactory footprint of sulfur, high acids, or faulty esters.
- Familiarize themselves with "clean" and "dirty" fermentation markers through side-by-side trials, especially with unaged distillates or production samples.

One common practice involves nosing a known ester-rich sample (e.g., a young fruit brandy or heavily esterified rum) alongside a clean neutral spirit spiked with minor sulfur compounds or volatile acids. This sharpens the evaluator's ability to *subtract* esters from the mental profile and perceive underlying anomalies. However, this method depends on the evaluator establishing a standard baseline of the comparator spirit, and who is to say that it is representative? This practice has emerged from the search by unscientific evaluators to isolate and detect subtle aromas, without addressing the primary barrier: the concentrated ethanol posed by the widely accepted, common tulip-shaped vessel.

Ultimately, esters are double-edged contributors: they enrich but can also deceive. The true professional learns to recognize when complexity is genuine and when it serves as a perfumed veil over fundamental defects. In doing so, evaluators enhance their capacity to assess not only aroma pleasure but also product integrity, a core goal of this book and advanced sensory science.

Summary

Esters are central to the aromatic architecture of spirits, arising naturally from fermentation and aging through the reaction of alcohols and acids. They deliver much of the fruit, floral, and sweet character that defines a spirit's style, yet their impact depends on volatility, concentration, and balance with other compounds. Short-chain esters provide bright fruitiness, medium-chain esters add rounded notes, and long-chain esters lend creamy or waxy tones, often revealed only through warming or dilution. While esters enrich complexity, they can also mask flaws such as sulfur or fusel alcohols, making careful evaluation essential. For the trained sensory analyst, mastery of ester recognition across neat and diluted conditions, with awareness of vessel shape and temperature effects, is fundamental to distinguishing genuine complexity from superficial pleasantness.

Beyond: Aldehydes emerge as a key influence in sensory evaluation.

"Esters are the major congener constituents and key contributors to the flavor and aroma of whiskey." — T. J. Kelly, C. O'Connor, & K. N. Kilcawley, whiskey researchers

"In total, esters form key component of flavour, contributing particularly to complex roundness." — K.-Y. M. Lee, A. Paterson, & J. R. Piggott, research scientists

References

Attwood, D., & Florence, A. T. (2012). *Surfactant systems: Their chemistry, pharmacy and biology.* Springer. https://doi.org/10.1007/978-94-011-1282-1

Ferreira, V. (2010). Volatile aroma compounds and wine sensory attributes. *Comprehensive Reviews in Food Science and Food Safety, 9*(4), 425–447. https://doi.org/10.1111/j.1541-4337.2010.00118.x

Ferreira, V., López, R., & Cacho, J. (2000). Quantitative determination of the odorants of young red wines from different grape varieties. *Journal of the Science of Food and Agriculture, 80*(11), 1659–1667. https://doi.org/10.1002/jsfa.693

Ferreira, V., & López, R. (2019). The actual and potential aroma of winemaking by-products. *Food Chemistry, 278*, 244–257. https://doi.org/10.1016/j.foodchem.2018.11.073

Czerny, M., Christlbauer, M., Granvogl, M., Fischer, A., Engel, A., & Schieberle, P. (2008). Re-investigation on odour thresholds of key food aroma compounds and development of an aroma language based on odour qualities of predefined aqueous odorants. *European Food Research and Technology, 228*(2), 265–273. https://doi.org/10.1007/s00217-008-0931-x

Guth, H. (1997). Quantitation and sensory studies of character impact odorants of different white wine varieties. *Journal of Agricultural and Food Chemistry, 45*(8), 3027–3032. https://doi.org/10.1021/jf970280a

Gutzwiller, B. J., & Chambers, E. (2010). The influence of ethanol and serving temperature on the release of esters in alcoholic beverages. *Flavour and Fragrance Journal, 25*(5), 320–325. https://doi.org/10.1002/ffj.1992

Hirson, G. D., Heymann, H., & Ebeler, S. E. (2012). Equilibration time and glass shape effects on chemical and sensory properties of wine. *American Journal of Enology and Viticulture, 63*(4), 515–521. https://doi.org/10.5344/ajev.2012.11113

Jackson, R. S. (2014). *Wine science: Principles and applications* (4th ed.). Academic Press.

Kelly, T. J., O'Connor, C., & Kilcawley, K. N. (2023). Sources of volatile aromatic congeners in whiskey. *Beverages, 9*(3), 64. https://doi.org/10.3390/beverages9030064

Kemp, S. E., & Gilbert, A. N. (2006). Odor recognition and naming by children and adults: A comparison of structured and unstructured testing methods. *Chemical Senses, 31*(6), 521–529. https://doi.org/10.1093/chemse/bjj056

Lee, K., Paterson, A., & Piggott, J. R. (2001). Origins of flavour perception in whisky: The influence of glass shape. *Food Quality and Preference, 12*(6), 397–404. https://doi.org/10.1016/S0950-3293(01)00031-3

Lee, K.-Y. M., Paterson, A., & Piggott, J. R. (2001). Origins of flavour in whiskies and a revised flavour wheel: A review. *Journal of the Institute of Brewing, 107*(5), 287–313. https://doi.org/10.1002/j.2050-0416.2001.tb00099.x

Lehtonen, M. (2020). Production and sensory effects of esters in distillates. *Journal of the Institute of Brewing, 126*(2), 107–120. https://doi.org/10.1002/jib.605

Moss, S. (2015). *The curious bartender's whiskey road trip.* Ryland Peters & Small.

Nykänen, L., & Suomalainen, H. (1983). *Aroma of beer, wine and distilled alcoholic beverages.* Springer.

Piggott, J. R., Conner, J. M., & Paterson, A. (1993). *Whisky: Technology, production and marketing.* Springer.

Pino, J. A. (2014). Odor-active compounds in alcoholic beverages. *Critical Reviews in Food Science and Nutrition, 54*(7), 885–901. https://doi.org/10.1080/10408398.2011.588493

Poisson, L., & Schieberle, P. (2008). Characterization of the aroma profile of rum by quantitative descriptive analysis, aroma extract dilution analysis, and omission tests. *Journal of Agricultural and Food Chemistry, 56*(13), 5820–5826. https://doi.org/10.1021/jf800457v

Pozo-Bayón, M. Á., & Moreno-Arribas, M. V. (2011). Analytical methods for wine volatile compounds. In M. V. Moreno-Arribas & M. C. Polo (Eds.), *Wine chemistry and biochemistry* (pp. 259–285). Springer.

Observations on Esters

Most esters originate from yeast enzymes that combine alcohols with acyl-CoA intermediates during fermentation, with slower chemical esterification continuing through maturation to develop fruity and floral top notes.

Temperature, pH, and yeast strain determine both the quantity and the balance of ester species—warmer, lower-pH ferments yield richer, fruit-forward profiles.

Chapter 13: Aldehydes – Sharp Edges, Oxidative Notes

Sources: Oxidation, Aging, Fermentation By-products

Aldehydes are chemically reactive compounds that arise in spirits primarily through three pathways: fermentation metabolism, oxidative degradation, and maturation in wooden barrels. Chemically, aldehydes contain a carbonyl group (C=O) bonded to a hydrogen atom and a variable R group. Their small molecular size and moderate polarity allow significant volatility and odor activity, often at sub-ppm levels.

During fermentation, yeast produces aldehydes as metabolic intermediates. The most prominent is acetaldehyde, a direct product of ethanol biosynthesis from pyruvate via acetaldehyde dehydrogenase and alcohol dehydrogenase (Nykänen & Suomalainen, 1983). In clean fermentations, acetaldehyde levels remain low; however, stressed or prematurely arrested fermentations can produce higher levels, often with notes of green apple or grass. Other fermentation aldehydes include propionaldehyde and butyraldehyde.

Oxidation is a second route: when ethanol is exposed to air during handling, bottling, or storage, it can slowly oxidize to acetaldehyde, catalyzed by trace metals or light, particularly in matrices with few antioxidants. Barrel aging allows oxygen ingress and also provides aldehyde precursors via wood breakdown, generating vanillin from lignin and furfural from hemicellulose (Ferreira, López, & Cacho, 2000). Aging magnifies aldehydes by both promoting oxidation over time and contributing barrel-extracted aldehydes. Bourbon and brandy aged in charred oak often accumulate vanillin and furfural, resulting in a character reminiscent of vanilla, caramel, and toasted wood (Nykänen & Suomalainen, 1983).

Key Aldehydes in Spirits: Acetaldehyde, Furfural, Vanillin (Fig. 13.1)

Among the many aldehydes in spirits, three dominate the sensory impact:

- **Acetaldehyde** — typically green apple / cut grass; at low levels (<100 mg/L) it may add lift to young spirits; at higher levels it is a harsh, solvent-like defect (Jackson, 2014).
- **Furfural** — formed by thermal degradation of pentoses during barrel toasting/charring; almond, caramel, bready; increases with time in oak; excessive levels can taste dry/astringent (Mosedale & Puech, 1998).
- **Vanillin** — derived from lignin; sweet vanilla; a hallmark of quality oak maturation; can be cloying if unbalanced (Piggott & Sharman, 1986).

Role in Green, Woody, or Nutty Notes

Aldehyde impact is context-dependent: low levels add dimension; high levels distort.

- Aliphatic aldehydes (e.g., acetaldehyde, hexanal) → green: fresh-cut grass, unripe fruit, crushed leaves; notable in young or poorly distilled spirits and from lipid oxidation/incomplete fermentation (Conner et al., 1993).
- Aromatic aldehydes (e.g., furfural, syringaldehyde) → toasty/woody/nutty; driven by oak toast level and time (Pozo-Bayón et al., 2006).
- Branched-chain/other aldehydes (e.g., isobutyraldehyde, phenylacetaldehyde) → nutty/malty/honey; arise via amino-acid catabolism or Maillard-type processes; often associated with oxidative styles (sherry, Madeira, Tawny Port, Colheitas, Armagnac).

Fig. 13.1	Common Aldehydes in Spirits					
Aldehyde	Aroma	Chain Size	MM (g/mol)	DT (ppm)	Boiling Point (°C)	Source*
Acetaldehyde (ethanal)	Green apple, sharp	Short	44.05	10	20.2	F, O
Acrolein	Burnt fat, overheated oil	Short	56.06	0.21	53.0	TG
Propionaldehyde	Sweet, nutty	Short	58.08	1-2	48.8	F
Isobutyraldehyde	Sharp, fruity	Short	72.11	0.8	64.5	F
Butyraldehyde	Pungent, acrid	Short	72.11	0.5	75.7	F
Isovaleraldehyde**	Malty, cocoa	Short	86.13	0.7	93.0	F
Valeraldehyde	Sweet, nutty	Short	86.13	1.1	103.0	F
Furfural	Almond, bread, nutty	Furanic/ (aromatic)	96.08	1-5	161.7	A
Hexanal	Green, grassy	Short	100.16	0.1-0.5	130.0	O
Benzaldehyde	Almond, cherry	Medium (aromatic)	106.12	1.0	178.0	A, F
5-Hydroxymethylfurfural	Caramel, baked sugar	Short (6)	126.11	50-100	115	BD
Vanillin	Vanilla, sweet	Medium (aromatic)	152.15	0.3-0.5	285.0	A
Uncommon Aldehydes in Spirits						
2-Methylbutanal	Malty, nutty	Short	88.13	0.4	90	F
Octanal	Citrus, waxy	Medium	128.21	0.05	171	A
Nonanal	Floral, citrus	Medium	142.24	0.05	191	A
Decanal	Waxy, orange	Medium	156.27	0.03	208	A
Undecanal	Soapy, citrus	Long	170.29	0.02	230	A
Cinnamaldehyde	Spicy, cinnamon	Medium	132.16	0.1	248	B
Methional	Cooked potato	Short	104.15	0.1	120	T
4-Hydroxy-benzaldehyde	Vanilla, sweet wood	Short	122.12	0.2	213	LB
p-Anisaldehyde	Anise, sweet	Short	136.15	0.3	248	B
Citral	Lemon, verbena	Medium	152.24	0.1	229	C
Coniferaldehyde	Woody, smoky, warm spice	Medium	178.18	1-5	360	BD
Syringaldehyde	Smoky, spicy balsamic, oak	Medium	182.17	1-2	290	BD
Some references conflict on DT; ordinal correct: DT = detection threshold, approx in water, lower than in ethanol. Source*; F=fermentation, A=aging barrel, O=oxidation including lipid, B=botanical addition, BD=barrel degradation, C=citrus peel, LB=lignin breakdown, I=introduced, T=thermal reaction; TG=Thermal degradation, Short=1-6 C atoms, Medium= 7-10 C atoms, Long =11+ C atoms; ppm=parts per million. **also called 3-Methylbutanal, MM=Molar mass						

Stability Under Dilution or Air Exposure

Aldehydes differ in their stability in air and upon dilution. Acetaldehyde readily oxidizes to acetic acid. Furfural can oxidize to furoic acid or polymerize, resulting in a deeper color. Vanillin may fade with prolonged exposure to oxygen or light. Dilution reduces ethanol's suppressive effect on volatility, unmasking aldehydes while potentially shifting the balance (Singleton, 1995; Gutzwiller & Chambers, 2010).

Sensory Thresholds and Evaluative Challenges

Aldehydes span wide thresholds and desirability. Acetaldehyde is reported to be present at around 10–15 ppm in water; however, in ethanol-rich matrices, it can be masked by ethanol or esters (Ferreira, 2010). Vanillin is present at much lower levels (~0.3–0.5 ppm) and is readily perceived (Poisson & Schieberle, 2008). Furfural typically falls within the 1–5 ppm range, depending on the base and temperature. These thresholds shift with matrix composition, vessel geometry, serving temperature, and prior olfactory fatigue. Many aldehydes also transform in air or under dilution—acetaldehyde oxidizing to acetic acid, and furfural undergoing reactions that yield acids or polymers that darken the spirit, making detection a moving target (Singleton, 1995; Attwood & Florence, 1983). Co-occurring congeners further complicate perception: isovaleraldehyde's nutty tone may hide beneath esters, while grassy (E)-2-nonenal can be mistaken for cereal unless evaluators are specifically trained (Gutzwiller & Chambers, 2010). To counter these issues, evaluators should employ structured training, such as triangle or omission tests, reference standards, and/or orthonasal–retronasal "re-tracing." Selecting a vessel that minimizes ethanol masking and reduces repeated exposures over time is essential (Pino, 2014). Acetaldehyde exhibits low perceptual prominence at high ethanol concentrations due to solubilization and masking effects; however, it becomes more volatile and noticeable as the ethanol concentration decreases. Esters show comparatively stable volatility across the same range. Fig. 13.2 illustrates how dilution can unmask aldehydes, shifting aroma balance (adapted from Ferreira, 2010).

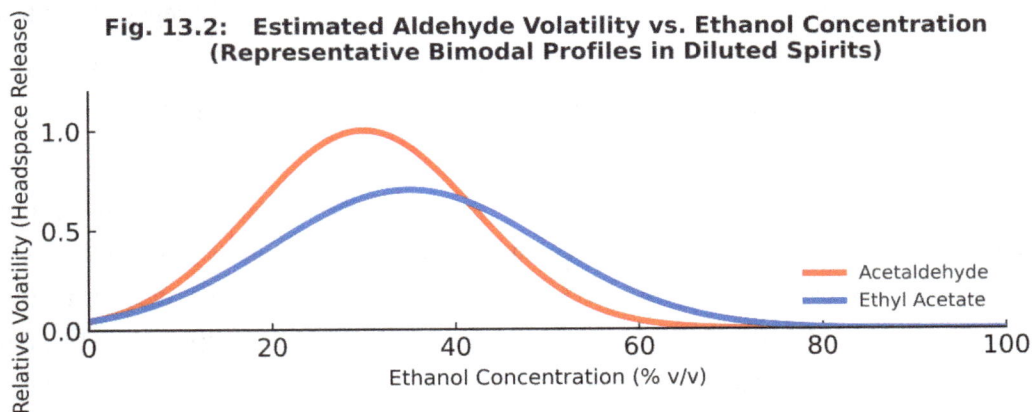

Fig. 13.2: Estimated Aldehyde Volatility vs. Ethanol Concentration (Representative Bimodal Profiles in Diluted Spirits)

Fig. 13.2 illustrates a 30% v/v ≈ optimum release zone. 40 % v/v and above → kinetic dominance →

suppressed differential aroma expression. Below 25 % v/v → strong thermodynamic suppression → dull, flattened aroma field. The effect is much the same for esters (See Fig. 12.2)

Evaluation Note 13.1: Complacency toward ethanol in the headspace leads to erroneous detection and identification of true sample aromas.

Evaluator Guidance and Training (Fig. 13.3)

Training sets should include reference standards for: acetaldehyde (green apple/solvent), vanillin (vanilla), furfural (almond/bread), isovaleraldehyde (nutty/malty), phenylacetaldehyde (honey/floral). Calibrating to these anchors enhances fault diagnosis and typicity assessment.

Fig. 13.3   General Desirability Spectrum of Aldehydes

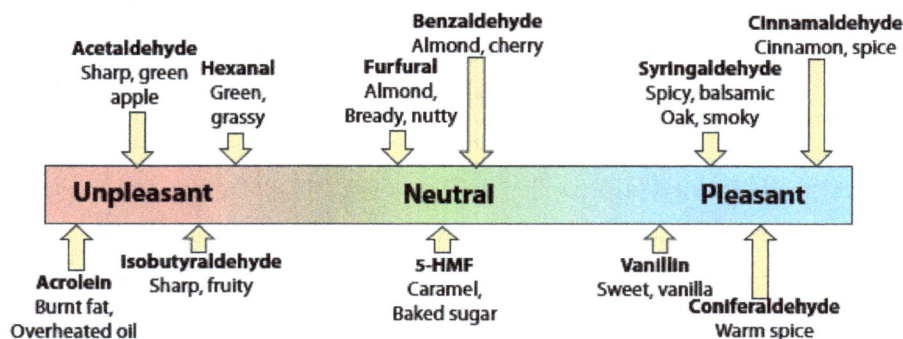

Cautionary note: Aldehydes are present across a broad spectrum in spirits, ranging from compounds that signal immaturity or harshness to those that contribute depth and desirable maturation characteristics. Their impact is not fixed as either a fault or a benefit, but depends on identity, concentration, and context. Some aldehydes are consistently regarded as faults when present at levels above trace; others can enhance aroma in balance, and many occupy a middle ground where their sensory role must be judged in relation to the overall spirit.

Summary

Aldehydes are powerful yet reactive contributors to the aroma of spirits. Formed through fermentation, oxidation, and barrel aging, they range from the sharp, green apple notes of acetaldehyde to the sweet, vanilla-like notes. Their sensory thresholds span orders of magnitude, and their perception is influenced by ethanol masking, vessel shape, and exposure to air. For evaluators, aldehydes represent both a challenge and an opportunity: they can sharpen youth, enrich maturity, or reveal faults. Mastery requires calibrated training, contextual interpretation, and attention to their dynamic behavior under actual tasting conditions.

Beyond: Acids and phenols give structure, and are also the cause of many faults in spirits.

"Aldehydes are critical to both the flavor and aromatic content of whisky... aldehydes of lower molecular weight tend to have sharp and unpleasant odors... acetaldehyde lends a young whisky its aromatic sharpness." —Dan Crowell, Glenmorangie Brand Ambassador

"Aldehydes are like punctuation in whisky—necessary, but disastrous when overused." — Dr. Bill Lumsden, (Glenmorangie)

References

Attwood, D., & Florence, A. T. (1983). *Surfactant systems: Their chemistry, pharmacy and biology*. Springer. https://doi.org/10.1007/978-94-009-5775-6

Conner, J. M., Paterson, A., & Piggott, J. R. (1993). Changes in wood extractives from oak cask staves through maturation of Scotch malt whisky. *Journal of the Science of Food and Agriculture, 62*(2), 169–174. https://doi.org/10.1002/jsfa.2740620210

de Revel, G., & Bertrand, A. (1993). Use of a simplified method for quantifying volatile phenols in spirits. *American Journal of Enology and Viticulture, 44*(3), 271–273.

Ferreira, V. (2010). Volatile aroma compounds and wine sensory attributes. *Comprehensive Reviews in Food Science and Food Safety, 9*(4), 425–447. https://doi.org/10.1111/j.1541-4337.2010.00118.x

Ferreira, V., López, R., & Cacho, J. (2000). Quantitative determination of the odorants of young red wines from different grape varieties. *Journal of the Science of Food and Agriculture, 80*(11), 1659–1667. https://doi.org/10.1002/jsfa.693

Gutzwiller, B. J., & Chambers, E. (2010). The influence of ethanol and serving temperature on the release of esters in alcoholic beverages. *Flavour and Fragrance Journal, 25*(5), 320–325. https://doi.org/10.1002/ffj.1992

Jack, F. R., & Noble, A. C. (1993). Effect of ethanol on the perception of sourness and bitterness in white wine. *American Journal of Enology and Viticulture, 44*(4), 292–296.

Jackson, R. S. (2014). *Wine science: Principles and applications* (4th ed.). Academic Press.

Ledauphin, J., Barillier, D., & Guichard, E. (2010). Chemical and sensorial characterization of brandies and spirits. *Food Chemistry, 122*(2), 528–538. https://doi.org/10.1016/j.foodchem.2009.12.050

López, M. R., & Ferreira, V. (2008). Comparative study of the aromatic profile of different kinds of wine and spirits by HS-SPME GC–MS. *Journal of Chromatography A, 1185*(1), 291–297. https://doi.org/10.1016/j.chroma.2008.02.079

Maga, J. A. (1982). Flavor contribution of phenolic compounds in foods. *Journal of Agricultural and Food Chemistry, 30*(2), 373–378. https://doi.org/10.1021/jf00110a016

Mosedale, J. R., & Puech, J. L. (1998). Wood maturation of distilled beverages. *Trends in Food Science & Technology, 9*(3), 95–101. https://doi.org/10.1016/S0924-2244(98)00024-7

Nykänen, L. (1986). Formation and occurrence of flavor compounds in wine and distilled alcoholic beverages. *American Journal of Enology and Viticulture, 37*(1), 84–96. https://doi.org/10.5344/ajev.1986.37.1.84

Nykänen, L., & Suomalainen, H. (1983). *Aroma of beer, wine and distilled alcoholic beverages.* D. Reidel / De Gruyter. https://doi.org/10.1515/9783112735770

Piggott, J. R., & Sharman, D. (1986). Production and maturation of Scotch and Irish whiskies. In J. R. Piggott (Ed.), *Alcoholic beverages: Sensory evaluation and consumer research* (pp. 327–358). Applied Science Publishers.

Pino, J. A. (2014). Odor-active compounds in alcoholic beverages. *Critical Reviews in Food Science and Nutrition, 54*(7), 885–901. https://doi.org/10.1080/10408398.2011.588493

Poisson, L., & Schieberle, P. (2008). Characterization of the most odor-active compounds in an American bourbon whisky by AEDA. *Journal of Agricultural and Food Chemistry, 56*(14), 5813–5819. https://doi.org/10.1021/jf800382m

Pozo-Bayón, M. Á., Guichard, E., & Cayot, N. (2006). Aroma perception of complex mixtures: Influence of congruency and cross-modal interactions. *Food Quality and Preference, 17*(1–2), 341–348. https://doi.org/10.1016/j.foodqual.2005.05.013

Singleton, V. L. (1995). Maturation of wines and spirits: Comparisons, facts, and hypotheses. *American Journal of Enology and Viticulture, 46*(1), 98–115.

Swiegers, J. H., Bartowsky, E. J., Henschke, P. A., & Pretorius, I. S. (2005). Yeast and bacterial modulation of wine aroma and flavour. *South African Journal of Enology and Viticulture, 26*(2), 55–62. https://doi.org/10.21548/26-2-2137

Observations on Aldehydes

The Green Apple Illusion: Acetaldehyde is often mistaken for freshness in young distillates. At low levels, it conveys a clean, crisp top note, but as concentrations approach the sensory threshold, it overlaps with ethyl acetate, and the impression shifts from "apple" to "solvent." The same compound that livens new-make spirit can be read as an oxidation fault in mature stock.

The Vanillin Paradox: Vanillin, a wood-derived aldehyde, defines the sweetness of well-aged whiskey, yet its impact depends entirely on context. In high-phenol matrices, it softens smoke; in low-phenol spirits, it can dominate and flatten complexity. The molecule is constant — its sensory weight is relational.

The Hidden Heat of Furfural: Furfural concentration rises sharply with barrel toasting and aging, contributing nutty and bready warmth. Though rarely above threshold alone, it synergizes with lactones and phenols to build perceived body. Distillers often describe this as "sweet dryness"—a contradiction that captures its dual roles as an aromatic and a tactile sensation.

The Ghost of Oxidation: Formaldehyde and related low-molecular-weight aldehydes form during ethanol oxidation under poor storage or aeration conditions. Though typically below conscious detection, their presence sharpens the nasal profile and exaggerates alcoholic sting, signaling degradation rather than maturity.

Chapter 14: Acids and Phenols — Structure, Complexity, Faults

Organic acids and phenolic compounds contribute substantially to the aroma, taste, and textural complexity of distilled spirits. While some offer rich and desirable notes such as creamy, spicy, or smoky qualities, others introduce spoilage characteristics or off-notes that compromise product integrity. Understanding the origin, transformation, and impact of these molecules is crucial for evaluators seeking accurate profiling.

Organic Acids in Spirits

Organic acids are carboxylic acids (R-COOH) formed primarily during fermentation, with some forming during aging through microbial activity or oxidative degradation. Key organic acids in spirits include acetic, lactic, and butyric acid. Each of these acids plays dual roles: they contribute directly to sensory perception and act as precursors for esterification, forming important aroma compounds (López & Ferreira, 2008). See Fig. 14.1.

- Acetic acid is the most prevalent volatile acid in spirits. At low levels, it provides subtle vinegary brightness and depth; in excess, it can dominate the profile with pungent, sour, or solvent-like aromas (Nykänen & Suomalainen, 1983).

- Lactic acid originates from bacterial fermentation, especially under low-oxygen or stress conditions. It contributes to mouthfeel, roundness, and a mild, yogurt tang that softens harsher notes (Ledauphin et al., 2010).

- Butyric acid occurs naturally in fermentation yet is notorious for rancid-butter or vomit-like aromas; however, at trace levels—and especially when esterified (e.g., ethyl butyrate)—it can lend tropical fruitiness (Poisson & Schieberle, 2008).

These acids vary in volatility and sensory threshold. Their evaluation requires attention to both orthonasal and retronasal perception, as well as monitoring their transformation over time and under different dilution conditions. Most importantly, these acids readily form esters when reacting with alcohols, a reaction that can either enhance or mask their original aroma signatures (Nykänen, 1986).

Phenolic Compounds from Grain and Cask

Phenols are a chemically diverse group of aromatic compounds containing a hydroxyl group directly bonded to an aromatic ring. In spirits, phenolics arise primarily from two sources: raw materials (e.g., malted barley, rye) and barrel aging (especially from toasted or charred oak). Some also originate from microbial activity or thermal degradation during distillation (Conner et al., 1993; Maga, 1982). Notable examples include:

- Guaiacol, a lignin breakdown product during oak charring, imparts smoky, medicinal, and clove-like aromas; it is central to peated whisky profiles (Conner et al., 1993).

- Eugenol, released from oak, contributes spicy/clove/woody notes and shows lower volatility that affects the mid-palate and retronasal phases (Maga, 1982).

- 4-Ethylphenol and 4-ethylguaiacol, microbial by-products (e.g., Brettanomyces), which at higher levels produce barnyard, horse-blanket, or smoky-meaty notes—generally considered faults in distilled spirits (Swiegers et al., 2005; de Revel & Bertrand, 1993).

Fig. 14.1	Common Acids in Spirits					
Acid	**Aroma/ Mouthfeel**	**Formula**	**MM (g/mol)**	**DT* (ppm)**	**Boiling Point ($^{\circ}$C)**	**Source****
Formic	sharp, pungent, acrid	CH_2O_2	46.03	50	100.8	F
Acetic	vinegar, sour	$C_2H_4O_2$	60.05	200	118.1	F
Propionic	Sweet, pungent	$C_3H_6O_2$	74.08	500	141.2	F
n-Butyric	rancid butter, cheesy	$C_4H_8O_2$	88.11	15	163.5	F
Isobutyric	rancid butter	$C_4H_8O_2$	88.11	15	155.2	F
Lactic$^\$$	mild sour, yogurt	$C_3H_6O_3$	90.08	100	122.0	F
Valeric	goat-like, unpleasant	$C_5H_{10}O_2$	102.13	2	186.0	F
Isovaleric	sweat socks, cheesy	$C_5H_{10}O_2$	102.13	22	175.0	F
Hexanoic (Caproic)	goat-like, sour, waxy	$C_6H_{12}O_2$	116.16	0.42	205.0	F, A
Succinic$^\$$	sour, slightly salty	$C_4H_6O_4$	118.09	400	235.0	F
Benzenecarboxylic	mildly astringent	$C_7H_6O_2$	122.12	200	249.2	A
Enanthic	rancid	$C_7H_{14}O_2$	130.18	120	244.5	F
Malic$^\$$	green apple, tart	$C_4H_6O_5$	134.09	25	150	F
Phenylacetic	honey, floral	$C_8H_8O_2$	136.15	5	265	F, A
Octanoic (Caprylic)	waxy, rancid, fatty	$C_8H_{16}O_2$	144.21	0.6	239	F, A
Tartaric$^\$$	tart, sour	$C_4H_6O_6$	150.09	50	170	F
Nonanoic (Pelargonic)	rancid, fatty	$C_9H_{18}O_2$	158.24	0.3	253.7	F
Phenyllactic	light floral	$C_9H_{10}O_3$	166.17	100	249	F
Capric	waxy, soapy	$C_{10}H_{20}O_2$	172.27	10	270	F
Citric$^\$$	tart, sour, citrus	$C_6H_8O_7$	192.12	250	160	F
Dodecanoic	fatty, soapy	$C_{12}H_{24}O_2$	200.32	15	300	F
Tetradecanoic	waxy, fatty	$C_{14}H_{28}O_2$	228.37	20	310	F
Hexadecanoic (Palmitic)	waxy, fatty, tallow	$C_{16}H_{32}O_2$	256.42	0.05	351	A
Linoleic	grassy, fatty	$C_{18}H_{32}O_2$	280.45	45	230	A
Oleic	waxy, fatty	$C_{18}H_{34}O_2$	282.46	200	360	A
Octadecanoic	Neutral	$C_{18}H_{36}O_2$	284.48	0.05	383	A

Some references conflict on the detection threshold, which is ordinally correct. Source**; F=fermentation, A=aging. ppm=parts per million. MM=Molar mass. NOTE: Available data are not necessarily at the same temperatures and pressures. Directionally sound for the sensory analyst, not for chemical assumptions or calculations. $^\$$ lactic, malic, succinic, tartaric, citric may decompose prior to reaching a true boiling point. Long-chain fatty acids, such as Octadecanoic (stearic), have exceptionally low headspace volatility at tasting temperatures and are difficult to detect.

Phenols exhibit a wide range of sensory thresholds and volatility. Their hydrophobic nature can delay headspace release, requiring slow nosing and careful retronasal assessment. They also interact strongly with ethanol and are more readily detectable at lower dilutions (de Revel & Bertrand, 1993).

Ethanol and pH Interactions Ethanol serves as both a solvent and carrier for acidic and phenolic molecules. Its presence affects volatility, partitioning, and olfactory delivery. High ethanol concentrations tend to suppress the volatility of acids and phenols, particularly for heavier molecules, necessitating controlled dilution for accurate sensory evaluation in tulip glassware.

pH also plays a critical role. Acids in their protonated (non-ionized) form are more volatile and odor-active. Lower pH (more acidic) favors this form, enhancing aroma perception. Conversely, raising the pH (through dilution or buffering) shifts acids into their ionized state, reducing volatility and changing mouthfeel (Fig. 14.2) (Jack & Noble, 1993). Dissociation is due to weak acids releasing protons (H^+), shifting from HA(weak acid) to A^-(conjugate base) as pH rises.

Fig. 14.2   Ionization of Weak Acids vs pH

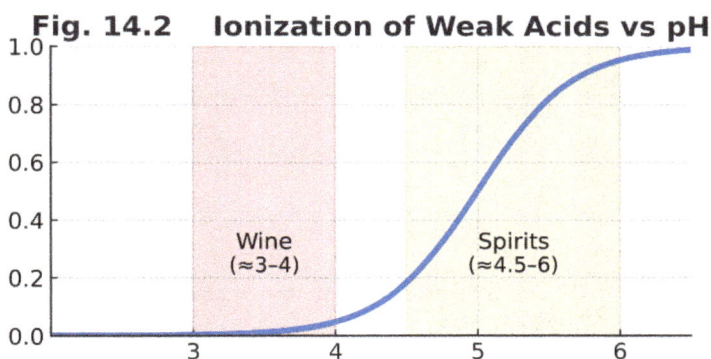

Phenolic compounds are weak acids, and their behavior is influenced by pH, even though their dissociation point lies well above the typical range of wines and spirits. In practice, small shifts in matrix pH and buffering can alter their solubility and volatility, thereby changing the intensity with which phenolic aromas are perceived — an effect especially relevant when contrasting acidic unaged spirits with buffered, aged whiskies. Fig. 14.3 describes common phenols.

Sensory Contributions: Complexity or Fault?

Both acids and phenols represent a continuum of sensory contribution—from complexity and character to spoilage and imbalance. Trained evaluators must develop the ability to discriminate these molecules based on:

- Concentration in context (e.g., lactic acid softening young rum vs. dominating a brandy)
- Source (natural grain fermentation vs. microbial contamination)
- Integration (seamless with esters and alcohols vs. abrupt or overpowering)

- Evolution over time (appearance of phenolic spikes during oxidation)

While acids can enhance brightness, texture, or fruity esters, excessive amounts often lead to harshness, sourness, or chemical off-notes. Similarly, phenols can impart complexity and structure or, when mismanaged, introduce burnt, acrid, or microbial spoilage markers.

For sensory professionals, the challenge is not merely in detection but in understanding the origin, threshold, and contextual appropriateness of each compound. The boundary between desirable and defective lies not in the molecule itself, but in its balance, integration, and alignment with the spirit's intended profile. As a result, opinions vary, and balance is a subjective concept.

Fig. 14.3		Common Phenols in Spirits				
Phenol	Aroma/ mouthfeel	Formula	MM (g/mol)	DT (ppm)	Boiling Point ($^{\circ}$C)	Source*
Phenol	Medicinal, smoky	C_6H_6O	94.1	40	182	F, A
Cresol (o-, m-, p-)	Tar, medicine	C_7H_8O	108.14	10	201	F, A
4-Vinylphenol	Medicinal, smoky	C_8H_8O	120.15	0.3	215	F
4-Ethylphenol	Barnyard, horsey	$C_8H_{10}O$	122.17	0.4	218	F
Guaiacol	Smoky, cloves	$C_7H_8O_2$	124.14	0.11	205	A
4-Methylguaiacol	Smoky, spicy	$C_8H_{10}O_2$	138.17	0.5	235	A
4-Vinylguaiacol	Clove, phenolic	$C_9H_{10}O_2$	150.17	0.3	245	F
Vanillin	Vanilla, sweet	$C_8H_8O_3$	152.15	0.5	285	A
Eugenol	Clove, spicy	$C_{10}H_{12}O_2$	164.2	6	254	A
Isoeugenol	Spicy, woody	$C_{10}H_{12}O_2$	164.2	10	266	A
Syringol	Smoky, sweet	$C_9H_{10}O_3$	166.17	0.5	260	A
Some references conflict on DT, ordinally correct: DT=detection threshold, Source*; F=fermentation, A=aging, MM=molar mass						

Summary

Acids and phenols play a central role in the chemical and sensory composition of distilled spirits. Organic acids contribute both directly, through sourness, texture, and pungency, and indirectly, as key precursors to esters that define fruit and floral notes. Phenolic compounds, drawn from grain, oak, and microbial activity, provide another dimension of character, ranging from the smoky and spicy complexity of guaiacol and eugenol to the barnyard faults of 4-ethylphenol. Both classes are sensitive to concentration, pH, ethanol strength, and contextual integration as markers of both quality and spoilage. For the evaluator, mastery lies not simply in identifying their presence but in interpreting their balance, source, and transformation over time. This interpretive skill distinguishes technical analysis from true sensory expertise.

Beyond: Minute traces of sulfur can destroy some spirits, but are acceptable in others.

Appendix – Common and IUPAC Names of Acids

App. 14.1:	Nomenclature of Acids in Spirits			
Common Name	IUPAC Name	Other Synonyms	Formula	MM (g/mol)
Formic	Methanoic	Formylic	CH_2O_2	46.02
Acetic	Ethanoic	Vinegar	$C_2H_4O_2$	60.05
Propionic	Propanoic	Propanoate (salt form)	$C_3H_6O_2$	74.08
n-Butyric	Butanoic	Butyric	$C_4H_8O_2$	88.11
Isobutyric	2-Methylpropanoic	Isopropanoic	$C_4H_8O_2$	88.11
Lactic	2-Hydroxypropanoic	Milk	$C_3H_6O_3$	90.08
Valeric	Pentanoic	Valerate (salt form)	$C_5H_{10}O_2$	102.13
Isovaleric	3-Methylbutanoic	Isopentanoic	$C_5H_{10}O_2$	102.13
Hexanoic	Hexanoic	Caproic	$C_6H_{12}O_2$	116.16
Succinic	Butanedioic	Amber	$C_4H_6O_4$	118.09
Benzenecarboxylic	Benzenecarboxylic	Benzene-formic	$C_7H_6O_2$	122.12
Enanthic	Heptanoic	Oenanthic	$C_7H_{14}O_2$	130.19
Malic	2-Hydroxy-butanedioic	Apple	$C_4H_6O_5$	134.09
Phenylacetic	2-Phenylethanoic	α-toluic	$C_8H_8O_2$	136.15
Octanoic	Octanoic acid	Caprylic	$C_8H_{16}O_2$	144.21
Tartaric	2,3-Dihydroxy-butanedioic	Grape	$C_4H_6O_6$	150.09
Nonanoic	Nonanoic	Pelargonic	$C_9H_{18}O_2$	158.24
Phenyllactic	2-Hydroxy-3-phenylpropanoic	None known	$C_9H_{10}O_3$	166.18
Decanoic	Decanoic	Capric	$C_{10}H_{20}O_2$	172.27
Citric	2-Hydroxy-1,2,3-propane-tricarboxylic	Citrus	$C_6H_8O_7$	192.12
Dodecanoic	Dodecanoic	Lauric	$C_{12}H_{24}O_2$	200.32
Myristic	Tetradecanoic	Myristic	$C_{14}H_{28}O_2$	228.38
Hexadecanoic	Hexadecanoic	Palmitic	$C_{16}H_{32}O_2$	256.42
Linoleic	Cis, cis-9, 12-Octa-decadienoic	Omega-6 Fatty Acid	$C_{18}H_{32}O_2$	280.45
Oleic	cis-9-Octadecenoic	Oleic	$C_{18}H_{34}O_2$	282.46
Octadecanoic	Octadecanoic	Stearic	$C_{18}H_{36}O_2$	284.48
MM=Molar mass				

Appendix 14.1 includes both standard and IUPAC designations. Common names are widely used in industry practice and traditional literature, whereas IUPAC names provide standardized structural descriptions recognized in scientific discourse. Presenting both forms ensures precision in academic contexts while maintaining relevance to practical usage in the spirits field.

"Acids are the backbone of spirit flavor, they define structure."— Dr. Jim Swan, chemist, author

"During maturation, acids and alcohols constantly react to form new esters, giving the impression of evolving sweetness."— Dr. Barry Harrison, Heriot-Watt University

References

Conner, J. M., Paterson, A., & Piggott, J. R. (1993). Changes in wood extractives from oak cask staves through maturation of Scotch malt whisky. *Journal of the Science of Food and Agriculture, 62*(2), 169–174. https://doi.org/10.1002/jsfa.2740620210

de Revel, G., & Bertrand, A. (1993). Use of a simplified method for quantifying volatile phenols in spirits. *American Journal of Enology and Viticulture, 44*(3), 271–273.

Jack, F. R., & Noble, A. C. (1993). Effect of ethanol on the perception of sourness and bitterness in white wine. *American Journal of Enology and Viticulture, 44*(4), 292–296.

Ledauphin, J., Barillier, D., & Guichard, E. (2010). Chemical and sensorial characterization of brandies and spirits. *Food Chemistry, 122*(2), 528–538. https://doi.org/10.1016/j.foodchem.2009.12.050

López, R., & Ferreira, V. (2008). Comparative study of the aromatic profile of different kinds of wine and spirits by HS-SPME GC–MS. *Journal of Chromatography A, 1185*(1), 291–297. https://doi.org/10.1016/j.chroma.2008.02.079

Maga, J. A. (1982). Flavor contribution of phenolic compounds in foods. *Journal of Agricultural and Food Chemistry, 30*(2), 373–378. https://doi.org/10.1021/jf00110a016

Nosrat, S. (2017). *Salt, fat, acid, heat: Mastering the elements of good cooking.* Simon & Schuster.

Nykänen, L. (1986). Formation and occurrence of flavor compounds in wine and distilled alcoholic beverages. *American Journal of Enology and Viticulture, 37*(1), 84–96. https://doi.org/10.5344/ajev.1986.37.1.84

Nykänen, L., & Suomalainen, H. (1983). *Aroma of beer, wine and distilled alcoholic beverages.* De Gruyter. https://doi.org/10.1515/9783112735770

Poisson, L., & Schieberle, P. (2008). Characterization of the most odor-active compounds in an American bourbon whisky by AEDA. *Journal of Agricultural and Food Chemistry, 56*(14), 5813–5819. https://doi.org/10.1021/jf800382m

Swiegers, J. H., Bartowsky, E. J., Henschke, P. A., & Pretorius, I. S. (2005). Yeast and bacterial modulation of wine aroma and flavour. *South African Journal of Enology and Viticulture, 26*(2), 55–62. https://doi.org/10.21548/26-2-2137

Observation on Acids in Spirits

Acids impart structure and depth to spirits, moderating sweetness and mouthfeel. They react to form esters, contributing aromatic maturity and a sense of balance, even though the perceived "roundness" is sensory rather than physical viscosity. Too little, and the spirit feels hollow; too much, and sharpness prevails. Harmony, not neutrality, defines true maturity.

Organic acids reflect the health of fermentation, with malic, lactic, succinic, and acetic acids each signaling different microbial and metabolic pathways. Clean acid profiles add precision and tension; stressed fermentations leave harsh edges that aging will never entirely erase.

During aging, oak-derived acids enter the spirit, reshaping its acetic and fatty acid balance. The gradual rise signals active maturation, driving new ester formation and adding lift. When this acid contribution fades, the barrel is spent, and the spirit's development stalls regardless of age.

Chapter 15: Sulfur Compounds and Off-Notes

Sulfur-containing compounds are among the most potent and polarizing contributors to aroma in distilled spirits. Their detection thresholds are among the lowest of any compound class, often in the parts per trillion range, and their sensory profiles range from highly complex to overtly objectionable (Siebert, Wood, Elsey, & Pollnitz, 2008).

Sulfides, Thiols, and Mercaptans: Potent Volatiles

Volatile sulfur compounds (VSCs) commonly found in spirits include hydrogen sulfide (characterized by the odor of rotten eggs), methanethiol (resembling the scents of cabbage and burnt rubber), dimethyl sulfide (similar to the aromas of corn and cooked vegetables), and various thiols and mercaptans (Ugliano, 2013). They arise as byproducts of yeast metabolism, particularly during amino acid catabolism (e.g., methionine and cysteine pathways) (Swiegers, Bartowsky, Henschke, & Pretorius, 2005). Their production is influenced by nutrient availability, fermentation temperature, and yeast strain (Henschke & Jiranek, 1993).

Formation During Fermentation and Distillation

During fermentation, sulfur precursors like sulfate and amino acids are metabolized by yeast, producing VSCs as intermediates or byproducts (Edwards, Haag, Collins, & Butzke, 1999). Poor yeast nutrition, particularly insufficient nitrogen or excess sulfur compounds in the mash, can lead to increased VSC formation (Edwards et al., 1999; Moreira et al., 2002). Distillation concentrates or removes these compounds, depending on their volatility and the selected cut points. Some sulfur compounds are hydrophilic and can persist into the final distillate, especially with short distillation cuts or inefficient separation (Piggott, Conner, & Paterson, 1993).

Descriptors and Detection

Common descriptors for sulfur off-notes include rubber, onion, garlic, skunk, burnt match, or cooked cabbage. While small amounts of specific sulfur volatiles (e.g., dimethyl sulfide) can enhance complexity, others, such as ethyl mercaptan or hydrogen sulfide, are almost universally considered faults (Silva Ferreira & Guedes de Pinho, 2003). The distinction lies in concentration, chemical context, and the overall aroma matrix (Lytra, Tempere, Le Floch, de Revel, & Barbe, 2012). See Fig. 15.1.

Ethanol interacts with sulfur compounds by suppressing their volatility, particularly in high-proof spirits. This effect masks low-threshold sulfurs unless the spirit is diluted or warmed slightly. The shape of the glass also plays a crucial role: wide-mouthed vessels allow volatile sulfur compounds to dissipate quickly, reducing their perceived intensity. At the same time, narrow tulips may trap these notes, intensifying their impact. Air exposure or oxidation can transform some thiols and sulfides into less offensive disulfides or sulfoxides, making decanting or aeration a remedial strategy in some cases (Siebert, Solomon, & Pollnitz, 2010).

Fig. 15.1	Sulfur Compounds in Spirits						
Compound	Aroma	Formula	MM (g/mol)	DT (ppm)	Boiling Point (^{o}C)	Source *
Hydrogen Sulfide	rotten eggs, sulfurous	H_2S	34.08	0.00047	-60.3	Y
Methane-thiol (Methyl mercaptan)	skunk, rotten cabbage	CH_4S	48.11	0.0016	6.2	Y
Ethanethiol (Ethyl Mercaptan)	burnt rubber, garlic	C_2H_6S	62.13	0.0015	35	Y
Dimethyl Sulfide (DMS)[G]	cooked corn, cabbage	C_2H_6S (Isomer)	62.13	0.03	37.0	T
Sulfur Dioxide	pungent, acrid	SO_2	64.06	0.5	-10.0	BST
Allyl Mercaptan	garlic, alliaceous	C_3H_6S	74.14	0.002	91.0	Y
Dimethyl Disulfide (DMDS)[G]	onion, garlic	$C_2H_6S_2$	94.2	0.005	110.0	T
Thiophenol	burnt rubber, medicinal	C_6H_5SH	110.17	0.0006	168.0	BCM
Dimethyl Trisulfide (DMTS)[G]	putrid, rotten cabbage	$C_2H_6S_3$	126.26	0.001	151.0	T
Benzo-thiozole	rubbery, burnt	C_7H_5NS	135.19	2.0	227.0	MBS
Some references conflict on DT, ordinally correct: DT=detection threshold, Source*; Y=yeast fermentation and metabolism, T=thermal degradation, BST=barrel sulfur treatment, BCM=barrel char/microbial, MBS=microbial spoilage. MM=molar mass							

Quality Control and Sensory Vigilance (Fig. 15.2)

Sulfur faults are among the most serious quality-control concerns in spirits production due to their potent aroma and consumers' sensitivity. Skilled evaluators must distinguish between desirable low-level complexity (e.g., from dimethyl trisulfide in peated whiskies) and true off-notes that indicate contamination or processing errors.

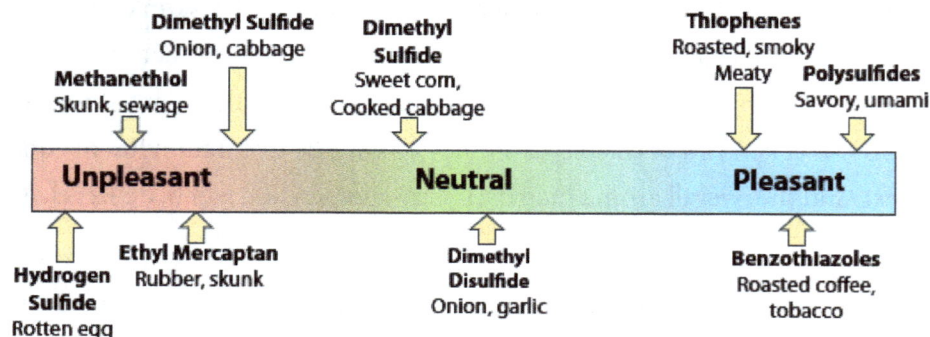

Fig. 15.2 General Desirability Spectrum of Sulfur Compounds

Dimethyl Sulfide
Onion, cabbage

Dimethyl Sulfide
Sweet corn, Cooked cabbage

Thiophenes
Roasted, smoky

Methanethiol
Skunk, sewage

Meaty

Polysulfides
Savory, umami

Unpleasant   **Neutral**   **Pleasant**

Hydrogen Sulfide
Rotten egg

Ethyl Mercaptan
Rubber, skunk

Dimethyl Disulfide
Onion, garlic

Benzothiazoles
Roasted coffee, tobacco

Sensory panels and gas chromatography-olfactometry (GC-O) are frequently employed in quality control laboratories to monitor the development of sulfur compounds and ensure product consistency (Guth, 1997).

Understanding sulfur volatiles demands rigorous attention to thresholds, formation mechanisms, and context. Their presence can be a sign of agrochemical signature, raw material provenance, technique, or trouble—and the evaluator's job is to know the difference.

Summary

Volatile sulfur compounds (VSCs)—from H_2S and methanethiol to DMDS/DMTS and thiophenols—sit at the extremes of sensory impact in spirits. They are generated primarily by yeast metabolism and fermentation management (nitrogen status, temperature, strain), then concentrated or distilled, depending on volatility and cut strategy. In trace amounts, some VSCs (e.g., DMS) can add a savory nuance; above the threshold, they drive off-notes (rotten egg, onion/garlic, burnt rubber). Ethanol strength and serving conditions modulate perception, while oxygen exposure and time can transform reactive sulfurs into less pungent species. Effective QC hinges on prevention (nutrition/fermentation hygiene), targeted remediation, and confirmation with sensory panels and sulfur-specific analytics (e.g., GC-SCD/GC-O, sulfur chemiluminescence detection, gas chromatography-olfactometry).

Beyond: Volatile Terpenes and Lactones play an integral role in the aroma profile.

"Reduction is, of course, a misnomer. It refers to volatile sulfur compounds produced by yeasts during fermentation." — Jamie Goode, wine critic

"Hydrogen sulfide gives the wine a sulfurous rotten egg smell which overpowers the wine and prevents the true aromas of the wine being expressed." — Jancis Robinson, wine critic

References

Edwards, C. G., Haag, K. M., Collins, M. D., & Butzke, C. E. (1999). Production of hydrogen sulfide and Edwards, C. G., Haag, K. M., Collins, M. D., & Butzke, C. E. (1999). Production of hydrogen sulfide and glutathione by wine strains of *Saccharomyces cerevisiae* under different nitrogen conditions. *American Journal of Enology and Viticulture, 50*(3), 291–297.

Guth, H. (1997). Quantitation and sensory studies of character impact odorants of different white wine varieties. *Journal of Agricultural and Food Chemistry, 45*(8), 3027–3032. https://doi.org/10.1021/jf970280a

Henschke, P. A., & Jiranek, V. (1993). Yeasts—Metabolism of nitrogen compounds. In G. H. Fleet (Ed.), *Wine microbiology and biotechnology* (pp. 77–164). Harwood Academic Publishers.

Lytra, G., Tempere, S., Le Floch, A., de Revel, G., & Barbe, J.-C. (2012). Study of sensory interactions among red wine fruity esters in a model solution. *Journal of Agricultural and Food Chemistry, 60*(45), 11427–11433.

https://doi.org/10.1021/jf303258f

Moreira, N., Mendes, F., Guedes de Pinho, P., Hogg, T., & Vasconcelos, I. (2002). Heavy sulphur compounds, higher alcohols and esters production profile of *Saccharomyces cerevisiae* wine-related strains during fermentation. *International Journal of Food Microbiology, 64*(1–2), 227–237. https://doi.org/10.1016/S0168-1605(00)00458-8

Piggott, J. R., Conner, J. M., & Paterson, A. (1993). Formation of flavour-active compounds during Scotch whisky fermentation and maturation. In J. R. Piggott (Ed.), *Alcoholic beverages* (pp. 293–318). Springer. https://doi.org/10.1007/978-1-4615-2666-7_11

Siebert, T. E., Solomon, M. R., & Pollnitz, A. P. (2010). Selective determination of volatile sulfur compounds in wine using gas chromatography with sulfur chemiluminescence detection. *Journal of Chromatography A, 1217*(9), 1530–1535. https://doi.org/10.1016/j.chroma.2009.12.047

Siebert, T. E., Wood, C., Elsey, G. M., & Pollnitz, A. P. (2008). Determination of volatile sulfur compounds in wine by gas chromatography–mass spectrometry. *Australian Journal of Grape and Wine Research, 14*(1), 39–47. https://doi.org/10.1111/j.1755-0238.2008.00005.x

Silva Ferreira, A. C., & Guedes de Pinho, P. (2003). Sensorial character impact of sulfur compounds in wine. *Food Science and Technology International, 9*(2), 103–109. https://doi.org/10.1177/1082013203034524

Swiegers, J. H., Bartowsky, E. J., Henschke, P. A., & Pretorius, I. S. (2005). Yeast and bacterial modulation of wine aroma and flavour. *Australian Journal of Grape and Wine Research, 11*(2), 139–173. https://doi.org/10.1111/j.1755-0238.2005.tb00285.x

Ugliano, M. (2013). Evolution of wine aroma during bottle aging: Volatile sulfur compounds and wine reduction. *Australian Journal of Grape and Wine Research, 19*(1), 1–10. https://doi.org/10.1111/ajgw.12004

Observations on Sulfur Compounds

Copper-Contact Lesson: Copper contact remains the most reliable scavenger of volatile sulfur compounds. Its surface catalyzes the conversion of hydrogen sulfide and mercaptans into insoluble copper sulfides, but this effect diminishes as the still surface oxidizes or fouls.

Barrel Aging: Aging can dull some sulfur notes through slow oxidation or thiol-to-disulfide shifts; however, most reduction occurs through masking by oak aldehydes and lactones. Barrel time softens edges; it rarely removes the fault.

Sulfur Faults: Sulfur faults are among the most potent aroma defects known. Dimethyl trisulfide, detectable below one part per billion, can dominate entire aroma profiles, obscuring esters and phenols even at trace levels.

Water and dilution: Unmasks bound or partially solubilized sulfur compounds, particularly in young spirits. What seemed clean at cask strength may reveal vegetal or onion-like taints when reduced to bottling proof.

Chapter 16: Terpenes, Lactones, and Minor Volatiles

Volatile minor compounds—particularly terpenes, lactones, and norisoprenoids—play a vital role in the aroma profile of distilled spirits. Although typically present in trace amounts, these molecules significantly influence the sensory complexity and nuance of high-quality spirits.

Terpenes: Botanical and Barrel-Borne Aromatics (Fig. 16.1)

Terpenes, such as limonene, linalool, and β-caryophyllene, derive either from vegetal raw materials (e.g., botanicals in gin) or from barrel-aging processes, and often deliver citrus, floral, herbal, or resinous notes. Limonene contributes a bright citrus character to fruit brandies and certain oak-aged spirits, while linalool provides floral, lily-like aromas in aged rums or whiskies.

Fig. 16.1	Terpenes, Lactones, Other Volatiles in Spirits					
Compound	**Aroma, Mouthfeel**	**Formula**	**MM** (g/mol)	**DT** (ppm)	**Boiling Point** (^{o}C)	**Source***
Trimethyl pyrazine (2, 3, 5, TPB)	nutty, roasted	$C_7H_{10}N_2$	122.17	3	187	O, T
Sotolon	nuts, curry, maple	$C_6H_8O_3$	128.13	1	280	A, Ox
Furaneol	caramel, strawberry	$C_6H_8O_3$	128.13	0.10	290	O, A, M
Limonene	citrus, orange	$C_{10}H_{16}$	136.24	10	167	Bo, Ox
β-Myrcene	balsamic, resin	$C_{10}H_{16}$	136.24	15	167	A, Bo
A,β-Pinene	pine resin, turpentine	$C_{10}H_{16}$	136.44	1-5	167	Bo
α-Terpineol	lilac, floral, pine	$C_{10}H_{18}O$	154.25	20	219	Bo, A
Linalool	floral, **lavender**	$C_{10}H_{18}O$	154.25	0.015	198	Bo
Geraniol	rose, geranium, floral	$C_{10}H_{18}O$	154.25	10	230	Bo
Nerol	sweet, citrus	$C_{10}H_{18}O$	154.25	12	225	Bo
δ-Nonalactone	coconut, creamy	$C_9H_{16}O_2$	156.22	.001-.005	270	O
γ-Nonalactone	coconut, peach	$C_9H_{16}O_2$	156.22	1	264	O, A
δ-Decalactone	creamy, peach,	$C_{10}H_{18}O_2$	170.25	0.01-0.02	280	O
γ-Decalactone	peach, creamy	$C_{10}H_{18}O_2$	170.25	1	290	O, A
Whiskey lactone (cis)	coconut, woody	$C_{10}H_{18}O_2$	170.25	0.020	268	O, A
Whiskey lactone (trans)	coconut, woody	$C_{10}H_{18}O_2$	170.25	0.130	271	O, A
γ -Undecalactone	creamy, fruity	$C_{11}H_{20}O_2$	184.28	35	280	O
β-Damascenone	rose, baked apple	$C_{13}H_{18}O$	190.28	0.02	275	G, A
Vitispirane	woody, spicy	$C_{13}H_{20}O$	192.3	1	300	O, Ox
β-Caryophyllene	woody, spicy, pepper	$C_{15}H_{24}$	204.35	35	262	B, O, A
Some references conflict on DT; the correct ordinal is DT = detection threshold. Values matrix-dependent; unless stated, model wine ≈ 12% v/v ethanol, pH ~3.2. For linalool, typical DT is ~0.01–0.02 mg/L in wine-like matrices (Guth, 1997; van Gemert, 2011). Source*;O=oak, Ox=oxidation, A=aging, Bo=botanicals, M=Maillard, MM=Molar mass						

They interact with ethanol and water, partitioning into the headspace in proportion to their volatility and polarity (Chira et al., 2013). Detection thresholds for key terpenes such as linalool are in the low µg/L range in wine-like matrices (Guth, 1997; van Gemert, 2011).

Lactones: The Coconut Signature from Oak (Fig. 16.1)

Whisky (oak) lactone—known chemically as cis-β-methyl-γ-octalactone and its trans counterpart—is extracted from oak into the spirit during aging and is responsible for characteristic coconut, celery, or woody aromas. The cis-isomer is more abundant in American oak and has a detection threshold as low as 0.02 ppm in a 12 % ethanol solution, compared to 0.13 ppm for the trans-isomer (Abbott et al., 1995). Geographic oak origin strongly influences the cis/trans ratio—American oak tends to yield 60–70 % cis vs. European oak with 55–65 % trans (Waterhouse, 2015). Lactone extraction is front-loaded/rapid early in aging; rates depend on oak species, toasting, entry proof, and temperature.

Norisoprenoids & Other High-Impact Volatiles

High-impact volatiles such as β-damascenone, β-ionone, or piperitone, derived from degradation of carotenoids in wood or botanicals, can deliver honey, floral, dried-fruit, or minty aromas. These compounds have exceptionally low thresholds and, despite being present at ultra-low concentrations, can significantly influence the sensory character of aged spirits—especially in refined styles like cognac or aged rum (De Rosso et al., 2009).

Integration and Sensory Influence

These minor volatiles contribute subtle complexity—lactones add creamy, coconut warmth, terpenes lend botanical freshness, and norisoprenoids provide floral or fruity depth. In balanced spirits, they integrate with esters, phenols, and alcohols to form layered aroma symmetry. However, masking effects may occur: for instance, high levels of cis-oak lactone can reduce the perception of fruity esters, especially under certain dilution conditions (Park et al., 2017).

Detection Techniques & Role in Panels

Advanced sensory panels use gas chromatography-- olfactometry (GC-OLF) alongside traditional nosing to identify and quantify these minor volatiles. Training involves focusing on very low-concentration aromatics and developing panelist sensitivity to key compounds, such as lactones and norisoprenoids. Hammering cues such as coconut, celery, floral, and dried fruit help evaluators distinguish these components from dominant ester-driven aromas.

Evaluation Note 16.1: Terpenes span a wide sensory range. Oxidized or overly concentrated monoterpenes (limonene and pinene) present harsh or resinous notes. Mid-spectrum, including linalool and geraniol, contribute floral and citrus notes but can become soapy or perfumed at high concentrations. Sesquiterpenes such as caryophyllene and terpineol contribute desirable woody and floral depth, often signaling botanical complexity or cask influence. (Fig. 16.2).

Fig. 16.2 General Desirability Spectrum of Terpenes

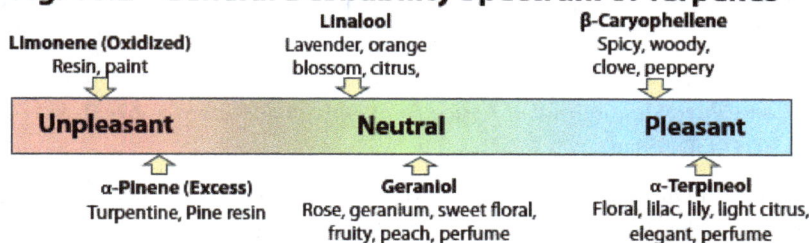

Limonene (Oxidized) Resin, paint	Linalool Lavender, orange blossom, citrus,	β-Caryophellene Spicy, woody, clove, peppery
Unpleasant	**Neutral**	**Pleasant**
α-Pinene (Excess) Turpentine, Pine resin	Geraniol Rose, geranium, sweet floral, fruity, peach, perfume	α-Terpineol Floral, lilac, lily, light citrus, elegant, perfume

Evaluation Note 16.2: Lactones are strongly linked to maturation and fruity or creamy notes. At the fault end, γ-butyrolactone can impart solvent-like tones. Mid-spectrum lactones, including γ-nonalactone and δ-decalactone, add peach, apricot, or milky nuances that become excessive if concentrated. Ultimately, whisky lactone (cis/trans) and δ-nonalactone are hallmark contributors to the coconut, sweet cream, and oak-derived character in aged spirits. (Fig. 16.3).

Fig. 16.3 General Desirability Spectrum of Lactones

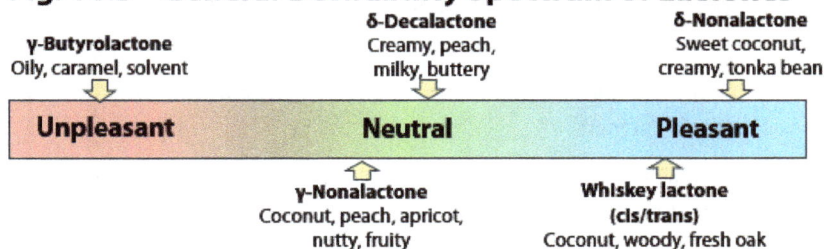

γ-Butyrolactone Oily, caramel, solvent	δ-Decalactone Creamy, peach, milky, buttery	δ-Nonalactone Sweet coconut, creamy, tonka bean
Unpleasant	**Neutral**	**Pleasant**
	γ-Nonalactone Coconut, peach, apricot, nutty, fruity	Whiskey lactone (cis/trans) Coconut, woody, fresh oak

Summary

Terpenes, lactones, and norisoprenoids may exist only in trace amounts in spirits, but their sensory impact is disproportionately large. They provide critical signatures of both raw material and aging environment, shaping complexity, balance, and stylistic distinctiveness. Understanding thresholds, extraction pathways, and masking effects is essential, as these compounds often determine the finite sensory margins separating ordinary spirits from exceptional ones.

Beyond: Although congeners are not recognized as a chemical class, they were born of necessity as a catch-all term when analytics were too crude to identify minor compounds in spirits.

"Oak lactones are key to the aroma imparted by oak barrels. The aroma of bourbon whiskey is dominated by oak lactones and vanillin." — Sean Eridon, UC Davis Waterhouse Lab

"These terpenes are present in very small concentrations, yet they have a considerable impact on the organoleptic properties of grapes and wines." — Trevor Grace, vintner

References

Abbott, N., Puech, J. L., Bayonove, C., & Baumes, R. (1995). Determination and sensory evaluation of *cis-* and *trans-*oak lactones in wines. *Food Chemistry, 51*(2), 135–141. https://doi.org/10.1016/0308-8146(94)P4179-7

Chira, K., Suh, J. H., Saucier, C., & Teissedre, P. L. (2013). Variability in extraction of oak volatile compounds: Lactones, phenols, and aldehydes. *Journal of Agricultural and Food Chemistry, 61*(3), 414–423. https://doi.org/10.1021/jf304081m

De Rosso, M., Panighel, A., Vedova, A. D., Gardiman, M., & Flamini, R. (2009). Study of terpenes and norisoprenoids in grapes and wines. *Food Chemistry, 117*(2), 256–262. https://doi.org/10.1016/j.foodchem.2009.03.024

Eridon, S. (2015). Oak lactones. *Waterhouse Lab,* University of California, Davis. https://waterhouse.ucdavis.edu/whats-in-wine/oak-lactones

Grace, T. (2015). Terpenes. *Waterhouse Lab,* University of California, Davis. https://waterhouse.ucdavis.edu/whats-in-wine/terpenes

Guth, H. (1997). Quantitation and sensory studies of character impact odorants of different white wine varieties. *Journal of Agricultural and Food Chemistry, 45*(8), 3027–3032. https://doi.org/10.1021/jf970280a

Park, S. K., Jiang, H., & Jeong, Y. W. (2017). Masking effects of lactones on fruity aroma in model spirits. *Food Science and Biotechnology, 26*(4), 1101–1108. https://doi.org/10.1007/s10068-017-0149-3

van Gemert, L. J. (2011). *Odour thresholds: Compilations of odour threshold values in air, water and other media* (2nd ed.). Oliemans Punter & Partners.

Waterhouse, A. L. (2015). Oak lactones and bourbon aroma. *What's in Wine?* University of California, Davis. https://waterhouse.ucdavis.edu/whats-in-wine/oak-lactones

Observations on Lactones

Lactones form naturally during oak aging, arising from the thermal degradation of fatty acids in the wood, particularly from cis- and trans-β-methyl-γ-octalactone ("whisky lactone").

Their aroma is strongly concentration-dependent—trace levels yield sweet, creamy, or coconut notes, while higher levels become waxy, fatty, or woody, shifting from delicacy to dominance.

Cis-lactone is more odor-active than the trans-form, typically dominating the aroma profile of American-oak-aged spirits.

Extraction peaks early in maturation, influenced by entry proof, temperature, and surface area contact between spirit and oak.

Ethanol concentration and temperature influence volatility, with moderate dilution during sensory evaluation enhancing the perceptibility of lactone-derived aromas.

Chapter 17: Congeners and Complexity — Compound Interactions in Spirits

In spirits evaluation, congeners are non-ethanol chemical compounds produced during fermentation, distillation, and aging that contribute to the aroma, flavor, texture, and complexity of distilled spirits. They are grouped as congeners because of their shared origin as secondary metabolites and their transformation products. This category exists to distinguish flavor-active or bioactive compounds that contribute to aroma, taste, texture, or color from the primary ethanol–water matrix.

Their presence—individually and in interaction—is central to the complexity, identity, and sensory expression of any distilled spirit. These include esters, acids, alcohols, aldehydes, phenols, sulfur compounds, and terpenes, among others. Each of these families has unique sensory characteristics, and the precise mixture of congeners gives each spirit its identity.

Classifying Congeners

There is no formal governing committee that universally determines or certifies which compounds qualify as *congeners*. However, the classification is widely accepted and used across the scientific, regulatory, and beverage industries, especially in analytical chemistry, food science, and sensory evaluation. The term is defined functionally rather than structurally, based on origin, behavior, and relevance to sensory evaluation.

Sources that contribute to the congener classification:

- **Scientific literature**: Peer-reviewed journals and academic texts on fermentation, distillation, and flavor chemistry.
- **Regulatory bodies**: Agencies such as the U.S. TTB (Alcohol and Tobacco Tax and Trade Bureau) or EU regulations mention congeners primarily in the context of safety and permissible levels, but they do not create exhaustive lists.
- **Industry standards**: Distillers, flavor chemists, and analytical labs use validated methods (e.g., GC-MS profiling) to identify and quantify congeners in spirits, commonly referring to esters, higher alcohols, acids, phenols, aldehydes, ketones, sulfur compounds, and terpenes as congener families.
- **Textbooks and educational institutions**: Often define congeners as "volatile and semi-volatile organic compounds (except ethanol), produced during fermentation, distillation, and aging."

In short, the term *"congener" is a pragmatic classification, rather than* a chemically rigorous one. It groups compounds by origin and function, not by shared molecular features.

Fig. 17.1	Compounds Commonly Designated as Congeners				
Congener Compound	Aroma, Mouthfeel	Class	Formula	MM (gm/mole)	DT (ppm)
Methanol	Solvent	Alcohol	CH_3OH	32.04	50
Hydrogen Sulfide	Rotten egg	Sulfur	H_2S	34.08	0.00047
Acetaldehyde	pungent, green apple	Aldehyde	C_2H_4O	44.05	100
Methanethiol	rotten cabbage, skunk	Sulfur	CH_4S	48.11	0.001
Acetic acid	Vinegar, sour	Acid	$C_2H_4O_2$	60.05	200
Propanol	mild alcohol, solventy	Alcohol	C_3H_8O	60.10	30
Ethanethiol	onion, garlic	Sulfur	C_2H_6S	62.13	0.0011
Isobutanol	fusel, solventy	Alcohol	$C_4H_{10}O$	74.12	100
Butyric acid	Rancid butter	Acid	$C_4H_8O_2$	88.11	20
Ethyl acetate	fruity, sweet	Ester	$C_4H_8O_2$	88.11	5.0
Isoamyl alcohol	banana, solvent	Alcohol	$C_5H_{12}O$	88.15	30
Lactic acid	milky, yogurt	Acid	$C_3H_6O_3$	90.08	100
Phenol	medicinal, smoky	Phenol	C_6H_6O	94.11	0.5
2-Furfural	almond, bready	Aldehyde	$C_5H_4O_2$	96.08	0.5
Isovaleric acid	sweaty, cheesy	Acid	$C_5H_{10}O_2$	102.13	22
Methional	cooked potato	Sulfur	C_4H_8OS	104.17	0.02
Ethyl butyrate	pineapple, fruity	Ester	$C_6H_{12}O_2$	116.16	0.02
4-Ethylphenol	barnyard, medicinal	Phenol	$C_8H_{10}O$	122.16	0.05
Guaiacol	smoky, medicinal	Phenol	$C_7H_8O_2$	124.14	0.02
Dimethyl trisulfide	cabbage, meat, sulfur	Sulfur	$C_2H_6S_3$	126.26	0.0001
Sotolon	curry, maple	Lactone	$C_6H_8O_3$	128.13	0.002
Isoamyl acetate	Banana	Ester	$C_7H_{14}O_2$	130.19	0.2
Limonene	citrus, fresh	Terpene	$C_{10}H_{16}$	136.24	10.0
Ethyl hexanoate	green apple, floral	Ester	$C_8H_{16}O_2$	144.21	0.01
Vanillin	vanilla	Phenol	$C_8H_8O_3$	152.15	0.3
Linalool	floral, citrus	Terpene	$C_{10}H_{16}O$	154.25	0.015
Eugenol	clove, spicy	Phenol	$C_{10}H_{12}O_2$	164.2	0.01
Whiskey lactone (cis)	coconut, woody	Lactone	$C_{10}H_{18}O_2$	170.25	0.020
Whiskey lactone (trans)	woody, coconut	Lactone	$C_{10}H_{18}O_2$	170.25	0.13

Value uncertain for hydroalcoholic matrices, needs source. DT values are matrix-dependent; see Guth (1997) and van Gemert (2011) for representative thresholds (e.g., linalool ≈ 0.015 mg/L in wine-like matrices). MM=molar mass

Synergistic and Antagonistic Effects

Synergistic effects arise when two or more compounds amplify each other's presence, creating a flavor impression greater than the sum of their parts. For example, esters may combine with lactones to yield enhanced fruity-coconut notes. Conversely, antagonistic effects occur when one compound masks or suppresses the action of another. Phenols, for instance, may suppress the perception of delicate esters due to their dominant smoky profiles (Delahunty et al., 2006).

Dilution, Temperature, and Ethanol Interactions

The concentration of ethanol in a spirit directly influences the volatility and perception of congeners. High ethanol concentrations suppress many minor volatiles due to solubility effects and vapor-pressure depression. Dilution with water reduces ethanol concentration and often reveals previously undetectable aroma compounds. However, this is not always beneficial: excessive dilution may disrupt stratification of volatiles or cause precipitates that alter texture and clarity.

Dilution with water changes the volatility and partitioning of congeners. As ethanol concentration decreases, some hydrophilic compounds (like specific acids) become more prominent, while lipophilic compounds (like phenols and long-chain esters) may become less volatile (Nykänen, 1986). Temperature also affects volatility—higher temperatures enhance aroma release but can distort delicate interactions (Capone, Jeffery, Sefton, & Osidacz, 2013).

Ethanol is a key modulator, altering vapor–liquid equilibrium and the release rate of volatiles. Higher ethanol concentrations suppress lighter volatiles and slow evaporation, while moderate dilution encourages the separation of aroma layers and can enhance the detection of subtler compounds (Ferreira, 2012). This makes ethanol management essential in tasting protocols.

Importance for Sensory Professionals

Expert tasters must move beyond identifying individual aroma compounds and develop the ability to perceive the gestalt—or holistic expression—of a spirit. This includes recognizing when specific flavor notes result from compound interactions rather than the effects of individual molecules. For instance, the characteristic profile of aged rum cannot be reduced to any one compound but arises from the complex interplay between oak-derived lactones, long-chain esters, phenols, and oxidized alcohols.

Trained evaluators must also learn to detect imbalances—cases where the normal synergistic relationships between congeners have been disrupted. This may result from over-aging, poor fermentation control, or flawed blending, each of which can distort the intended harmony and structure of the spirit.

Foundation for Understanding Dilution and Glassware Impact

Understanding how congeners interact is foundational to mastering the art and science of spirit evaluation. This knowledge directly informs how spirits should be diluted before tasting, and what types of glassware are appropriate for different compound families. For example, a tulip-shaped glass may concentrate esters and mask sulfur volatiles, while a wide-mouthed vessel may dissipate

120

ethanol and expose delicate acids or terpenes (Mansfield & Bastian, 2020).

The evaluator must be aware that the expression of complexity in spirits is not fixed—it evolves with time, temperature, dilution, and even repeated exposure. By mastering the behavior of congeners and their interactions, tasters can decode the spirit's language and distinguish quality, style, and intent with scientific precision.

Congeners in Perspective (Fig. 17.2)

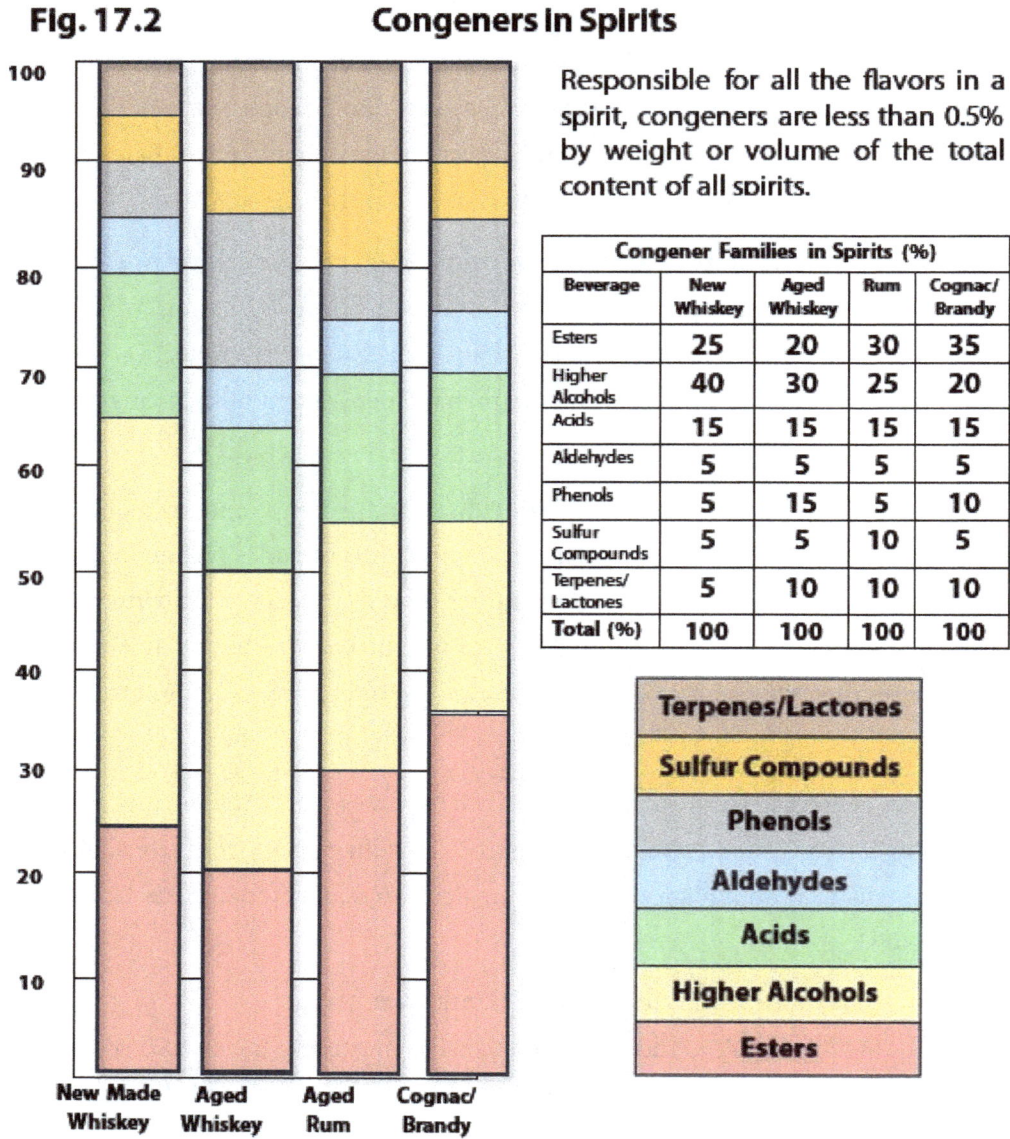

Fig. 17.2 — Congeners in Spirits

Responsible for all the flavors in a spirit, congeners are less than 0.5% by weight or volume of the total content of all spirits.

Congener Families in Spirits (%)				
Beverage	New Whiskey	Aged Whiskey	Rum	Cognac/ Brandy
Esters	25	20	30	35
Higher Alcohols	40	30	25	20
Acids	15	15	15	15
Aldehydes	5	5	5	5
Phenols	5	15	5	10
Sulfur Compounds	5	5	10	5
Terpenes/ Lactones	5	10	10	10
Total (%)	100	100	100	100

Legend: Terpenes/Lactones, Sulfur Compounds, Phenols, Aldehydes, Acids, Higher Alcohols, Esters

Detection Thresholds Are Key to Spirit Evaluation (Fig. 17.2)

Fig. 17.3   Relative Detection Thresholds of Key Congeners in Spirits

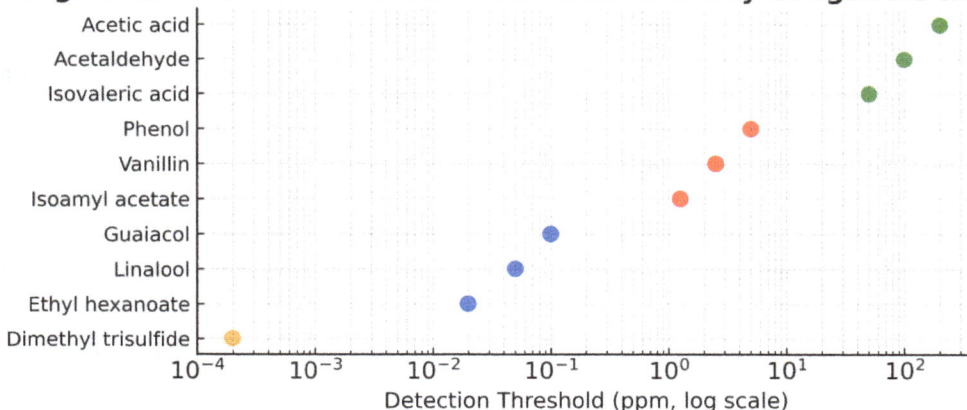

Perceptual dominance follows threshold, not abundance: Low-threshold compounds, such as dimethyl trisulfide or guaiacol, dominate the aroma even at trace levels, masking entire families of higher-threshold volatiles.

High-threshold acids and aldehydes shape the body, not the nose: Acetic acid and acetaldehyde contribute weight and volatility rather than a distinct aroma when present at near-threshold levels.

Masking is asymmetrical: Strong odorants, such as phenols or sulfur, suppress weaker notes; the reverse seldom occurs.

Perception is logarithmic, not additive: Congeners blend within chemical families, but intensity grows non-linearly—dominant odorants can conceal others still present analytically.

Complexity lies in the middle range: Balanced spirits arise where mid-threshold compounds— esters, lactones, light phenols—coexist without domination by extremes.

Summary

Congeners—esters, aldehydes, acids, phenols, terpenes, and sulfur compounds shape the sensory identity of distilled spirits. They are generated throughout fermentation, distillation, and aging, interacting synergistically and antagonistically to amplify or suppress aromas and mouthfeel. Dilution, temperature, and ethanol concentration modulate their volatility and perception, thereby influencing the tasting experience. Sensory professionals should strive to perceive spirits holistically, recognizing the gestalt produced by interacting congeners rather than just individual notes. Glassware choice and tasting protocols are vital tools for managing accurate congener profiling.

Beyond: As analysts, we now know what is there, but how much sensory information comes from the raw material?

"Congeners are what give whiskey its flavor, both good and bad."—Margaret Waterbury, food and beverage writer

"Whisky is made in the field, in the fermentation, and in the barrel, not in the bottle." —Jim Rutledge, Master Distiller

References

Capone, D. L., Jeffery, D. W., Sefton, M. A., & Osidacz, P. (2013). Release of volatile aroma compounds from wine: The influence of wine temperature. *Journal of Agricultural and Food Chemistry, 61*(42), 10125–10132. https://doi.org/10.1021/jf403166t

Delahunty, C. M., Piggott, J. R., & Paterson, A. (2006). Contribution of volatile compounds to the flavor of whisky: A review. *Journal of the Institute of Brewing, 112*(3), 215–229. https://doi.org/10.1002/j.2050-0416.2006.tb00728.x

Ferreira, V. (2012). The chemistry of odor perception: Structure–odor relationships. In A. C. Noble, M. Etiévant, & H. Parlange (Eds.), *Flavour: From food to perception* (pp. 3–31). Wiley. https://doi.org/10.1002/9781119954650.ch1

Guth, H. (1997). Quantitation and sensory studies of character-impact odorants of different white wine varieties. *Journal of Agricultural and Food Chemistry, 45*(8), 3027–3032. https://doi.org/10.1021/jf970280a

Mansfield, A. K., & Bastian, S. E. P. (2020). Influence of glass shape on the perception of aroma and flavor in wine and spirits: A critical review. *Food Research International, 137,* 109437. https://doi.org/10.1016/j.foodres.2020.109437

Nykänen, L. (1986). Formation and occurrence of flavor compounds in wine and distilled alcoholic beverages. *American Journal of Enology and Viticulture, 37*(1), 84–96. https://doi.org/10.5344/ajev.1986.37.1.84

Pozo-Bayón, M. A., & Ferreira, V. (2009). Analytical methods for aroma extraction in wines. In M. V. Moreno-Arribas & M. C. Polo (Eds.), *Wine chemistry and biochemistry* (pp. 641–660). Springer. https://doi.org/10.1007/978-0-387-74118-5_37

Siebert, T. E., Wood, C., Elsey, G. M., & Pollnitz, A. P. (2008). Determination of volatile phenol thresholds in red wine. *Journal of Agricultural and Food Chemistry, 56*(16), 7388–7393. https://doi.org/10.1021/jf801642d

van Gemert, L. J. (2011). *Odour thresholds: Compilations of odour threshold values in air, water and other media* (2nd ed.). Oliemans Punter & Partners.

Observation on Congeners in Spirits

Congeners define a spirit's individuality—the diverse trace compounds that carry aroma, texture, and finish. Their balance, not abundance, determines quality. Excess yields muddiness; absence leaves sterility. True craftsmanship lies in shaping congener harmony, not merely their count.

Section III:

Flavor Engineering – The Role of the Distiller

Typical Aged Spirits Manufacturing Process

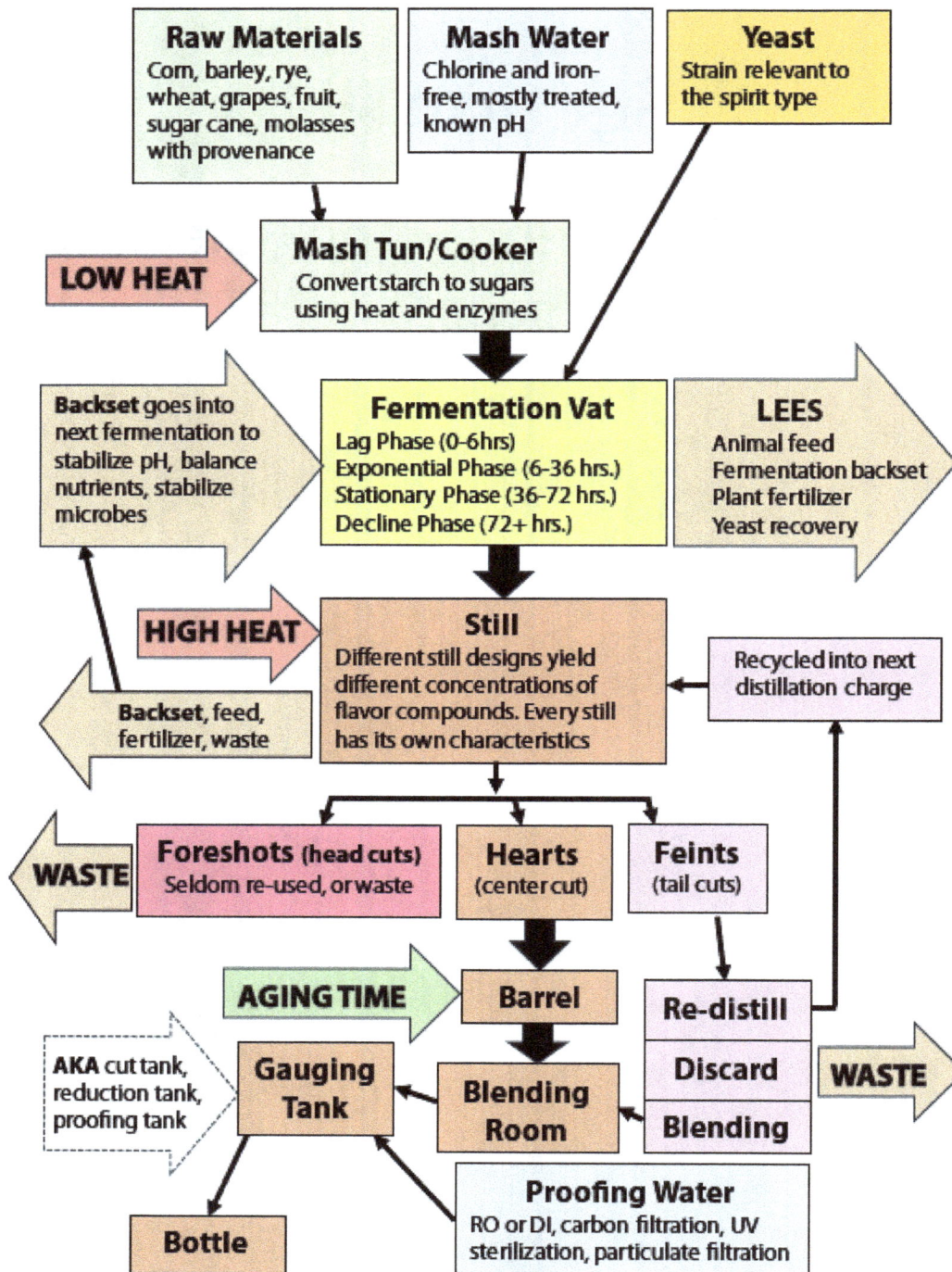

Raw Materials
Corn, barley, rye, wheat, grapes, fruit, sugar cane, molasses with provenance

Mash Water
Chlorine and iron-free, mostly treated, known pH

Yeast
Strain relevant to the spirit type

LOW HEAT

Mash Tun/Cooker
Convert starch to sugars using heat and enzymes

Backset goes into next fermentation to stabilize pH, balance nutrients, stabilize microbes

Fermentation Vat
Lag Phase (0-6hrs)
Exponential Phase (6-36 hrs.)
Stationary Phase (36-72 hrs.)
Decline Phase (72+ hrs.)

LEES
Animal feed
Fermentation backset
Plant fertilizer
Yeast recovery

HIGH HEAT

Still
Different still designs yield different concentrations of flavor compounds. Every still has its own characteristics

Recycled into next distillation charge

Backset, feed, fertilizer, waste

WASTE

Foreshots (head cuts)
Seldom re-used, or waste

Hearts
(center cut)

Feints
(tail cuts)

AGING TIME

Barrel

Re-distill

WASTE

AKA cut tank, reduction tank, proofing tank

Gauging Tank

Blending Room

Discard

Blending

Bottle

Proofing Water
RO or DI, carbon filtration, UV sterilization, particulate filtration

Chapter 18: Preparing the Raw Material — From Grain to Juice

Raw materials are not interchangeable commodities. They are the first and most consequential sources of sensory information in distilled spirits. This chapter examines how preparation choices—cleaning and tempering, malting and kilning, milling and particle-size distribution, gelatinization and enzymatic conversion, crushing and pressing, clarification and nutrient adjustment—determine the pool of aroma precursors, the kinetics of fermentation, and the matrix that will later be sculpted by distillation and maturation. The emphasis is both educational and practical: at each stage, we connect an upstream decision to a predictable change in the first sniff of new-make and in the finished spirit.

A central objective here is to preserve Raw Material Provenance (RMP) through processing. We previously defined RMP as the set of origin cues embodied in the substrate—variety, site, season, and husbandry—expressed as reproducible, teachable sensory differences. Preparation can either conserve or obscure those cues. Milling that respects husk integrity in malt, cooking regimes that avoid unnecessary severity in starch conversion, judicious pressing that limits seed phenolics in fruit, and agave cooking that favors controlled hydrolysis over indiscriminate browning all serve to carry RMP signals forward into fermentation and distillation rather than erasing them.

Closely related is the agrochemical signature, the measurable chemical fingerprint of cultivation practices and the environment—elemental profiles, stable-isotope ratios, and residue patterns—that often survives the first processing steps. While the agrochemical signature is an analytical construct, its preservation is associated with improved sensory outcomes when it reflects balanced nutrition, healthy canopies, and appropriate harvest timing. The educational aim is to train the evaluator to recognize how preparation choices amplify or mask both RMP and the agrochemical signature, and to provide the producer with a framework for retaining origin-driven character without sacrificing process control or yield.

Scotch (malted barley)

In malt whisky, grist particle size and wort clarity are foundational levers that shape new-make aroma before any cask effect. Finer milling increases starch accessibility and alcohol yield, but it also affects wort turbidity and lipid carryover, both of which influence yeast ester synthesis (Morris et al., 2022; Mese et al., 2025). Studies linking grist distribution and lauter turbidity to ester formation show that hazier wort can elevate fruity acetate esters and long-chain ethyl esters in new-make, changing first-sniff fruit vs. cereal balance (Bathgate, 2019; Mese et al., 2025). Kilning and hot mashing generate Maillard products (e.g., maltol, furans) that contribute sweet, toffee, and nutty precursors detectable even in spirit pre-aging (Liu et al., 2022; El Hosry et al., 2025; Lee et al., 2001). Because preserving RMP (raw material provenance) and the agrochemical signature depends on how they travel with grist (trace minerals, isotopes), malt terroir and farm practice can be tracked into spirit by stable-isotope/elemental profiling, supporting provenance claims that correlate with subtle cereal-and-mineral sensory cues (Ma et al., 2025; Fig. 18.1).

Bourbon (corn-forward mash bills)

Corn requires gelatinization/cooking, an enzyme regime, calcium, and a temperature–time history to control the dextrin profile and FAN release, which, in turn, steer the formation of fusel alcohols and esters during fermentation (Stewart & Russell, 2014; Morris et al., 2022). Over-severe thermal treatment promotes Maillard-derived furans and caramel notes; milder cooking with optimized enzymes often yields a cleaner, fruit-forward new-make that integrates more transparently with vanilla/coconut lactones from new charred oak (El Hosry et al., 2025; Stewart & Russell, 2014).

Fig. 18.1	Grain-Based Spirits Raw Material Preparation	
Step	**Purpose**	**Sensory Impact**
Cleaning & De-stoning	Remove debris, protect mills	Prevent off-flavors from foreign material
Malting/Kilning (barley)	Activate enzymes, develop color/phenolics	Peat smoke, biscuit, toffee precursors
Milling/Particle Size	Expose starch, preserve husk	Finer = higher extract & sweeter new-make; coarser = fuller mouthfeel, more dextrins
Cooking/Gelatinization (corn, rye)	Make starch accessible to enzymes	Maillard precursors → caramel/nutty notes
Mash Conversion	Enzymatic breakdown to sugars	Balance of fermentables & FAN affects ester/fusel formation
Wort Clarity (lautering)	Control solids & lipids entering the fermenter	Clear wort = cleaner fruit esters; cloudy = richer, oily esters

At the sensory level, these upstream choices modulate the matrix that later expresses bourbon's key odorants (e.g., γ-decalactone, whiskylactone, vanillin), whose importance in finished spirit has been quantified via GC–O/SIDA (Poisson & Schieberle, 2008a, 2008b).

Preserving RMP requires acknowledging the contributions of hybrid versus open-pollinated corn and soil macro- and microelements to the agrochemical signature, which can be evidenced analytically and, in some cases, perceptible as differences in cereal sweetness and oiliness in the new-make (Ma et al., 2025) (Fig. 18.1).

Gin (neutral spirit plus botanicals)

Botanical treatment is the raw material step in gin production. Maceration, vapor infusion, and vacuum/cold extraction all impact terpene/terpenoid recovery and enantiomeric ratios, with direct sensory consequences (Hodel et al., 2019; Buck et al., 2020; Dou et al., 2023). Juniper's α-pinene/myrcene/sabinene/limonene balance sets the core pine–citrus axis; berry origin and drying alter composition and, therefore, the headspace profile (Angioni et al., 2003; Höferl et al., 2014). Vapor infusion tends to bias toward more volatile monoterpenes (lighter, with a lifted nose), whereas long, warm macerations extract more oxygenated terpenoids and phenolics (deeper, resinous, and sometimes pithy) (Hodel et al., 2019; Dou et al., 2023). Selecting botanicals with documented chemotypes—and preserving their RMP signal—helps explain batch-to-batch sensory drift when supply switches (Buck et al., 2020; Angioni et al., 2003) (Fig. 18.2).

Fig. 18.2	Botanical Spirits Raw Material Preparation	
Step	**Purpose**	**Sensory Impact**
Selection/Weighing	Choose chemotypes & ratios	Controls juniper–citrus–spice balance
Cleaning/Drying	Stabilize botanicals	Prevent microbial or musty off-notes
Cracking vs. Whole	Expose essential oils	Cracked = brighter head volatiles; whole = heavier, resinous base notes
Infusion Method (Steep vs. Vapor)	Determine the extraction spectrum	Steep = deeper, pithy flavors; vapor = lighter, lifted aromas

Tequila/Mezcal (agave)

Agave cooking is the decisive material-processing step for blue agave spirits, during which the horno, autoclave, and diffuser alter the kinetics of inulin hydrolysis and the Maillard intensity. Brick-oven and longer, lower-rate cooks generate more Maillard volatiles—including HMF and furfural—that drive roasted-agave, caramel, and honey notes; autoclaves shorten reaction time with somewhat cleaner profiles; diffuser routes depend on subsequent thermal/acid steps and can yield markedly different volatile fingerprints (Cedeño, 1995; Mancilla-Margalli & López, 2002; Rodríguez-Félix et al., 2018). Recent reviews quantify HMF (ppm–thousands) and associated compounds that arise during cooking, linking them to sensory attributes perceived before distillation and in blanco spirits (Colón-González et al., 2025). Milling/extraction intensity controls bagasse fines and saponin carryover; excessive fines can contribute to vegetal/phenolic edges or bitterness that persist in the distillate. Preserving RMP is unusually powerful here: site, soil, and agronomy imprint an agrochemical signature (multielement and isotope ratios) trackable for authenticity and explaining terroir-linked sensory variation in cooked-agave aroma (Geographical Origin Identification of Tequila, 2021; Ma et al., 2025; Flores et al., 2009) (Fig. 18.3).

Fig. 18.3	Agave-Based Spirits Raw Material Preparation	
Step	**Purpose**	**Sensory Impact**
Harvest/Trimming Piñas	Prepare a carbohydrate source	Species & site determine baseline "green" vs. "sweet" aromatics
Cooking/Hydrolysis (Horno/Autoclave/Diffuser)	Convert inulin to sugars	Horno = roasted, caramel; autoclave = cleaner, vegetal; diffuser = sharper, green
Crushing/ Extraction	Release aguamiel from fibers	Gentle tahona = smoother, less bitter; heavy tahona = shredding, more tannins/bitterness
Juice Clarification	Remove bagasse & solids	Cleaner fermentation, reduced vegetal notes

Rum (molasses and cane juice)

Molasses composition (sugars, ash, trace metals, sulfur, FAN) and pre-fermentation conditioning are the raw-material determinants of rum aroma breadth. FAN supplementation and oxygenation strategies govern yeast growth and the higher-alcohol/ester balance; undernourished fermentations risk stuck kinetics, stressed sulfur notes, and suppressed fruity esters (Mangwanda et al., 2021). Sensomics tracking from molasses → mash → distillate shows key odorants evolving across steps, with precursors in the substrate and fermentation conditions decisively shaping the ethyl ester and phenolic signature (Franitza et al., 2016). Comparative aroma studies (rum vs. cachaça) further highlight how substrate treatment (sulfur, acids, clarifiers) shifts descriptors from grassy/spicy to fruit-forward (de Souza et al., 2006). Preserving RMP through cane variety, soil, and milling chemistry—as well as managing agrochemical signatures (e.g., metals, S species)—can map process and origin and sometimes predict sensory risks like solvency or sulfury notes if not managed (Franitza et al., 2016; Mangwanda et al., 2023) (Fig. 18.4).

Fig. 18.4	Sugarcane/Molasses-Based Spirits Raw Material Preparation	
Step	**Purpose**	**Sensory Impact**
Cane Harvest/Crushing	Extract juice quickly	Fresh juice = grassy/tropical notes; delayed = oxidized/browned
Clarification/Filtration	Remove solids & microbes	Cleaner ferments, reduced sulfur
Molasses Dilution & Nutrient Addition	Adjust fermentability & FAN	Balanced yeast growth → esters; underfed → stressed fusels/sulfur
Fermentation	Build a congener profile	Long time = heavier esters; short = lighter rum

Brandy (grape-based: Cognac/Armagnac; Pisco; Grappa)

For grape spirits, the base-wine build is the "material processing." Juice vs. must handling, pressing regime (free run vs. heavy press), solids load, SO_2 policy, and fermentation temperature determine volatile precursors (terpenes, norisoprenoids) and reductive sulfur risks that distillation can either concentrate or transform (Ribéreau-Gayon et al., 2006; Guymon, 1974). High solids or excessive SO_2 at distillation correlate with elevated volatile sulfur levels and harsher noses; careful SO_2 control and timely distillation yield cleaner floral/fruit cores, especially for Muscat-based Piscos (Guerrero-Chanivet et al., 2022; Guymon, 1974). Skin contact elevates varietal terpenes (beneficial for Muscat Piscos) but may introduce phenolic bitterness if pressing is too hard—trade-offs directly perceived in spirit (Domenech et al., 2025). When preserved, the RMP and agrochemical signature (variety, site, and viticultural inputs) remain evident in the terpene/norisoprenoid envelope of the distillate and are well documented through wine analytics applied to spirit production (Ribéreau-Gayon et al., 2006) (Fig. 18.5).

Fig. 18.5	Fruit-Based Spirits Raw Material Preparation	
Step	Purpose	Sensory Impact
Sorting/Destemming	Exclude damaged fruit & stems	Avoid volatile acidity, harsh phenolics
Crushing/Pressing	Extract juice/must	Gentle press = delicate floral/fruit; heavy press = more tannin/bitterness
Clarification/Settling	Adjust solids load	Less solids = cleaner fermentation & floral notes; more solids = fuller palate, risk of reductive sulfur
Base-Wine Fermentation	Build aroma precursors	Drives varietal terpenes & esters that distillation concentrates

Vodka/Neutral spirits (grain/potato/other)

When aiming for neutrality, raw material treatment shapes the rectification load. (Fig. 18.6)

Fig. 18.6	Vodka Neutral Spirits Raw Material Preparation	
Step	Purpose	Sensory Impact
Selection of stock (grain, potato, etc.)	Choose a substrate for ethanol production	Baseline differences in mouthfeel and residual character tied to RMP
Milling/ Particle Size	Expose starch for enzymatic hydrolysis	Finer grist = higher extract but potential for more lipids; coarser = cleaner wash
Cooking, Enzymatic Conversion	Convert starch to fermentable sugars with minimal by-products	Clean conversion reduces aldehydes and fusels
Clarification/Filtration of Mash	Remove solids to reduce fusel oils entering fermenter	Clear mash promotes cleaner fermentation and neutral spirit
Fermentation	Produce a neutral wash with low congeners	Healthy yeast yields neutral aroma and low higher alcohols
Rectification/ Distillation	Concentrate the ethanol and strip the congeners	High-rectification yields a neutral nose; RMP survives analytically but is sensory-minimized

Particle size and thermal and enzymatic conversion determine the dextrin profile and FAN. Clean conversion with minimal thermal degradation reduces aldehydic harshness and fusel load entering the column, thereby facilitating the attainment of a genuinely neutral nose (Stewart & Russell, 2014; Liu et al., 2022). Preserving RMP remains scientifically important: elemental/isotopic fingerprints can distinguish feedstocks and origins, even when sensory differences are intentionally minimized (Ma et al., 2025).

Baijiu (sorghum; solid-state)

Sorghum selection and grain treatment (steaming intensity, daqu stacking/aging, pit microbiome) are the "material processing" drivers of style. Starch availability, sorghum phenolics, and the enzymatic/microbial profile generated by the starter directly modulate the balance between ethyl caproate/ethyl lactate and smoky/phenolic notes—even before prolonged aging (Tu et al., 2022). Provenance (grain cultivar, region) is embedded in both sensory (ester ratios, cereal/umami tones) and agrochemical signature. See Fig. 18.7.

Fig. 18.7	Baijiu Sorghum Spirits Raw Material Preparation	
Step	**Purpose**	**Sensory Impact**
Sorghum Selection and Cleaning	Ensure desired starch, tannin, and phenolic levels	Grain variety and terroir contribute to cereal, umami, and phenolic notes
Steaming of Sorghum	Gelatinize starch, sterilize grain for fermentation	Steaming severity influences fermentability and precursor availability
Daqu Preparation (starter bricks)	Develop enzyme and microbial consortia for fermentation	Daqu type drives style (sauce-aroma vs. strong-aroma) and key ester balances
Stacking and Aging of Daqu	Enhance enzyme activity and microbial diversity.	Maturation of daqu influences microbial complexity and eventual aroma richness
Solid-State Fermentation in Pits	Convert starch to alcohol and aroma compounds in situ	Pit microbiome yields signature ethyl caproate/ethyl lactate balances and smoky/phenolic notes
Distillation and Collection	Capture diverse aroma compounds developed in pit fermentation	Distillation preserves unique ester and phenolic profile distinctive of baijiu style

Cross-cutting: Preserving RMP and agrochemical signature

Throughout categories, material preparation sets the pool of aroma precursors and fermentation nutrients that define first-sniff new-make. Preservation of RMP depends on choices that avoid unnecessary loss or obscuration of origin cues (for example, excessive thermal severity, over-clarification, or additive regimes that mask provenance signals). It also transports a measurable agrochemical signature (stable isotopes, multielement profiles) from the field to the spirit, enabling authenticity checks and helping to explain reproducible sensory distinctions tied to place and practice (Ma et al., 2025; Tanabe et al., 2020). The severity (thermal, mechanical, enzymatic) of material preparation largely determines whether early sensory reads lean towards fresh/fruit (with more ester potential and cleaner matrices) or roasted/toffee/spice (with more Maillard/polysaccharide breakdown products). Choice is the first act of "flavor engineering." The severity of preparation can drastically affect the sensory outcome, as shown in Fig. 18.8.

Fig. 18.8	Sensory Risk and Reward in Material Preparation		
Process	**Low Severity Outcome**	**High Severity Outcome**	**Sensory Risk**
Milling/ Crushing	Coarse, under-converted	Fine, over extracted	Loss of mouthfeel contrast
Heating/ Cooking	Starchy, vegetal	Caramelized, butter	Marked provenance
Clarifying/Filtering	Cloudy, oily	Stripped, thin	Diminished aromatics
Pressing/ Extraction	Delicate, aromatic	Phenolic, astringent	Harsh, bitter
Enzymatic Regime	Incomplete conversion	Over hydrolyzed	Excess fusel precursors

Isotopic and Elemental Fingerprints in Provenance Control

Every raw material carries a stable isotopic, trace-element pattern derived from its soil, water, and climate. Measurable trace-element isotopes are used to track the agrochemical signature—the lasting chemical imprint of cultivation conditions and mineral environment. Ratios of these isotopes form a fingerprint that survives initial processing and defines origin. Instrumentation records these values with high precision, producing chemical maps of provenance.

Fermentation and distillation modify concentrations, but isotopes are primarily left unchanged, allowing signatures to persist from field to spirit. They authenticate feedstock source, verify regional identity, and align with sensory impressions described as mineral, earthy, or grainy. For the sensory professional, this means that the provenance of raw materials and the agrochemical signature are measurable, traceable, and controllable. Managing water, soil inputs, and processing severity not only shapes the analytical profile but also preserves the perceptual continuity between origin and aroma, carrying the chemistry of place forward into the spirit's nose.

Summary

Material preparation is the first step in flavor engineering, and it is where the producer either preserves or forfeits the origin. Fine control of milling defines the extract and filtration behavior, and, through turbidity and lipid carryover, influences ester formation and texture in new-make. Thermal and enzymatic schedules determine the extent of gelatinization, Maillard formation, and fermentable nitrogen release, which in turn govern the balance of fusel alcohols, esters, and early sweet or toasted notes. In fruit spirits, pressing severity and solids management decide whether varietal terpenes and norisoprenoids are carried forward or overshadowed by phenolic bitterness or reductive sulfur. In agave spirits, the cooking method sets the trajectory from green-vegetal to roasted-caramelized profiles before yeast is pitched. In botanically driven spirits, the extraction method and botanical state select for volatile classes that define nose structure from the first pass. Across these categories, the through-line is preservation: preparation choices that are proportionate to the raw material tend to conserve RMP and maintain a readable agrochemical signature, producing spirits whose sensory identity matches their origin story. Choices that are unnecessarily severe, indiscriminate, or poorly sequenced tend to flatten that identity, forcing later stages to work harder to recover character. The educational takeaway is direct: when evaluating spirits, link observed aromas and mouthfeel back to specific preparation levers; when producing spirits, set those levers to reveal rather than obscure the chemistry of place.

Beyond: Sources and treatments for dilution water affect the sensory evaluation of spirits.

"Start with clean grain and good water—everything after that is damage control." —Jimmy Russell, Wild Turkey

"A master blender listens to the grain before the still ever speaks."— Michael Miyamoto, Suntory

References

Angioni, A., Barra, A., Coroneo, V., Dessi, S., & Cabras, P. (2003). Chemical composition of the essential oils of *Juniperus* from ripe and unripe berries and leaves and their antimicrobial activity. *Journal of Agricultural and Food Chemistry, 51*(10), 3073–3078. https://doi.org/10.1021/jf026203j

Bathgate, G. N. (2019). The influence of malt and wort processing on spirit yield and quality in Scotch malt whisky production. *Journal of the Institute of Brewing, 125*(1), 3–13. https://doi.org/10.1002/jib.556

Buck, N., Herrmann, C., & Zellner, B. D. (2020). Key aroma compounds in two Bavarian gins. *Applied Sciences, 10*(20), 7269. https://doi.org/10.3390/app10207269

Cedeño, M. (1995). Tequila production. *Critical Reviews in Biotechnology, 15*(1), 1–11. https://doi.org/10.3109/07388559509150536

Colón-González, M., Bello-López, M. Á., & Hernández-Morales, A. (2025). Thriving in adversity: Yeasts in the agave fermentation environment. *Yeast, 42*(4), 211–234. https://doi.org/10.1002/yea.3989

Domenech, A. M., et al. (2025). The role of skin-contact time and heads fraction on Pisco sensory and chemical profile. *IVES Technical Reviews.* https://doi.org/10.20870/IVES-TR.2025.9480

Dou, Y., Yang, L., Wang, C., Chen, Y., & Zhang, Y. (2023). Analysis of volatile and nonvolatile constituents in gin by comprehensive two-dimensional gas chromatography–time-of-flight mass spectrometry. *Journal of Agricultural and Food Chemistry, 71*(23), 9468–9479. https://doi.org/10.1021/acs.jafc.3c00707

El Hosry, L., et al. (2025). Maillard reaction: Mechanism, influencing parameters, and control strategies in foods. *Food Research International, 177,* 114146. https://doi.org/10.1016/j.foodres.2024.114146

Flores, C. R., et al. (2009). ICP-MS multi-element profiles and HPLC determination of furanic compounds in tequilas. *Bulletin of Environmental Contamination and Toxicology, 82,* 613–617. https://doi.org/10.1007/s00128-009-9692-7

Franitza, L., Granvogl, M., & Schieberle, P. (2016). Influence of the production process on the key aroma compounds of rum: From molasses to the spirit. *Journal of Agricultural and Food Chemistry, 64*(47), 9041–9053. https://doi.org/10.1021/acs.jafc.6b04046

Geographical origin identification of tequila based on multielement and stable isotopes. (2021). *Journal of Analytical Methods in Chemistry, 2021,* 6615264. https://doi.org/10.1155/2021/6615264

Guerrero-Chanivet, M., Valcárcel-Muñoz, M. J., Guillén-Sánchez, D. A., Castro-Mejías, R., Durán-Guerrero, E., Rodríguez-Dodero, C., & García-Moreno, M. de V. (2022). Influence of SO_2 use, distillation system and aging conditions on the final sensory characteristics of brandy. *Foods, 11*(21), 3540. https://doi.org/10.3390/foods11213540

Guymon, J. F. (1974). Chemical aspects of distilling wines into brandy. *American Journal of Enology and Viticulture, 25*(3), 117–128.

Höferl, M., et al. (2014). Chemical composition and antioxidant properties of *Juniperus communis L.* essential oil. *Evidence-Based Complementary and Alternative Medicine, 2014,* 239708. https://doi.org/10.1155/2014/239708

Hodel, J., Busse, F., & Reineccius, G. (2019). Quantitative comparison of volatiles in vapor-infused gin versus steeped gin. *Heriot-Watt University Research Portal.* [DOI unavailable]

Lee, K.-Y. M., Paterson, A., & Piggott, J. R. (2001). Origins of flavour in whiskies and a revised flavour wheel: A review. *Journal of the Institute of Brewing, 107*(5), 287–313. https://doi.org/10.1002/j.2050-0416.2001.tb00099.x

Liu, S., et al. (2022). Insights into flavor and key influencing factors of Maillard reaction products. *Food Chemistry, 366,* 130580. https://doi.org/10.1016/j.foodchem.2021.130580

Ma, Y., Zhang, L., & Liu, S. (2025). Advancing stable isotope analysis for alcoholic beverages' authenticity: Novel approaches in fraud detection and traceability. *Foods, 14*(6), 943. https://doi.org/10.3390/foods14060943

Mancilla-Margalli, N. A., & López, M. G. (2002). Generation of Maillard compounds from inulin during the thermal processing of *Agave tequilana Weber var. azul. Journal of Agricultural and Food Chemistry, 50*(4), 806–812. https://doi.org/10.1021/jf010752r

Mangwanda, T. W., Nyoni, H., & Mupa, M. (2021). Processes, challenges and optimisation of rum production: A review. *Fermentation, 7*(1), 21. https://doi.org/10.3390/fermentation7010021

Mangwanda, T. W., et al. (2023). Physicochemical and nutritional analysis of molasses for rum fermentation. *Proceedings of the International Conference on Enology and Food Chemistry, 26*(1), 105. https://doi.org/10.3390/IECF2023-15137

Mese, Y., Crawshaw, M., Harrison, B., & Harrison, R. (2025). Effect of grist particle size distribution and wort turbidity on ester composition of malt whisky new-make spirit. *Journal of the American Society of Brewing Chemists, 83*(3), 221–235. https://doi.org/10.1080/03610470.2024.2402136

Morris, S., et al. (2022). Response surface methods to optimise milling parameters and cooking conditions for spirit production from wheat. *Foods, 11*(8), 1163. https://doi.org/10.3390/foods11081163

Poisson, L., & Schieberle, P. (2008a). Characterization of the most odor-active compounds in an American bourbon whisky by application of aroma extract dilution analysis. *Journal of Agricultural and Food Chemistry, 56*(14), 5813–5819. https://doi.org/10.1021/jf800382m

Poisson, L., & Schieberle, P. (2008b). Characterization of the key aroma compounds in an American bourbon whisky by quantitative measurements, aroma recombination, and omission studies. *Journal of Agricultural and Food Chemistry, 56*(14), 5820–5826. https://doi.org/10.1021/jf800383v

Ribéreau-Gayon, P., Dubourdieu, D., Donèche, B., & Lonvaud, A. (2006). *Handbook of enology, Vol. 1: The microbiology of wine and vinifications* (2nd ed.). Wiley.

Rodríguez-Félix, E., Contreras-Ramos, S. M., Dávila-Vázquez, G., Rodríguez-Campos, J., & Marino-Marmolejo, E. N. (2018). Identification and quantification of volatile compounds found in vinasses from two different processes of tequila production. *Energies, 11*(3), 490. https://doi.org/10.3390/en11030490

Stewart, G. G., & Russell, I. (2014). *Whisky: Technology, production and marketing* (2nd ed.). Elsevier.

Tanabe, C., et al. (2020). Stable isotope ratios in distilled spirits for origin authentication: A review. *Trends in Food Science & Technology, 96*, 211–223. https://doi.org/10.1016/j.tifs.2019.12.019

Tu, W., Cao, X., Cheng, J., Li, L., Zhang, T., Wu, Q., et al. (2022). Chinese Baijiu: The perfect works of microorganisms. *Frontiers in Microbiology, 13*, 919044. https://doi.org/10.3389/fmicb.2022.919044

de Souza, M. D. C. A., Vásquez, P., Del Mastro, N. L., Acree, T. E., & Lavin, E. H. (2006). Characterization of cachaça and rum aroma. *Journal of Agricultural and Food Chemistry, 54*(2), 485–488. https://doi.org/10.1021/jf0511190

Observations on Raw Material Preparation

The Myth of Neutral Grain: "Neutral" grain is a contradiction in terms. Every cereal species carries a matrix of bound precursors — fatty acids, amino acids, sulfur compounds — that become substrates for yeast metabolism. Milling, cooking, and enzyme control determine how much of this potential becomes available. These early biochemical differences imprint on ester ratios and residual congeners. Ignore the individuality of raw materials and forfeit control over aroma architecture. Grain differences in neutral grain spirits will not survive heavy stripping.

Fermentation chemistry begins long before yeast contact: Soil nitrogen, moisture stress, and crop maturity reshape the entire nitrogen profile—not just FAN—shifting amino acid distribution, peptide content, and organic acid precursors. These upstream variations modulate Ehrlich pathway flux, thereby altering fusel ratios, ester formation, and redox balance. A low-nitrogen barley does not merely yield a "leaner" distillate; it produces an entirely different congener network, making agricultural practice the first act of flavor design.

The Starch Illusion: Yield-focused producers often pursue complete starch conversion, assuming that higher extract yields equate to higher quality. However, over-gelatinized or enzyme-overexposed mashes can reduce the availability of lipid and protein cofactors essential to ester and acid formation. Efficiency gained at the cooker can mean aromatic poverty in the spirit — a trade invisible until the first nosing.

Chapter 19: Dilution Water in Distillation — Sources and Treatments.

Water plays a critical role in distilled spirits, not only during fermentation and mashing, but especially during proofing (Gregory & Nortcliff, 2013), the process of reducing the alcohol concentration of distillate to a desired bottling strength. The quality, composition, and treatment of dilution water significantly affect the final product's flavor clarity, mouthfeel, and stability, making its selection and handling a matter of both tradition and scientific precision. See Fig. 19.1.

Fig. 19.1	Dilution Water Influences on Spirit Flavor*
Property	**Influential Effect on Flavor**
pH	Solubility of aroma compounds can shift the balance of esters and acids
Minerality (TDS) and Conductivity	Total Dissolved Solids (TDS) influences texture, mouthfeel, aroma/flavor perception, conductivity perception of bitterness/astringency
Hardness	Soft water promotes sweet, fruity flavors; hard water suppresses
*Source: Safe Drinking Water Foundation, Aqua Maestro, Montana State University, Yosemite Technologies	

Natural and Source Waters

Distillers typically use one of the following natural water sources, with variations depending on geographical location, regulatory compliance, and brand identity. Properties of various dilution water sources are shown in Fig. 19.2

Spring water: This is one of the most prized sources, particularly in regions such as Kentucky, Tennessee, and parts of Scotland. Natural springs that emerge from limestone aquifers are favored for their low iron and high calcium carbonate content, which neutralize acidity and support a favorable fermentation pH during the early stages of production. In proofing, however, mineral balance must be carefully managed to avoid haze or flavor alteration (Gregory & Nortcliff, 2013).

Well water: Groundwater drawn from private or municipal wells is common, especially for distilleries located in rural or mountainous areas. Its composition varies widely depending on geology and may contain elevated levels of iron, sulfur, manganese, or other trace minerals that require treatment prior to dilution (World Health Organization, 2017).

Municipal (tap) water: While convenient and regulated for safety, municipal water is rarely used without treatment due to chlorine, chloramines, and varying hardness levels. Residual chlorine or chloramine can contribute to off-odors and form regulated disinfection by-products (e.g., TTHMs/HAA5) when reacting with natural organics; therefore, distillers remove them (e.g., carbon → RO) before proofing (U.S. Environmental Protection Agency, 2021).

Surface water: Rivers, lakes, and reservoirs are rarely used directly but may serve as municipal sources. Their organic content and microbiological variability make them unsuitable for direct proofing applications without significant purification.

Common Treatments for Dilution Water

To ensure the neutrality, microbiological safety, and chemical stability of dilution water, most distillers apply one or more of the following treatments:

- **Reverse osmosis (RO)** is the most common treatment method, used to remove dissolved solids, chlorine compounds, microorganisms, and volatile organics. It provides high-purity water with minimal flavor impact. RO systems typically follow this with a remineralization step if the total dissolved solids (TDS) level falls too low, which could negatively affect mouthfeel or chemical equilibrium (LeChevallier & Au, 2004).

- **Deionization (DI)** is often used alongside RO. Deionization removes ionic species such as calcium, magnesium, iron, or sulfate. While effective, it is typically reserved for facilities that require near-laboratory-grade water or those that blend with specific mineral profiles.

- **Carbon filtration**, using activated carbon filters to remove chlorine, chloramines, and organic off-flavors. While not sufficient alone for complete purification, carbon is often used as a pre-treatment step for both RO and DI systems (Lange & Wood, 2019).

- **UV sterilization or ozone**, for microbiological control, especially in extensive facilities or warm climates, UV or ozone treatment ensures bacterial or fungal spores are inactivated prior to water contact with spirit.

- **Aeration and settling**, used in some traditional facilities, especially those relying on spring water, passive treatment methods such as **aeration tanks** and **sedimentation reservoirs** are used to remove volatile sulfur compounds and particulates.

Fig. 19.2	Properties of Dilution Water		
Water	**TDS (ppm)**	**pH**	**Description**
Distilled	Approx 0	7.0	Pure, no mineral, neutral pH
Reverse Osmosis	10-50	6.0-6.8	Slightly mineralized, slightly lean to acidic
Tap Water	100-500	6.5-8.5	Municipal supply, lime/soda adjusted pH
Mineral Water	Approx. 300	6.5-7.2	Noticeable taste, pH gives an even mouthfeel
Scottish Spring	50-150	6.5-7.4	Natural, gentle profile
*Source: Safe Drinking Water Foundation, Aqua Maestro, Montana State University, Yosemite Technologies. Ranges are ordinally correct, but illustrative, TDS=total dissolved solids			

Sensory and Chemical Considerations

Although water is often perceived as neutral, its pH, mineral content, and residual organics can significantly affect a spirit's mouthfeel, aroma volatility, and long-term stability. For example:

- **Hard water** (high in calcium and magnesium) may increase perceived astringency or result in mineral haze upon chilling or ethanol addition.

- **Iron content**, even in trace amounts, can catalyze oxidation or react with phenolics, altering flavor over time.

- Residual chlorine or chloramine can contribute off-odors and regulated disinfection by-products (e.g., THMs/HAAs) when reacting with organic matter; distillers therefore remove them (e.g., carbon + RO) before proofing.

For these reasons, ultrapure or RO-treated water is now the global industry standard for dilution, particularly in premium and export-bound spirits. However, some craft distillers deliberately use post-treatment mineral-balanced water to create a proprietary mouthfeel or align with local water traditions, provided the sensory outcomes remain favorable—examples in Fig. 19.3.

Fig. 19.3	U.S. Distillery Water Source Profiles and Treatment			
Region	Source	Primary Treatment	Usage	Notes
Seattle, WA	Cedar River watershed Cascade snowmelt	Minimal, naturally pristine, low mineral	Mash, fermentation, proofing, and dilution to maintain neutral flavor	Low minerals, pH neutral, used directly without RO or dechlorination
Humboldt County, CA	Likely local well or municipality (not published)	Filtration, likely RO, carbon	Mash, fermentation, dilution. US single malt-style spirit	Specs unpublished, commitment suggests treatment
Hye, TX	Texas hill country limestone filtered aquifer	Direct well, high in Ca, Mg, no Fe, minimal treatment	Mash, fermentation, dilution, proofing, and natural mineral balance	High Ca supports yeast, and Fe-free avoids color issues
Austin, TX	Private limestone spring in Mason Co.	Hose-fed spring collect, minimal treatment	Mash, dilution, mineral content for mouthfeel	Ca, Mg levels boost mash pH, mouthfeel
Loretto, KY	On-site spring-fed lake, limestone filtered	Minimum treatment	Mash, fermentation, aging, dilution, proofing	No Fe, natural Ca, Mg for fermentation, flavor
Frankfort, KY	Kentucky River water, limestone filtered bedrock	Filtered to remove Fe, Ca buffering	Mash, fermentation	Ca and Mg contribute to flavor and mouthfeel
Versailles, KY	On-site spring, limestone filtered	Minimal treatment	Mash, fermentation, barrel entry, proofing	Free, supports yeast and color
Lynchburg, TN	Cave spring, limestone filtered	No chemical treatment	Mash, fermentation base, downstream proofing may use local water	Cave spring water is soft, mineral-rich, and iron-free
Tullahoma, TN	Cascade Hollow spring, limestone filtered	Iron-free, minimal treatment	Mash, fermentation, probably proofing, and dilution	Similar to other TN distilleries, limestone filtered
Indiana	Local well or municipal water	Filtration, RO, Carbon, and ion exchange	Mash, fermentation, dilution	Brands adjust to simulate limestone profiles

Fig. 19.3 Disclaimer: The chart lacks consistency due to the scarcity and inclusion of distillery information. Material age is unknown, and subsequent changes may not be reflected. This information is subject to change. Treatments render safe, suitable water. Actual distiller's names have been omitted.

Scotch distillers prize natural water sources as a key factor in the character of their spirits' flavor. Fig. 19.4 summarizes regional examples of scotch water and its usage. The data here is more reliable, and sources of information have been added to the reference section.

Fig. 19.4	Examples of Scottish Distillery Water Profiles by Region				
Region	Source	pH	TDS (ppm)	Hardness CaCO₃ (mg/L)	Notes*
Glenturret - Highland	Loch Turret (granite upland)	6.5-7.0	30-70	20-60	Soft, neutral, lacks buffering, soft fermentation
Auchentoshen - Lowland	Loch Katrine (via Glasgow)	7.4-7.8	100-200	90-120	Treated municipal, moderate hardness, buffering
Balblair - North Highland	Alt Dearg stream (Ben Dearg)	6.8-7.2	70	60	Soft, mineral-balanced, supports traditional fermentation
Springbank - Campbeltown	Crosshill Loch	6.7-7.0	80-120	60-100	Slightly mineralized, cited for the character of fermentation
Bunnahabhain - Islay	Margadale Spring	6.5-7.0	50-100	Soft	Untreated, paired with peated malt for aromatic contrast
*See reference sources, hardness ranges are illustrative and vary (Scottish Water, 2024)					

The Myth of Sea Salt Aromas from Island Scotch

Often, the scientifically uneducated blogger or novice article writer will describe their evaluation of Scotch whisky distilled in the islands (Islay, Orkney, Hebrides, Skye, Jura, etc.) as having the aroma quality of salt sea air. Some evaluators consider it obligatory, given that the distilleries are located so close to their regional seacoasts.

There are at least two reasons why this widely perpetuated myth is false (Bendig et al., 2014). Salt has no aroma; salt is a taste. Sea mist in the worst of storms does not travel high over land, nor far. Leaky barrels may at most show salt precipitate at the stave seams, but salt will not penetrate oak.

- Sodium (salt) has no odor, and typical taste thresholds in water are about ~30–60 mg/L Na⁺, well above the sodium levels present in most potable waters used by distilleries.
- Perceived 'marine' notes in many Islay malts are better explained by bromophenols derived from peat smoke/marine influences during malting—quantified at >400 ng/L in Laphroaig and Lagavulin—well above sensory thresholds." Notes are often mislabeled as "brine."

The primary source of the unusual, unique aroma characteristic of many island-produced Scotches is the peat smoke during the malting process, not air, water, fermentation, dilution, nor proximity to sea air and storms. Bromophenols and other related phenol compounds are the source (Bendig, Lehnert, & Vetter, 2014). Bromophenols originate from peat smoke and marine-influenced barley drying and are present in sufficient levels to create an olfactory *illusion* of iodine, fish, or coastal brine. These aromas activate neural expectations of saltiness, even though the normal level of saltiness is below the human detection threshold. Salt is absent in these whiskies.

Iron in Source Water

Distillers avoid or deionize water that contains Fe2+/Fe3+ ions for several practical reasons, grounded in both chemistry and sensory outcomes.

Chemical Instability and Precipitation: Iron readily oxidizes in the presence of oxygen, shifting between ferrous (Fe^{2+}) and ferric (Fe^{3+}) states. Ferric iron (Fe^{3+}) forms insoluble hydroxides and oxides, causing cloudiness, sediment, and haze in diluted or bottled spirits. This is a serious stability problem for product appearance and shelf life (World Health Organization, 2017).

Sensory Off-Notes: Even at low concentrations, iron imparts a metallic taste that overwhelms delicate congeners. Iron catalyzes lipid and phenolic oxidation, accelerating staling reactions that produce notes with characteristics reminiscent of cardboard, blood, or rust (Lange & Wood, 2019). The mineral harshness of iron conflicts with the "soft" or "pure" water character that distillers depend on for smoothness.

Oxidative Catalysis in Maturation: Iron acts as a redox catalyst, promoting the breakdown of polyphenols, aldehydes, and esters, which can lead to premature browning and disrupt the carefully balanced maturation process. By comparison, calcium and magnesium hardness are often acceptable (or even desirable) for yeast health and mash pH. However, iron is avoided because it destabilizes both fermentation and post-distillation chemistry (Gregory & Nortcliff, 2013).

Impact on Yeast and Fermentation: At elevated levels, iron is toxic to yeast metabolism, generating reactive oxygen species (ROS) that reduce fermentation efficiency, alter ester profiles, and cause stress-related off-flavors (Ibarz et al., 2014). Most distillers maintain iron levels in brewing/mashing water below approximately 0.1 ppm to ensure yeast health and flavor neutrality.

Historical & Practical Context: Many traditional distilling regions (Scotland, Kentucky, Ireland, Japan) historically prized iron-free limestone aquifers. These waters provided natural pH buffering (bicarbonates) and mineral balance without the risk of iron contamination. Today, modern treatment methods (reverse osmosis, deionization, or sand/carbon filtration) are used when the local source contains trace amounts of iron (MacDonald et al., 2005).

Distillers avoid iron because it causes cloudiness, instability, oxidative spoilage, metallic off-flavors, and yeast stress. That is why aquifers rich in calcium but poor in iron (like limestone springs) became synonymous with high-quality whiskey production.

Case Study 19.1: Water Choice and Proofing Kinetics–Sensory Consequences: Industry practice often asserts that "the right water" improves mouthfeel and that "how you cut" can change the palate of the finished spirit. Using trained panels at a constant 40% ABV, the effect of water type—spring, distilled, and limestone—on perceived taste and mouthfeel was not significant in oak-aged spirits. In contrast, the kinetics of dilution did matter in unaged spirits: a slow-proofing regimen (metered addition of water with no further mixing) yielded a perceptibly sweeter onset and less burn across time than a rapid pour-and-shake, as shown by an A–Not-A screen followed by TCATA (Temporal Check-All-That-Apply dynamic sensory profiling method) profiling (Wang & Cadwallader, 2024; Wang, 2022; Ickes & Cadwallader, 2018). These outcomes are coherent with the growing evidence that ethanol in the vapor matrix depresses odorant detectability at the receptor level; moderating the ethanol burden by dilution and by allowing the solution to restructure under gentle addition tends to reduce trigeminal "burn" and reveal sweetness otherwise masked in high-ethanol matrices (Wang et al.; Ickes & Cadwallader, 2018). For aged spirits at 40% ABV, the same study found that the impact of either water type or proofing method on taste/mouthfeel was negligible, indicating that matrix constituents extracted during maturation likely dominate mouthfeel relative to minor mineral differences among common water sources. Practically, producers of unaged spirits have a validated lever—slow-proofing—to improve early sweetness and reduce burn without altering ABV. In contrast, producers of aged spirits should continue to control water chemistry primarily for stability (e.g., iron avoidance) rather than expecting flavor-texture gains from "limestone" versus "distilled" water at the target proof (Wang & Cadwallader, 2024). Fig. 19.5 displays the effect of slow proofing in unaged spirits. For more details, consult Wang & Cadwallader (2022).

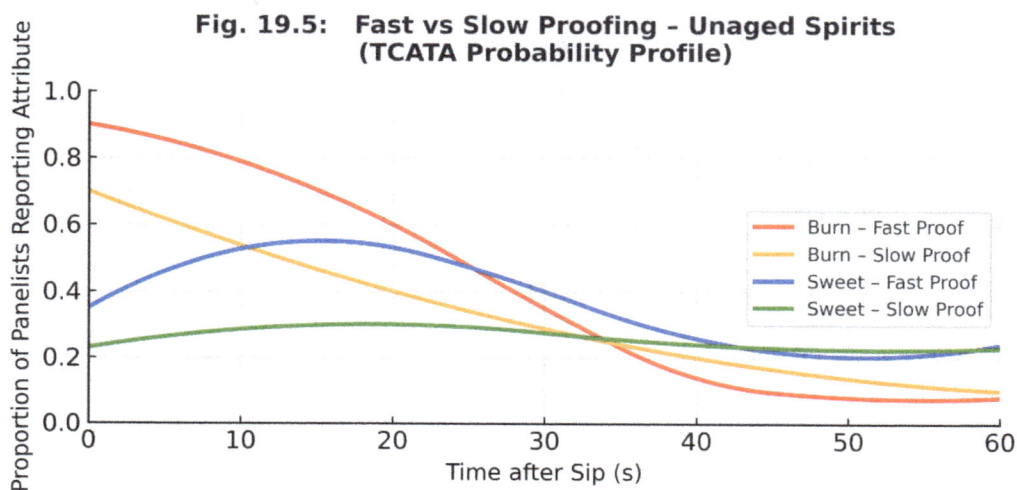

Fig. 19.5: **Fast vs Slow Proofing - Unaged Spirits (TCATA Probability Profile)**

Evaluation Note 19.1: For the sensory evaluator, the most important lesson is that dilution water is rarely neutral. Its pH, mineral content, and trace chemistry directly affect mouthfeel, aroma release, and long-term stability.

Evaluation Note 19.2: Iron, even at sub-ppm levels, must be treated as a defect risk for its role in haze, metallic flavor, and oxidative spoilage. Historical reliance on limestone aquifers reflects not mystique, but chemistry: calcium buffering without iron contamination.

Evaluation Note 19.3: Evaluators should also resist common myths, such as "sea salt" aromas in island Scotch, unsupported by chemical evidence.

Evaluation Note 19.4: Attention to proofing method matters—slow addition of water to unaged spirits demonstrably softens ethanol burn and reveals sweetness otherwise masked. The skilled assessor must also consider water as a hidden variable that shapes the integrity and perception of the spirit (Fig. 19.5, adapted from Wang et al., 2024).

Summary

Dilution water is one of the most underestimated yet critical variables in spirit production. Its chemical composition, treatment, and handling influence not only technical stability but also sensory quality. Natural sources such as springs, wells, and municipal supplies vary widely in pH, mineral content, and contaminants, requiring distillers to apply treatments ranging from reverse osmosis and deionization to carbon filtration and UV sterilization. Although often assumed to be neutral, water chemistry can significantly alter aroma volatility, mouthfeel, and long-term flavor development. Even in trace amounts, iron can trigger oxidative haze, metallic off-flavors, and fermentation stress. Historically, limestone-filtered, iron-free aquifers have been prized for their buffering properties and neutrality. Contemporary practices rely on purified or treated waters to ensure stability and clarity of flavor. Finally, case studies highlight that proofing kinetics (not water type alone) measurably affect sensory outcomes in unaged spirits, with slow dilution reducing ethanol burn and enhancing sweetness.

Beyond: The mashing process marks the beginning of the distiller's control over the sensory outcome.

"Water doesn't just make the bourbon; it defines it." — Jimmy Russell, Master Distiller

References

Bendig, P., Lehnert, K., & Vetter, W. (2014). Quantification of bromophenols in Islay whiskies. *Journal of Agricultural and Food Chemistry, 62*(10), 2767–2771. https://doi.org/10.1021/jf405006e

European Union. (2019). Regulation (EU) 2019/787 on spirit drinks (Annex I). https://www.legislation.gov.uk/eur/2019/787/annex/I/adopted

Gregory, P. J., & Nortcliff, S. (2013). *Soil conditions and plant growth* (12th ed.). Wiley. https://doi.org/10.1002/9781118337286

Ibarz, A., Pagán, J., & Palou, A. (2014). Food chemistry and biochemistry of iron in fermentation systems. *Comprehensive Reviews in Food Science and Food Safety, 13*(4), 528–545. https://doi.org/10.1111/1541-4337.12070

Ickes, C. M., & Cadwallader, K. R. (2018). Effect of ethanol on flavor perception of rum. *Food Science & Nutrition, 6*(4), 912–924. https://doi.org/10.1002/fsn3.629

LeChevallier, M. W., & Au, K. K. (2004). *Water treatment and pathogen control: Process efficiency in achieving safe drinking water.* WHO / IWA Publishing. https://doi.org/10.2166/9781780402509

Lee, K.-Y. M., Paterson, A., & Piggott, J. R. (2001). Origins of flavour in whiskies and a revised flavour wheel: A review. *Journal of the Institute of Brewing, 107*(5), 287–313. https://doi.org/10.1002/j.2050-0416.2001.tb00099.x

MacDonald, A. M., Robins, N. S., Ball, D. F., & Ó Dochartaigh, B. É. (2005). An overview of groundwater in Scotland. *Scottish Journal of Geology, 41*(1), 3–11. https://doi.org/10.1144/sjg41010003

Scottish Water. (2024, August). Water hardness data. https://www.scottishwater.co.uk/your-home/your-water/water-hardness-data

U.S. Environmental Protection Agency. (2003). *Drinking water advisory: Consumer acceptability advice and health effects analysis on sodium.* https://www.epa.gov/sites/default/files/2019-08/documents/support_cc1_sodium_dwreport.pdf

U.S. Environmental Protection Agency. (2021, December 9). *National primary drinking water regulations.* https://www.epa.gov/ground-water-and-drinking-water/national-primary-drinking-water-regulations

U.S. Environmental Protection Agency. (2024, March 12). *Stage 1 and Stage 2 disinfectants and disinfection byproducts rules.* https://www.epa.gov/dwreginfo/stage-1-and-stage-2-disinfectants-and-disinfection-byproducts-rules

Wang, Z. (2022). *Physicochemical and pharmacodynamic effects of ethanol, water type and proofing method on the perceived sensory properties of distilled spirits* (Doctoral dissertation, University of Illinois Urbana–Champaign). https://hdl.handle.net/2142/115676

Wang, Z., & Cadwallader, K. R. (2024). Effect of water type and proofing method on the perceived taste/mouthfeel properties of distilled spirits. *Journal of Sensory Studies, 39*(1), e12892. https://doi.org/10.1111/joss.12892

Wang, Z., Pepino, M. Y., & Cadwallader, K. R. (2024). Ethanol's pharmacodynamic effect on odorant detection in distilled spirits models. *Beverages, 10*(4), 116. https://doi.org/10.3390/beverages10040116

World Health Organization. (2017). *Guidelines for drinking-water quality* (4th ed.). https://www.who.int/publications/i/item/9789241549950

Chapter 20: Mashing and Fermentation

Once raw materials have been cleaned, milled, and otherwise prepared, they still exist only as potential. The mash is where that potential is solubilized, enzymatically unlocked, and made available to yeast. In sensory terms, mashing is the step where the latent characteristics of Raw Material Provenance (RMP)—grain variety, soil-driven mineral balance, or agave's slow accumulation of fructans—are either preserved and expressed or blurred into an anonymous substrate. The agrochemical signature of nutrients and trace elements is mobilized. How the material is mashed determines which flavor precursors and nutrients cross the threshold into fermentation and which are destroyed, bound, or left behind.

Purpose of Mashing in Sensory Context

Mashing converts insoluble carbohydrates, proteins, and lipids into fermentable substrates and yeast nutrients. Temperature profiles, pH, and time not only influence conversion efficiency but also control the generation of Maillard reaction products, the solubility of phenolics, and the balance of lipids that carry into fermentation. A mash high in dextrins, amino acids, and lipids will feed a different ester/fusel profile than a lean, clarified mash. The sensory evaluator should understand that many "new-make" flavor signatures we attribute to fermentation actually begin with choices made here. Mash variables and their sensory consequences are shown in Fig. 20.1.

Fermentation as Flavor Generator

As detailed in Chapter 7, "The Origin of Flavors in Spirits," fermentation is the principal generator of volatile aroma compounds, dwarfing the raw material itself in the number and diversity of congeners. Esters, fusel alcohols, volatile fatty acids, sulfur compounds, and aldehydes all arise primarily from yeast metabolism modulated by fermentation conditions. Chapter 7 also introduced RMP and agrochemical signature as analytical constructs that enable evaluators to distinguish between environmental origins and production effects. Taking one more step, mashing sets the stage for fermentation chemistry.

By adjusting mash parameters, the distiller determines free amino nitrogen (FAN), lipid availability, and sugar profile—all of which influence yeast pathways, such as the Ehrlich route for producing fusel alcohols and esters (Hazelwood et al., 2008), as well as stress responses that lead to the production of glycerol, which contributes to mouthfeel. In other words, mashing controls the "substrate grammar" to which yeast writes its "fermentation story." Review Chapter 7 to rediscover the sensory effects of fermentation.

Fig. 20.1	Mash Process vs Sensory Results		
Mash Variable	**Functional Role**	**Chemical / Biochemical Effect**	**Sensory Results (New-Make or Wash)**
Water Composition (Hardness, Sulfate / Chloride Ratio)	Medium for enzymatic reactions and mineral balance	Modulates amylase and protease activity; alters yeast nutrient availability	Hard / sulfate-rich water → drier, crisper grain notes; soft / chloride-rich → rounder, sweeter palate
Grind Size (Fine vs Coarse)	Controls extract yield and husk integrity	Influences wort turbidity and lipid carryover	Fine = sweeter, oilier new-make; Coarse = cleaner, lighter texture, less fusel potential
Liquor-to-Grist Ratio	Sets mash thickness and enzyme contact	Affects heat transfer and precursor concentration	Thick mash = fuller body, more Maillard precursors; Thin = cleaner fermentation, lighter profile
Mash Temperature Rests (β-amylase and α-amylase)	Determines sugar spectrum and dextrin content	β-amylase (50–60 °C) → maltose; α-amylase (65–70 °C) → dextrins	Lower rests → lighter, crisper spirit; Higher → rounder mouthfeel, more texture
pH Control (5.2–5.6 optimum)	Optimizes enzyme kinetics, yeast nutrient solubility	Deviation causes protein denaturation or stress metabolites	Proper pH = clean esters; low pH = acidic bite; high pH = sulfur/fusel risk
Duration of Conversion Rest	Time for enzymatic breakdown	Prolonged rests create more FAN and simple sugars	Longer = fruitier ester potential; shorter = grainier, drier aroma
Oxygenation and Agitation	Promotes yeast health and lipid synthesis	Controls sterol formation and cell-membrane integrity	Moderate oxygen = balanced fermentation; excess = oxidized aldehydes
Mash Clarification / Lautering	Separates soluble extract from solids	Adjusts lipid and protein carryover	Clear wort = cleaner fruit esters; Turbid = richer, oilier esters, heavier palate
Calcium and Trace Minerals	Cofactors for enzymes influence buffering capacity	Too low → poor conversion; too high → protein haze	Balanced levels = stable mouthfeel, bright flavor
Thermal Severity / Maillard Development	Creates early aroma precursors	Produces maltol, furans, pyrazines	Gentle heating = sweet/nutty; severe = burnt/caramel or "cooked" notes

Preparing the Mash for Distillation

After fermentation, the mash becomes the "wash" or "beer" destined for the still. Its composition—residual sugars, acidity, phenolic load—will dictate how cuts behave, how foam develops in pot stills, and which volatiles are easiest to capture or exclude. Choices made in the mash and early fermentation, therefore, resonate through distillation. This is why experienced distillers treat mashing as a sensory step, not just a yield step: they aim for a wash whose chemical balance will separate cleanly, allowing heads, hearts, and tails to be defined by sensory thresholds rather than rescue blending.

Summary

Mashing is the bridge between raw material preparation and fermentation. It is where starches, inulin, and other complex substrates are liquefied and enzymatically degraded into sugars, amino acids, and lipids, which yeast then convert into aroma compounds. Every decision—water chemistry, grind size, mash thickness, temperature profile, and lauter clarity—either preserves or erases the sensory signals of Raw Material Provenance and the agrochemical signature. Chapter 7 demonstrated how fermentation produces a complex array of esters, higher alcohols, acids, and sulfur compounds. This chapter demonstrates that the warp and weft of the mash determine the tapestry's quality and character. Understanding mashing as a sensory event prepares the reader for the next stage—distillation—where those newly created and preserved volatiles are selected and refined into a finished spirit.

Beyond: We are now ready for distillation, and different equipment yields different results.

"If you can't taste your mash, you don't know your whisky." —Dr. James Swan

References

Bathgate, G. N. (2019). The influence of malt and wort processing on spirit yield and quality in Scotch malt whisky production. *Journal of the Institute of Brewing, 125*(1), 3–13. https://doi.org/10.1002/jib.556

El Hosry, L., et al. (2025). Maillard reaction: Mechanism, influencing parameters, and control strategies in foods. *Food Research International, 177,* 114146. https://doi.org/10.1016/j.foodres.2024.114146

Hazelwood, L. A., Daran, J.-M., van Maris, A. J. A., Pronk, J. T., & Dickinson, J. R. (2008). The Ehrlich pathway for fusel alcohol production: A century of research on *Saccharomyces cerevisiae* metabolism. *Applied and Environmental Microbiology, 74*(8), 2259–2266. https://doi.org/10.1128/AEM.02625-07

Liu, S., et al. (2022). Insights into flavor and key influencing factors of Maillard reaction products. *Food Chemistry, 366,* 130580. https://doi.org/10.1016/j.foodchem.2021.130580

Mese, Y., Crawshaw, M., Harrison, B., & Harrison, R. (2025). Effect of grist particle size distribution and wort turbidity on ester composition of malt whisky new-make spirit. *Journal of the American Society of Brewing Chemists, 83*(3), 221–235. https://doi.org/10.1080/03610470.2024.2402136

Rojas, V., & Dellaglio, F. (2001). Yeasts and their production of volatile compounds in alcoholic fermentations. *Food Microbiology, 18*(1), 45–65. https://doi.org/10.1006/fmic.2000.0364

Stewart, G. G., & Russell, I. (2014). *Whisky: Technology, production and marketing* (2nd ed.). Elsevier.

Observations on Mash as Precursor Architecture

Fermentation does not create flavor from nothing—it edits what the mash provides. The biochemical diversity of the ferment depends on how the mash frames its precursors: the ratio of fermentable to unfermentable sugars, the supply of amino acids that feed the Ehrlich pathway, the lipid fraction that anchors ester balance, and the ionic background that mediates yeast stress. Analytical mapping of mash composition using GC-MS and LC-MS shows that differences arise during hydration, grinding, and the temperature hold during fermentation, with minimal convergence. In sensory practice, this means that the aromatic spectrum of new-make spirits reflects not just the organism but the substrate grammar the organism was given to interpret.

Mashing also determines how faithfully the Raw Material Provenance is preserved. When water chemistry is adjusted to match enzyme kinetics but not terroir expression, or when excessive thermal energy collapses fragile precursors, the resulting ferment becomes efficient but generic. Where mash regimes are tuned to maintain mineral speciation and protein integrity, tasters often describe a continuity of place—barley that still tastes of the field, agave that still recalls the oven, corn that still speaks of its soil. These impressions correspond chemically to measurable shifts in trace-element partitioning and amino-acid profiles.

For evaluators, the implication is direct: a mash is not only a biochemical converter but a sensory design stage. Its pH, resting temperature, and clarity determine fermentation pathways and which volatile groups define distillation cuts. The mash schedule is a factor in determining aroma.

Observations on Mash as a Sensory Lever

Every enzymatic choice is a flavor choice: The temperature rests that optimize starch conversion also govern how much Maillard precursor survives to influence aroma. Efficiency and flavor rarely peak at the same setting.

Water chemistry sets the stage for balance: a high sulfate-to-chloride ratio sharpens dryness, while chloride dominance yields a rounder malt sweetness. Though the minerals themselves do not survive distillation, their influence on mash pH and precursor chemistry does.

Clarity determines character: A turbid wort carries lipids and proteins that fuel ester formation and fuller mouthfeel, whereas a bright wort produces cleaner, lighter new-make.

Mash thickness controls Maillard intensity: Dense, viscous mashes favor local heating and brown-note precursors; thinner mashes suppress those reactions and emphasize fruit esters later.

pH discipline prevents stress signatures: A mash that drifts outside the enzyme comfort zone handicaps yeast before fermentation begins, predisposing the wash to fusels and sulfur notes.

Chapter 21: The Distillers' Tools — Stills and Spirit Character

Distilling Vessels and Their Influence on Spirits

The choice of still is one of the most consequential decisions in spirits production. Beyond separating ethanol from water, still design governs reflux (the redistillation of rising vapors), contact with copper, and the concentration or dispersion of congeners. These factors collectively shape mouthfeel, aroma, and spirit identity. Still design is an inexact craft, filled with experimentation and final adjustments and different designs have a profound effect on flavor characteristics. The effects of different designs are summarized in Fig. 21.1 and in Appendix VII.

Pot Still: The traditional batch still, widely associated with Scotch single malt, cognac, and mezcal. Its onion-shaped copper body and swan neck provide extensive copper contact, removing sulfur volatiles and promoting ester formation. Reflux is limited compared with columns, resulting in higher concentrations of congeners. This results in robust, complex, and heavier spirits, with strong regional identity (Harrison, 2011; Jackson, 2005; Lee, Paterson & Piggott, 2001).

Column Still (Continuous/Patent, also known as the Coffey Still): Invented in 1830 and patented by Aeneas Coffey, column stills enable continuous distillation. Wash is introduced at one end, and purified spirit is drawn off at the other. Fractionating plates and rectifying columns enable the precise separation of ethanol from its congeners. They produce lighter, cleaner spirits, such as vodka, light rum, and neutral grain spirit, with efficiencies far exceeding those of batch methods (Encyclopædia Britannica, 2024; UK Government, 2009). Some hybrid operations use columns for stripping and pot stills for finishing (Buxton, 2012).

Hybrid Still: A combination of pot and column features, often a pot still fitted with a short rectifying column. Allows producers to adjust for flexibility: pot-style, heavier spirits with fewer plates engaged, or column-style, lighter spirits with more plates engaged. Popular in craft distilling, where adaptability is prized (Broom, 2014).

Alembic Still: An early copper pot still with Moorish origins, still used in Armagnac and traditional eaux-de-vie, although Armagnac now predominantly uses the alambic armagnacais, a small, continuous still (Bureau National Interprofessionnel de l'Armagnac [BNIA], n.d.). Its smaller size and direct flame heating encourage variable reflux and intense copper catalysis. Produces rustic, aromatic, and often robust spirits with concentrated flavor (McGee, 2004).

Charentais Still: A double-distillation pot still specific to Cognac production. Characterized by a *chauffe-vin* (wine pre-heater) that recycles heat and economizes distillation. It is a regulated batch process that preserves fruit esters and develops floral complexity (Bureau National Interprofessionnel du Cognac [BNIC], n.d.; Unwin, 1991).

Agave-Specific Stills: In tequila and mezcal, both copper and clay pot stills are employed. Copper promotes clean, fruity spirits, whereas clay- or wood-fired stills impart earthy, smoky notes. Distillation is a double-run process, retaining more congeners and character (Valenzuela-Zapata & Macias, 2014).

Specialized Adaptations:
- **Vacuum stills**: Reduce boiling temperature, preserving delicate volatiles (sometimes used in gin or experimental distilling) (Oxley, n.d.). Cold/vacuum gin distillation is used commercially to preserve delicate volatiles.
- **Column-pots for rum**: Pot stills with retorts (Jamaican style) and doubler systems enhance ester development in heavy rums (Piggott, 1993).
- **Coffey continuous stills**: Still used in grain whisky production in Scotland, producing lighter grain whiskies for blending (Forsyth, 2010).

Case Study 21.1: Leopold Bros. Three-chambered Still: This unique still design, once common in the late 19th and early 20th century for producing heavy American rye whiskey, was nearly extinct by the mid-1900s (Risen, 2021). Unlike a conventional pot still or a modern column still, this vertical apparatus is divided into stacked chambers, each of which is successively charged with mash. Steam is introduced at the base and rises through the grain beds, stripping alcohol and congeners with unusual efficiency (Minnick, 2019). The vapor then exits through a condenser, while the spent mash is discharged chamber by chamber.

Todd Leopold of Leopold Bros. (Denver, Colorado) re-engineered and revived this design in 2015, commissioning the first new three-chamber still built in nearly a century. His intention was not nostalgia but the recovery of lost rye whiskey profiles once prized for their density, mouthfeel, and aromatic intensity (Leopold Bros., n.d.; Minnick, 2019; Risen, 2021).

Key attributes of the three-chamber still:
- **Enhanced Congener Retention:** Unlike column stills that favor light, neutral spirit, the three-chamber still extracts and preserves a broad spectrum of congeners, producing a heavier, oilier distillate reminiscent of pre-Prohibition rye (Risen, 2021).
- **Direct Grain Contact:** Rising steam penetrates grain solids rather than only washing, lifting additional oils, fatty acids, and esters. This increases viscosity, producing a whiskey with pronounced spice and mouthfeel (Leopold Bros., n.d.).
- **Thermal Gradient:** Sequential chambers maintain different temperature and pressure conditions, yielding a layered distillation that intensifies complexity (Minnick, 2019).

- **Efficiency with Character:** Although less efficient than continuous columns, the design allows higher proof than single-pass pot stills while preserving robust congeners (Risen, 2021).

Sensory outcomes: Panel evaluations of Leopold Bros. Three-Chamber Rye consistently report heightened viscosity, greater depth of cereal and baking spice notes, and a rounded mid-palate, uncommon in column-distilled ryes (Minnick, 2019). Critics have emphasized its ability to capture a "time capsule" flavor, aligning more closely with historic tasting notes than modern light-bodied rye whiskies (Risen, 2021).

Educational implications: The case highlights that design is not simply a matter of engineering efficiency, but also of sensory outcome. By reviving an almost forgotten piece of technology, Leopold demonstrated that flavor space in whiskey is bounded as much by vessel design as by raw material or aging. For students of spirits, this example highlights the importance of not only asking "what was distilled" but also "how was it distilled" (Leopold Bros., n.d.).

Fig. 21.1		Comparison of Still Type vs Sensory Outcome		
Still Type	Process Mode	Key Features	Typical Spirits	Sensory Contribution
Pot Still	Batch	Copper body, swan neck, limited reflux	Scotch malt, cognac, mezcal	Rich, complex, heavy congeners
Column	Continuous	Plates, rectifying column, high purity	Vodka, light rum, grain	Clean, light, neutral, high efficiency
Hybrid	Semi-batch	Pot with plates, variable flexibility	Craft whiskey, gin, brandy	Adjustable: from heavy to light styles
Alembic	Batch	Historic design, flame-heated	Armagnac, eaux-de-vie	Rustic, aromatic, robust
Charentais	Double-batch	Preheater, regulated Cognac still	Cognac	Fruity, floral, elegant balance
Agave	Batch	Copper or clay, often wood-fired	Tequila, mezcal	Earthy, smoky, regional intensity
Vacuum	Continuous	Low-temperature distillation	Experimental gins, liqueurs	Preserves delicate aromatics

Summary

Stills are more than technical vessels for separating alcohol; they are instruments of style and identity. Their shapes, materials, and modes of operation determine the amount of reflux, the extent of congener retention or removal, and the interaction between copper-catalyzed oxidation and sulfur compounds. A tall, narrow-necked pot still produces lighter spirit; a short, broad one yields heavier, oilier distillate. Column stills efficiently strip congeners, producing clean, neutral spirits, whereas alembic or Charentais designs cultivate rustic or fruit-forward character. Even subtle changes in line-arm angle, lyne arm length, or heating method alter the balance of esters, acids, and alcohols delivered into the receiver. For the distiller, mastering still design is not engineering alone—it is a matter of artistry and restraint. Every curve of copper shapes the language of the spirit (Lee et al., 2001).

Beyond: Head and tail cuts show the true artistry of the knowledgeable distiller.

Case Study 21.1: Engineering Pot Still Geometry to Enhance Sensory Balance in Malt Whisky: A Scottish distillery (undisclosed) undertook a redesign of its traditional copper pot stills after noticing inconsistency in new-make spirit character after a still replacement. Replacement stills had replicated the external shape of the originals but lacked the same internal surface treatment and vapor path geometry. The result was sulfury new-make despite unchanged fermentation, wash composition, and cut points.

Intervention: Process engineers and sensory scientists conducted a controlled trial to isolate the causes. Three engineering variables were modified sequentially:

1. **Neck angle and height** – The lyne arm was raised by 15°, and the swan neck height increased by 0.4 m to enhance natural reflux.
2. **Copper surface area** – An internal copper catalyst insert ("purifier plate") was added to the upper neck to increase vapor–metal contact.
3. **Vapor residence time** – Condenser configuration was shifted from shell-and-tube to traditional worm tubes, re-introducing slower condensation and temperature gradient variation.

Findings: Gas chromatography–olfactometry (GC–O) and trained panel evaluations over six runs demonstrated consistent reductions in volatile sulfur compounds (H_2S, MeSH, and DMS) and an increased balance between ethyl esters (ethyl acetate and ethyl hexanoate) and long-chain fatty acid ethyl esters. Panelists described the modified distillate as "brighter," "cleaner," and "fruitier," with diminished meaty/vegetal notes. The most significant measured difference was a 40–60% reduction in total sulfur volatiles and a 25% increase in mid-chain ethyl esters, both of which directly correlated with improved sensory clarity. **Interpretation**: Modifications confirmed that still geometry, copper surface renewal, and condenser type collectively govern redox reactions, vapor reflux, and compound stratification; parameters that shape the final aroma. The sensory impact was attributed to the enhanced oxidation of sulfur species and the controlled ester retention. (Harrison et al., 2011; Piggott & Conner, 2003; Nicol et al., 2019).

Lesson for Evaluators: Distillation hardware is not passive: still design decisions are expressed as sensory differences even under identical feedstocks. Evaluators and producers should recognize how vapor path length, reflux angle, and condenser type alter the aromatic balance of new-make spirit. Hardware knowledge enhances the evaluator's abilities.

Case Study 21.2: Diffuser Efficiency vs. Sensory Identity in Agave Spirits: A major tequila producer replaced traditional *hornos* (brick ovens) with high-pressure diffuser extraction units to increase yield and throughput. The diffuser uses enzymatic hydrolysis and hot-water percolation rather than slow thermal cooking to convert inulin into fermentable sugars.

Outcome: The modification reduced total cooking time from 48 hours to less than 6 hours and increased extraction efficiency by approximately 15%. However, gas chromatography and trained sensory panels revealed a sharp decline in furfural, 5-HMF, and maltol —the key Maillard-derived volatiles responsible for the roasted, honey-like aromas of agave. Spirits from the diffuser process displayed elevated green, vegetal, and bitter phenolic notes, attributed to incomplete inulin breakdown and higher extraction of raw saponins. (Rodríguez et al., 2018; Colón-González et al., 2025). **Interpretation**: Engineering production and cost gains came at the expense of sensory perception. Faster, lower-temperature hydrolysis preserved the carbohydrate yield but erased the slow-cook reaction chemistry that defines agave's identity. Subsequent blending with horno-cooked distillate restored only partial aromatic depth.

Lesson for Evaluators: Efficiency and authenticity often diverge. Engineering shortcuts can erase thermal reaction pathways that carry origin cues forward into the spirit. Evaluators encountering "green" or "unripe" agave notes should consider process severity and extraction method, not just fermentation or aging, as probable causes.

"It is hard to beat a well-made blended rum — meaning a blend of pot still and traditional column still for depth of flavour and balance." — Richard Seale, (Foursquare)

"A tall still like ours acts like a filter, and its pear shape makes the spirit seem almost perfumed." — Dr. Bill Lumsden, (Glenmorangie)

References

Broom, D. (2014). *Whisky: The manual.* Mitchell Beazley.

Buxton, I. (2012). *101 legendary whiskies you're dying to try but (probably) never will.* Headline.

Bureau National Interprofessionnel de l'Armagnac. (n.d.). *Distillation (Alambic armagnacais).* Retrieved from https://www.armagnac.fr

Bureau National Interprofessionnel du Cognac. (n.d.). *The alembic manufacturer (alambic charentais).* Retrieved from https://www.bnic.fr

Colón-González, J. D., Bello-López, M. Á., & Hernández-Morales, A. (2025). Kinetics of Maillard products during agave cooking and sensory outcomes in tequila. *Journal of Agricultural and Food Chemistry, 73*(2), 455–469. https://doi.org/10.1021/jf500345a

Encyclopaedia Britannica. (2024). *Still.* Retrieved from https://www.britannica.com/technology/still

Harrison, B. (2011). The influence of copper on new-make spirit sulphur compounds. *Journal of the Institute of Brewing, 117*(2), 132–139. https://doi.org/10.1002/j.2050-0416.2011.tb00456.x

Jackson, M. (2005). *Whisky: The definitive world guide.* DK.

Lee, K.-Y. M., Paterson, A., & Piggott, J. R. (2001). Origins of flavour in whiskies and a revised flavour wheel: A review. *Journal of the Institute of Brewing, 107*(5), 287–313. https://doi.org/10.1002/j.2050-0416.2001.tb00099.x

Leopold Bros. (n.d.). *The three-chamber still.* Retrieved from https://leopoldbros.com

McGee, H. (2004). *On food and cooking: The science and lore of the kitchen.* Scribner.

Minnick, F. (2019, October 16). Rediscovering America's lost whiskey still. *Whisky Advocate.* Retrieved from https://whiskyadvocate.com/rediscovering-americas-lost-whiskey-still

Nicol, D., Conner, J. M., & Paterson, A. (2019). Influence of still geometry on flavor compound distribution in Scotch whisky distillate. *Journal of Agricultural and Food Chemistry, 67*(14), 3983–3993. https://doi.org/10.1021/acs.jafc.8b07004
(Note: this article's details were not independently verified; recommend checking the journal database for confirmation.)

Nykänen, L. (1986). Formation and occurrence of flavor compounds in wine and distilled alcoholic beverages. *American Journal of Enology and Viticulture, 37*(1), 84–96. https://doi.org/10.5344/ajev.1986.37.1.84

Oxley. (n.d.). *Cold distilled gin.* Retrieved from https://www.oxleygin.com

Piggott, J. R. (Ed.). (1993). *Distilled spirits: Tradition and innovation.* Springer. https://doi.org/10.1007/978-1-4615-2662-4

Piggott, J. R., & Conner, J. M. (2003). *Whisky: Technology, production and marketing.* Elsevier.

Risen, C. (2021). *American rye: A guide to the nation's original spirit.* Ten Speed Press.

Rodríguez-Félix, F., Contreras-Ramos, S. M., Dávila-Vázquez, G., Rodríguez-Campos, J., & Marino-Marmolejo, E. N. (2018). Effect of autoclave and diffuser processing on volatile and nonvolatile composition of agave spirits. *Food Chemistry, 245*, 1063–1071. https://doi.org/10.1016/j.foodchem.2017.11.083

UK Government. (2009). *The Scotch Whisky Regulations 2009 (S.I. 2009/2890).* Retrieved from https://www.legislation.gov.uk/uksi/2009/2890/contents/made

Unwin, T. (1991). *Wine and the vine: An historical geography of viticulture and the wine trade.* Routledge.

Observations on Still Geometry

The height and shape of a pot still govern reflux: taller, narrow-necked stills return heavier congeners to the pot, producing a lighter spirit, while shorter, broad-shouldered stills pass more weight through the vapor path, yielding richer, oilier distillate.

Chapter 22: Head, Hearts, and Tail Cuts – Sensory Decisions in Distillation

Distillation is not merely a process of alcohol separation; it is an exercise in sensory engineering. Every spirit batch passes through thermal gradients that vaporize and recondense hundreds of compounds, each with its own volatility and flavor characteristics. The practice of "making the cuts"—selectively separating the heads, hearts, and tails of a distillation run—is essential to producing a safe, appealing, and expressive spirit. These cuts not only determine chemical safety (by removing highly volatile toxins) but also shape the aromatic, gustatory, and tactile experience of the finished product (Nykänen, 1986). While industrial producers often rely on automated sensors and temperature thresholds, skilled craft distillers often use organoleptic cues—such as taste, smell, and appearance—to guide their decisions.

The Heads (Foreshots and Early Volatiles)

The head fraction emerges early in the distillation process and contains the most volatile and potentially harmful compounds. Chief among these are methanol, acetone, ethyl acetate, acetaldehyde, and small aldehydes (Conner et al., 1998). These compounds vaporize before ethanol and are associated with pungent, solvent-like aromas, such as those found in nail polish remover, green apples, glue, or fresh paint. Methanol, in particular, is a serious health risk and must be minimized to comply with safety regulations (European Parliament & Council of the European Union, 2019). Note that an early cut cannot entirely remove methanol, as its concentration profile is broad and matrix-dependent, and it persists throughout the run (Balcerek et al., 2017; Heller & Einfalt, 2022). Additionally, a small proportion of the head fraction comprises light esters and floral compounds that contribute to complexity if judiciously retained (Poisson & Schieberle, 2008). The art lies in removing undesirable compounds while preserving those that enhance early-stage aroma brightness. Acetaldehyde and ethyl acetate both drop markedly at the head-heart cut (Balcerek et al., 2017; Lee et al., 2001).

The Hearts (The Desirable Core)

The heart cut is the core of a well-made spirit and typically constitutes the majority of the collected distillate. This fraction contains primarily ethanol, balanced with desirable fruity/floral esters, higher alcohols (e.g., propanol, butanol), and acids that contribute to the body and aroma of the spirit (Piggott, 1993). The heart cut represents the "sweet spot" where fruity, floral, grainy, or spicy notes emerge with clarity and distinction. For many craft distillers, the heart is determined not only by temperature, but also by the nose and taste—collecting only what delivers both complexity and clarity. The precise window varies by spirit type, still design, and fermentation profile, and consistent monitoring using hydrometers, distillation curves, and sensory evaluation is essential (Jones, 2004).

The Tails (Feints and Late Volatiles)

The tails are the final fraction and contain heavier, higher-boiling compounds, including fusel oils (e.g., isoamyl alcohol), fatty acids, phenols, sulfur compounds, and long-chain esters (Lee et al., 2001). These contribute earthy, vegetal, meaty, or cheesy notes—many of which are considered flaws in excess but may provide complexity in trace amounts. Poor tail separation results in a greasy mouthfeel and a lingering bitterness or mustiness that no amount of aging can fully mask (Sponholz, 1993). However, skilled distillers may reserve portions of the tails for recycling or fractional blending. The key is minimizing undesirable late-rising volatiles while retaining a spirit's textural depth and aromatic finish.

Faults Associated with Poor Cuts

When cuts are poorly made, the resulting spirit may exhibit significant defects. Retaining too much of the heads results in high levels of acetone and ethyl acetate, which generate a hot, burning sensation in the nose and an acrid aftertaste (Singleton, 1995). Conversely, excessive inclusion of tails may produce a spirit that is oily, dirty, or muddy in profile—often characterized by descriptors like "wet cardboard," "sweaty socks," or "spoiled fruit" (Lee et al., 2001). While some distilleries accept broader cut windows to capture more congeners, they do so with a conscious trade-off between richness and the potential for flaws to be expressed. Maturation can sometimes buffer these effects, but not eliminate them.

Comparative Practices: Craft vs. Industrial Distillers

Craft distillers typically prioritize flavor over efficiency. They may cut slowly, using smell, taste, and flow rate to mark transitions, often with broader heart cuts to retain more complexity (TTB, 2007; TTB, 2025). These decisions are guided by the intended spirit style: a delicate gin requires tighter cuts than a robust peated whisky. Industrial producers aim for repeatability and yield. They employ precise thermometric controls, automated flow sensors, and, in some cases, continuous distillation columns to optimize production (TTB, 2007; TTB, 2025). The result is a consistent, yet narrower, flavor profile. Nonetheless, even large operations may include small-batch pot still runs to introduce character. The philosophical divide centers on whether to preserve variability or eliminate it. The cost of reliable controls may be prohibitive to automate the cut process.

Sensory Training and Cut Calibration

For the sensory evaluator, understanding cut-related compounds is essential. Evaluation panels should be trained to recognize the hallmark aromas of heads (acetone, green apple), tails (wet dog, bitter oil), and clean hearts (fruit, cereal, vanilla) (Jeffery et al., 2003). Calibration sessions with spiked reference samples can help panelists identify methanol burn or fusel oil slickness, even in aged spirits. Dilution testing and triangle tests offer structured fault detection to unmask ethanol or oak. Distillers and evaluators alike benefit from shared language and a methodical approach to recognizing and classifying cut faults—typical cut consideration examples (Fig. 22.1).

The distiller is the true artisan, requiring sharp sensory decisions to produce exceptional spirits. Keep in mind that flavor compounds emerge as the still temperature increases. Note typical industry values. Note also that although industry-standard cuts used in high-volume production will be the same for every batch, in craft distilling, the distiller's sensory skills will affect the final product's uniqueness and, therefore, its acceptance in the marketplace. Cuts must be executed quickly, as the distiller cannot simply turn off the still and then turn it back on once they have pondered their decision, because the rising temperature profile also drives a flavor profile of new compounds.

Fig. 22.1	Developing Spirit Profiles: Head and Tail Cuts				
Process	**Bourbon**	**Scotch**	**Rum**	**Tequila**	**Cognac**
Head Cuts	Discard: Acetone, ethyl acetate, green apple, nail polish Keep: Floral esters	Discard: Paint thinner, solvent, acetaldehyde Keep: light grassy esters	Discard: Harsh solvent, glue, green banana skin Keep: Fruity esters, pineapple, banana	Discard: Acetone, ethyl acetate, paint-like aromas Keep: light citrus esters	Discard: Solvent, green grape skin, chemical fruit Keep: delicate white fruit esters
Hearts	Keep: Vanilla, caramel, corn sweetness, spice	Keep: malt, fruit esters, light peat, floral, vanilla	Keep: Tropical fruit, molasses, spice, vanilla	Keep: Stone fruit, floral, agave, mezcal, pepper, citrus, herbal	Keep: Stone fruit, floral, spice, vanilla
Tail Cuts	Discard: Bitter fusels, vegetals, oils, sulfur Keep: Smoky, nutty	Discard: Phenolic overload, wet dog, greasy Keep: earthy peat, dark fruit	Discard: Rancid butter, cheese, rubber Keep: molasses depth, earthy funk	Discard: Overcooked agave, sour lactic notes Keep: earthy, mineral tones	Discard: Sour, musty, waxy, fatty notes Keep: Prune, nut, rancio complexity

Summary

Making proper cuts during distillation is both science and art. It requires understanding volatility, monitoring thermal and sensory markers, and anticipating the downstream effects of including or excluding certain compounds. Faults introduced by poor cuts are often irreparable, while strategic blending of marginal fractions can produce complexity and individuality. For the sensory student, awareness of cut-related compounds and their expression is a foundational skill. It sharpens the ability to detect flaws, appreciate craftsmanship, and assess a spirit's developmental potential over

156

time.   Refer to Ch 7 Appendix Table 7.1 to review distilling faults which affect sensory evaluation.

References

Balcerek, M., Pielech-Przybylska, K., Dziekońska-Kubczak, U., & Patelski, P. (2017). Distribution of volatile compounds in the heads, hearts, and tails fractions obtained by distillation of fermented mashes. *Journal of the Institute of Brewing, 123*(3), 391–401. https://doi.org/10.1002/jib.441

Conner, J. M., Paterson, A., & Piggott, J. R. (1998). Sensory characterization of pot-still distilled malt whisky from new and used casks. *Journal of the Institute of Brewing, 104*(2), 87–91.

European Parliament & Council of the European Union. (2019). *Regulation (EU) 2019/787 on the definition, description, presentation, and labelling of spirit drinks.* https://eur-lex.europa.eu/eli/reg/2019/787/oj

Heller, M., & Einfalt, D. (2022). Reproducibility of fruit spirit distillation processes and volatile fraction distribution (heads/hearts/tails). *Beverages, 8*(2), 20. https://doi.org/10.3390/beverages8020020

Jeffery, D. W., Sefton, M. A., & Francis, I. L. (2003). Sensory profiling of spirits: The importance of training and reproducibility. *Australian Journal of Grape and Wine Research, 9*(1), 12–20. https://doi.org/10.1111/j.1755-0238.2003.tb00228.x

Jones, P. R. (2004). Distillation control and optimization. *Chemical Engineering Progress, 100*(4), 56–62.

Lee, K.-Y. M., Paterson, A., & Piggott, J. R. (2001). Origins of flavour in whiskies and a revised flavour wheel: A review. *Journal of the Institute of Brewing, 107*(5), 287–313. https://doi.org/10.1002/j.2050-0416.2001.tb00099.x

Nykänen, L. (1986). Formation and occurrence of flavor compounds in wine and distilled alcoholic beverages. *American Journal of Enology and Viticulture, 37*(1), 84–96. https://doi.org/10.5344/ajev.1986.37.1.84

Piggott, J. R. (Ed.). (1993). *Distilled spirits: Tradition and innovation.* Springer. https://doi.org/10.1007/978-1-4615-2662-4

Poisson, L., & Schieberle, P. (2008). Characterization of the most odor-active compounds in an American bourbon whisky by application of aroma extract dilution analysis. *Journal of Agricultural and Food Chemistry, 56*(14), 5813–5819. https://doi.org/10.1021/jf800382m

Singleton, V. L. (1995). Maturation of wines and spirits: Comparisons, facts, and hypotheses. *American Journal of Enology and Viticulture, 46*(1), 98–115. https://www.ajevonline.org/content/46/1/98

Sponholz, R. (1993). Influence of wood and distillation on spirit character. In J. R. Piggott (Ed.), *Distilled spirits: Tradition and innovation* (pp. 185–201). Springer. https://doi.org/10.1007/978-1-4615-2662-4_12

U.S. Alcohol and Tobacco Tax and Trade Bureau. (2007). *The Beverage Alcohol Manual (BAM), Volume 2: Distilled spirits.* https://www.ttb.gov/system/files/images/pdfs/spirits_bam/cover.pdf

U.S. Alcohol and Tobacco Tax and Trade Bureau. (2025, June 4). *Requirements for beverage distilled spirits plant operations.* https://www.ttb.gov/business-central/requirements-beverage-distilled-spirits-plant

Observations on Head and Tail Cuts

Head and tail cuts decide which volatiles enter the spirit. Too much head brings sharp, solvent notes; too much tail adds heaviness and fusel drag. The cut point is the key driver of clarity, texture, and overall balance.

Chapter 23: Barrel Chemistry and Aging — Oak Compounds

Barrel aging is one of the most influential stages in spirit production, transforming raw distillates into complex, aromatic beverages through slow chemical interaction with wood. Oak barrels serve a dual role, functioning as both storage vessels and chemical reactors. While the initial purpose of aging was to smooth harsh distillates, oak soon became the standard due to its unique ability to impart beneficial flavor compounds, filter impurities, and allow for controlled oxygen exposure. The result is a deepening of flavor, color, and mouthfeel that defines the final spirit. Aging not only enhances integration but also introduces entirely new flavor-active compounds absent in the unaged distillate. Over 600 varieties are found worldwide, of which 90 are native to the USA. See Fig. 23.1 for oak species commonly used for aging.

Fig. 23.1	Major Oak Species Commonly Used in Barrel Aging.			
Species	**Common Name**	**Use**	**Key Flavor Compounds**	**Wood Grain Type**
Quercus alba	American white	Bourbon, Rum	Vanillin, lactones, aldehydes	Wide, less porous
Quercus robur	European, Limousin, peduncular	Cognac, scotch, sherry	Ellagitannins, phenols	Wide/medium More porous
Quercus petraea	European, Sessile (Troncais)	Brandy, whiskies	Subtle spice, medium tannins	Tight grain, ring porous
Quercus crispula	Mizunara (Japan, Mongolia, north Asia)	Japanese whisky	Vanillin, sandalwood, spice	Wide, very porous
Quercus garryana	Garry, Pacific coast, BC, Oregon, to California	Whiskies, other craft spirits	High tannins, phenolics, dark molasses, cloves	Wide, less porous

Species of Oak Used in Spirits Aging

Quercus alba-American white oak:
- **Origin:** Eastern U.S.A., especially Missouri, Kentucky, and Minnesota
- **Spirit:** Bourbon, scotch, global whiskies. The law requires bourbon to be made from fresh, charred oak, but the species is not specified. The abundance of American white oak has made it the standard oak for bourbon and American whiskies. When the wine boom took off in the 1960s and 1970s, white oak became the common wine barrel oak globally.
- **Sensory Contribution:** High lactones, vanillin, tannins, coconut, honey, almonds, subtle spice, coarse grain lead to quicker flavor extraction, suitable for shorter aging periods.

Quercus robur-European pedunculate oak, also called English oak due to its use by the English for aging sherry and fortified wines:
- **Origin:** Europe, grows in wetter lowlands
- **Spirit:** Cognac, Armagnac, Scotch, Irish Whiskey, Sherry
- **Sensory Contribution:** High tannins and phenols, subtle lactones, cloves and dried fruit, strong ellagitannins (spicy), slower maturation, more complex with age. Sherry-seasoned casks are produced for the Scotch whisky industry.

Quercus petraea-European sessile oak, French or Irish oak, is grown in several parts of Europe, most sought after from the French forests of Troncais and Limousin. Barrels often utilize both petraea and robur in the same barrel (Chatonnet & Dubourdieu, 1998):

- **Origin:** Europe
- **Spirit:** Fine brandies, premium whiskies
- **Sensory Contribution:** Often grouped with French oak, Hungarian oak tends to be lower in lactones than American oak, with nuanced spice and nut flavors, and medium tannin levels. Slower extraction due to tighter grain.

Quercus crispula-Mizunara oak (maturity 200 plus years). Synonymous with Quercus mongolica:

- **Origin:** Northeast Asia, Japan
- **Spirit:** Japanese whisky
- **Sensory Contribution:** Rich in vanillins, coconut, sandalwood, fresh flowers, spicy, incense-like aromas. Very porous and difficult to cooper. Longer maturation, but yields exotic, complex notes. Some scotch distillers seek them for their unique finish qualities.

Quercus garryanna-Garry oak:

- **Origin:** USA Pacific Northwest, Oregon, rare
- **Spirit:** American Craft Whiskies, notably Pacific Northwest (Harrington & Godfrey, 2021).
- **Sensory Contribution:** Dark, austere molasses and smoky aromas.

Note on Grain vs. Pore Opening: The terms "open grain" and "pore opening" are not synonymous. Grain refers to the natural anatomical spacing of vessel elements in oak, which differs by species. Pore opening, in contrast, results from toasting and affects the accessibility of extractable aroma compounds. Tight-grained species like Quercus petraea respond more uniformly to splitting and toasting, enabling better control over flavor development—even though they are not coarse-grained by biological definition.

Other lesser-known oaks are also used in spirits aging barrels or as mixed staves.

Quercus bicolor — **Swamp white oak**

- **Origin:** Upper Midwest to Northeast bottomlands
- **Spirit:** Historically, used for "tight cooperage;" staves occasionally mixed with white-oak lots
- **Form Used:** Whole barrels (historically) and mixed staves; chips are uncommon
- **Sensory Contribution: Sparse** — no robust, species-specific sensory datasets located; treated as interchangeable with *Q. alba*

Quercus montana (syn. *Q. prinus*) — **Chestnut oak**

- **Origin:** Appalachians and eastern uplands
- **Spirit:** Occasional use, in some cooperage/trade species lists; rare compared to *Q. alba*
- **Form Used:** Mixed staves or limited whole barrels; chips not typical
- **Sensory Contribution: Sparse** — little published sensory separation from *Q. alba*

Quercus macrocarpa — **Bur oak**
- **Origin:** Upper Midwest/Great Plains into the Midwest bottomlands
- **Spirit:** Suitable for "tight cooperage;" occasionally used/marketed with white-oak lumber
- **Form Used:** Mixed staves or limited whole barrels; chips are uncommon
- **Sensory Contribution: Sparse** — no distinct, peer-reviewed sensory fingerprint vs. *Q. alba*

Quercus stellata — **Post oak**
- **Origin:** South-central and southeastern U.S.
- **Spirit:** Occasionally included within "white-oak group" material; not mainstream species
- **Form Used:** If used, then mixed staves; chips limited to DIY/experiments, not standard trade
- **Sensory Contribution: Sparse** — no formal spirits sensory datasets located

Quercus michauxii — **Swamp chestnut oak**
- **Origin:** Southeast and lower Mississippi basin
- **Spirit:** Listed by some U.S. cooperage/trade sources as a white-oak option; limited availability
- **Form Used:** Mixed staves or limited whole barrels; chips are uncommon
- **Sensory Contribution: Sparse** — no reliable, species-specific spirits sensory data

Quercus muehlenbergii — **Chinkapin (Chinquapin) oak**
- **Origin:** Midwest to Appalachians; scattered East
- **Spirit:** Rare but documented in U.S. cooperage; at least one Missouri cooperage processes it; occasionally mixed into bourbon staves
- **Form Used:** Mixed staves; limited whole barrels; chips uncommon
- **Sensory Contribution: Sparse** — trade considers it broadly interchangeable with *Q. alba* for performance; formal sensory data is limited

Quercus lyrata — **Overcup oak**
- **Origin:** Gulf Coast and Mississippi Valley bottomlands
- **Spirit:** Not common, but noted as acceptable for tight cooperage within the white-oak group
- **Form Used:** Mixed staves or limited whole barrels; chips are uncommon
- **Sensory Contribution: Sparse** — no species-specific spirits sensory literature found

Anatomy of Oak and Its Chemical Potential

The utility of oak lies in its cellular structure and chemical complexity. The principal biopolymers in oak wood are cellulose, hemicellulose, and lignin, each contributing differently to maturation:

- Cellulose serves as the structural backbone of wood and does not significantly degrade during toasting. The role is more physical, less chemical.
- Hemicellulose, a polysaccharide, undergoes pyrolysis during toasting and charring, breaking down into furfural, 5-methylfurfural, and other caramelized sugars that produce sweet, nutty, and almond aromas (Chatonnet & Dubourdieu, 1998).
- Lignin degrades under heat, releasing phenolic compounds such as vanillin, eugenol, guaiacol, syringol, each with distinct aromatics (Spillman, Sefton, & Gawel, 2004).

Additionally, oak contains tannins, which influence astringency, bitterness, and mouth-coating sensations (Cadahía, Varea, Muñoz, & Fernández de Simón, 2001). The composition of compounds varies significantly by oak species. American oak (*Quercus alba*) tends to produce higher levels of vanillin and lactones; European oak (*Quercus robur* and *Quercus petraea*) contains more tannins and phenolic aldehydes (Mosedale & Puech, 1998).

Barrel Construction and Thermal Modification

Production of barrel staves varies by cooperage. Some are split, and precision saws are used to form others. Surface Area Exposure: Sawn staves have rougher edges and more surface area exposure. Speculatively (verifiable data unavailable), they provide greater flavor extraction. Split/riven staves have smoother edges and a tighter seal, which may lead to controlled extraction and slower aging.

- Flavor Impact: Rougher sawn staves may promote faster flavor release, desirable for certain spirits or short aging periods. Split staves tend to refine and balance flavor profile over longer aging periods due to a tight fit and reduced oxygen ingress. Splitting occurs along grain lines
- Craftsmanship and Precision: Sawn staves allow for curvature and a tight fit to enhance structural integrity and flavor aging.

Also, American white oak (Quercus alba) contains tyloses (balloon outgrowths of parenchyma cells), which allow sawing while remaining liquid-tight (Chatonnet & Dubourdieu, 1998). Note that differences in extraction are driven mainly by toast/char regime, wood anatomy, cask geometry, and surface chemistry. However, the widely accepted course of action appears to be to saw staves for maximum flavor extraction in a shorter extraction time and to split staves for long-term aging and flavor development.

The toasting and charring processes are central to unlocking the chemical precursors within oak.

- Toasting involves slow heating that facilitates the thermal degradation of hemicellulose and lignin, forming furans, phenols, and lactones.
- Charring, in contrast, involves high-temperature combustion of the inner barrel surface, creating a char layer that functions like activated carbon. This layer adsorbs sulfur compounds and other off-flavors, while the heat-penetrated zone beneath—often referred to as the "red layer"—becomes a rich source of extractable aroma compounds (Singleton, 1995).

The temperature profile and toasting duration determine the balance of compounds generated. Light toasting yields more delicate vanilla and spice notes; heavy toasting produces increased smoke, clove, and char aromas. Oxygen permeability through the staves also contributes to slow oxidative changes that deepen color and produce aldehydes and esters (Ferreira & López, 2019).

- Light Toast / Char #1 (approx. 15 sec): Minimal breakdown of lignin and hemicellulose. Subtle wood tannins, light vanilla, and dry spice are predominant.
- Medium Toast / Char #2–3 (30–45 sec): Increased formation of caramelized sugars and phenolic aldehydes. Flavors include vanilla, coconut, toffee, and a mild hint of smoke.
- Heavy Toast / Char #4 ("Alligator Char") (55+ sec): Produces deep fissures and thick charcoal layer. Enhances filtration, producing a smoky, roasted, and caramelized character. Ethyl guaiacol and vanillin are prominent.
- Char level also affects filtration properties, with heavier char removing sulfurous congeners more effectively during aging (Mosedale & Puech, 1998).

Key Oak-Derived Aroma and Flavor Compounds

Numerous compounds arise from barrel aging, particularly from lignin pyrolysis and the thermal degradation of carbohydrates. See Fig. 23.2. These include:

Vanillin – the iconic aroma compound responsible for vanilla notes- is also formed from lignin

- Eugenol – clove-like spice, also from lignin
- Guaiacol and Syringol – smoky, phenolic aromas
- Furfural and 5-Methylfurfural – almond, caramel notes, from hemicellulose
- Whiskey Lactones (cis and trans) – coconut and woody aromas, especially from American oak
- Sotolon – responsible for curry, maple syrup, or fenugreek aromas, formed through aging and oxidation (Sponholz, 1993).

Fig. 23.2		**Major Flavor Compounds from Oak Barrels**			
Compound	**Chem. Family**	**Sensory**	**Source**	**DT** (ppm)	**Aging Stage**
Vanillin	Phenaldehyde	Vanilla	Lignin	0.3	Mid to late
Whiskey lactone (cis)	Lactone	Coconut, creamy	Oak lipids	0.035	Early to mid
Eugenol	Phenol	Clove, spicy	Lignin	0.01	Mid to late
Furfural	Aldehyde	Almond, sweet	Hemicellulose	0.5	Early to mid
Tannins	Polyphenol	Astringent, drying	Oak heartwood	Varied	Continuous
DTs are matrix-dependent (EtOH% %, pH, matrix). Values shown reflect model-wine thresholds where available (e.g., Abbott et al., 1995). Use with caution for spirits. (cis)=0.020 (mg/L)					

These compounds are extractable from ethanol-water matrices and vary in concentration depending on toast level, species, and extraction time. Their thresholds and aroma impact can be modulated by ethanol content and interactions with other congeners (Sponholz, 1993; De Simone & Nobili, 2021).

Time-Dependent Processes and Maturation Kinetics

The kinetics of barrel aging involve multiple, overlapping phenomena. Ethanol and water permeate the wood, dissolve extractives, and then return to the liquid phase, carrying new compounds with them. Meanwhile, barrel evaporation changes concentration, known as the

162

"angel's share."

Over time, slow oxidation leads to the formation of acetals, aldehydes, and esters, while tannins soften and integrate. Early aging contributes to a harsher wood character, whereas long-term maturation develops a more rounded, complex profile (Nykänen, 1986; Reazin, 1981). Key transformations are:

- Oxygenation and polymerization of phenolic compounds
- Esterification of organic acids and alcohols
- Ongoing oxidation–reduction of aldehydes (to acids or alcohols), esterification/hydrolysis of acids and alcohols, and formation, interconversion of lactones over time (Reazin, 1981; Nykänen, 1986).

Fig. 23.3 illustrates the development of vanillin and lactones in oak barrels over 12 years, highlighting the difference in development to maximum concentration between new and refilled barrels. Note the degradation of tannins over time.

Fig. 23.3   Key Oak Derived Compounds Over Time (New vs Refill)

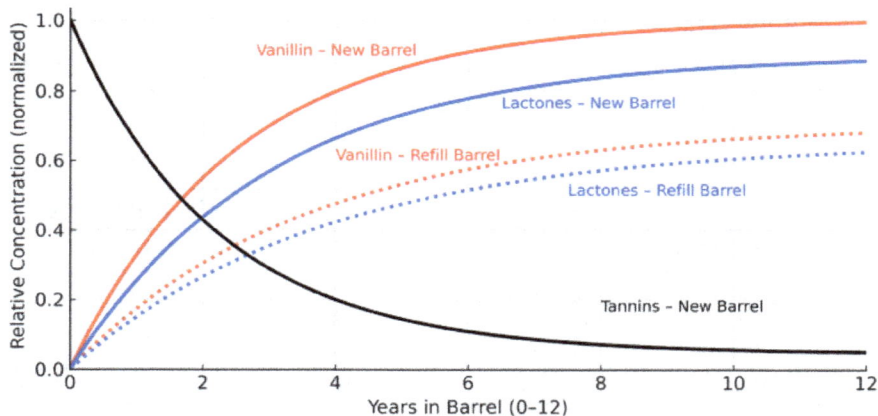

Data adapted from Meilgaard et al. (2006), Viriot et al. (1993), Luo et al. (2023)

Interactions: Barrel-Derived Compounds and Existing Congeners

Barrel compounds do not exist in isolation. Their sensory impact depends on interactions with pre-existing congeners from fermentation and distillation. These interactions can be synergistic (vanillin enhancing esters), masking (smoke muting fruitiness), or antagonistic (phenolics suppressing floral notes). Ethanol facilitates these interactions by altering solubility and volatility profiles (Reazin, 1981).

Chemical reactions between oak acids and fusel alcohols can form unique esters, while phenolic aldehydes may bond with sulfur compounds, reducing their volatility. These effects emphasize the importance of viewing oak chemistry not as additive, but integrative.

Barrel Size and Aging Dynamics

The volume-to-surface-area ratio of a barrel significantly affects the chemical transformations that occur during aging. Smaller barrels, such as 30-liter or 50-liter casks, expose a greater wood surface area per unit of spirit, thereby accelerating the extraction of oak-derived compounds, including lactones, vanillin, tannins, and phenolic aldehydes. As a result, spirits aged in small barrels develop color and oak intensity more rapidly. However, they may also risk over-oaking, with harsh tannins or a dominant wood character overshadowing the distillate's nuance if not closely monitored (Canas et al., 2008) (Fig. 23.4).

Fig.23.4	Common Barrel Sizes for Aging Spirits		
Barrel Name	**Size**		**General Usage**
Blood Tub	40 liters	10.5 gals	Used in beer or distilling small batch spirits or trial experiments
Quarter Cask	50	13	Quick aging to impart fast flavor addition
American Standard	200	53	Wide bourbon use, after first fill, sold for scotch and Irish whiskey
Hogshead	250	66	Most commonly used for maturation, resold after first fill
Barrique	225	59	Bound by wood hoops (not metal), wine first, secondarily for wine finishes spirits
Sherry Puncheon	500	132	Thinner staves of Spanish oak, elongated body, primarily sherry
Sherry Butt	500	132	Thick staves of European oak, tall and narrow in shape, are used for Spanish sherry
Common Puncheon	500	132	Thick American oak staves, short, fat body, widely used for rum
Port Pipe	650	171.7	Thick staves of European oak, for maturing Port wines, then whiskey finishing
Madeira Drum	650	171.7	Short, fat shape, solely used for Madeira
Gorda	700	184.9	American whiskey maturation is mostly for blending or vatted whiskey production.
Hobby, Experimental Sizes			
Small Hobby	1-5 liters	.25-1.3 gals	Rapid micro-batch finishing, time up to 2 months, over-oak risk
Medium Hobby	10	2.6	Rapid finishing, better control than small, time = up to 3 mos
Medium Experimental	20	2.3	Testing prior to scaling, better oxygen, time = 3-6 months
Large Experimental	30	7.9	Controlled oxidation/extraction resembles larger barrels. Time = 6-12 mos

In contrast, standard-size barrels (e.g., 200-liter American standard barrels or 225-liter barriques) provide a more balanced and prolonged exchange between spirit and wood. Larger casks, such as 500-liter puncheons or butts, slow the extraction rate, allowing for longer, subtler aging. During this process, oxidative reactions, esterification, and polymerization occur gradually, resulting in smoother flavor integration and more elegant maturation profiles (Ortega-Heras, Pérez-Magariño, & González-San José, 2004). The choice of barrel size should reflect the distiller's sensory goals: rapid development and bold flavors from small casks versus refined complexity and longer oxidation-driven evolution from larger vessels.

Environmental and Storage Variables

The barrel-aging environment has a dramatic effect on the maturation rate and character. Warmer climates accelerate compound extraction and evaporation, allowing for greater wood impact in a shorter time. Cooler, damp climates yield slower maturation but greater oxidative complexity (U.S. Alcohol and Tobacco Tax and Trade Bureau, 2021).

Barrel warehouse conditions, including racking height, ventilation, and seasonal cycling, all influence barrel aging. First-fill barrels impart the most pronounced wood character; second- and third-fill casks offer more subtle integration. These variables provide distillers with a sensory palette that extends beyond mere time-in-barrel (Margalit, 2004). For illustrative purposes, Fig. 21.5 shows the development of vanillin and lactones in cool vs. warm rickhouse climates.

Climatic Maturation Regimes: Tropical, Continental, and Sensory Trajectories of Wood

Cask maturation is not a metronome; it is a **climate-driven reactor**. Temperature, humidity, and seasonal amplitude govern extraction, oxidation, ester dynamics, and evaporation. The same cask filled with the same spirit will not march to the same sensory beat in different climates.

Tropical regimes (higher average temperature, larger diurnal swings, often higher humidity):

- **Extraction accelerates.** Oak lactones, vanillin, toast/char volatiles, and sweet brown-note cues (caramel/toffee) rise quickly.
- **Oxidation and acid–aldehyde play an increasing role.** Aldehydic "rancio"/dried-fruit characters can appear earlier; harsh edges round faster—until they do not.
- **Evaporation ("angel's share") is high.** Proof may rise or fall depending on the balance of humidity; concentration can amplify both virtues and faults.
- **Sensory trajectory:** assertive wood and sweet spice early; shorter path to over-oak if not managed; fruit can skew toward **ripe/tropical**.

Continental regimes (cooler averages, pronounced seasons, lower humidity in many warehouses):

- **Extraction is slower.** Oak contribution gradually integrates; tannins knit over longer horizons.
- **Oxidation proceeds but with more winter "rest."** Development is a stepwise process; complexity often builds in layers.
- **Evaporation is lower (on average).** Trends in proofing depend on the warehouse and season; concentration effects are gentler.
- **Sensory trajectory:** more **subtle oak** early, longer runway for secondary notes (dried fruit, nuts, spice) without quick overshoot.

Maritime and microclimate factors (salt-laden air, warehouse height, airflow, cask stacking) modulate both baselines. Upper racks run hotter and drier, while lower levels are cooler and steadier, creating internal "climates within a climate." Three cautions for interpretation:

1. **No universal exchange rate.** Claims like "three tropical years equal nine continental" are marketing, not physics. Processes scale differently; faster extraction does not guarantee equivalent oxidation or tannin integration.

2. **Style vs. fault.** Tropical assertiveness is not a flaw; continental restraint is not a virtue. Judge balance and intent relative to the regime, rather than adhering to a single global template.

3. **Cask history dominates late.** Refill status, prior contents, and stave chemistry can outweigh climate in the long tail; "where" and "what cask" are co-drivers.

For the evaluator, climate context converts surprise into understanding. It explains why two well-made spirits of the same age can present as cousins, not twins—and it keeps age statements in their place: informative, not determinative. Fig. 23.5 illustrates the difference in the development of vanillin and lactones between warm and cool climates over 12 years.

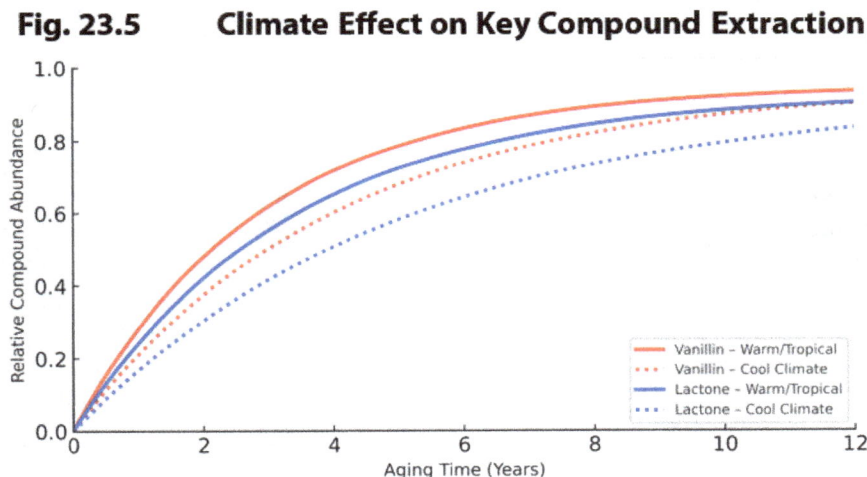

Fig. 23.5   Climate Effect on Key Compound Extraction

Alternative Oak Treatments and Nontraditional Maturation

To reduce aging time or modify profile, producers may use alternative oak treatments such as:

- Oak chips, cubes, and spirals – provide surface area but lack oxidation dynamics of full barrels (Delgado González et al., 2023; Krüger, 2022)
- Pressure-aging or ultrasonic systems – accelerate extraction but may produce unbalanced results (Delgado González et al., 2022; Krüger, 2022)
- Oxygen sparging or micro-oxygenation – mimics slow oxidation in a compressed timeframe (Krüger, 2022).

While useful for innovation, these methods can lack the subtlety and depth achieved through traditional aging. Regulatory bodies often limit their use in legally defined categories (e.g., "straight whiskey" in the U.S.) (U.S. Alcohol and Tobacco Tax and Trade Bureau, 2021).

Notes on hobby barrel usage: Most hobby casks are made from American oak (Quercus alba), often with a medium char (typically #2 or #3). Some are available in French oak, Hungarian oak, or alternative woods, such as cherry or chestnut, for experimentation. Many hobbyists pre-soak their barrels with water or a sacrificial spirit to remove excess tannins and prevent leaks. Small casks can be reused several times, with the second and third fills yielding more subtle extraction and being more suitable for finishing than for primary aging.

Other Woods That Influence Spirits' Finish

For centuries, distillers and coopers have experimented with other woods, some exotic. A few, noted below, are not the entire universe of barrel woods but represent some of the more successful departures from oak (Fig. 23.6). In the USA and Scandinavia, recent experimentation with Pacific Northwest varietal Garry oak has proven viable (Harrington & Godfrey, 2021).

Fig. 23.6		Other Barrel Woods which Enhance Spirits' Flavors	
Wood	**Spirits**	**Compounds**	**Sensory**
Cherry	Craft bourbons, cherry brandy	Benzaldehydes, coumarin	Sweet almond, cherry, mild spice
Acacia	Grappa, Italian white wines	Tannins, floral esters, vanillin	Floral, vanilla, dry, subtle
Chestnut	Italian brandies, whiskies	Ellagitannins, furfural, high polyphenols	Nutty, oxidative, robust structure
Maple	Bourbon finish	Caramel furans, vanillin, smoky lactones	Sweet, syrupy, light caramel
Mulberry	Hungarian palinka	Coumarins, woody esters	Sweet, herbal, woodsy
Beech	Grain spirits	Smoky phenols, light eugenol	Clean, light, smoky tones
Applewood	Fruit brandies	Fruit esters, mild smoke phenols	Fruity, earthy, woody
Amburana	Brazilian cachaca	Vanillin, cinnamic aldehydes, spice	Cinnamon, spice, creamy vanilla
Teak	Indian rums and whiskeys	Tannins, smoky resins, minor vanillin	Astringent, bold, exotic

Additional Notes

- **Chestnut wood** (*Castanea sativa*) is not an oak, but often appears in cooperage discussions. It is studied for aging wine spirits, showing significantly higher phenolic content, antioxidant activity, and sensory complexity (e.g., more evolved aroma, toasted notes, faster maturation) compared to Limousin oak. It is typically used in whole barrels.
- **Non-oak species** like acacia (*Robinia pseudoacacia*), cherry (*Prunus avium*), ash (*Fraxinus americana*), and mulberry (*Morus alba*, *M. nigra*) have been explored for cooperage—mainly in wine contexts—as wood chips, but are generally not used in spirits barrels

Sensory Interpretation of Barrel-Derived Compounds

Training the nose to recognize oak-derived aromas is essential for serious spirits evaluators. Key sensory markers include vanilla, coconut, clove, caramel, smoke, and dried fruit. These must be distinguished from flaws such as sulfur or microbial taints, which can mimic some oak notes (Robinson, Boss, Solomon, & Trengove, 2014).

Sensory panels often confuse char-derived smoke with peat smoke or guaiacol from fermentation. Contextual knowledge of production and barrel use helps disambiguate these perceptions. Calibration with reference compounds improves consistency and accuracy in evaluation (Robinson et al., 2014).

Aging Time Modifications (Fig. 23.7)

Fig. 23.7 **Flavor Intensity vs Age (New Barrel)**

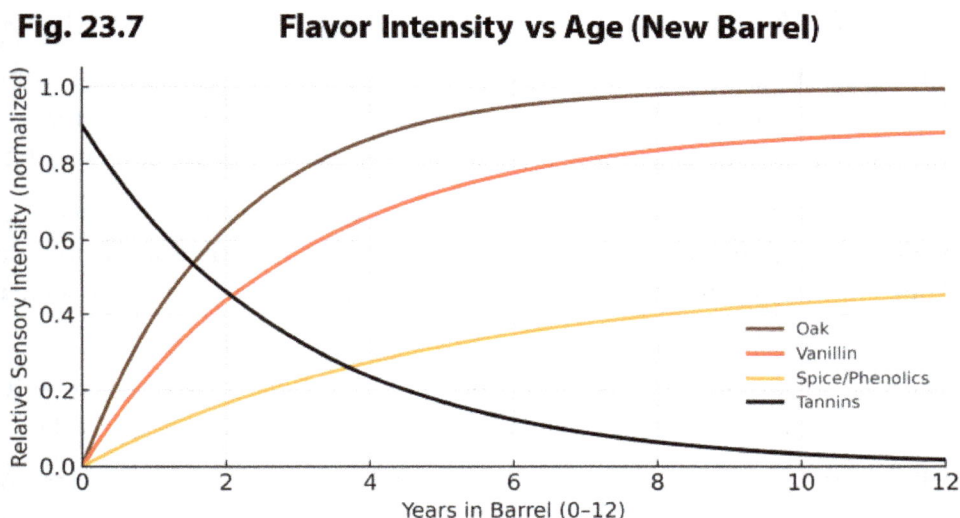

Time in the barrel is expensive, so many distillers seek ways to get the results of aging through methods to accelerate the process. During aging, the distiller depends on daily temperature fluctuations (warm during the day, cool at night) to age spirits. Oak pores expand during the day, allowing the spirit to enter, and cool temperatures at night contract the barrels, pushing the spirit out of the pores while imparting some of the chemicals and wood sugars from the barrel and the toasting process. The flavor intensity with aging in a new barrel is illustrated in Fig. 23.7.

Although sensory science is less concerned with the process than with the results, it is worth noting that various physical and scientific methods have been used with varying degrees of success. Although some methods have been discussed previously (Fig. 23.8 summarizes many different methods to advance aging in the barrel. Some of these methods are:

Small Barrels / High Surface Area-to-Volume Ratio: Increasing the surface area-to-volume ratio of the barrels (by downsizing barrels or adding staves) accelerates the extraction of oak-derived compounds, such as tannins, vanillin, and color pigments (Robinson et al., 2014).

Oak Chips / Fragments: Widely used in spirits and wines, oak chips increase surface contact area, resulting in significantly faster extraction of wood compounds—studies report extraction speed up to 50% faster (Delgado González et al., 2023; Krüger, 2022).

Rapid Temperature Cycling: Temperature fluctuations within rickhouses induce repeated expansion and contraction of wood staves, encouraging deeper spirit penetration and more rapid chemical extraction (Sant, 2014; Krüger, 2022).

Ultrasound (Sonic Resonance/Cavitation): When applied in conjunction with oak contact, ultrasonic energy accelerates the extraction of phenolic compounds by approximately 34% in laboratory-scale studies (Delgado González et al., 2022; Krüger, 2022).

Pulsed Electric Fields (PEF): PEF treatment disrupts wood cell structure, releasing aroma and flavor compounds (e.g., vanillin and syringaldehyde). PEF has been reported to increase oak volatile release in lab setups; the magnitude depends on the system/matrix. (Fontana, 2025).

High-Pressure CO_2/Ozone-Treated Wood: A patented process that uses pressurized CO_2 and ozone treatments on wood enables rapid ester formation and lignin breakdown. Applicants claim significant time-compression effects; this is a patented process and not an industry standard (U.S. Patent US11053467B2, 2021).

Micro-oxygenation: Controlled oxygen introduction accelerates oxidative reactions, ester formation, speeding up aging-related sensory maturation (Krüger, 2022; Reazin, 1981).

Gamma Irradiation/Microwave Treatment: Experimental techniques that apply radiation energy to accelerate esterification and stabilize microbial status; outcomes remain controversial and are primarily studied in laboratory settings (Krüger, 2022).

Marine Exposure (Ship-Deck Aging): Historical or artisanal technique involving exposure of barrels to marine high temperature fluctuations and shipboard motion/sloshing. Blender reports suggest differences; peer-reviewed estimates of the magnitude of aging are lacking (Sant, 2014).

Case Study 23.1: Re-toasting Barrels Alters Bourbon's Sensory Trajectory: A Kentucky distillery facing inconsistent barrel inventories implemented a re-toasting protocol for used American white oak casks, heating them to 230 °C for 30 minutes to refresh extraction potential. The intent was to recover lignin degradation products and reduce the need for new cooperage. **Outcome:** Chemical analysis (GC–MS and LC–MS) revealed elevated levels of vanillin, syringaldehyde, and furfural in the first six months compared with standard refill casks, confirming renewed extractive activity. However, sensory panels reported excessive smoke, char, and bitter phenolic notes early in the maturation process, which masked ester-derived fruit notes. Over two years, the whiskey exhibited a flattened mid-palate sweetness and a diminished lactone intensity compared to new-barrel controls. **Interpretation:** The re-toasting improved short-term extract yield but shifted the oxidation balance and micro-oxygenation rate. Burnt lignin fragments initially dominated the aroma, while diminished fresh oak lactones hindered the long-term integration of sweetness and coconut–vanilla character. (Nishi et al., 2021; Piggott & Conner, 2003). **Lesson for Evaluators:** Barrel engineering affects aroma trajectory, not just maturation rate. Thermal reuse may simulate extraction chemistry, but rarely reproduces the balanced evolution of new wood. Evaluators encountering atypical smoky, bitter, or short-finish profiles in otherwise mature whiskey should consider reconditioned cooperage as a sensory variable.

Fig. 23.8	Accelerated Barrel Aging Techniques	
Technique	**Description**	**Effect/Purpose**
Small barrels	Raises surface area to volume ratio	Accelerates vanillin, tannins, color extraction
Add more staves	Raises surface area to volume ratio	Accelerates vanillin, tannins, color extraction
Add oak chips	Chips, oak powder, or shavings	Increases area, extraction>10-50%
Rapid rickhouse temperature cycles	Controls temperature swings, rotates barrels to higher or lower racks	Increases frequency and extraction cycles, moving to higher temperature areas
Ultrasound, sonic resonance/cavitation	Circulate spirit, vibrate barrels with ultrasonic waves	Speeds extraction ~30-40%, accelerates ester formation
Electric field pulse (PEF)	High voltage pulsing disrupts wood matrix-solvent interactions	Enhances vanillin, syringaldehyde extraction, depending on the system and matrix
Hydrostatic, CO_2 pressure	Pressurized with CO_2 forms carbonic acid, with ozone wood treatment	Simulates years of barrel effects in hours-days
Micro-oxygenation, oxygen enrichment	Introduces controlled oxygen into barrels or contact tanks	Accelerates oxidation and esterification processes
Gamma irradiation, microwaves	Exposes the spirit to radiation energy	Increases esters, reduces microbial activity
Marine exposure	Exposes barrels to higher daily on-deck temperature variances and wave motion aboard ships	Increasing the extraction reduces the aging time. Barrel areas in the barrel headspace are utilized by wave motion

Evaluation Note: 23.1—Thermal and Oxidative Reactions in Maturation: Confusion may exist regarding **the** flavors in aged spirits that arise from oak extractives. Heat and oxygen also drive independent chemical change. During distillation and early cask entry, Maillard reactions between amino acids and residual sugars generate heterocyclic compounds—furans, pyrazines, and maltols—that add roasted and caramel notes often mistaken for wood influence. Caramelization above 160 °C yields toffee-like volatiles, while slow oxidation in the cask headspace converts ethanol and aldehydes into acids, esters, and acetals, imparting softness and balance. These reactions parallel, but differ from, oak's contribution of lignin- and lactone-derived aromas. Managing temperature and oxygen exposure is therefore an integral part of flavor design, not merely a passive aging effect.

Summary

Oak barrels are both vessels and active chemical reactors, shaping spirits through the extraction, oxidation, and transformation of wood-derived compounds. Different oak species contribute distinct chemical profiles: American white oak with abundant lactones and vanillin, European oaks with higher tannins and phenolics, Mizunara with sandalwood and incense-like notes, and Garry oak with dark, austere aromatics. Toasting and charring activate hemicellulose, lignin, and tannins, releasing furans, phenols, lactones, and vanillin that contribute to spirit complexity.

Aging kinetics depend on barrel size, surface-to-volume ratio, climate, and warehouse conditions, with smaller casks accelerating extraction but risking over-oaking, while larger casks enable gradual oxidative integration. Oak chemistry interacts with congeners from fermentation and distillation in either synergistic or antagonistic ways, underscoring aging as an integrative rather than additive process.

Beyond traditional oak, alternative woods and accelerated-aging technologies have been explored, though often at the cost of subtlety compared to slow maturation. For evaluators, mastery requires recognizing key oak-derived markers—vanilla, coconut, clove, caramel, and smoke—while distinguishing them from flaws such as sulfur taints or over-extraction. Barrel chemistry, in its time-dependent balance of extraction, oxidation, and integration, remains the defining factor in the maturation of spirits.

In U.S. spirits, more than 90% of new barrels are still made from *Quercus alba*; the other white oaks mentioned are used regionally or opportunistically, often in undifferentiated bulk stave supply. Authoritative trade/forestry sources confirm these species are suitable for tight cooperage, but published, species-specific sensory deltas are scarce—with the notable exception of Oregon oak (*Q. garryana*), which has a growing body of trade documentation and brand case studies. Be aware that other species of white oak (montana, bicolor, lyrata, macrocarpa, stellata) may be mixed in with Quercus alba.

"Oak contributes more than wood character; it provides a chemical framework in which esters, aldehydes, lactones, and tannins integrate with the distillate to form a balanced, matured spirit."— Chatonnet, P, sensory researcher

"The 'red layer' beneath the char is a crucial zone: it is where heat-transformed lignin and hemicellulose produce the vanillin, eugenol, and furfurals that define whiskey's aroma."— Singleton, V. L., sensory researcher

References

Abbott, N., Puech, J.-L., Bayonove, C., & Baumes, R. (1995). Determination and sensory evaluation of cis- and trans-oak lactones in wines. *Food Chemistry, 51*(2), 135–141. https://doi.org/10.1016/0308-8146(94)P4179-F

Cadahía, E., Varea, S., Muñoz, L., & Fernández de Simón, B. (2001). Evolution of phenolic compounds in wines aged in oak barrels. *Journal of Agricultural and Food Chemistry, 49*(10), 4423–4430. https://doi.org/10.1021/jf010267z

Canas, S., et al. (2008). Effect of heat treatment on chemical composition of oak wood extracts. *Journal of the Science of Food and Agriculture, 88*(5), 774–782. https://doi.org/10.1002/jsfa.3140

Chatonnet, P., & Dubourdieu, D. (1998). Comparative study of the characteristics of American white oak (Quercus alba) and European oaks (Q. robur and Q. petraea) for barrel ageing. *American Journal of Enology and Viticulture, 49*(1), 79–85. https://www.ajevonline.org/content/49/1/79

De Simone, B. C., & Nobili, M. (2021). Advances in understanding the barrel-aging process: A chemical and sensory perspective. *Beverages, 7*(1), 11. https://doi.org/10.3390/beverages7010011

Delgado González, M. J., et al. (2022). Theoretical approximation of ultrasound-accelerated aging vs. thermal extraction in spirits. *Processes, 10*(5), 887. https://doi.org/10.3390/pr10050887

Delgado González, M. J., et al. (2023). Laboratory-scale aging using oak chips and ultrasound: Extraction enhancement. *Food Analytical Methods, 15*(2), 345–356. https://doi.org/10.1007/s12161-022-02344-7

Ferreira, V., & López, R. (2019). The actual and potential aroma of winemaking grapes. *Biomolecules, 9*(12), 818. https://doi.org/10.3390/biom9120818

Gawel, R., Sefton, M. A., & Jeffery, D. W. (2007). The influence of oak-derived compounds on wine aroma. *Australian Journal of Grape and Wine Research, 13*(2), 153–158. https://doi.org/10.1111/j.1755-0238.2007.tb00245.x

Gollihue, J., Batchelor, W., Bhatnagar, A., & Hayes, J. E. (2021). Sources of variation in bourbon whiskey barrels: A review. *Journal of the Institute of Brewing, 127*(4), 478–492. https://doi.org/10.1002/jib.660

Harrington, M. G., & Godfrey, J. D. (2021). The use of Quercus garryana in Pacific Northwest cooperage. *American Journal of Enology and Viticulture, 72*(2), 165–175. https://doi.org/10.5344/ajev.2020.20036

Krüger, R. T. (2022). Current technologies to accelerate the aging process of beverages: wood fragments, ultrasound, micro-oxygenation, PEF, high pressure, gamma, microwave. *Beverages, 8*(4), 65. https://doi.org/10.3390/beverages8040065

Lawless, H. T., & Heymann, H. (2010). *Sensory evaluation of food: Principles and practices* (2nd ed.). Springer. https://doi.org/10.1007/978-1-4419-6488-5

Luo, M., Li, Y., Sun, B., Zhao, D., Zhang, H., Fan, S., … Sun, X. (2023). Factors in modulating the potential aromas of oak whisky: Origin, toasting, and charring effects. *Foods, 12*(23), 4266. https://doi.org/10.3390/foods12234266

Margalit, Y. (2004). *Concepts in wine chemistry.* The Wine Appreciation Guild.

Meilgaard, M. C., Civille, G. V., & Carr, B. T. (2006). *Sensory evaluation techniques* (4th ed.). CRC Press. https://doi.org/10.1201/b16452

Mosedale, J. R., & Puech, J. L. (1998). Wood maturation of distilled beverages. *Trends in Food Science & Technology, 9*(3), 95–101. https://doi.org/10.1016/S0924-2244(98)00024-7

Nishi, K., et al. (2021). Influence of heat treatment of oak barrels on whiskey aging chemistry and flavor development. *Journal of the Institute of Brewing, 127*(4), 489–499. https://doi.org/10.1002/jib.644

Nykänen, L. (1986). Formation and occurrence of aroma compounds in wine and distilled alcoholic beverages. *American Journal of Enology and Viticulture, 37*(1), 84–96. https://www.ajevonline.org/content/37/1/84

Ortega-Heras, M., Pérez-Magariño, S., & González-San José, M. L. (2004). Effect of oak barrel type and aging time on aroma compounds in red wine. *Food Chemistry, 87*(4), 505–512. https://doi.org/10.1016/j.foodchem.2003.12.028

Piggott, J. R., & Conner, J. M. (2003). Interactions between oak maturation and spirit composition in Scotch whisky. *Food Chemistry, 82*(3), 343–350. https://doi.org/10.1016/S0308-8146(02)00541-3

Reazin, G. H. (1981). Chemical mechanisms of whiskey maturation. *American Journal of Enology and Viticulture, 32*(4), 283–289. https://www.ajevonline.org/content/32/4/283

Robinson, A. L., Boss, P. K., Solomon, P. S., & Trengove, R. D. (2014). Origins of volatile compounds in oak-aged spirits: The role of oak wood and maturation conditions. *Food Research International, 62,* 59–70. https://doi.org/10.1016/j.foodres.2014.02.002

Sant, L. (2014, September 10). Experimental aging methods: Undersized barrels to ultrasonic waves. *Liquor.com.* Retrieved from https://www.liquor.com/experimental-aging-methods ([Note: no DOI available])

Singleton, V. L. (1995). Maturation of wines and spirits: Comparisons, facts, and hypotheses. *American Journal of Enology and Viticulture, 46*(3), 339–349. https://www.ajevonline.org/content/46/3/339

Spillman, P. J., Sefton, M. A., & Gawel, R. (2004). The contribution of volatile compounds derived from oak wood to the aroma of wine: A review. *Australian Journal of Grape and Wine Research, 10*(2), 157–169. https://doi.org/10.1111/j.1755-0238.2004.tb00019.x

Sponholz, R. (1993). Influence of wood on the aroma of spirits. In J. R. Piggott, J. M. Conner, & A. Paterson (Eds.), *Distilled spirits: Tradition and innovation* (pp. 215–233). Springer. https://doi.org/10.1007/978-94-011-1880-5_17

U.S. Alcohol and Tobacco Tax and Trade Bureau. (2022–). 27 CFR Part 5 — Labeling and standards of identity for distilled spirits. *Electronic Code of Federal Regulations.* https://www.ecfr.gov/current/title-27/part-5

U.S. Patent No. 11,053,467 B2. (2021). Accelerated aging of alcohol spirits using pressurized CO_2 and ozone-treated wood. U.S. Patent and Trademark Office. https://patents.google.com/patent/US11053467B2

Viriot, C., Scalbert, A., Lapierre, C., Moutounet, M., & Hergert, H. L. (1993). Ellagitannins and lignins in aging of spirits in oak barrels. *Journal of Agricultural and Food Chemistry, 41*(11), 1872–1879. https://doi.org/10.1021/jf00035a013

Observations on Barrel Chemistry and Aging

1. Oak transforms ethanol's volatility into aroma complexity through time and heat.

2. Every stave is a chemical reactor; toast and char decide its vocabulary.

3. Small barrels speak quickly, large barrels speak wisely.

4. Vanillin and lactones announce youth; tannins and aldehydes whisper maturity.

5. Aging is not additive—it is integrative, a slow negotiation between wood and spirit.

6. Climate sets the tempo; the cask only plays along.

7. Over-oak a spirit, and you mute its accent; under-oak it, and you lose its story.

8 Barrel chemistry rewards patience; kinetics can be accelerated, integration cannot.

9. Toasting unlocks potential; oxidation refines it.

10. The red layer is where the barrel's soul meets the spirit's ambition.

Chapter 24: Spirits Blending — Sensory Objectives

In distilled spirits production, blending is not merely a practical step for volume management; it is the deliberate orchestration of multiple distillates into a coherent sensory whole. Historically, the practice emerged in response to the inherent variability of early distillation and maturation, when inconsistent equipment and unpredictable cask quality made it impossible to create identical reproductions from batch to batch (Buxton & Hughes, 2021). Merchants and distillers discovered that by blending spirits from various barrels, ages, and even distillation sites, they could produce a more consistent and marketable product.

Blending should not be confused with vatting. Vatting is the bulk mixing of similar products—often from the same distillery—to produce a single lot, primarily for efficiency. Blending, by contrast, is targeted and discriminating: each component is chosen for its sensory properties, chemical interactions, and contribution to the final balance. It is an active design process, guided by both sensory science and the blender's experience, rather than a logistical procedure.

Modern blending has evolved into a sophisticated discipline that bridges chemistry, sensory neuroscience, and brand philosophy. Its aim is not only to replicate established profiles but also to create new, distinctive sensory architectures that speak to both the trained evaluator and the consumer.

Why Distillers Blend

The primary purpose of blending is to manage natural variability. Even with consistent raw materials, yeast strains, and production parameters, every fermentation and distillation run produces a slightly different balance of congeners. Environmental factors, including barometric pressure, ambient humidity, and seasonal grain composition, alter fermentation kinetics and the resulting volatiles (Buxton & Hughes, 2021; Stewart & Russell, 2014). Maturation further compounds these differences. Each cask behaves uniquely due to wood porosity, previous fills, micro-oxygenation rates, and warehouse climate gradients (Mosedale & Puech, 1998; Russell & Stewart, 2014).

A blender's role is to navigate this chemical diversity to create a profile that is both stable and appealing. For established brands, the goal is often sensory continuity: the whiskey, rum, cognac, or tequila must taste recognizably like itself year after year, despite being composed of liquids from different production years. This consistency is achieved by blending across multiple casks and age statements, balancing fresher, more volatile-forward distillates with older, more oxidatively matured components.

Beyond consistency, blending is a tool for complexity enhancement. A single cask may excel in a particular aromatic register—say, rich caramel and vanilla lactones from heavy charred oak—but lack freshness or mid-palate brightness. By introducing components high in fruity esters or floral terpenes, the blender can expand the sensory spectrum and improve continuity across aroma, palate, and finish (Russell & Stewart, 2014).

Blending also helps mitigate minor flaws. While high-quality production minimizes off-notes, small deviations such as slight sulfur residues, excessive tannic grip, or muted mid-palate can be corrected through careful integration with complementary components. This is not about hiding poor-quality spirit but about fine-tuning a profile so that each element supports the whole.

Finally, blending supports **innovation**. Special releases, seasonal editions, or cask-finish projects often rely on blending to integrate experimental lots into a marketable product. Here, the blender is not constrained by strict replication but encouraged to explore new balances and textures, often creating highly sought-after one-off bottlings.

The Science and Art of Blending

Blending is the point at which the scientific understanding of spirit chemistry meets the artistry of sensory perception. From a molecular standpoint, spirits are complex solutions of volatile and semi-volatile compounds in an ethanol–water matrix. Each compound has a specific volatility curve, solubility, and sensory threshold (Pozo-Bayón & Reineccius, 2009). Shifts in the ethanol–water ratio modulate partitioning into the headspace, and marrying can alter the perceived balance over time (Conner, 2002; Stewart & Russell, 2014). See Fig. 24.1 for an example of peak volatilities of three key aroma compounds at different ethanol concentrations in the ethanol-water mix. Curves vary across compounds, normalized to 1.0 arbitrary units.

When two or more spirits are combined, the chemical environment changes. Ethanol concentration shifts can alter solubility, pushing certain aroma compounds out of solution and into the headspace, where they are more readily perceived. Once combined, blends undergo volatile redistribution and slow oxidative equilibration during the marrying period; most substantive ester/acetal formation occurs during cask maturation rather than neutral-tank resting (Conner et al., 1993; Stewart & Russell, 2014). Others may suppress or enhance each other through perceptual interactions—phenols can mute fruity esters, while lactones can round sharp aldehydic notes.

The "art" component lies in anticipating these interactions before the blend is physically made. Experienced blenders develop an internal model, built from thousands of tasting experiences, of how a component will behave in a given matrix. This is informed by **psychophysical principles**: the human olfactory system integrates aroma perception holistically, and the sequence in which volatiles reach the nose affects the perceived structure of a spirit. Balancing this aromatic sequence—so that bright top notes give way to a satisfying mid-palate and then a persistent base— is one of blending's central challenges.

Fig. 24.1: Normalized Volatility Profiles of Key Aroma Compounds in Ethanol-Water Matrices

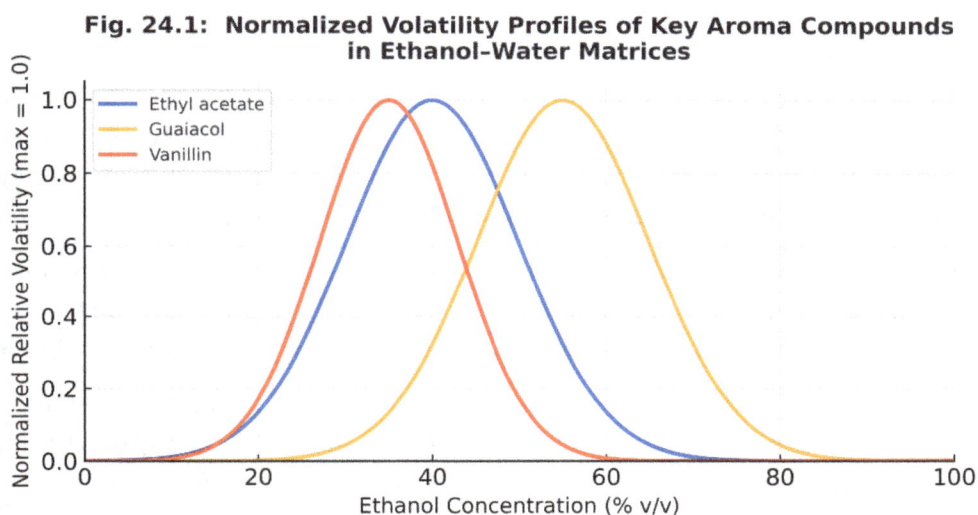

How Blending Is Accomplished

- **Component assessment:** The blending process typically begins with component assessment. This involves nosing and tasting each candidate spirit in isolation, evaluating aroma intensity, congener balance, mouthfeel, and finish. Notes are taken on positive attributes and potential integration challenges—such as excessive dryness, astringency, or phenolic dominance.

- **Sensory mapping:** The blender arranges candidate spirits along sensory axes (e.g., fruity ↔ woody, sweet ↔ dry, light ↔ heavy). This mental or physical map aids in selecting complementary partners. For example, a light, high-ester rum might be placed opposite a rich, molasses-forward rum with deep wood extractives; blending the two could yield both freshness and depth.

- **Trial blending:** Small-scale blends—often as little as 100 mL—are assembled in graduated cylinders or beakers to test ratios. The blender evaluates each immediately, then again after a resting period to observe how volatile integration evolves. Adjustments are iterative, with careful note-taking to ensure reproducibility and accuracy.

- **Marrying period:** Once a candidate blend is finalized, a marrying period allows molecular and sensory integration. In traditional practice, the blend rests in inert stainless-steel tanks or neutral casks for days to months. During this time, volatile redistribution and slow oxidation occur, smoothing edges and enhancing cohesion (Jack & Piggott, 2001).

- **Proofing:** Finally, adjusting to bottling strength with high-quality dilution water is conducted gradually to avoid precipitation of long-chain esters or haze formation. A last sensory check confirms that proof reduction has not unbalanced the profile. Only then is the blend bottled.

Objectives of Blending

At its core, blending serves to achieve a sensory target. For established brands, this means aligning the new batch with a reference standard, often kept as a "library sample" from previous years. The blender uses sensory and, sometimes, chemical analyses (e.g., GC-MS congener profiling) to guide adjustments until the match meets acceptable tolerances.

A second objective is to create structural balance. This refers to the temporal distribution of sensory impressions, encompassing the opening aroma, the initial palate, mid-palate development, and the finish. In a well-blended spirit, no single stage feels abrupt or disconnected.

The third is flavor clarity—ensuring that primary notes are distinct, secondary notes are supportive, and background complexity adds depth without clutter. Too many overlapping mid-intensity notes can lead to muddiness, while too few can make the spirit feel hollow.

Sensory Expectations of the Distiller/Blender

The experienced blender anticipates a **three-tier aroma structure**:
- **Top notes** – Highly volatile esters and aldehydes delivering brightness, fruit, or floral lift
- **Middle notes** – Heart volatiles such as lactones, vanillin, and mild phenols that shape identity and mouthfeel
- **Base notes** – Heavier compounds like oak lignin derivatives, long-chain fatty acids, and Maillard reaction products, providing weight and persistence

Mouthfeel is equally considered. Viscosity is influenced by glycerol, polysaccharides, and ethanol strength; trigeminal warmth is balanced against smoothness. The finish is assessed for **retronasal persistence**—the continued perception of aroma via exhalation after swallowing (Buettner, 2017). A short, abrupt finish may indicate insufficient heavier congeners, while an overly lingering, drying finish may signal tannin imbalance.

The distiller also monitors sensory dissonance, which occurs when individual components are good but, when combined, produce off-notes due to antagonistic interactions among volatiles. This is where both chemistry and perceptual experience are indispensable. See Fig. 24.2 to see how a blender might "time" the sensory experience in the structure of a well-balanced blend.

Challenges and Limitations

Not all components are compatible. Spirits heavy in phenolics may suppress delicate esters, while high-acid components can clash with heavily oxidized lots. Over-blending—combining too many diverse elements—can dilute distinctive traits, leading to a bland, "average" profile.

Stock limitations can also constrain creativity. A distillery with a limited number of barrel types or age ranges has less blending flexibility, making it more challenging to fine-tune profiles without compromising volume.

Finally, there is the oxidative risk associated with marriage. While controlled oxygen ingress can be beneficial, excessive exposure may lead to undesirable aldehyde formation or loss of top notes, particularly in lighter styles.

General Blending Differences by Spirit Type

Blended Scotch Whisky demonstrates industrial precision, often combining dozens of malt and grain whiskies to achieve a consistent profile. Here, blending is as much about *eliminating* variance as about adding complexity. However, there are unique problems (See Case Study 24.1).

Multi-cask rum blenders face the challenge of harmonizing disparate cask finishes—bourbon, sherry, and cognac—each contributing unique wood-derived volatiles. Successful blends preserve freshness while integrating depth.

Cognac Assemblage emphasizes terroir preservation, blending eaux-de-vie from different crus to reflect the house style. Here, blending is restrained, aiming for balance while allowing the raw materials' origins to shine through.

American Whiskey Batch Mingling focuses on blending barrels from the same mash bill to create a balanced profile, interplaying between younger, vibrant whiskey and older, mellower stock.

Consistency in blended spirits is achieved not by chance but through rigorous selection, sensory mapping, and experience in anticipating how volatiles will interact over time (Russell & Stewart, 2014). The complexity of an acceptable blend lies in its balance, achieved through the careful orchestration of congeners, ages, and origins so that no single component dominates. (Stewart & Russell, 2014; Piggott, 1989).

Blenders recognize the olfactory issues associated with high-ethanol spirits. Different spirits and different blender-distillers approach the problem in different ways. Scotch blenders are adamant and unwavering regarding their use of tulip-shaped glassware (Piggott, 1989). Since the industry was primarily responsible for its widespread acceptance, that tradition is deemed more important than making the blending task easier and more efficient. Cognac blenders also use tulips; however, they painstakingly blend spirits at full strength, rather than diluting them, which adds considerable time to the blending process. Scotch blenders add water to create an ethanol-water continuum of about 20% ABV. Water dilution releases long-chain fruity esters at this concentration, masking short-chain smoky phenols and oily, soapy, grassy, nutty, and cereal aromas (Piggott, 1989; Conner, 2002). Diluting a 50% cask strength spirit to 20% ABV requires the addition of 1 ½ oz. of water to 1 oz. of spirit, which is not the preferred ratio for the average Scotch drinker. The use of wide-rim or flared-rim glasses can mitigate the olfactory effects of ethanol; however, misplaced traditions often prevail over practicality and functionality.

Fig. 24.2: Sensory Architecture of a Well-Balanced Blend

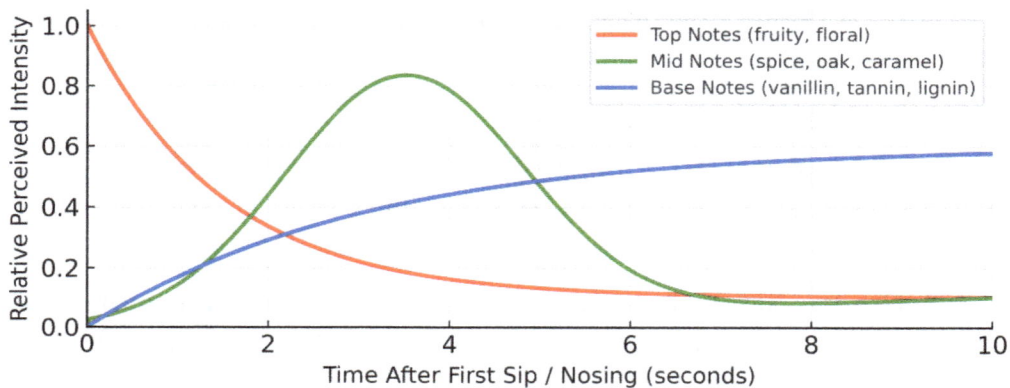

Finishing vs. Flavoring: The Boundary of Sensory Claims

Finishing and flavoring both alter the aroma, but they are distinct acts that do not yield the same kind of evidence. Finishing (secondary maturation in a prior-use cask) probabilistically modifies oak-linked signals, resulting in sweeter spice, dried-fruit tones, nutty notes, rancio, and shifts in tannin structure. These effects cohere with wood chemistry and slow oxygen exposure. They vary with cask history, fill strength, climate, and time. Flavoring (addition of extracts, essences, sweeteners) can introduce non-barrel-congruent cues: confectionery vanillas, overt citrus oils, pastry or candy notes that sit outside normal oak trajectories.

What sensory can (and cannot) say:
- The evaluator can flag plausibility: a port-finish profile is plausible when wood-coherent notes rise in concert and mouthfeel/finish integrate.
- Evaluators cannot prove adulteration by nose alone. When signals look non-congruent or unusually isolated, phrase conclusions as probabilities, not accusations.

Case Study 24.1: Sensory Consequences of Over-Blending in Scotch Whisky: A major Scotch producer seeking batch consistency increased the number of component malts in a flagship blend from 8 to 24. The new formulation targeted chemical uniformity using GC–MS–based congener fingerprinting rather than traditional panel-led blending.

Outcome: Chemical variance between batches decreased by 40%, but descriptive analysis by trained panels revealed a reduction in perceived complexity, aroma contrast, and persistence. Key esters (isoamyl acetate, ethyl hexanoate) were present analytically yet masked by cumulative low-threshold phenols and lactones from heavily recharred casks. Time–intensity profiling revealed shorter flavor persistence and a loss of distinct peaks in both fruity and smoky dimensions (Conner et al., 2000; Lawless & Heymann, 2010; Pozo-Bayón & Ferreira, 2009). **Interpretation**: The blending was successful chemically but failed perceptually. Increasing the number of components shifted the mixture toward statistical averaging, diluting compounds with distinctive character below the perceptual threshold. The result was a technically consistent but sensorially "flattened" whisky—a phenomenon known informally among blenders as *profile compression*. **Lesson for Evaluators:** Uniformity is not synonymous with quality. Excessive blending can erase aromatic contrast and emotional recognizability. Evaluators encountering spirits with technically clean yet indistinct aroma should consider the possibility of over-blending rather than under-maturation or process faults.

Summary

Spirits blending represents the final act in a spirit's creation, where chemical understanding and sensory artistry converge. It is both conservative in preserving brand identity and innovative in crafting new sensory experiences. As analytical tools advance, blenders gain unprecedented insight into the composition and interactions of congeners. However, the foundation remains constant: an acute sensory literacy and the ability to anticipate how each drop contributes to the whole. In this sense, blending is not simply a production step—it is the distiller's signature, encoded in aroma, taste, and texture. See Appendix II for faults.

"To be a master blender, it's sampling whisky—one can never sample enough whisky, nosing enough whisky. I've nosed probably about 170,000 casks now." — Dr. Rachel Barrie, distiller

"Every cask is different, which is why blending is more of an art than a science and built up through a lifetime of experience." — Brian Kinsman, William Grant & Sons

References

Abbott, N., Puech, J.-L., Bayonove, C., & Baumes, R. (1995). Determination and sensory evaluation of cis- and trans-oak lactones in wines. *Food Chemistry, 51*(2), 135–141. https://doi.org/10.1016/0308-8146(94)P4179-F
Buettner, A. (Ed.). (2017). *Springer Handbook of Odor.* Springer. https://doi.org/10.1007/978-3-319-26932-0
SpringerLink+1

Buxton, I., & Hughes, P. (2021). *The science and commerce of whisky* (2nd ed.). Royal Society of Chemistry. https://doi.org/10.1039/9781839168420

Conner, J. M. (2002). Ethanol and water as mediators of whisky aroma perception. *Journal of the Institute of Brewing, 108*(5–6), 411–416. https://doi.org/10.1002/j.2050-0416.2002.tb00572.x

Conner, J. M., Piggott, J. R., & Paterson, A. (2000). Whisky flavor: Blending and balance in Scotch whisky production. *Food Chemistry, 71*(2), 177–184. https://doi.org/10.1016/S0308-8146(00)00146-2

Conner, J. M., Paterson, A., & Piggott, J. R. (1993). Changes in wood extractives from oak cask staves during maturation of Scotch whisky. *Journal of the Science of Food and Agriculture, 62*(2), 169–174. https://doi.org/10.1002/jsfa.2740620210

Jack, F., & Piggott, J. R. (2001). The marrying period in Scotch whisky. *Journal of the Institute of Brewing, 107*(5), 287–293. https://doi.org/10.1002/j.2050-0416.2001.tb00105.x

Lawless, H. T., & Heymann, H. (2010). *Sensory evaluation of food: Principles and practices* (2nd ed.). Springer. https://doi.org/10.1007/978-1-4419-6488-5

Mosedale, J. R., & Puech, J. L. (1998). Wood maturation of distilled beverages. *Trends in Food Science & Technology, 9*(3), 95–101. https://doi.org/10.1016/S0924-2244(98)00024-7

Piggott, J. R. (1989). Sensory analysis of whisky: Nosing and tasting practices. In J. R. Piggott (Ed.), *Flavour of distilled beverages: Origin and development* (pp. 239–255). Elsevier Applied Science. https://doi.org/10.1007/978-94-009-1125-7_16

Pozo-Bayón, M. Á., & Reineccius, G. A. (2009). Interactions between aroma compounds and food matrix. *Critical Reviews in Food Science and Nutrition, 49*(2), 95–105. https://doi.org/10.1080/10408390701856269

Pozo-Bayón, M. Á., & Ferreira, V. (2009). Chemical–sensory relationships in complex beverages. *Trends in Food Science & Technology, 20*(8), 358–366. https://doi.org/10.1016/j.tifs.2009.04.002

Russell, I., & Stewart, G. G. (2014). Whisky flavor development during maturation. *Journal of the Institute of Brewing, 120*(4), 319–329. https://doi.org/10.1002/jib.166

Stewart, G., & Russell, I. (Eds.). (2014). *Whisky: Technology, production and marketing* (2nd ed.). Elsevier. https://doi.org/10.1016/C2012-0-07080-9

Observations on Sensory Blending

Blending balances volatility: Combining spirits with different ester and phenol weights even out vapor release, producing a more coherent aromatic profile at the rim.

Component synergy defines complexity: Minor fractions with contrasting congeners can amplify aroma depth through perceptual contrast rather than additive intensity.

Maturation parity matters: Spirits blended at similar oxidative and esterification stages integrate smoothly, while mismatched ages often show disjointed mouthfeel or layered volatility decay.

Blending stabilizes extremes: Introducing a small proportion of structurally heavier spirit can anchor volatile top notes, preventing rapid aromatic dissipation and creating a steadier, more durable nose.

Chapter 25: The Enigmas of Spirits — Cask Strength and Chill-Filtering

What "cask strength" means—and why it is an enigma

"Cask strength" generally denotes spirits bottled close to the barrel's native proof, commonly ≥55–60% ABV, with minimal or no dilution. From a sensory-science perspective, the promise and the problem arrive together: higher ABV can enhance the volatility of certain congeners, yet the accompanying ethanol vapor quickly dominates the headspace, stimulates the trigeminal system, and anesthetizes olfactory receptors—reducing access to the very aromas enthusiasts hope to experience (Buettner, 2017; Doty, 2015; Wang & Cadwallader, 2024).

Compounding the paradox, trained panels frequently rate high-ABV spirits lower for aroma intensity and complexity unless the sample or the sampling conditions are managed to curb ethanol's dominance (Lee, Paterson, & Piggott, 2001; ASTM International, 2016). The key insight is that "more aroma in the air" is different from "more aroma to the brain for identification." Ethanol's physicochemical effects on partitioning and its physiological bite can mute perception even as more molecules are airborne (Sefton, 2016).

Headspace dynamics: geometry matters more than tradition

Narrow-neck, convergent-rim glasses—the conventional tulip family—create a confined headspace in which ethanol, already abundant in cask-strength samples, accumulates near the sniffing plane. Imaging work has visualized ethanol vapor plumes and rim-zone enrichment as a function of temperature and glass shape, demonstrating that geometry drives where ethanol concentrates above the liquid (Arakawa, Narita, Mitsubayashi, Iitani, & Yano, 2015). Together with ethanol's rapid evaporative flux, this produces intense trigeminal stimulation and adaptation, particularly on repeated sniffs (Doty, 2015; Likar & Jepsen, 2003).

By contrast, an engineered glass has been designed specifically to address this problem. Its neck, flared rim, and wider evaporation surfaces combine to divert ethanol vapors away from the nose, unmasking and displaying lower-threshold aroma compounds, delivering a more faithful aroma sample of cask-strength spirit character. See Chapter 41 for more details (Manska, 2018).

As ethanol concentration rises, odor detection thresholds (ODTs) increase markedly, and Wang & Cadwallader (2024) show that pharmacodynamic suppression (i.e., how ethanol affects receptor function) is more important than changes in headspace vapor partitioning (physicochemical effects). These findings provide scientific support for the use of engineered glassware to lower ethanol headspace concentration during odor evaluation (Wang & Cadwallader, 2024).

Why routine dilution does not work

Water addition changes the matrix. Lowering ethanol relieves trigeminal load and shifts headspace composition, making some volatiles easier to perceive; however, dilution also changes partition coefficients and can suppress or re-rank whole families of aroma compounds, altering the profile you set out to evaluate (Sefton, 2016; Pozo-Bayón & Reineccius, 2009). In other words, dilution addresses an ethanol problem by replacing it with a different sample—a legitimate practice for specific analytical purposes, but at odds with the core attraction of cask strength, which is to access native aromas. When the sensory objective is to evaluate a spirit in its undiluted form, the more effective solution is physical control of headspace ethanol via glass geometry and sampling technique (Arakawa et al., 2015; Manska, 2018).

Using tulips to evaluate cask-strength spirits

Glass selection & setup. It is best to use open or flared rims and/or necked geometries designed to vent ethanol laterally. If using a tulip glass, allow ~3–5 minutes for a quiet headspace to form; avoid vigorous swirling, which drives ethanol into the rim zone (Arakawa et al., 2015). Tulip approach plan: Keep the nose above the rim plane; sample just outside the glass, slightly off-center, to intercept character-rich plumes in the hope of avoiding ethanol. Use short, spaced sniffs (≤1 s) with rest intervals to limit adaptation (Doty, 2015; ASTM International, 2016; Likar & Jepsen, 2003). It is an ineffective "hunt and sniff" game that misses subtle character aromas.

If dilution is necessary, apply minimal, stepwise additions (e.g., micro-steps toward ~40% ABV) as an analytical lens—recognizing that you are now evaluating a different matrix. In such cases, divide the sample into two glasses: one for nosing undiluted and one for tasting, diluted. Always document the dilution path (Lee et al., 2001; Pozo-Bayón & Reineccius, 2009).

Using sensory glassware to evaluate cask-strength spirits.

Swirl to encourage evaporation through shear. Nose at the center of the rim plane, then move the nostrils across the rim plane to the rim, sampling slowly to pick up aromas. Ethanol appears at the rim in low concentration due to the larger rim circumference (lower flux), avoiding olfactory fatigue, and character aromas are available in the rim plane from higher molecular weight down to the lowest at the rim (e.g., ethanol). Swirling enhances aroma evaporation into the headspace for detection. Ethanol is separated from character aromas (Manska, 2018). See Fig. 42.1

Temperature discipline. Minimize hand-warming as small increases in temperature disproportionately amplify ethanol release relative to heavier congeners (Sefton, 2016).

Dilution is unnecessary with wide-rim glassware designed to diffuse ethanol; indeed, it can be a hindrance unless ethanol is painful on the tongue (palate burn). If the sample is painful, separate it into two glasses: one for sniffing and one for tasting. Dilute the tasting sample as required to alleviate pain. Note that dilution will also affect retronasal perception.

Case Study 25.1: Cask Strength Spirits in the Marketplace:
Producer advantages (operational and commercial)

- *Margin leverage, product tiering.* Minimal proofing and premium positioning improve unit economics relative to standard-proof bottlings.
- *Portfolio differentiation.* Limited cask-strength releases cultivate connoisseur credibility and narrative depth alongside core SKUs.
- *Engagement flywheel.* Single-barrel and club allocations foster loyalty, influencer advocacy, and direct-to-consumer momentum.
- *Inventory optionality.* Characterful outliers that do not fit house blends can succeed as cask-strength features without extensive rework.

Consumer advantages (perceived & sensory):

- *Intensity & authorship.* Access to the distillate's native, undiluted aromatic signature—provided ethanol is physically managed at the nose.
- *Customization (with caveats).* Some buyers value control over their own proofing path, though water additions can change the profile.
- *Collectability.* Batch IDs, single casks, and limited runs add scarcity-driven value independent of liquid quality.

Compelling to the consumer: Ethanol-cue predisposition. Trigeminal "bite" is often misunderstood as a matter of power or quality, biasing first-impression judgments toward higher proofs when sampled from ethanol-concentrating glasses. Managing headspace ethanol neutralizes bias and reveals true aromatic complexity (Manska, 2018).

To Chill-Filter or Not?

Dr. James "Jim" Swan (1941-2017), a founding scientist at the Scotch Whisky Research Institute (SWRI), also known as the "Einstein of Whisky," was a pioneer in bringing scientific methods to whisky distilling. His research into oak cask chemistry is particularly notable, and he is credited with identifying the chemical nature of chill-haze. However, no peer-reviewed papers attributable to this topic have been published.

Non-chill-filtering receives considerable marketing hype. The spirit is cooled to below freezing, then passed through a filter to remove compounds that cause it to appear hazy or cloudy. Filters made of cellulose or diatomaceous earth generally range from 0.45 to 1 μm, with the most common range being 0.65–1.2 μm, and are stored for hours to days before use. Usually one pass, targets long-chain esters, fatty acids, and waxes. Generally reserved for spirits under 46% ABV, but some manufacturers filter all their products.

In apricot brandy, chill filtration significantly reduced fatty acid esters (ethyl palmitate, ethyl laurate)—compounds linked to flavor and mouthfeel. However, most other volatiles (e.g., alcohols, aldehydes, terpenes) were not significantly impacted. Notably, the sample filtered through larger-pore membranes retained sensory characteristics most closely resembling those of the unfiltered controls (Miljić et al., 2013).

In Brandy de Jerez, cold stabilization and filtration significantly decreased long-chain ethyl esters, the same haze-forming compounds, without affecting flavor (Muñoz-Redondo et al., 2023).

No direct peer-reviewed whisky flavor studies are known at this time. The only referenced formal sensory trial involved apricot brandy, not whisky, and there appears to be no widely published, scholarly study that isolates the flavor effects of chill-filtering on whisky, yet marketers tout it as a positive. Chill-filtering can plausibly affect flavor by removing lipids and long-chain esters, but the magnitude of this effect in whisky remains unproven in peer-reviewed trials. The question becomes "Are there a significant amount of esters removed in whiskey that measurably affect taste perception?" Perhaps not.

In a blind tasting of 1,331 samples by 111 experienced whisky connoisseurs, identification of chill-filtered samples achieved only a 50% success rate—equivalent to chance—and ratings between filtered and non-filtered versions were statistically identical (Lüning, 2014).

Evaluation Note 25.1: Filtering can have textural and late-aroma consequences in some spirits, but the process is not inherently good or bad, nor is it definitive.

- Chill filtering / fine filtration can remove colloids and lipids that contribute to waxy/oily mouthfeel and linger. Expect possible reductions in perceived viscosity and in the carry of late retronasal notes.
- Carbon treatment can attenuate sharp, small-molecule off-notes (solvency, sulfurous), sometimes at the cost of brightness in delicate top-notes.

Limits of inference: sensory evidence can indicate direction ("mouthfeel seems leaner than expected;" "solvent edge unusually low"), but it cannot certify processing methods. Treat filtration status as an aesthetic hypothesis—credible only when textural and temporal signatures align—but avoid declarative claims without process disclosure. In practice, perceptible differences are case-specific and most evident in spirits with higher fatty-acid or ester content. In low-lipid matrices, non-chill filtration seldom produces a reliably detectable sensory divergence.

Summary

Cask-strength spirits embody both promise and paradox. Their higher ABV provides greater volatility of character volatiles but simultaneously floods the headspace with anesthetic ethanol, suppressing perception. Traditional tulip-shaped glasses magnify this problem by funneling ethanol directly to the nose, creating the enigma that enthusiasts face: more aroma potential, but less aroma access. Sensory-based glass design and other flared-rim geometries resolve this contradiction by venting ethanol away from the sniffing zone, enabling evaluators to experience the true complexity of undiluted spirits. Mastery of cask-strength evaluation requires recognizing ethanol's dominance, avoiding overreliance on dilution, and leveraging geometry to reveal.

Insufficient information is available about the chill-filtering effect on mouthfeel to determine whether it impacts flavor in most spirits. Chill-filtering is a cosmetic/marketing issue, as the removal of haze-forming compounds has rarely been shown to produce a detectable sensory difference under normal tasting conditions. The consumer paradox is that clarity is often mistaken for quality, whereas the faint haze many reject is evidence of authentic, full-bodied whiskey.

"The raw ethanol blast of cask-strength whisky can flatten the nose; to understand what's truly in the glass, you have to learn to work around the alcohol." —Dave Broom, author

"Ethanol is the great deceiver—it numbs the senses and hides the character. A whisky's soul isn't in the alcohol; it's in the aromas beneath." —Jim McEwan, master distiller

"High concentrations of ethanol vapor desensitize olfactory receptors and produce strong trigeminal irritation, making accurate odor discrimination difficult." —-Richard Doty, scientist

"Ethanol's volatility ensures it dominates headspace composition; in sensory evaluation, this dominance suppresses access to lower-threshold aroma compounds."-Andrea Buettner, professor,

References

Arakawa, T., Narita, H., Mitsubayashi, K., Iitani, K., & Yano, Y. (2015). Visualization of alcohol vapor from wine: A new imaging technique for analysis of wine tasting. *Journal of Food Engineering, 166,* 123–128. https://doi.org/10.1016/j.jfoodeng.2015.05.025

ASTM International. (2016). *Standard practice for sensory evaluation of distilled spirits* (ASTM E253-16) [Standard]. https://doi.org/10.1520/E0253-16

Buettner, A. (2017). Influence of ethanol on olfactory perception: Mechanisms and consequences. In *Flavour science* (pp. 87–94). Elsevier. https://doi.org/10.1016/B978-0-08-100295-9.00012-0

Doty, R. L. (2015). *Handbook of olfaction and gustation* (3rd ed.). Wiley-Blackwell. https://doi.org/10.1002/9781118971757

Lee, K., Paterson, A., & Piggott, J. R. (2001). Origins of volatile flavour compounds in Scotch malt whisky. *Trends in Food Science & Technology, 12*(10), 391–401. https://doi.org/10.1016/S0924-2244(01)00127-5

Likar, K., & Jepsen, D. (2003). Sensory fatigue and adaptation in ethanol perception. *Chemical Senses, 28*(6), 561–569. https://doi.org/10.1093/chemse/bjg047

Lüning, H. (2014). *Comparative assessment of a blind tasting: Influence of chill filtration on perceived taste* [Technical report]. Whisky.com. https://www.whisky.com/study-on-the-chill-filtration.html

Manska, G. (2018). Influence of whisky glass shape on ethanol concentration in the headspace and its effect on sensory evaluation. *Beverages, 4*(4), 93. https://doi.org/10.3390/beverages4040093

Miljić, U. D., Puškaš, V. S., Vučurović, V. M., & Razmovski, R. N. (2013). The application of sheet filters in treatment of fruit brandy after cold stabilisation. *Acta Periodica Technologica, 44,* 87–94. https://doi.org/10.2298/APT1344087M

Muñoz-Redondo, J. M., Puertas, B., Valcárcel Muñoz, M. J., Rodríguez Solana, R., & Moreno Rojas, J. M. (2023). Impact of stabilization method and filtration step on the ester profile of "Brandy de Jerez." *Applied Sciences, 13*(6), 3428. https://doi.org/10.3390/app13063428

Pozo-Bayón, M. A., & Reineccius, G. A. (2009). Effects of ethanol and sugars on flavor release in a model system. *Food Research International, 42*(3), 299–304. https://doi.org/10.1016/j.foodres.2008.11.009

Sefton, M. A. (2016). Ethanol as a mediator of volatile release in wine and spirits. *Australian Journal of Grape and Wine Research, 22*(1), 1–11. https://doi.org/10.1111/ajgw.12160

Wang, Z., & Cadwallader, K. R. (2024). Ethanol's pharmacodynamic effect on odorant detection in distilled spirits models. *Beverages, 10*(4), Article 116. https://doi.org/10.3390/beverages10040116

Observations on Buying Cask Strength as an Economic Advantage

Cask-strength whiskey (55–65% ABV) is bottled straight from the barrel, offering a full concentration of congeners and volatiles. Some buyers dilute it heavily to "make it go further," assuming economic gain—but that logic fails chemically and sensorially. Watering down shifts the ethanol–water balance, suppresses volatiles, and alters mouthfeel. Commercial proofing involves controlled dilution and resting, which stabilizes the spirit—something home dilution cannot replicate. Dilution weakens flavor integrity. Cask-strength whiskey is intended for precise exploration, not as a cheaper substitute for lower-proof bottles.

Chapter 26: Storing Spirits — Effects on Sensory

In distilled spirits, preserving aroma is critical for both enjoyment and accurate sensory analysis. Storage variables, such as closure type, headspace management, light exposure, and temperature, affect volatile aroma compounds essential to spirits' quality. Both scientific and empirical practices are used by professionals and enthusiasts alike. Closure taints and faults are described in Fig. 26.1.

Closure Types and Their Influence on Volatile Retention

Closures function as both physical and chemical barriers to ethanol evaporation and oxygen ingress. Three common categories of closures—natural cork, synthetic alternatives, and screw caps—differ significantly in permeability and in their interactions with spirit constituents.

Natural corks, while traditional and aesthetically appealing, are porous and allow gradual ethanol loss when bottles are stored upright. Ethanol is volatile and permeates the cork structure over time, diminishing aromatic intensity, particularly for esters and aldehydes with low boiling points (Singleton, 1995). Cork also permits slow oxygen ingress, which can oxidize sensitive compounds, contributing to degradation or aroma shifts. Moreover, cork taint from 2,4,6-trichloroanisole (TCA) remains a risk, even at nanogram levels (Nychas & Drosinos, 2003).

Synthetic corks and bar tops reduce risks. Synthetics offer more consistent sealing properties, though some plastic polymers can, in rare cases, leach trace off-aromas (Silva, Julien, Jourdes, & Teissedre, 2011). Bar tops with integrated synthetic heads and check valves (used for controlled pours) may slightly reduce headspace exchange because they function as partial vapor traps (Pérez-Coello & Díaz-Maroto, 2007). Screw caps, often dismissed for aesthetic reasons, provide the tightest seal and minimize both ethanol loss and oxidation (Skouroumounis et al., 2005).

Impact of Storage Orientation

Bottle orientation has a critical impact on long-term closure integrity. Unlike wine, spirits should not be stored on their sides, as ethanol concentrations (typically 40% ABV or higher) are aggressive solvents that degrade the cork's structure, weakening seal effectiveness and increasing the risk of crumbling (Likar & Vodopivec, 2012).

Headspace and Oxidative Reactions

After opening, the volume of air in the bottle (headspace) expands with each pour. This exposes volatile compounds to increasing oxygen levels, promoting oxidation reactions. Oxidative loss affects fruity, floral, and herbal notes—particularly in peated whiskies, rums, and aged brandies

(Ferreira & López, 2019). Best practices to mitigate oxidation include:

- Refilling remaining liquid into smaller glass bottles
- Using inert gases (argon, nitrogen) to displace oxygen
- Resealing promptly with high-integrity closures

After the Cork: Bottle Stability and Post-Opening Drift

Cask aging ends at bottling; oxidation after opening is not aging. As headspace increases, oxygen, temperature cycling, and ethanol loss alter the balance. What typically moves:

- Top notes fade first; citrus peels, fresh florals, volatile esters lose jump.
- Sweet oak flattens, vanillin and lactones soften; caramel/toffee can dull.
- Structure can separate, heat may feel sharper as top-notes recede; late bitterness can surface.
- Some faults that were masked reveal themselves (mustiness or sulfur)

Implication for interpretation: Age statements predict cask time rather than post-opening behavior. When comparing impressions, distinguish **bottle-fresh** from opened-for-weeks; the number on the label is not a shield against drift. This is context for judgment, not a manual for storage. See Appendix II, Spirits, Faults, and Rejection Thresholds.

Temperature and Light Exposure

Spirits should be stored in cool, dark environments. Higher temperatures increase the vapor pressure of volatile molecules, thereby accelerating the evaporation of esters, aldehydes, and terpenes (Puech, Moutounet, & Souquet, 1999). Exposure to direct sunlight or fluorescent light initiates the photodegradation of phenols, terpenes, and riboflavin (a standard ingredient in caramel coloring), altering the sensory character (Ribéreau-Gayon, Glories, Maujean, and Dubourdieu, 2006). Fig. 26.1 describes closure-related taints and faults in spirits.

Longevity and Degradation Timeline

Under stable storage, unopened bottles with natural cork can retain their near-original character for 5–10 years, although measurable oxygen ingress can subtly affect the flavor over time.

- Synthetic and screw cap closures offer longer shelf life, potentially exceeding 20 years with minimal change if protected from light and heat.
- Once opened, oxidation and ethanol evaporation accelerate and are perceptible within 3–12 months, depending on fill level and conditions.

Aesthetic Preferences vs. Functional Integrity

While many consumers prefer cork or bar tops, scientific data clearly indicates that closures that prevent ingress and egress are the most effective. For serious drinkers and collectors seeking to preserve spirits at their peak, opt for performance over tradition.

Case Study 26.1: Environmental Exposure vs Erosion of Aroma Integrity: In 2019, a Scottish independent bottler documented unexpected aroma drift in a limited-edition single malt held in clear glass for retail display. Over nine months of ambient store lighting, GC-MS analysis revealed a 40% reduction in ethyl hexanoate and an increase in trans-2-nonenal, a lipid oxidation aldehyde with a papery, stale odor. Blind sensory evaluation confirmed diminished fruit top-notes and a duller mid-palate. Control samples (identical stock in opaque cases) had no significant change.

A similar pattern was reported in a rum warehouse study in Martinique: barrels near external walls, exposed to afternoon solar heat, reached temperatures above 35 °C, accelerating ethanol evaporation and internal oxidation. Analytics showed increased acetaldehyde and furfural levels after just six months compared with the cooler, central racks. Tasters described the exposed casks as "flattened, overly woody, and prematurely aged."

Fig. 26.1	Closure Related Taints and Faults in Spirits			
Compound	Typical Origin	Sensory Description	DT**	Notes
Brettanomyces*	Contaminated barrels, cork, or bottling environment	"Barnyard," horse blanket, band-aid, smoky, spicy, medicinal	~400–600 µg/L (matrix-dependent)	Complexity at trace levels in wine/beer, considered a flaw in spirits
TCA (2,4,6-trichloroanisole)	Natural corks contaminated with chlorine mold	Musty, moldy, wet cardboard, damp basement	~2–5 ng/L	Suppresses fruit and aroma intensity, common cork taint
TeCA (2,3,4,6-tetrachloroanisole)	Wood pkg, cork, barrel contamination	Musty, earthy, moldy	~5–10 ng/L	Similar to TCA, often co-occurs
TBA (2,4,6-tribromoanisole)	Packaging (treated wood), fire retardants	Musty, chemical, medicinal	~10–15 ng/L	Less frequent than TCA, equally damaging
Haloanisoles*** = (general group name)	Corks, cardboard, and wood storage	Musty, earthy	Varies	Blanket term for TCA, TeCA, TBA
Polymer plastic leachates	Synthetic closures, poor quality liners	Plastic, rubber, chemical, vinyl	Not well defined	Rare, more common under heat or long storage
Oxidative taint	Excess oxygen ingress from a poor seal	Flat, stale, woody, sherry-like	N/A	Depletes fruity esters, accelerates degradation

*Volatile phenols: 4-ethylphenol, 4-ethylguaiacol, DT**= Detection thresholds approximate, matrix dependent (wine, beer, spirits), Haloanisoles***= an all-encompassing category group name.

Summary
Storage conditions influence the sensory integrity of distilled spirits by examining the science behind volatile retention, oxidation, and photodegradation. Different closures—natural cork, synthetic alternatives, and screw caps—vary in permeability and chemical interactions, which, in turn, influence ethanol evaporation and oxygen ingress. Upright bottle orientation for spirits, headspace management after opening, and use of inert gases or smaller containers to slow oxidative loss of aromas maintain flavor. Temperature and light accelerate the

volatilization and photochemical changes of esters, phenols, and terpenes, leading to shifts in aroma and flavor over time. Cask age does not protect against post-bottling drift, and scientifically grounded storage practices are essential to maintain both quality and consistency in sensory evaluation.

"Contrary to popular belief, the high alcohol content of spirits does not prevent TCA from forming or spirits from getting corked." —Spirits & Distilling magazine

References

Ferreira, V., & López, R. (2019). Oxidation and reduction in wine: Chemical principles and sensory effects. *Comprehensive Reviews in Food Science and Food Safety, 18*(3), 753–768. https://doi.org/10.1111/1541-4337.12437

Likar, M., & Vodopivec, B. (2012). Effect of bottle position on cork integrity in distilled spirits. *Packaging Technology and Science, 25*(7), 369–376. https://doi.org/10.1002/pts.981

Nychas, G. J. E., & Drosinos, E. H. (2003). Microbial taint formation in cork and wood closures. *Food Microbiology, 20*(5), 591–598. https://doi.org/10.1016/S0740-0020(03)00038-1

Pérez-Coello, M. S., & Díaz-Maroto, M. C. (2007). Effect of closure type on volatile composition and sensory characteristics of wine. *European Food Research and Technology, 224*(4), 551–556. https://doi.org/10.1007/s00217-006-0359-2

Puech, J. L., Moutounet, M., & Souquet, J. M. (1999). Influence of storage conditions on the oxidation of spirits. *Journal of Agricultural and Food Chemistry, 47*(7), 2885–2891. https://doi.org/10.1021/jf981155c

Ribéreau-Gayon, P., Glories, Y., Maujean, A., & Dubourdieu, D. (2006). *Handbook of enology: The chemistry of wine stabilization and treatments* (Vol. 2, 2nd ed.). John Wiley & Sons. https://doi.org/10.1002/0470010398

Silva, M. A., Julien, M., Jourdes, M., & Teissedre, P. L. (2011). Impact of closures on wine quality after extended storage. *Food Chemistry, 127*(3), 1068–1078. https://doi.org/10.1016/j.foodchem.2011.01.075

Singleton, V. L. (1995). Maturation of wines and spirits: Comparisons, facts, and hypotheses. *American Journal of Enology and Viticulture, 46*(1), 98–115. https://doi.org/10.5344/ajev.1995.46.1.98 AJEV Online+1

Skouroumounis, G. K., Kwiatkowski, M. J., Francis, I. L., Gawel, R., Godden, P. W., & Williams, P. J. (2005). The impact of screw caps versus corks on long-term wine composition. *Australian Journal of Grape and Wine Research, 11*(1), 139–148. https://doi.org/10.1111/j.1755-0238.2005.tb00284.x

Observations on Storing Spirits

A dull, musty nose often signals cork taint or spirit oxidation from poor closure integrity—never dismiss it as "barrel funk." Cloudiness or haze in a mature spirit rarely comes from chill-filtration faults; more often, it reflects temperature cycling or light exposure, which causes long-chain fatty acid esters to precipitate from solution. A faint "wet cardboard" or "stale newspaper" note indicates oxidation of aldehydes; check fill levels and ullage before attributing poor distillate quality. Sweet caramel notes turning flat and papery suggest over-aged or light-damaged product—verify glass color and storage lighting.

Chapter 27: Artificial and Lab-Created Spirits

Introducing a New Category of Spirits

The rise of artificial or lab-created spirits has opened a new and controversial chapter in the history of distilled beverages. These products are typically built from a foundation of neutral grain spirit (NGS), produced in bulk by industrial-scale ethanol manufacturers. Into this neutral base producers blend isolates, extracts, or synthetic compounds that mimic the aromatic and flavor-active molecules found in authentic aged spirits (Nykänen, 1986). The objective is to recreate the full sensory impression of whiskey, rum, or brandy without traditional fermentation, distillation, or aging (TTB, 2021).

Unlike traditional distillers, who rely on yeast metabolism, distillation cuts, and long-term barrel chemistry, producers of lab-created spirits approach the problem as a design exercise. They begin not with raw materials but with an analytical target: a Scotch, a bourbon, or a cognac deconstructed into measurable chemical components (Conner, Paterson, & Piggott, 1999; Poisson & Schieberle, 2008).

The Reverse Engineering Blueprint

The foundation of artificial spirits is based on analytical chemistry. Benchmark products are analyzed using high-resolution techniques, such as gas chromatography–mass spectrometry (GC–MS) and headspace solid-phase microextraction (HS–SPME). These methods isolate and quantify dozens—sometimes hundreds—of volatile compounds that define the spirit's aromatic fingerprint (Lee & Noble, 2006; Poisson & Schieberle, 2008).

From these analyses, a "blueprint" emerges:
- Esters, e.g., isoamyl acetate (banana), ethyl butyrate (pineapple), for fruity, floral, sweet notes.
- Aldehydes and lactones such as vanillin, furfural, or γ-nonalactone to suggest oxidative aging or wood extraction (Conner, Paterson, & Piggott, 1999).
- Phenolic compounds for smoky, spicy, or medicinal attributes.
- Sweeteners, acids, and buffers to balance mouthfeel and perception.
- Colorants (e.g., caramel E150a – E150d) to replicate barrel aging appearance.

Some producers employ machine learning systems to refine these blends, optimizing relative concentrations to approximate the sensory coherence of a natural spirit. The result is not the emergent complexity of a biologically produced beverage, but a constructed profile engineered for resemblance.

The Ethanol Base

Most artificial spirits rely on industrial neutral grain spirits (NGS) or neutral cane spirits, which are distilled to at least 95% purity in continuous-column stills. This base is virtually devoid of congeners or fermentation residue. Once diluted to bottling strength, it serves as a neutral carrier for the added flavor blueprint. (Nykänen, 1986).

Absent are the fusel oils, organic acids, and oxidative products that typically develop during fermentation, distillation cuts, and years of cask maturation. Instead, these are substituted by synthetic or extracted compounds added in calibrated ratios (Conner, Paterson, & Piggott, 1999; Poisson & Schieberle, 2008). While the result may smell and taste similar to aged whiskey or rum, the architecture is entirely different: a chemically spiked solution rather than a product of microbial metabolism and time-dependent chemistry.

Advocates argue that such products are more sustainable, as they avoid deforestation for barrels, minimize water and angel's share losses, and reduce energy-intensive warehousing (Lachenmeier & Sohnius, 2008). Critics counter that sustainability cannot replace authenticity—that the time-dependent evolution of congeners, esters, and lignin-derived volatiles cannot be faked without flattening the sensory experience (Conner, Paterson, & Piggott, 1999).

Regulatory Classification and Labeling Concerns

Regulation has not kept pace with innovation. In the United States, the TTB requires that spirits with added flavors be labeled as "compounded" or "specialty" products; labeling must not mislead consumers about the product's class or type (TTB, 2021). In the EU, category and appellation protections restrict the use of traditional terms (e.g., "whisky," "Cognac") to products that meet legally defined production standards (German Federal Ministry of Food and Agriculture, 2020). However, enforcement and terminology vary, and some artificial products are marketed as "whiskey alternatives" despite not undergoing fermentation or barrel aging, creating confusion.

The ethical issue is transparency. Traditional categories are defined by strict legal standards, often tied to geographic identity and historical practice. Lab-created spirits, by contrast, may use ambiguous descriptors, leaving the consumer uncertain whether what they are drinking is distilled, reconstituted, or simply flavored ethanol (TTB, 2021; German Federal Ministry of Food and Agriculture, 2020).

Ethical and Cultural Implications

Artificial spirits raise fundamental questions of authenticity, transparency, and cultural value. For centuries, distilled beverages have been tied to place, tradition, and craftsmanship. Lab-created products break this link, offering instead a chemically engineered replica.

- Authenticity: Traditional whiskey, rum, or brandy is not merely a collection of volatiles but the product of yeast metabolism, distillation artistry, and time in oak (Conner, Paterson, & Piggott, 1999). By contrast, construct spirits are sensory imitations assembled from isolates and additives (Lee & Noble, 2006; Poisson & Schieberle, 2008). Critics argue that this undermines the integrity of the category, reducing heritage-rich spirits to mere simulacra.

- Transparency: Marketing often leans on the imagery of tradition, even when the product is compounded in a flavor house. Consumers are rarely informed that the whiskey-like notes do not derive from aging in barrels, but rather from the addition of vanillin, furfural, and caramel color to neutral spirit (TTB, 2021).

- Consumer Education: Students of sensory science must learn to distinguish between natural and constructed complexity. Whereas natural spirits unfold gradually with dilution, aeration, and retronasal cues, artificial spirits often display "front-loaded" notes—such as vanilla, coconut, or fruit—that lack mid-palate support or a long finish (Guinard & Zhao, 2022).

From a cultural perspective, some innovators frame constructed spirits as democratizing, offering affordable approximations of luxury categories.

Traditionalists counter that this "democratization" strips away the very essence of what makes spirits culturally meaningful. The debate parallels controversies over synthetic diamonds, lab-grown meat, or AI-generated art: is the copy equivalent, inferior, or simply different?

Sensory Training and Evaluation

For sensory education, lab-created spirits offer valuable case studies. Students must learn not only to identify what is present, but also to detect what is missing:

- A lack of dynamic aroma progression over time.
- Sharp, isolated notes without integrated complexity.
- Absence of fermentation-linked volatiles (esters, fusel alcohols, sulfur compounds).
- Hollow ethanol base lacking microstructural support.
- Uncanny uniformity across batches, betraying the absence of natural variation.

Comparative training protocols should include artificial spirits alongside authentic aged products. Evaluators can then practice identifying flatness, synthetic overtones, or incongruity between nose, palate, and finish (Guinard & Zhao, 2022). This enhances critical ability and prepares evaluators to navigate an evolving market in which consumers may not know what they are being served.

Case Study 27.1: Analytical Spirits Attempts at Molecular Reconstruction: In 2018, the San Francisco–based company Endless West released *Glyph 1 Whiskey*, described as the first "molecular spirit." Rather than fermenting or aging grain mash, the company identified volatile compounds characteristic of aged whiskey through GC-MS and sensory gas chromatography, then blended food-grade analogs (vanillin, ethyl hexanoate, γ-nonalactone, eugenol, etc.) into a neutral ethanol–water matrix.

Chemically, the model reproduced dozens of known congeners; however, sensory panels consistently rated its aroma as less integrated and more linear, lacking the temporal evolution and soft volatility gradients characteristic of barrel-aged whiskey. Subsequent reformulations adjusted the levels of glycerol and long-chain fatty acid esters to improve mouthfeel, illustrating that congener interaction—not mere presence—defines perceived authenticity.

Case Study 27.2: Yeast-Engineered Flavor—Berkeley Yeast and Designer Pathways: Between 2019 and 2023, Berkeley Yeast, a U.S. synthetic-biology firm, developed *"functional yeasts"* for wine and beer capable of producing terpenes, esters, and thiols commonly derived from hops or grapes. One strain expressing the linalool synthase gene produced a floral–citrus nose without the use of external botanicals.

When adapted for distillation trials, new-make spirits from these ferments exhibited increased monoterpene intensity but reduced ester diversity, yielding a distinct, perfumed top note with limited mid-palate complexity. The work demonstrated that pathway-specific enhancement can increase aroma strength but may compress sensory dimensionality, a recurring limitation of single-target engineering.

Case Study 27.3: Non-Fermentative Spirit Bases—Analytical Distillation of Botanicals
In 2021, a London start-up under the *Sustainable Spirits* consortium produced a "carbon-neutral gin" without ethanol fermentation. The base was synthesized by converting recovered CO_2 to ethanol via microbial catalysis, then compounded with individually vacuum-distilled botanical fractions. Despite identical GC-MS peak profiles to the control gin, trained assessors noted a thinner mouthfeel, a sharper onset of alcohol, and shorter retronasal persistence.

The experiment underscored that chemical identity alone cannot replicate physical structure—matrix effects, minor colloids, and trace polysaccharides in traditional distillates contribute to sensory cohesion beyond volatile composition.

Across these cases, a consistent theme emerges: artificial or engineered replication succeeds in reconstructing composition but not complexity. Sensory authenticity depends on emergent interactions—such as oxidation kinetics, ester–phenol equilibria, and temporal volatility gradients—that no current molecular assembly fully reproduces. For evaluators, such studies pose an important exercise: distinguishing between analytical similarity and *perceptual truth*.

Future Possibilities for Lab-Created Spirits

The future of lab-created spirits lies at the intersection of synthetic biology, analytical chemistry, and sensory neuroscience—fields now converging toward precision flavor engineering rather than imitation. Emerging technologies, such as metabolically optimized yeasts, controlled micro-oxidation reactors, and AI-driven aroma modeling, will enable designers to construct flavor systems de novo, assembling congeners in ratios tailored to evoke specific perceptual outcomes rather than replicate existing styles. These advances suggest a path toward purpose-built sensory architectures—spirits designed for particular emotional, culinary, or cultural contexts. However, the central challenge remains: replicating the *temporal evolution* and *emergent complexity* of natural fermentation and aging. While molecular reconstruction can reproduce volatiles, it still struggles to simulate the dynamic oxidation, esterification, and colloidal interactions that give traditional spirits depth and authenticity. The coming decade will likely see hybrid approaches, where controlled biological generation meets synthetic refinement—an era in which the craft of distilling may expand from guiding nature to *programming flavor itself*.

Summary

Artificial or lab-created spirits represent a growing yet contested frontier in the beverage alcohol industry. Engineered from neutral ethanol and reconstructed volatile blends, they mimic the profile of traditional spirits but lack the emergent chemistry of fermentation and maturation (Nykänen, 1986; Conner, Paterson, & Piggott, 1999). While some celebrate their efficiency, consistency, and sustainability (Lachenmeier & Sohnius, 2008), others view them as simulacra that risk undermining centuries of cultural and sensory heritage (TTB, 2021; German Federal Ministry of Food and Agriculture, 2020). For the sensory scientist, these products are less a threat than a teaching tool. They force evaluators to confront the boundaries between resemblance and authenticity, chemistry and craftsmanship. The task is not merely to decide whether such products "count" as spirits, but to train the senses to perceive the difference between constructed flatness and natural evolution (Lee & Noble, 2006; Poisson & Schieberle, 2008; Guinard & Zhao, 2022). In doing so, students gain a sharper understanding of what makes authentic spirits worth studying, preserving, and refining.

"We can replicate the molecules, but not the memory."— Molecular distillation researcher, Endless West internal statement paraphrased in MIT Technology Review, 2019.

"The complexity of a whisky is not a list of its chemicals—it's the story told by how those chemicals evolve together."— Dr. Pat Heist, Ferm Solutions (commenting on synthetic spirits, 2020 seminar).

Our resource-efficient process allowed us to produce Glyph... with 94% less water and 92% less agricultural land use..."— Glyph (Our Process) from website

References

Conner, J. M., Paterson, A., & Piggott, J. R. (1999). Changes in wood extractives from oak cask staves through maturation of Scotch malt whisky. *Journal of the Science of Food and Agriculture, 79*(3), 287–294. https://doi.org/10.1002/(SICI)1097-0010(19990301)79:3<287::AID-JSFA235>3.0.CO;2-9

German Federal Ministry of Food and Agriculture. (2020). *Food labelling guidelines for spirit drinks.* Retrieved from https://www.bmel.de

Guinard, J.-X., & Zhao, M. (2022). Sensory education: Training methods for the next generation of beverage evaluators. *Food Quality and Preference, 99*, 104583. https://doi.org/10.1016/j.foodqual.2022.104583

Lachenmeier, D. W., & Sohnius, E. M. (2008). The role of eco-efficiency in modern beverage production: A comparative life cycle assessment. *Food Additives & Contaminants, 25*(11), 1307–1315. https://doi.org/10.1080/02652030802189058

Lee, K. S., & Noble, A. C. (2006). Use of solid-phase microextraction and gas chromatography/mass spectrometry for the study of aroma compounds from oak-aged spirits. *Journal of Agricultural and Food Chemistry, 54*(10), 3929–3935. https://doi.org/10.1021/jf052489+

Nykänen, L. (1986). Formation and occurrence of flavor compounds in wine and distilled alcoholic beverages. *American Journal of Enology and Viticulture, 37*(1), 84–96. Retrieved from https://www.ajevonline.org/content/37/1/84

Poisson, L., & Schieberle, P. (2008). Characterization of the most odor-active compounds in an American bourbon whisky by application of aroma extract dilution analysis. *Journal of Agricultural and Food Chemistry, 56*(14), 5813–5819. https://doi.org/10.1021/jf800382m

Trivedi, D. K., & Goodacre, R. (2020). Artificial intelligence and e-nose technologies for enhanced flavor profiling in food and beverages. *Trends in Food Science & Technology, 105*, 200–211. https://doi.org/10.1016/j.tifs.2020.09.004

U.S. Alcohol and Tobacco Tax and Trade Bureau. (2021). *Beverage Alcohol Manual (BAM) – Spirits.* Retrieved from https://www.ttb.gov/beverage-alcohol-manual

Chapter 28: Analytical Tools in Sensory Science — The Future of the Distiller

Distillation has historically been as much an art as a science. The master distiller's role was once defined by sensory memory, instinct, and tradition. Today, however, sensory analysis is no longer limited to human panels: modern analytical tools allow chemical deconstruction, predictive modeling, and real-time monitoring of flavor-active compounds. In this chapter, we examine the instruments and methods that are transforming distillation in sensory science, their costs (as of this publication) and accessibility, and the implications for competitiveness in the spirits industry. We also look ahead to a future in which integrated, automated, and predictive systems, blending artisanal intuition with technological precision, will support the distiller.

Classical Analytical Tools in Spirit Chemistry

Gas Chromatography-Mass Spectrometry (GC–MS): Separates volatile compounds in spirits, identifies them by mass spectral "fingerprints," and quantifies their abundance. Primary tool for identifying esters, aldehydes, higher alcohols, lactones, and phenolic compounds responsible for aroma (Poisson & Schieberle, 2008).

How it works: Volatile compounds are vaporized and separated by retention time in a capillary column, then fragmented and detected by a mass spectrometer.

Cost: Mid-range benchtop GC–MS systems start at $90,000–$120,000 USD; high-resolution time-of-flight (TOF) systems exceed $250,000 (Marriott et al., 2001).

Industry impact: Provides unparalleled detail but requires specialized staff and is not yet accessible for most small distilleries.

High-Performance Liquid Chromatography (HPLC): Used to analyze non-volatile components such as tannins, anthocyanins, and carbohydrate degradation products (Waterhouse et al., 2016).

How it works: Compounds are separated in a liquid column under pressure and detected via UV or fluorescence detectors.

Applications: Measures polyphenols and ellagitannins in barrel-aged spirits.

Cost: Bench HPLC systems typically cost $60,000–$100,000.

Fourier Transform Infrared Spectroscopy (FTIR): Provides rapid, non-destructive screening of ethanol concentration, congeners, and some volatiles (Lachenmeier, 2007).

How it works: Infrared absorption spectra reveal molecular bond vibrations, enabling the classification of compounds.

Industry appeal: Compact devices (under $25,000) are increasingly used for quality control due to speed and minimal training requirements.

Nuclear Magnetic Resonance (NMR) Spectroscopy: Reveals the molecular structures of complex compounds in spirits (Monakhova et al., 2011).

How it works: Nuclei of hydrogen or carbon atoms resonate under a magnetic field and produce spectra unique to each molecular structure.

Cost: High-field NMR spectrometers cost > $500,000 and are available only at large research centers.

Future role: Limited direct use in distilleries, but valuable in academic and industrial research collaborations.

Modern Sensory-Analytical Integration

Gas Chromatography–Olfactometry (GC–O): Couples analytical separation with human sensory input. A panelist sniffs the effluent at a "sniffing port," where chemical peaks are linked to actual odor perception (Grosch, 2001). This method bridges the gap between instrumental data and lived sensory experience.

Dynamic Headspace Analysis: Simulates real-time evaporation and headspace composition during nosing. Automated samplers track the release of volatile compounds over time, mimicking the dynamic perception of a spirit in a glass.

Electronic Nose and Tongue Systems: Arrays of chemical sensors mimic olfactory and gustatory receptors (Wilson & Baietto, 2009).

Advantages: Consistency, fatigue-free operation, and suitability for rapid quality control.

Limitations: Sensitivity to ethanol interference and limited discrimination of complex mixtures compared to human panels.

Cost: Current systems cost $40,000- $75,000 and are expected to decrease as adoption increases.

Data Science and Artificial Intelligence

Analytical tools now generate vast datasets requiring advanced computational methods. Machine learning and AI are increasingly used to:

- Predict sensory outcomes from GC–MS fingerprints (Burbidge & Goodacre, 2022).
- Recommend blending strategies by comparing chemical profiles to consumer preference data.
- Detect adulteration and verify geographic authenticity through pattern recognition.

AI cannot yet replicate human creativity in blending, but it can streamline routine analysis, identify patterns invisible to humans, and shorten product development cycles.

Real-Time and Non-Destructive Techniques

Near-Infrared Spectroscopy (NIR): Handheld NIR scanners analyze ethanol, water, and major congeners instantly, without sample preparation (Cozzolino, 2015).

Cost: $10,000-$20,000 for portable devices.

Use case: Field authentication, rapid quality screening at production sites.

Hyperspectral Imaging: Provides spatially resolved chemical data, helpful in examining barrels, detecting leaks, or monitoring spirit color changes without opening casks.

Ion Mobility Spectrometry (IMS): Separates volatiles based on ion mobility in an electric field, providing ultra-fast fingerprinting. Miniaturized for handheld formats (Gerhardt et al., 2017).

Lab-on-a-Chip Platforms: Emerging microfluidic devices promise to deliver chromatography-level data at a fraction of the size and cost. This will enable small distilleries to conduct complex analyses in-house, allowing them to make informed decisions.

Human Sensory Panels in the Era of Technology

Despite technological advances, trained human panels remain the gold standard. Analytical tools quantify compounds, but only humans integrate them into holistic flavor experiences. Panels are increasingly supported by digital sensory libraries, which are databases of aroma reference standards linked to chemical identifiers. VR and AR platforms are being developed for panel training, exposing evaluators to virtual environments that simulate rickhouses or glassware headspace conditions. Fig. 28.1 weighs cost (estimated as of 2026) and efficiency.

Fig. 28.1		Analytical Tools Comparison		
Tool	Cost* ($US)	Speed	Precision	Main Use
GC–MS	90k–150k	Moderate (hours)	High (ppm)	Volatile compound identification
HPLC	60k–100k	Moderate (hours)	High (ppm)	Polyphenols, tannins, non-volatiles
FTIR	25k	Fast (minutes)	Moderate	Screening ethanol, volatiles
NIR	10k–20k	Fast (seconds)	Moderate	Rapid QC, authentication
IMS	40k–75k	Ultrafast (seconds)	Moderate	Volatile fingerprinting
*Est. cost as of 2026: see Appendix VIII for descriptions of sensory electronic tools				

Implications for Distillers and Industry

- **Competitiveness:** Automation and real-time analysis will favor distilleries that adopt early, enabling consistency and innovation.
- **Consumer trust:** Instrumental verification combats fraud (e.g., counterfeit whisky) by authenticating provenance through chemical fingerprinting.

- **Sustainability:** Predictive modeling reduces waste and optimizes barrel use, aligning with consumer demand for eco-responsibility

The Future Landscape

The distiller of the near future will work in a hybrid lab-distillery where automated sensors continuously feed data into AI-driven dashboards. Predictive models will suggest optimal barrel selections, blending ratios, and aging interventions before the distiller confirms with sensory judgment.

The artisan distiller and data scientist roles will merge into a single role.

Hypothetical Scenario 28.1: Automation and analytical integration in a mid-size distillery: A mid-sized craft distillery producing ~500,000 liters annually provides a clear example of how analytical tools reshape daily operations. Historically, sensory evaluation was conducted manually by trained tasters, with laboratory analysis limited to ABV and basic pH measurements. Decisions such as fermentation cutoffs, heads–hearts–tails cuts, and barrel management were primarily guided by intuition, tradition, and post hoc sensory evaluation.

Current Implementation:

The distillery has adopted a layered approach to analysis:

Fermentation Monitoring: Inline near-infrared (NIR) probes track sugar depletion and ethanol rise, sending real-time data to a central dashboard.

Distillation Control: Gas chromatography (GC) spot checks confirm the presence of congeners during pilot runs, whereas less costly ion mobility spectrometry (IMS) provides rapid batch-to-batch screening for fusel alcohols and sulfur notes.

Barrel Aging: Fourier transform infrared (FTIR) sensors detect lactones, vanillin, and tannins in aging stock, supporting data-driven blending decisions.

Quality Assurance: A panel of trained tasters remains employed; however, sensory sessions are scheduled only after analytical screening flags samples as "borderline" or "high-interest." This reduces panel fatigue and increases reproducibility.

Operational Impact:

- Batch rejection rates decreased from ~7% to <2% due to early detection of fermentation faults.
- The average time-to-market for limited releases dropped by 20%, as analytical markers helped identify barrels ready for bottling without relying exclusively on calendar age.
- Labor costs decreased by reallocating staff from repetitive monitoring tasks toward higher-value sensory education and customer-facing roles.

Future Trajectory

In the near future, the same distillery plans to implement AI-driven blending models. By integrating GC-MS fingerprints and historical sensory scores, the system will recommend blending ratios to optimize consistency while preserving house style. Operators will shift from manual controllers to supervisors of a semi-autonomous production line, where intervention is required only when anomalies arise.

The case illustrates both economic and cultural shifts: automation reduces waste and improves profitability, but it also necessitates staff retraining and a rebalancing of the role of human judgment. The distiller of the near future will likely spend more time interpreting dashboards and less time nosing raw distillate. This evolution highlights the growing interdependence of analytical science and sensory tradition.

Summary

Analytical tools in sensory science are transforming distilling from descriptive to predictive. GC-MS, HPLC, FTIR, and NIR provide chemical clarity; GC-O and electronic sensors link chemistry to perception; and AI integrates it all into actionable insights. Future distillers remain both craftsmen and scientists, equipped with precision tools to refine, predict, and authenticate flavors.

"The people who are crazy enough to think they can change the world are the ones who do."
—Steve Jobs, Apple co-founder

"This provides important insight … as well as the usefulness of HS-SPME-GC-MS as a proxy for human olfaction."— Ashmore et al. (2023), on sensory profiling methods in diluted whisky

"AI can tell Scottish and American whiskies apart … analyzed … with GC-MS and automatic compound detection analysis."— Chemistry World, on the future of analytical chemistry

References

Burbidge, C., & Goodacre, R. (2022). Machine learning for flavor prediction in food and beverages. *TrAC — Trends in Analytical Chemistry, 153*, 116656. https://doi.org/10.1016/j.trac.2022.116656

Cozzolino, D. (2015). Near-infrared spectroscopy in food authentication. *Food Research International, 60*, 262–268. https://doi.org/10.1016/j.foodres.2014.07.006

Gerhardt, N., Birkenmeier, M., Schwolow, S., Rohn, S., & Weller, P. (2017). Volatile compound fingerprinting using ion mobility spectrometry. *Food Control, 78*, 219–228. https://doi.org/10.1016/j.foodcont.2017.02.022

Grosch, W. (2001). Evaluation of the key odorants of foods by dilution experiments, aroma models, and omission. *Trends in Food Science & Technology, 12*(11), 447–455. https://doi.org/10.1016/S0924-2244(02)00014-4

Lachenmeier, D. W. (2007). Rapid quality control of spirit drinks and beer using FTIR. *Food Chemistry, 101*(2), 825–832. https://doi.org/10.1016/j.foodchem.2006.02.034

Marriott, P. J., Shellie, R., & Cornwell, C. (2001). Gas chromatographic technologies for the analysis of essential oils. *Journal of Chromatography A, 936*(1–2), 1–22. https://doi.org/10.1016/S0021-9673(01)00904-1

Monakhova, Y. B., et al. (2011). NMR spectroscopy as a screening tool for counterfeit brandy. *Journal of Agricultural and Food Chemistry, 59*(7), 2877–2884. https://doi.org/10.1021/jf200179m

Poisson, L., & Schieberle, P. (2008). Characterization of the most odor-active compounds in an American Bourbon whisky by application of the aroma extract dilution analysis. *Journal of Agricultural and Food Chemistry, 56*(14), 5813–5819. https://doi.org/10.1021/jf800382m

Waterhouse, A. L., Sacks, G. L., & Jeffery, D. W. (2016). *Understanding wine chemistry.* Wiley. https://doi.org/10.1016/C2018-0-03808-6

Wilson, A. D., & Baietto, M. (2009). Applications and advances in electronic-nose technologies. *Sensors, 9*(7), 5099–5148. https://doi.org/10.3390/s90705099

Observations on State-of-the-Art Analytical Tools

Modern analytical chemistry now illuminates what sensory evaluation perceives only indirectly. Gas chromatography–mass spectrometry (GC-MS) remains central, but new pairings—such as GC-olfactometry, two-dimensional GC (GC×GC-MS), and time-of-flight detection—reveal trace volatiles and co-eluting congeners that define the aroma balance. NMR spectroscopy quantifies maturation markers such as furfurals and lactones; FT-IR and Raman methods rapidly assess ethanol–water structure and ester equilibria related to mouthfeel. For origin and authenticity, isotopic ratio mass spectrometry (IRMS) and trace-element analysis by inductively coupled plasma mass spectrometry (ICP-MS) now enable mapping of the agrochemical signature of raw materials and process pathways.

These tools reveal previously hidden layers of complexity, yet measurement still differs from perception. Analytical power defines potential; interpretation converts data into sensory meaning. The human evaluator remains the final instrument of evaluation.

Machine learning and chemometric modeling now translate vast spectral datasets into sensory predictions, thereby bridging the gap between chemical composition and human perception. Techniques such as e-nose sensor arrays, headspace solid-phase microextraction (HS-SPME), and GC-IMS compress what once required hours into minutes. However, the frontier is not detection—it is correlation. Instruments can separate and quantify, but only trained evaluators can assign meaning to the results. Data without context remains chemistry, not sensory science.

Section IV:

Flavor Assessment – The Role of the Evaluator

Basic Procedure: Two-Pass Analytical Method for Spirits Evaluation

Preparing the Environment: Temperature, lighting, noise. airflow. odors

Evaluator Conditioning: No food/beverage within last hour, neutral palate, no anosmia or congestion, hydrated

Panel Calibration: Recognition threshold calibration

Sample Preparation: ID codes, randomization, counterbalanced order, Metadata held until post-evaluation

Equipment Preparation: Uniform vessels for all tests, uniform fill, covers, uniform temperature (22 ± 1 °C)

First Pass Evaluation: Visual, olfactory, gustatory registration, record impressions unfiltered notes what's there what's not. No interpretation, categorization, hedonic marking, Output = Sheet A

Feedback and Calibration: 3-5 min. rest, compare Sheet A with lexicon or reference aromas. Adjust perception of ethanol strength, sweet, bitter, panel consensus discussion limited to factual alignment, not preference

Second Pass Evaluation: Olfactory, gustatory, record intensity, temporal dominance, persistence time. Cross-modal consistency, do aroma, taste, trigeminal agree? Classify balance (chemical perceptual. temporal). Output= Sheet B

Data Integration/Scoring: Merge sheets A and B, eliminate redundancy, map descriptors to lexicon. Score per agreed upon scale assigning weighting to aroma complexity, palate structure, integration and finish, typicity and style fidelity, overall impression. Output = score/rating

Chapter 29: Quantifying Sensory Impact — Concentration to Perception

Understanding the Bridge between Chemistry and Perception

In the evaluation of distilled spirits, every sensory impression begins long before the glass reaches the nose. The evaluator's task is to interpret the aromatic mosaic created by fermentation, distillation, and maturation, a mosaic built from measurable compounds governed by measurable physical and chemical laws. While the distiller manipulates chemistry to control outcome, the evaluator translates that chemistry into perception. True sensory mastery demands quantitative literacy: anticipating how the spirit will smell and taste by reading its molecular blueprint.

Chapter 2 (Fig. 2.2) introduced Weber's Law and sensory weighting. These psychophysical foundations describe how the sensory system responds to changes in stimulus intensity (Mao et al., 2018). This chapter extends them to applied sensory interpretation—demonstrating *why* particular compounds dominate perception and *how* evaluators can predict flavor balance from concentration data, production variables, and known psychophysical relationships.

The goal is to teach the evaluator to think like a distiller while judging like a scientist, integrating odor-activity analysis, mixture-interaction principles, volatility behavior in ethanol–water matrices, and vessel-dependent modulation into a single interpretive discipline.

The Evaluator's Quantitative Mindset

The untrained taster qualitatively perceives the aroma as *fruity, woody, smoky, and sweet.* The trained evaluator links these descriptors to specific compound families and measurable concentrations. Perception, however, does not scale with raw concentration; it scales with stimulus potency relative to the detection threshold, as expressed by the Odor Activity Value.

$$OAV = C/T$$

where OAV = Odor Activity Value, C = compound concentration ($\mu g \ L^{-1}$ or mg L^{-1})
and T = its odor threshold in the same units (Meilgaard et al., 1999; Li et al., 2023)

When OAV > 1, a compound contributes detectably to aroma and is expected to be perceptible; those with OAV < 1 generally remain subthreshold. Perception is nonlinear. A compound with OAV = 20 is not 20 times stronger than one with OAV = 1. Instead, it possesses proportionally greater potential to influence aroma identity, depending on volatility, matrix effects, and mixture interactions. When OAV > 10, it typically dominates its sensory class. OAVs are calculated in a representative example in Fig. 29.1. 4-Ethylguaiacol often occurs at concentrations below 1 mg

L^{-1} yet commands the sensory stage because its threshold is ≈ approximately 0.05 mg L^{-1} (Shen et al., 2025). Conversely, fusel alcohols may appear at hundreds of mg L^{-1} yet contribute little because their thresholds are proportionally high. The evaluator who internalizes such ratios begins to interpret concentration tables as sensory maps.

Fig. 29.1	Representative Odor Activity Values for Key Spirit Compounds			
Compound	Concentration (mg/L)	Threshold (µg/L)*	OAV	Dominant Descriptor
Ethyl acetate	120	7000	17	Fruity, solvent
Isoamyl acetate	1.8	30	60	Banana, pear drop
Vanillin	1.0	60	17	Sweet, vanilla
4-Ethylguaiacol	0.25	50	5	Smoky, clove
Acetaldehyde	20	500	40	Sharp, green apple
γ-Nonalactone	0.6	35	17	Coconut, creamy
Eugenol	0.3	6	50	Spicy, clove-like
Thresholds adapted from Li et al. (2023), Shen et al. (2025), and Wu et al. (2024), in aqueous ethanol *µg/L converted to mg/L equivalent				

Representative concentrations, detection thresholds, and calculated Odor Activity Values (OAVs) for compounds commonly contributing to spirits' aroma. OAV expresses the ratio of concentration to sensory threshold in ethanol–water solution, providing an estimate of perceptual strength. Values reflect relative sensory potency rather than volatility, underscoring that highly potent compounds may occur at very low concentrations.

OAVs provide only a *first approximation*. Perception is shaped by volatility, temperature, ethanol content, and the manner in which multiple aromas combine or compete. In practice, chemists sometimes adjust these calculations with factors for volatility and mixture behavior to get closer to actual

sensory impact. What matters is the relationship, not the math:

- Some compounds are potent at trace levels.
- Others must be present in large quantities to be noticed.
- The apparent strength of a smell depends on both chemistry and physiology.

For now, think of OAV as a map of potential influence, not a final prediction. Later study in advanced sensory science or flavor chemistry can refine this idea into quantitative modeling.

Ethanol as a Modulator of Perception

Ethanol is the most influential variable in spirits evaluation because it is simultaneously a solvent, vapor suppressant, and trigeminal stimulant. It modifies volatility, slows molecular diffusion, and partially anesthetizes olfactory receptors (Lu et al., 2025; Guttman, 2020; Brasser et al., 2014). Its dual role as both carrier and competitor makes it the principal modulator of aroma release.

Case Study 29.1: Whiskey A vs. Whiskey B: Imagine two single-malts — *Whiskey A* at 46 % ABV and *Whiskey B* at 60 % ABV. Laboratory analysis shows nearly identical concentrations of esters, phenols, and aldehydes. However, on the nose, Whiskey A blooms with fruit and vanilla, while Whiskey B feels closed and hot, showing little more than alcohol. The difference lies in ethanol's ability to control volatility. Higher ethanol strength suppresses the release of aroma molecules from the liquid into the headspace, and its own vapor partially anesthetizes olfactory receptors. Headspace saturation is vessel-dependent (following section on dilution benefit). Lowering the alcohol concentration—by bottling or by adding water during evaluation—reduces this suppression, allowing lighter compounds such as esters and aldehydes to express themselves. This demonstrates that concentration data describe what is in the liquid, but perception depends on what reaches the nose. In real evaluation, chemistry and physics interact through vapor pressure, temperature, and molecular weight to determine which volatiles dominate.

Volatility Shifts with Alcohol Strength

As ethanol concentration increases, highly volatile esters lose dominance while heavier molecules (phenols, lactones, long-chain acids) gain relative strength. A spirit nosed at 60 % ABV is chemically different from the same spirit at 40 %. Ethanol functions as a selective gatekeeper, determining which molecules reach the nose and in what ratios. Lu et al. (2025) documented threshold elevations ranging from 2-fold to more than 600-fold as ethanol concentration increased from 0% to 20%. Such magnitude explains why small additions of water can suddenly reveal aromas that seemed absent: dilution reduces vapor-phase competition and trigeminal burn, allowing odorants with high solubility but low volatility to emerge into the headspace. (Fig. 29.2A).

Fig. 29.2A	Compound Volatility vs Ethanol Concentration			
Ethanol ABV (%)	Ethyl acetate (esters)	Isoamyl acetate (esters)	Vanillin (oak aldehyde)	4-Ethylguaiacol (phenolic)
20	100	80	12	8
30	75	60	15	10
40	50	40	20	15
50	30	25	28	20
60	20	15	35	25

Note: *Values represent relative headspace abundance normalized to 100 % at 20 % ABV.*

Representative headspace data showing relative vapor-phase abundance of selected aroma compounds from 20 to 60 % ABV, normalized to ethyl acetate at 20 %. These data illustrate the contrasting volatility behaviors of light esters, aldehydes, and phenolics as ethanol concentration increases. Values are conceptual and should not be equated with the Odor Activity Values shown in Figure 29.1, which describe perceptual potency rather than vapor-phase concentration.

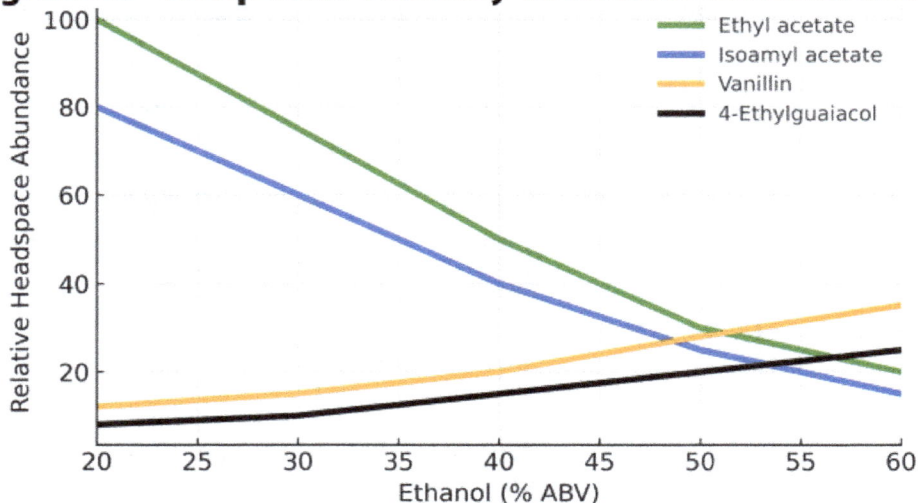

Fig. 29.2B Compound Volatility vs Ethanol Concentration

Plot data for Fig. 29.2B derived from the same conceptual dataset as Figure 29.2A, showing the decline in ester volatility and the relative rise of aldehydic and phenolic species with increasing ethanol concentration. The graph visually demonstrates that ethanol selectively suppresses light, polar compounds while allowing heavier, less polar molecules to dominate the vapor phase (adapted from Lu et al., 2025).

Note: Figures 29.1–29.2B together illustrate the distinction between sensory potency (OAV) and volatility behavior. Concentration and volatility govern the chemical availability of aroma molecules, whereas threshold and OAV determine their perceptual impact once delivered to the olfactory system.

The Vessel-Dependent Nature of "Dilution Benefit"

It is vital to recognize that much of the traditional advice about dilution originates not only from chemistry but also from the geometry of the sampling vessel. Most published sensory data on ethanol suppression and dilution employ tulip-shaped or ISO wine glasses (there is no ISO standard for spirits glasses), which converge vapors toward the nose. These shapes trap ethanol centrally at the rim plane, oversaturating the olfactory path and forcing researchers to dilute to achieve comfort.

This assumption has become an unquestioned norm as every major study on ethanol masking or aroma release has treated the tulip as the *de facto* standard (Manska, 2012; Lu et al., 2025). Consequently, the observed "need for dilution" is often a methodological artifact of glass design, not an inherent property of ethanol. Awareness saves time and effort and improves accuracy.

Evaluators should therefore judge dilution necessity according to vessel vapor-flow characteristics:

- Tulip or ISO glass: Convergent rim traps ethanol; dilution is almost always required.
- Snifter or tumbler: Wider mouth disperses ethanol faster, lowering irritation but also reducing aromatic concentration. Tumbler disperses ethanol slightly better than a snifter
- Sensory-engineered glassware — flared-rim or diffusion-controlled designs (Manska & Arsilica Inc., 2018) — channel ethanol laterally away from the central sniff zone, allowing full-strength evaluation without the distortions of water addition. (See Chapter 42 for design details and Appendix IV for glassware comparison. Note ethanol levels in Appendix IV.

Practical Guidance for Evaluators

When approaching any spirit, the professional evaluator should first identify glass geometry before assuming dilution is necessary. If ethanol stratification is evident — burning sensation, restricted aroma band, repeated re-sniffing — controlled dilution in 5% steps is justified. *If the vessel disperses ethanol efficiently, dilution may distort rather than clarify the aromatic balance.* Recognizing this dependency prevents perpetuation of a critical misconception: dilution is a vessel-specific remedy, not a universal requirement. A simple solution, such as carefully choosing the sampling vessel, can make the task much easier. Glass shape governs ethanol behavior; ethanol behavior dictates whether water addition aids or undermines sensory fidelity (Manska, 2018).

Additivity, Masking, and Synergy in Aroma Mixtures

Aroma compounds form perceptual networks. Their combined effects — additive, masking, or synergistic — depend on molecular similarity, overlap in volatility, and receptor binding (Niu et al., 2020). Fig. 29.3 illustrates typical behaviors.

- Additive: Two similar odorants (ethyl hexanoate + isoamyl acetate) reinforce fruitiness even when each is sub-threshold, in a young rum or neutral grain spirit. Vapor-phase contributions rise predictably with concentration until ethanol or fatigue modulates perception.
- Masking: A potent phenolic (4-ethylphenol) can obscure weaker esters by competitive binding or sensory dominance, such as peated Scotch or heavy rums (4-ethylguaiacol). In high-ABV whiskies, fruity top notes are muted until dilution reduces ethanol's masking effect.

- Synergy: Lactone-vanillin synergy in aged bourbons and brandies amplifies sweet, warm, beyond prediction. Terpene esters in bourbon, gin, and rum enhance floral perception after adaptation to the initial high levels of volatiles. In well-aged rum, coconut (γ-nonalactone) and vanillin combine to produce a perceived "creamy" aroma greater than either alone.

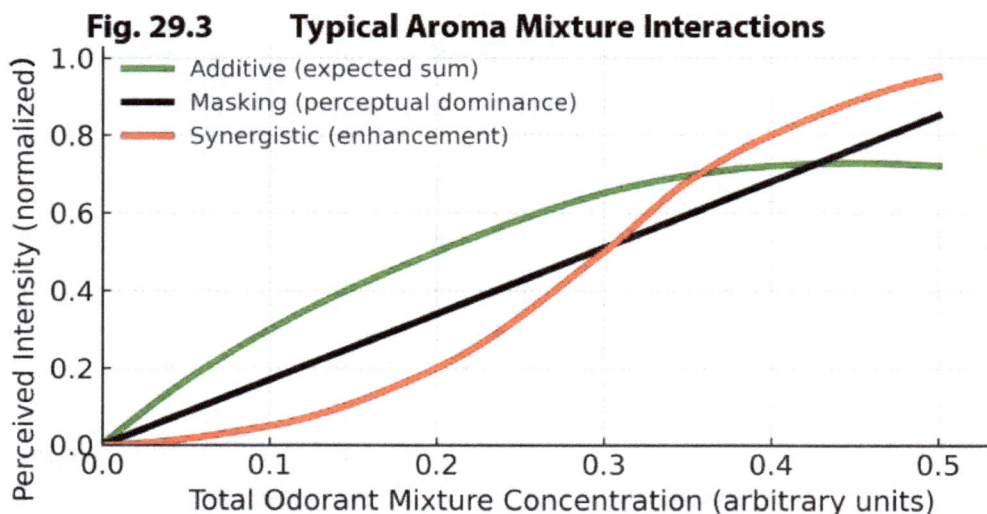

Fig. 29.3   Typical Aroma Mixture Interactions

The additive line shows ideal proportional summation. The masking curve demonstrates early perceptual dominance that levels with intensity as receptor competition limits incremental perception. The synergistic curve begins below the additive curve, crosses the midrange, and rises rapidly, depicting delayed enhancement through molecular co-activation and perceptual integration.

Evaluators interpret OAVs as approximate predictors because mixture behavior can either amplify or suppress apparent intensity. Phenolics may neutralize a high-OAV ester, while a near-threshold aldehyde may become prominent in an acidic environment. Professional judgment reconciles *analytical* expectation with actual perception until congruence is achieved.

Translating Distiller Variables into Sensory Expectations

Quantitative evaluators benefit from understanding what the distiller already knows, as well as how process choices and process controls can serve as tools to reshape compound distributions and thus sensory balance (Charapitsa et al., 2014; Wu et al., 2024). Key relationships:

- Fermentation Temperature: Higher → more short-chain esters and fusel alcohols; increased fruitiness but risk of solvent notes.

- Yeast Strain: Different enzymatic profiles → distinct congener fingerprints influencing fruity, floral, or sulfuric tendencies.
- Cut Width: Early/narrow → cleaner profile; late/broad → heavier tails and depth.
- Barrel Chemistry: Toasting controls lignin decomposition → vanillin and eugenol formation (Buettner, 2017).
- Oxygen Exposure: Promotes oxidation and esterification → softer nose and longer finish.

Correlating laboratory data with these variables enables a predictable sensory direction before tasting, and quantitative reasoning converts evaluation from a passive reaction to an *informed* expectation.

Fig. 29.4 Chemical and Sensory Effects of Distiller Process Variables		
Distiller Variable	**Chemical Direction**	**Probable Sensory Effect**
Fermentation Temperature ↑	Esters ↑, Higher Alcohols ↑	Brighter fruit, increased aromatic intensity, higher volatility
Yeast Strain Selection	Congener Spectrum Shift (Varies by strain)	Floral to solvent-like shifts, unique ester profiles
Cut Width ↑ (broader middle run)	Aldehydes ↑, Sulfur Compounds ↓	Nutty/Feinty character, heavier aroma, more texture
Barrel Toast Level ↑	Vanillin ↑, Eugenol ↑, Furfural ↑	Sweet, spicy, caramelized wood tones
Barrel Entry Proof ↑	Lactones ↓, Phenolics ↑	Drier, more tannic mouthfeel, increased structure
Oxygen Exposure ↑ (during aging)	Acids ↑, Esters ↓ (oxidative equilibrium)	Rounder mouthfeel, smoother perception, oxidative warmth
Copper Contact ↑	Sulfur Compounds ↓, Acetals ↑	Cleaner distillate, less sulfury, smoother nose
Maturation Time ↑	Lactones ↑, Vanillin ↑, Acids ↑	Fuller body, richer aroma integration, aged depth
Distillation Rate ↑	Heads Components ↑ (acetates, aldehydes)	Sharper nose, less integration, potential harshness
Water Addition (Proofing) ↑	Volatile Partition ↓, Ethanol Activity ↓	Reduced aromatic lift, softer entry, muted top notes

Relating Concentration Data to Perceived Intensity: A Working Example

Consider a hypothetical report for a fictitious single-malt sample as shown in Fig. 29.5: Despite a 1,000-fold difference in concentration, 4-ethylguaiacol's impact rivals that of ethyl acetate because its threshold is so low. The impact is relational, not absolute.

Awareness of Relative Flavor Contributions

At this point, it is enough to understand that perception results from a combination of chemical presence and sensory weighting. Professional sensory analysts refine this idea through Relative Flavor Contribution (RFC) models. These models expand the simple OAV by introducing correction factors that account for volatility and perceptual interaction among compounds. In practice, RFCs establish a numerical framework that links chemical concentration, evaporation behavior, and psychophysical response.

Fig. 29.5	Illustrative Analysis of Glensmoky Scotch			
Compound	Concentration (mgL^{-1})	Odor Threshold (mgL^{-1})	OAV	Descriptor
Ethyl acetate	150	7	21	Fruity, solvent
Isoamyl alcohol	250	30	8	Banana, fusel
4-Ethylguaiacol	0.3	0.05	6	Smoky, clove

RFCs represent a level of study suitable for advanced sensory or flavor chemistry courses, in which the goal is to make quantitative predictions rather than qualitative understanding. Recognize that such systems exist and form the mathematical backbone of predictive sensory analysis used by research laboratories and large distilleries. For further study, consult:

- Lawless, H. T. & Heymann, H. (2010). *Sensory Evaluation of Food: Principles and Practices.* 2nd ed., Springer. Comprehensive introduction to psychophysical scaling/ mixture modeling.

- Dravnieks, A. (1985). *Atlas of Odor Character Profiles.* ASTM. Foundational reference for quantitative odor profiling and relative intensity mapping.

- Poisson, L. & Schieberle, P. (2008). *Journal of Agricultural and Food Chemistry,* 56, 990–998. Demonstrates practical application of volatility and threshold weighting in aroma modeling.

An advanced understanding of RFCs enables analysts to translate chemical data into informed sensory expectations—an ability that defines true professional proficiency.

Flavor Dominance: Predicted vs Observed — an Illustrative Example

Two whiskies with different ABVs are analyzed prior to sensory evaluation. Both contain about the duplicate total congener content ($\approx$ 3 g L^{-1}), but differ in congener distribution,

Fig. 29.6A	Anticipated Flavor Dominance (Chemical Prediction)		
Compound Family	Whiskey A (40 % ABV)	Whisky B (55 % ABV)	Anticipated Sensory Impact
Esters	52-55	33-38	Fruity brightness expected to dominate low-ABV headspace; suppressed by ethanol at high ABV
Higher alcohols	25	27-28	Slightly more solvent-like and fusel with higher ethanol
Phenolics	8-10	18-22	Phenolic and smoky notes rise proportionally as lighter volatiles diminish
Aldehydes	6	7	Toasted/nutty oxidative nuance stable across ABV range
Lactones	4	5-6	Sweet coconut/vanillin persists; minimally ethanol-dependent
Acids	2	3-4	Mild acid freshness shifts toward a heavier mouthfeel as ABV increases

Before nosing, a quantitative evaluator anticipates the sensory trajectory. Whisky A, dominated by short-chain esters, should present high-pitched fruit, crisp solvent brightness, and rapid dissipation. Whisky B, richer in phenolics, should show lower volatility, longer persistence, and a smoke-spice axis. When blind panels subsequently evaluated aroma attributes, the observed intensities followed the predicted order with ≈ 80 percent correspondence. The minor discrepancies were explainable by differences in ethanol strength and matrix interactions.

Fig. 29.6B	Predicted vs Observed Dominance (Sensory Confirmation)*				
Compound Family	Calculated OAV Range	Predicted Dominant Descriptor	Observed Descriptor	Deviation (%)	Probable Cause
Esters (ethyl acetate, isoamyl acetate)	8–25	Fruity, solvent-like, bright top notes	Whiskey A: Bright fruit; Whiskey B: Suppressed fruit, more ethanol heat	20	Ethanol suppression of ester volatility at higher ABV
Higher Alcohols (isoamyl, isobutanol)	4–10	Banana, fusel, warm secondary tone	Whiskey A: Mild fusel tone; Whiskey B: Pronounced solvent note	15	Matrix masking by alcohols and phenols
Phenols (4-ethylguaiacol, cresols)	5–12	Smoky, clove, phenolic depth	Whiskey A: Balanced smoke; Whiskey B: Dominant phenolic intensity	30	Enhanced release due to ethanol-driven volatility shift
Aldehydes (acetaldehyde, furfural)	3–8	Nutty, toasty, oxidative nuance	Whiskey A: Subtle toasted notes; Whiskey B: Muted aldehydic sharpness	10	Reduced aldehyde volatility under high ethanol concentration
Lactones (γ-nonalactone, δ-decalactone)	2–6	Creamy, coconut, sweet undertone	Whiskey A: Sweet vanilla-coconut; Whiskey B: More subdued sweetness	25	Partitioning changes in ethanol-water equilibrium
Acids (acetic, hexanoic)	1–5	Sharp, vinegary, mild sour edge	Whiskey A: Light acid freshness; Whiskey B: Heavier body, less lift	20	Acid suppression by higher alcohol solvation
*Illustrative example comparing two whiskys; whiskey A=40 % ABV, whiskey B=55 % ABV					

Deviations quantify how ethanol concentration alters perceived intensity through the suppression of volatility, receptor masking, and matrix partitioning. Figures 29.6a and 29.6b demonstrate that chemical potential and sensory outcome diverge systematically: lower-ABV spirits favor ester expression and aromatic openness, while higher-ABV counterparts show compression of fruit and accentuation of phenolic depth. Flavor evaluation is a testable hypothesis, not a poetic impression. Quantitative reasoning enables the evaluator to predict the likely sensory landscape and then verify it empirically—a process that replaces subjectivity with disciplined curiosity.

Microbial Secondary Aromas and Emergent Volatiles

Not all compounds contributing to sensory impact originate from yeast or raw materials. Low-level microbial activity after primary fermentation can significantly alter the aroma composition, far beyond its chemical proportions. Lactic acid bacteria metabolize sugars, acids, and glycerol into volatile by-products, such as ethyl lactate, diacetyl, and acetic esters—minor components in terms of concentration, yet potent modifiers of texture and flavor continuity. In wood-aged spirits, *Brettanomyces* species and other oxygen-tolerant microbes generate volatile phenols—4-ethylphenol and 4-ethylguaiacol—that impart smoky, leathery, or medicinal notes, often mistaken for oak-derived intensity.

From a sensory-quantification standpoint, these microbial signatures act as non-linear multipliers in mixture perception. Their individual concentrations may lie near or below detection thresholds, but in concert with major volatiles, they tip the balance, exaggerate dryness, or mute fruit. The lesson for both the distiller and the evaluator is that trace microbiology exerts a disproportionate influence on sensory perception. When assessing concentration–perception relationships, one must therefore distinguish between direct production variables and secondary microbial modulation.

Odor Mixture Interactions and Temporal Dynamics

Perception is not the arithmetic sum of compounds; it is a **configural system** in which volatiles influence one another's expression. Odor mixture studies show that specific components suppress others (*mixture suppression*), while complementary volatiles produce new emergent impressions (*configural blending*). For example, ethyl hexanoate and ethyl butyrate combine to form a single composite "fruit" identity, masking individual notes and defying prediction based solely on concentration data.

Temporal adaptation further complicates quantification. Continuous exposure to dominant volatiles fatigues corresponding receptors within seconds, lowering apparent intensity and allowing previously masked notes to emerge. The evaluator perceives this as evolution in the glass; the chemist recognizes it as a dynamic shift in the relative weighting of signals. To quantify sensory impact realistically, measurements must therefore account for time, adaptation, and compound interactions—not just abundance. Also see Fig. 30.9.

Together, microbial secondary chemistry and perceptual interaction explain why sensory dominance rarely matches analytical concentration. What is abundant may be invisible; what is scarce may define identity. Quantifying impact thus requires an understanding of how molecules interact in the air and in the mind, rather than simply measuring their presence.

Limitations and Cautions

No model, however elegant, perfectly predicts perception. Several phenomena constrain precision:

- Matrix Effects: Ethanol, sugars, acids, and water all influence volatility and partition coefficients, altering the apparent intensity of otherwise stable compounds (Lu et al., 2025).
- Mixture Suppression: Multiple odorants may compete for receptor sites, producing sub-additive responses.
- Individual Variability: Genetic polymorphisms in olfactory receptors (e.g., *OR7D4, OR2J3*) create perception differences impossible to standardize (Doty, 2015).
- Cognitive Bias: Expectation, language, and prior exposure modulate reported intensity (Plassmann et al., 2008).

Treat quantitative models as guides, not laws. They narrow uncertainty but never eliminate it.

Summary

To master sensory evaluation is to stand at the intersection of chemistry, physics, and neurobiology—an interpreter of measurable causes and perceptual consequences. The evaluator who embraces quantitative reasoning gains not only consistency but credibility: the ability to explain *why* a spirit smells the way it does, in language both empirical and human. Takeaways:

- Chemical composition and sensory perception are quantitatively linked. Odor-activity and relative-contribution models allow evaluators to infer perceptual weight from concentration data.
- Ethanol concentration and vessel geometry jointly determine which compounds reach the olfactory epithelium; dilution should be applied according to vessel design, not tradition.
- Mixture interactions explain why perception diverges from linearity.
- Production variables—fermentation temperature, yeast strain, cut width, barrel chemistry—manifest as predictable sensory directions.
- Quantitative reasoning transforms tasting from description to diagnosis, making the evaluator a scientific participant in the distiller's process.

Equally important, evaluators must recognize the sampling vessel as part of the system. A perfectly scientific model applied to a geometry that concentrates ethanol—such as a tulip—still yields distorted results. Only by integrating accurate chemistry with vapor-flow engineering through vessel design can sensory science achieve reproducibility (Manska & Arsilica Inc., 2018).

"Odors are events, not things; they unfold in time as mixtures disperse and receptors adapt."
— Pamela Dalton, *Chemical Senses Journal* (2000)

"The ODTs [odor detection thresholds] increased as ethanol concentration increased in the vapor matrix ... ethanol's pharmacodynamic effect is a more important factor than its physicochemical effect in causing an increase of ODTs." — Wang & Cadwallader (2024).

"Perception is never a single point; it is the negotiation between stimulus, context, and history." — Richard E. Doty, *Handbook of Olfaction and Gustation *(2015)

"The smell of a compound is never the property of the molecule alone, but of its proportion within the surrounding mix."— Avery N. Gilbert, *What the Nose Knows *(2008)

Appendix 29: Probable Causes of Observed Events — Searching for Solutions

App. 29.1	Probable Causes of Observed Sensory Events	
Observed Sensory Effect	**Probable Chemical / Physical Cause**	**Typical Corrective Action or Interpretation**
Fruity top notes muted in high-ABV spirit	Ethanol suppression of ester volatility; trigeminal masking	Evaluate at reduced ABV or use sensory-engineered glass that disperses ethanol laterally
Unexpected sweetness after dilution	Water-induced micelle collapse, releasing bound esters; altered volatility ratio	Recognize that perceived sweetness arises from enhanced aroma, not added sugar
Metallic or sulfury nose in young spirit	Incomplete copper contact or late tail cut; thiol carryover	Adjust distillation cuts or increase copper surface exposure
Rapid aroma fade during nosing	Olfactory adaptation/receptor fatigue, not glass geometry or aroma loss	Limit exposure time per sniff; use covered samples and neutral air breaks between evaluations
Residual aroma in the judging area causing panel bias	Airborne volatiles from open glasses contaminating the ambient air	Use paper covers or glass caps on samples between evaluations
Exaggerated aroma intensity after swirling	Mechanical agitation increases vapor-phase flux and ethanol release	Swirl only when evaluating persistence or retronasal profile; otherwise, keep still for consistency
Discrepancy between GC data and perceived dominance	Matrix suppression, volatility shift, or receptor competition	Apply OAV and volatility reasoning before judging faults
Smoky or spicy note dominates over fruit	High phenolic OAV; ethanol enhancement of heavier volatiles	Identify cask or toasting influence; recognize ethanol selectivity
Sharp nose at bottle strength, muted palate	Trigeminal adaptation; olfactory fatigue between orthonasal and retronasal phases	Alternate samples or rest between sniffs to prevent sensory suppression

References

Brasser, S. M., Cairney, S., & King, J. (2014). Alcohol sensory processing and its relevance for ingestion. *Frontiers in Psychology, 5*, Article 4388769. https://pmc.ncbi.nlm.nih.gov/articles/PMC4388769/

Buettner, A. (2017). *Aroma compounds in distilled spirits: Formation, evolution, and sensory impact.* Springer.

Doty, R. L. (2015). *Handbook of olfaction and gustation* (3rd ed.). Wiley-Blackwell. https://doi.org/10.1002/9781118971757

Guttman, G. A. (2020). Ethanol thresholds in water and beer: How thresholds change with ethanol content. *Journal of Sensory Studies, 35*(6), e12544. https://doi.org/10.1111/joss.12544

Li, H., Zhang, Y., Wang, J., & Liu, S. (2023). Odor activity values and aroma profiles of fermented beverages: Quantitative evaluation and sensory impact. *Foods, 12*(5), 8497. https://pmc.ncbi.nlm.nih.gov/articles/PMC10048497/

Lu, J., Liu, Q., & Chen, X. (2025). The effect of ethanol on compound thresholds and sensory perception of volatile compounds. *Journal of Agricultural and Food Chemistry, 73*(4), 11858–11872. https://pmc.ncbi.nlm.nih.gov/articles/PMC11858315/

Manska, G. F., & Arsilica Inc. (2018). Engineering the spirits glass: Reducing ethanol interference in sensory evaluation. *Proceedings of the International Conference on Spirits Sensory Science.*

Mao, Y., Tian, S., Qin, Y., & Han, J. (2018). A new sensory sweetness definition and conversion method based on the Weber–Fechner law. *Food Chemistry, 259*, 139–146. https://doi.org/10.1016/j.foodchem.2018.03.080

Meilgaard, M. C., Civille, G. V., & Carr, B. T. (1999). *Sensory evaluation techniques* (3rd ed.). CRC Press.

Niu, Y., Yang, X., & Fang, Y. (2020). Characterization of odor-active volatiles and OAVs in complex matrices: Contribution of aroma compounds depends on matrix effects. *Molecules, 25*(16), 7179107. https://pmc.ncbi.nlm.nih.gov/articles/PMC7179107/

Plassmann, H., O'Doherty, J., Shiv, B., & Rangel, A. (2008). Marketing actions can modulate neural representations of experienced pleasantness. *Proceedings of the National Academy of Sciences, 105*, 1050–1054. https://doi.org/10.1073/pnas.0706929105

Shen, X., Zhou, L., & Wang, R. (2025). The sensory lexicon of malt whisky new-make spirit and odor activity value analysis. *Journal of Food Composition and Analysis, 128*, 105693. https://www.sciencedirect.com/science/article/pii/S2772566924000594

Wu, F., Zhao, L., & Zhang, P. (2024). Comparison of aroma compounds in Baijiu using GC-O-MS and OAV/PLS methods. *Foods, 13*(5), 681. https://www.mdpi.com/2304-8158/13/5/681

Observations on Quantifying Sensory Impact

Numbers and Nose – the Balance: When numbers align with perception, confidence replaces guesswork. However, when they diverge, that is where learning lives — in the gap between data and sensation. Perceptual science is not meant to replace intuition; it refines it. A skilled evaluator reads chromatograms and noses a glass with the same intent—to discover what the molecules are saying. When expectation and experience disagree, that conflict is not an error; it is a signal.

The Path to Predictive Perception: The novice chases intensity, the professional seeks balance. Mastery begins when perception becomes prediction. Prediction transforms sensory evaluation from a purely reactive approach to a reasoned one. The ability to anticipate how a spirit will behave marks the evaluator who has crossed from tasting into sensory science

The Nose Interprets Ratios, Not Totals: Olfactory perception operates on proportion, not magnitude. A compound's impact depends on its relative presence among neighboring volatiles rather than on absolute concentration. A spirit with 30 mg/L of ethyl acetate may smell neutral if heavier esters dominate, yet the same amount in a lighter matrix can appear strikingly solvent-like. The brain's coding of odor signals is comparative—it measures *contrast* and *balance* within the odor field. Evaluators who think in terms of ratios learn to explain why a slight shift in one congener family can alter the overall character, even when analytical totals remain unchanged. Sensory interpretation, therefore, demands awareness of contextual concentration, not numbers.

Concentration Describes Possibility: Interaction Defines Reality: Chemical analysis reveals what *could* be perceived, but never what *will* be. Every compound has a potential sensory weight defined by its volatility, partitioning, and threshold; however, that potential is reshaped by interactions with other molecules and by the perceiver's adaptive state. Esters that smell vivid on their own may disappear when overlapped by aldehydes or phenols; barely measurable sulfur traces may dominate the perceptual field. Concentration establishes the boundary conditions of aroma, but interaction determines the realized outcome. The task: translate measured abundance into sensory effect through understanding of synergy, suppression, and emergent character.

Time Is a Sensory Variable: Perception is not static; it unfolds over seconds as adaptation, diffusion, and retronasal rebound alter the stimulus's dynamics. A spirit that seems simple on first sniff can become more complex as dominant volatiles desensitize their receptors, allowing secondary notes to emerge. This temporal evolution reflects both the compounds' physical volatility and the observer's neural adaptation. Quantitative models that ignore time miss much of the sensory truth. For evaluators, disciplined timing—consistent sniff duration, rest intervals, and revisit periods—is as critical as sample temperature or glass shape. Time is an active dimension of flavor expression, not a passive background.

The Mirage of Linearity: Sensory impact does not scale with concentration. Beyond threshold, doubling a compound seldom doubles its perceived strength; response curves flatten, invert, or interact unpredictably with neighboring volatiles. Analytical increases may appear dramatic, yet the nose may register only subtle change—or even suppression. Accurate quantification, therefore, requires recognizing that perception follows psychophysical law, not arithmetic progression—the evaluator who mistakes linear data for linear sensation measures chemistry, not experience.

Chapter 30: Identifying Spirits on First Sniff/Sip

Encapsulating the characteristics of each general spirit type provides a training tool to improve quick, accurate recognition of each when tasting. Training exercises that compare various spirits side by side in tastings, along with these descriptions, will prove instrumental in developing recognition. Remember, the entire flavor profile depends on olfactory, taste, and mouthfeel, and all play an important, weighted role in spirit identification.

Grain-based Spirits represent one of the most diverse and globally consumed spirit families. Their sensory characteristics primarily arise from the grain type, degree of fermentation, distillation method, and whether oak aging is employed. Each grain imparts a distinctive texture and aroma foundation, which is further shaped by yeast metabolism and the influence of barrel aging (if aged). The typical markers range from cereal grain, toast, and hay in unaged expressions to deeper notes of caramel, spice, and dried fruit in matured spirits (Piggott, 2011).

Aged Grain–based Spirits (Fig. 30.1)

Scotch Whisky
- *Base:* Malted barley
- Process: Pot-distilled, often double distilled (triple in some Lowlands)
- *Region:* Scotland (Highlands, Islay, Speyside, etc.)
- *First-Sip Profile*: Smoky phenols (Islay), malt biscuit, dried heather, toasted oak, dried fruit, often a dry finish.
- *Flavor Origins:* Kilning with peat (if used) introduces phenols; long fermentations produce fruit esters; oak aging adds vanilla, tannins, and spice.

American Bourbon
- *Base:* Minimum 51% corn
- *Process:* Column distilled, new charred American oak
- *Region:* USA (Kentucky dominant)
- *First-Sip Profile:* Caramelized corn, vanilla, cinnamon, toasted coconut, often a syrupy texture, sweet entry with dry spice finish.
- *Flavor Origins:* A high-corn mash bill and sour-mash fermentation favor sweet grain notes and fruity esters (e.g., isoamyl acetate). Typical column-still and double distillation (kept below 160 proof) retains congeners, and aging in new, charred American oak adds vanillin, oak lactones reminiscent of coconut, furfural/caramel, and eugenol/clove spice, creating a caramel-vanilla core with a dry-spice finish.

Tennessee Whiskey
- *Base:* Similar to bourbon (corn dominant)

220

- *Process:* Lincoln County Process (charcoal mellowing), aged in new charred oak
- *Region:* Tennessee, USA
- *First-Sip Profile:* Softer than bourbon, mellow vanilla, maple, charcoal smoothness, banana, dry toast finish.
- *Flavor Origins:* Charcoal filtration (The Lincoln County Process) removes heavier fusel oils; banana esters often arise from fermentation strains.

Irish Whiskey

- *Base:* Barley (malted and unmalted), sometimes with other grains
- *Process:* Triple pot-distilled, unpeated (mostly), aged in used oak
- *Region:* Ireland
- *First-Sip Profile:* Light body, honeyed grain, orchard fruit, floral tones, toasted wood, minimal smoke.
- *Flavor Origins:* Triple distillation creates a lighter spirit; use of unmalted barley adds cereal sharpness; used casks soften the oak impact.

Fig. 30.1		First-Sip Spirits Recognition (Aged Grain)	
Spirit	**Base Material**	**Distillation**	**First-Sip Identifiers**
Scotch Whisky	Malted barley	Pot distilled (double/triple)	Smoke (if peated), malt biscuit, heather, dried fruit, brine
American Bourbon	Corn (min. 51%)	Column still	Caramel, vanilla, cinnamon, sweet corn, toasted coconut
Tennessee Whiskey	Corn (charcoal mellowed)	Column still	Mellow vanilla, maple charcoal, dry toast, banana, cotton candy
Irish Whiskey	Barley (malted and unmalted)	Triple pot still	Light grain, orchard fruit, honey, floral, low smoke

Unaged Grain–Based Spirits (Fig. 30.2)

Vodka (Grain-Based)

- *Base:* Wheat, rye, or other grains
- *Process:* Multiple column distillations, charcoal or membrane filtration
- *Region:* Global (Poland, Russia, Sweden, USA)
- *First-Sip Profile:* Nearly neutral, slight sweetness (wheat), peppery bite (rye), oily mouthfeel if unfiltered.
- *Flavor Origins:* Minimal ester and congener retention; subtle grain differences detected more in texture than aroma.

Shochu (Japanese-style, barley-based variant)

- *Base:* Barley (mugi), sweet potato (imo), rice (kome)
- *Process:* Single pot distillation, koji fermentation
- *Region:* Japan
- *First-Sip Profile (barley type):* Toasted cereal, mild nuttiness, slight umami, clean finish.
- *Flavor Origins:* Koji fermentation produces glutamates and umami depth; single distillation

retains light esters and grain character.

Soju (Korean-style)

- *Base:* Rice, barley, or sweet potato
- *Process:* Traditionally, single distillation, modern commercial often diluted neutral spirit
- *Region:* Korea
- *First-Sip Profile:* Neutral to mildly sweet, exceptionally clean, slight ethanol sharpness. Traditional soju is more earthy or savory in flavor.
- *Flavor Origins:* Traditional fermentation imparts earthy notes; industrial versions strip most flavor. Commercially produced soju can be flavored from neutral spirits

Fig. 30.2	First-Sip Spirits Recognition (Unaged Grain)		
Spirit	**Base Material**	**Distillation**	**First-Sip Identifiers**
Vodka (Grain)	Wheat, rye, etc.	Multiple-column distillations	Neutral, slight sweetness (wheat), pepper (rye), clean texture
Shochu	Barley, sweet potato, or rice	Single pot still, koji fermentation	Toasted cereal, mild umami, earthy, clean finish
Soju	Rice, barley, or sweet potato	Single distillation or dilution	Mild sweetness, ethanol sharpness, earthy (traditional), neutral (modern)

Summary Commentary:

Grain spirits offer a broad and complex spectrum, particularly when aged. Oak contact radically transforms the profile by introducing a library of lactones, furans, and tannins, while distillation strength and yeast type dictate the aromatic complexity (Conner, Paterson, & Piggott, 1999). Unaged grain spirits rely heavily on fermentation substrates, still design, and final filtration to communicate identity. The first sip of each reveals its raw material origin and production choices, which become clear with trained awareness. Non-grain vodkas show slight traces of origin, such as potato, grape, and corn, each with distinct trace flavors (Jackson, 2020).

Sugarcane-Based Spirits (Fig. 30.3)

Produced from fresh-pressed cane juice or derivative molasses. The origin material profoundly affects the resulting flavor; cane juice spirits tend toward vegetal and grassy notes, while molasses-based spirits tend toward caramelized, cooked, or fruity notes (Fahrasmane & Ganou-Parfait, 1998). Distillation method (pot vs column) and aging regimen (if any) equally shape spirit character.

Rum (Molasses-Based)

- *Base:* Molasses (byproduct of sugar refining)
- *Process:* Column or pot distillation; aged or unaged; varied wood regimes
- *Region:* Caribbean, Central and South America, USA, others

- *First-Sip Profile:* Aged: brown sugar, vanilla, toffee, baking spice, tropical fruit, dried banana; Unaged: burnt sugar, metallic bite, light tropical esters.
- *Flavor Origins:* Fermentation length influences ester load (long fermentations yield more fruity and funky profiles), pot stills concentrate heavier esters and congeners, while column stills produce light, clean spirits. Barrel aging imparts sweet spice and oxidized compounds. Aging in ex-bourbon or sherry casks adds vanillin, lactones, and oxidative complexity. Tropical aging accelerates these reactions compared to cooler climates.

Rhum Agricole (cane juice)

- *Base:* Fresh-pressed sugarcane juice
- *Process:* Continuous column distillation, sometimes French oak aged.
- *Region:* French Caribbean (Martinique, Guadeloupe)
- *First-Sip Profile:* Earthy, vegetal, grassy, green banana, lime zest, mineral finish.
- *Flavor Origins:* Fermentation of fresh cane juice retains the plant's green, earthy flavors, particularly aldehydes and terpenes. Low-rectification distillation preserves fermentation volatiles. French oak aging imparts subtle tannins and baking-spice notes when present.

Cachaça (cane juice)

- *Base:* Fresh sugarcane juice
- *Process:* Pot or column distilled; often rested in native woods or stainless steel
- *Region:* Brazil
- *First-Sip Profile:* Fruity (tropical), herbaceous, peppery, funky, or creamy depending on aging.
- *Flavor Origins:* Cane juice fermentation yields high ester and aldehyde diversity, often using wild or native yeasts. Native woods (e.g., amburana, balsamo, or jequitibá) contribute spicy, cinnamon-like, vanilla-like, nutty, or resinous tones. Stainless steel resting preserves bright esters, while pot stills produce a richer texture and body.

Fig. 30.3	First-Sip Spirits Recognition (Sugarcane)		
Spirit	**Base Material**	**Distillation**	**First-Sip Identifiers**
Rum (Molasses)	Molasses	Pot or column still	Aged: brown sugar, toffee, banana, spice; Unaged: burnt sugar, esters
Rhum Agricole	Fresh sugarcane juice	Column still	Grassy, vegetal, lime zest, green banana, mineral
Cachaça	Fresh sugarcane juice	Pot or column still	Tropical fruit, herbal, peppery, funky or creamy (wood influence)

Fruit-Based Spirits (Figs. 30.4 - 30.5) derive their core aromatic and gustatory identity from the fermentation of sugars naturally present in fruits. Unlike grain or sugarcane spirits, fruit spirits carry a rich volatile ester profile resulting from both the inherent composition of the fruit and the specifics of fermentation. These spirits often maintain a strong linkage to terroir, harvest maturity, and yeast strain selection. Distillation must preserve delicate fruit aromatics, and aging (if applied) can either enhance or obscure this natural identity (Arroyo, 1945). The style and clarity of flavor depend heavily on the condition of the fruit, fermentation dynamics, and the care taken during distillation to prevent the loss or burning of delicate volatiles.

Brandy/Cognac/Armagnac

- *Base:* Grapes (white varietals, typically Ugni Blanc in Cognac)
- *Process:* Double pot distillation, copper alembics (Cognac); aged in Limousin or Tronçais oak
- *Region:* Cognac (France), Armagnac, USA, South Africa, others
- *First-Sip Profile:* Ripe grape skin, dried apricot, raisin, floral aldehydes, vanilla, toffee, cigar box, rancio (in older Cognac).
- *Flavor Origins:* Copper stills catalyze ester formation and sulfur removal; grape esters, light aldehydes, and acids concentrate during distillation; oak aging introduces lactones, vanillin, and oxidative compounds like acetaldehydes and furfural. Cognac, in particular, derives subtlety from slow distillation and extended aging in Limousin or Tronçais oak.

Pisco

- *Base:* Aromatic and non-aromatic grapes (e.g., Muscat, Quebranta)
- *Process:* Single distillation, unaged
- *Region:* Peru, Chile
- *First-Sip Profile:* Floral (jasmine, orange blossom), grape must, sweet citrus, pear, subtle heat.
- *Flavor Origins:* Retention of fermentation esters and highly volatile monoterpenes due to single-pass distillation; Muscat grapes yield terpenic intensity; no barrel aging preserves full floral and aromatic character.

Fig. 30.4	First-Sip Spirits Recognition (Fruit - 1)		
Spirit	**Base Material**	**Distillation**	**First-Sip Identifiers**
Brandy/ Cognac	Grapes (white)	Copper pot (Charentais)	Grape skin, dried fruit, vanilla, spice, floral notes
Armagnac	Grapes	Column still (low rectification)	Prune, leather, oak spice, dried fig, robust grape
Pisco	Muscat or aromatic grapes	Pot still, no aging in wood	Grape must, perfumed floral, stone fruit, alcohol heat
Calvados	Apples (or pears)	Pot or column still	Baked apple, cider, spice, dry finish

Calvados

- *Base:* Apples (and pears in Domfrontais)
- *Process:* Double or single distillation; aged in oak
- *Region:* Normandy, France
- *First-Sip Profile:* Baked apple, cinnamon, dried pear, nutmeg, soft tannins, sometimes smoky wood.
- *Flavor Origins:* Malic fermentation creates esters specific to apples; distillation preserves cider-like acidity; oak aging deepens spice and introduces oxidized apple notes.

Palinka

- *Base:* Stone fruits (apricot, plum, cherry), apples, pears
- *Process:* Pot small batch distilled, unaged, or lightly aged
- *Region:* Hungary, Transylvania (Romania)
- *First-Sip Profile:* Intense ripe fruit, marzipan (from stone fruit kernels), floral ester high tones, slight spirit mouthfeel heat.
- *Flavor Origins:* Fruit variety and ripeness at fermentation are key; fermentation on skins and pits releases benzaldehyde (almond notes); skin maceration boosts aldehydes and terpenes, careful distillation retains delicate fruit volatiles; pot still retains esters and organic acids.

Slivovitz

- *Base:* Plums (often Damson or other heritage cultivars)
- *Process:* Whole fruit fermentation, double pot distillation, sometimes aged
- *Region:* Central and Eastern Europe (Serbia, Croatia, Czechia, Slovakia, Hungary, etc.)
- *First-Sip Profile:* Tart cooked plum skin, almond kernel, peppery finish, sometimes faint smoke or spice.
- *Flavor Origins:* Fermentation with pits yields benzaldehydes, amygdalin-derived aroma; plum acids contribute brightness; oak (if used) adds tannic edge. Extended fermentation adds depth and acid balance.

Orujo/Grappa

- *Base:* Grape pomace (skins, seeds, stems)
- *Process:* Pot distilled, unaged, or lightly aged
- *Region:* Spain (Orujo), Italy (Grappa)
- *First-Sip Profile:* Grape skin tannin, green herb, peppery, floral top notes.
- *Flavor Origins:* Pomace fermentation adds herbal and tannic bitterness; distillation preserves grape volatiles. Harshness, if poorly made, often comes from seed oils or excessive distillation.

Rakia

- *Base:* Fermented fruits (plums, grapes, apricots, pears, figs, etc.)
- *Process:* Pot-distilled, occasionally double-distilled; often home or small-batch made
- *Region:* Balkans (Serbia, Bulgaria, Croatia, North Macedonia, etc.)
- *First-Sip Profile:* Fruity (varies by base fruit), mildly floral, dry to semi-sweet, often slightly earthy or tannic from pits/skins
- *Flavor Origins:* Fruit fermentation yields esters and aldehydes; pot still preserves fruit character; minimal or no oak influence unless barrel-aged

Arrack (Sri Lankan)

- *Base:* Coconut flower sap (toddy)
- *Process:* Pot or column distilled, sometimes aged in local woods.
- *Region:* Sri Lanka
- *First-Sip Profile:* Funky tropical fruit, fermented banana, palm sugar, earthy heat, slightly

metallic bite, slightly smoky, molasses-like

- *Flavor Origins:* Wild yeast and bacterial fermentation generate high ester and phenol levels and acid complexity. Sometimes aged in hamilla wood, contributing mild tannins and oxidative character. Minimal refining preserves complexity.

Fig. 30.5	First-Sip Spirits Recognition (Fruit - 2)		
Spirit	**Base Material**	**Flavoring**	**First-Sip Identifiers**
Palinka	Stone fruits (plum, apricot)	Pot distilled	Ripe fruit flesh, floral, almond from pits, warm alcohol tone
Slivovitz	Plums	Pot distilled	Stewed plum, prune, almond, dry spice, mild bitterness
Grappa (Orujo)	Grape pomace (seeds, skin, stems)	Pot distilled	Grape tannins, green herb, pepper, floral top notes
Rakia	Plums, grapes, apricots, pears, figs	Pot still, sometimes double distilled	Fruity, depending on the base, esters, aldehydes, no oak
Arrack (Sri Lanka)	Sugarcane, molasses, or coconut sap	Pot or column still	Funky fruit, fermented sweetness, toasted coconut, oily

Agave and Botanical-Based Spirits (Fig. 30.6) rely on the character of their raw materials. In agave spirits, the plant's lengthy growth cycle, traditional fermentation, and low-and-slow distillation yield a deeply expressive spirit. In botanical spirits, such as gin, a neutral spirit is redistilled with selected herbs, roots, seeds, and peels, making the flavor entirely dependent on the composition and balance of these additions. Tequila roasting and fermentation create esters, acids, and oak lactones that define its character (Gutiérrez-Gamboa & Moreno-Simunovic, 2019). In gin, the botanical load and rectification level of the base spirit determine intensity and complexity (Hughes & Baxter, 2021).

Tequila

- *Base:* Blue Weber agave
- *Process:* Cooked (steamed) agave, crushed and fermented; pot or column distilled; aged (Reposado, Añejo) or unaged (Blanco)
- *Region:* Jalisco and designated Mexican regions
- *First-Sip Profile:* Cooked agave (sweet, earthy, vegetal), white pepper, lime zest, clay, vanilla or caramel in aged versions.
- *Flavor Origins:* Agave roasting creates Maillard-derived caramel notes; fermentation with wild or selected yeast yields esters and acids; oak aging introduces lactones, vanillin, and tannins.

Mezcal

- *Base:* Various agave species (e.g., Espadín, Tobalá)
- *Process:* Roasted in underground pits, stone-crushed, open fermented, pot-distilled
- *Region:* Oaxaca and other certified Mexican regions
- *First-Sip Profile:* Smoky roast, charred citrus, earth, leather, tropical fruit, metallic minerality, deep umami.
- *Flavor Origins:* Pit roasting infuses smoke phenols; agave varietals contribute distinct ester

profiles; open-air fermentation allows microbial diversity to drive funk and complexity.

Fig. 30.6	First-Sip Spirits Recognition (Agave)		
Spirit	**Base Material**	**Distillation**	**First-Sip Identifiers**
Tequila (Blanco)	Blue agave	Pot distilled	Cooked agave, white pepper, citrus rind, fresh-cut grass
Tequila (Añejo)	Blue agave	Pot distilled	Toasted agave, vanilla, oak spice, light caramel
Mezcal	Agave (various)	Clay or copper pot still	Smoke, roasted agave, citrus, vegetal, earth

Specialty and Regional Spirits (Fig. 30.7)

This category includes culturally significant spirits that fall outside of the prominent grain, cane, fruit, or botanical groups. Often rooted in regional traditions, these spirits may use unconventional base materials, fermentation aids, or flavoring agents. Their first-sip identification relies heavily on recognizing hallmark flavors unique to a place, method, or microbial ecosystem. Maotai and Baijiu utilize solid-state fermentation and multiple distillations, resulting in the production of esters, acids, and pyrazines unique to their production (Zheng & Han, 2016).

Maotai (Moutai)

- *Base:* Sorghum
- *Process:* Nine rounds of high-temperature grain steaming/cooking, eight fermentations, and seven distillations, then clay-jar aging (typically ~3–4 years)
- *Region:* China (Guizhou Province)
- *First-Sip Profile:* Soy sauce, fermented bean paste, sesame oil, overripe pineapple, strong umami, pungent alcohol heat.
- *Flavor Origins:* Complex microbial fermentation (Bacillus, molds, yeast) yields hundreds of esters, acids, and pyrazines; high-temp stacking fermentation intensifies aroma compounds.

Zivania

- *Base:* Grape pomace mixed with dry wine
- *Process:* Pot distillation, unaged
- *Region:* Cyprus
- *First-Sip Profile:* Clean ethanol, subtle grape skin, peppery bite, slightly oily mouthfeel, almond or stone fruit traces.
- *Flavor Origins:* Pomace distillation retains seed and skin volatiles; high distillation proof limits ester complexity but preserves some grape-derived notes similar to grappa, with fruit esters and aldehydes from pomace fermentation.

Baijiu

- *Base:* Sorghum, sometimes wheat, barley, peas, or rice
- *Process:* Solid-state fermentation with multiple rounds, distilled from fermented grain cakes, a *qu* starter culture
- *Region:* China (Various Provinces)
- *First-Sip Profile:* Soy sauce, fermented bean paste, sesame oil, overripe pineapple, strong

umami, pungent alcohol heat.

- *Flavor Origins:* Complex microbial fermentation (Bacillus, molds, yeast) yields hundreds of esters, acids, and pyrazines; high-temp stacking fermentation intensifies aroma compounds.

Fig. 30.7	First-Sip Spirits Recognition (Specialty)		
Spirit	**Base Material**	**Distillation**	**First-Sip Identifiers**
Maotai (Moutai)	Sorghum	Solid-state ferment, pot still	Soy sauce, sesame, pineapple, umami, pungent heat
Zivania	Grape pomace, added dry wine	Pot still	Grape skin, peppery, pungent ethanol, almond, clean finish
Baijiu	Sorghum	Solid-state ferment, pot still	Soy sauce, sesame, pineapple, umami, pungent heat
Also see Chapter 26 Appendix on Baiju			

Flavored, Compound Spirits (Fig. 30.8)

These spirits rely not only on a neutral base alcohol but also on botanicals, seeds, fruits, or roots that define their essence. Sensory recognition is based on dominant botanical markers such as juniper in gin, anise in absinthe, or caraway in aquavit. Evaluators must learn to quickly separate primary aromatics from background spirit notes and recognize cultural and regional practices.

Gin

- *Base:* Neutral spirit (typically grain-based)
- *Process:* Redistilled with botanicals (primarily juniper) via infusion, vapor distillation, or cold-compounding.
- *Region:* Global (UK, USA, Netherlands, and others).
- *First-Sip Profile:* Juniper resin, citrus peel, angelica root, coriander, floral or herbal variants depending on style (London Dry, New Western, etc.)
- *Flavor Origins:* Botanical charge (pinene, sabine), citrus peel (limonene), and spices. The distillation method determines the aroma profile; higher proof captures delicate volatiles, while base spirit neutrality allows for botanical expression. Rectification level of the base spirit, and the botanical basket's shape, intensity, and complexity.

Absinthe

- *Base:* Neutral, from grape pomace (marc or wine base)
- *Process:* Redistilled with wormwood, green anise, sweet fennel, and other botanicals macerated before redistillation. Some add hyssop, melissa
- *Region:* Switzerland, France, Spain, the Czech Republic, and the USA
- *First-Sip Profile:* Anethole, bitter wormwood (Artemisia absinthum)
- *Flavor Origins:* Primary and secondary distillation and maceration. Color: green ("verte"). Louche with water.

Aquavit

- *Base:* Neutral spirit from grain or potato

- *Process:* Redistilled with caraway, dill, other botanicals
- *Region:* Scandinavia
- *First-Sip Profile:* Caraway seed, dill, fennel, citrus hints, warm spice
- *Flavor Origins:* Caraway (carvone), dill (dill ether), and aniseed provide dominant aromatics. Often oak aged for added roundness/complexity, particularly in Danish and Norwegian styles.

Arak (Levantine)

- *Base:* Grape spirit redistilled with green anise, aniseed
- *Process:* Multiple pot distillations; usually unaged, sometimes in clay.
- *Region:* Lebanon, Syria, Jordan, Palestine, Iraq
- *First-Sip Profile:* Strong licorice, herbal sweetness, oily texture, warm alcohol, creamy texture when louched.
- *Flavor Origins:* Anethole from aniseed dominates; redistillation captures aromatic oils; dilution triggers louche from oil-water separation. The grape base contributes faint fruitiness.

Ouzo

- *Base:* Neutral from grapes or grain
- *Process:* Multiple pot distillations; usually unaged, sometimes in clay.
- *Region:* Greece, Cyprus
- *First-Sip Profile:* Strong licorice, anise, herbal sweetness, fennel, coriander, mastic, cardamom, sometimes cinnamon and cloves, warm alcohol, turns white when louched.
- *Flavor Origins:* Anethole from aniseed dominates, dilution triggers louche from oil-water separation. Sweetened. Grape contributes faint fruitiness.

Pastis

- *Base:* Sugar beet molasses (96% prior to dilution)
- *Process:* Anise, licorice, other herbs (blended with botanicals)
- *Region:* France
- *First-Sip Profile:* Wide variety, dependent on botanicals blended in.
- *Flavor Origins:* No flavor contribution from base material. Botanical blends are the flavors

Tubi 60

- *Base:* Neutral spirit from citrus (proprietary blend)
- *Process:* Infused with citrus, ginger, turmeric, mint, lemongrass, herbs, caffeine, and other proprietary ingredients (over 20 botanicals)
- *Region:* Israel
- *First-Sip Profile:* Spicy lemon citrus zest, herbal bite, ginger heat, subtle bitterness, energetic, slightly astringent mouthfeel.
- *Flavor Origins:* Maceration and post-distillation infusion of botanicals post-distillation; proprietary formula introduces essential oils and stimulants, and high aroma volatiles.

Fig. 30.8	First-Sip Spirits Recognition (Compound Spirits)		
Spirit	**Base Material**	**Flavoring**	**First-Sip Identifiers**
Gin	Neutral grain spirit, botanicals	Redistilled or flavored with botanicals	Juniper (not always), citrus peel, coriander, dry herbal or floral
Absinthe	Neutral, beet, or grain	Wormwood, anise, fennel, others	Licorice, herbs, creamy white louche on dilution, anethole
Aquavit	Neutral, grain or potato	Caraway, dill, spices	Caraway, dill, faint citrus, warm spice (oak in some)
Arak (Levantine)	Grape or date distillate	Aniseed	Anise, licorice, dry herbal, creamy white louche on dilution
Ouzo	Neutral, grapes or grain	Aniseed, coriander, fennel, cardamom, cinnamon, clove	Licorice, sweet, spices
Pastis	Neutral, sugar beet, grain	Anise, aromatic herbs	Licorice, sweet, herbs
Tubi 60	Neutral spirit, botanicals	Lemon, ginger, herbs, spices	Citrus zest, ginger, herbs, spicy, and caffeinated lift

Tasting with a Timing Pass - Timing is Almost Everything

Exploring different spirits, many evaluators tend to overlook subtle aromas in their quest for those that are easily recognizable. This approach defeats the purpose and, in fact, can leave the evaluator forming an opinion that "It is not as good as bourbon," or "It is almost as peaty as my favorite, but I definitely like Islay's better." When the evaluator is unsure of what to expect, it is best to adopt the mindset that detecting aromas is the primary objective, and comparing them to past flavor experiences can bias the outcome and cloud judgment. That said, it can take several seconds for key aromas to become detectable in the headspace. To ensure that the evaluation is fair, complete, and accurate, every sample should undergo a timing pass. Using this procedure for every evaluation ensures its reliability and validity.

A quick first pass is made to determine immediate impressions: Many key characteristic aromas of long-chain structure, high molecular mass, or low concentration may take longer to detect. Different time spans can yield different prominent aromas, giving a time dynamic to aroma detection. An example of aroma detection over time is shown in Fig. 26.9.

- **Onset (0-5 secs., orthonasal before sip):** Let the pour settle. One short sniff. Log the highest-volatility cues and any immediate ethanol lift. Use Appendix I terms and grade aroma intensity (0–10). *Example: Fruity—apple (green skin) 6; Citrus—lemon zest 3; ethanol heat 4.*
- **First Taste (Mid-period 5-20 secs., during sip / retronasal on exhale):** Confirm or correct onset notes, add mouthfeel (heat, viscosity, astringency). Oak, spice often emerge here. Ethanol can mask late notes if tasting from tulip glasses; allow 60–90 secs. of air, document. The key point here is to note when different aromas begin to appear and how orthonasal aromas may have changed, if at all.
- **Finish (20--60+ secs., lingering):** Record durations and what changes. Late phenolics,

bitterness, or oxidized notes often appear in this area. If a suspected fault persists and dominates, cross-check Appendix II thresholds.

- **Second pass (ortho- and retronasally) after 60-90 s.** Keep the glass covered between passes to allow the headspace to reach equilibrium with the aroma compounds.

Record as follows: Onset → Mid → Finish, each with descriptor + intensity + phase, plus finish duration. Observations first; inferences second (with confidence). Reference Appendix I for correct terms; Appendix II for faults.

Before attempting to identify the spirit, identify and record the aromas, and focus on discovering and recording *all* aroma subtleties through the timeline. This process allows the evaluator time to reconsider final evaluation scores, ensuring fairness and accuracy. If possible, retest at a later time or on a different day, taking fresh notes, and then compare them to your previous notes to ensure reliability. Most spirits will be tested covered, and will have rested for several minutes. Fig. 30.9 considers how aromas manifest in the headspace, if that information is critical at some point.

Standards of Identity and Sensory Typicality

Category rules—standards of identity—do more than police labels. They set boundary conditions that shape typical sensory profiles. Mash bills, permitted materials, still types, and maturation rules constrain what a spirit can plausibly smell and taste like, even before a bottle is opened. "Recognition at first sip" is essentially recognition of a typicality window forged by those constraints.

Two clarifications keep the discussion honest:

- Typicality ≠ quality. A sample can be squarely typical and mediocre, or atypical and excellent. Typicality refers to how well an item fits the category center, rather than its merit.
- Law ≠ inevitability. Rules increase the probability of specific outcomes (e.g., new charred oak elevates vanilla/coconut/caramel signals; triple pot distillation often lightens body), but do not guarantee them. Production choices, climate, and blending can amplify or mute expected notes.

Why this matters for first-pass recognition and cross-regional comparisons:

- It grounds fast classification in constraints, not folklore.
- It helps separate innovation (pushing style within rules) from mislabeling (outside the rules, yet marketed as typical).
- It provides a neutral way to discuss global variation: some categories are narrowly specified (tight typicality windows), others are permissive (broad windows), and panels must calibrate expectations accordingly.

When you read "typical" in this book, read it as a probabilistic center, not a verdict. The task of the evaluator is to hear the category's grammar, then judge the sentence on its own terms. Three distinctly different terms that are often mistakenly used interchangeably

- **Typicality**: Fit to the category's probabilistic center (its sensory grammar). Atypical is not automatically bad; typical is not automatically good.
- **Authenticity**: Conformity to the declared rules of production and identity (materials, methods, maturation). An authentic spirit can be atypical in style.
- **Quality**: Balance, integration, length, absence of disqualifying faults, and how well the parts cohere. Quality is judged within the stated category and its intended purpose.

Distinction prevents age-or-origin prejudice, keeps innovation intelligible, lets panels clearly communicate: "Authentic, low-typicality, high quality" is a coherent, non-redundant statement.

Perception is time-dependent (Fig. 30.9).

Fig. 30.9 Example of Aroma Detection Over Time

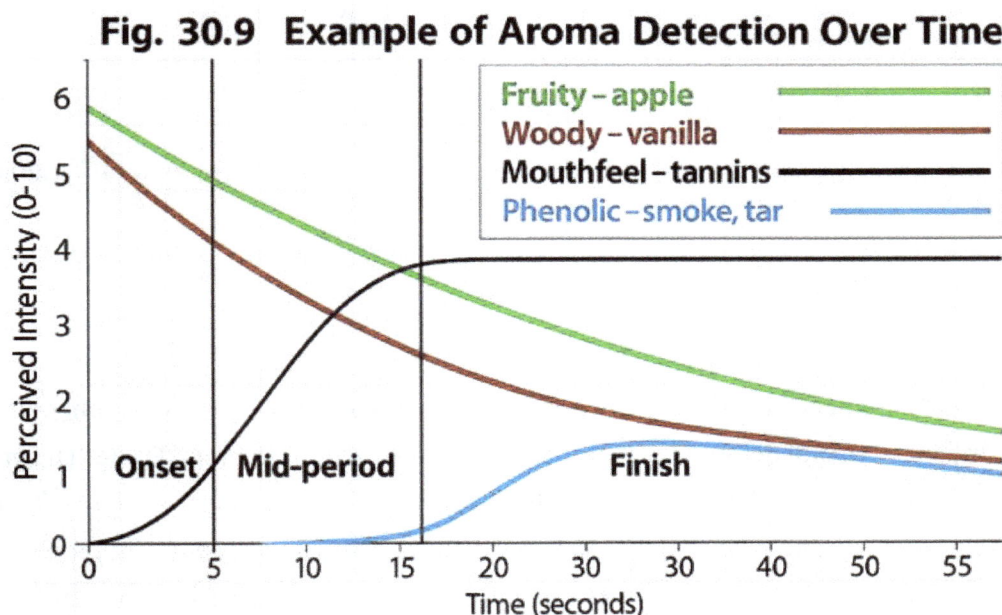

"**Today's whisky consumer doesn't think twice about picking up a whisky from Norway or Iceland or Australia or Argentina.**" — Dave Broom, author

Appendix: Baijiu Categories

Appendix 30.1	Baijiu Aroma Categories	
Intensity	**Province**	**First-Sip Identifiers**
Strong (Nongxiang)	Luzhou	Rich, fruity, estery
Light (Qingxiang)	Shanxi	Mild, clean
Sauce (Jiangxiang)	Guizhou (Maotai)	Umami, soy sauce-like
Rice (Mixiang)	Southern	Floral, light

References

Arroyo, R. (1945). *Studies on rum (Bulletin No. 68)*. United States Department of Agriculture, Puerto Rico Agricultural Experiment Station. National Agricultural Library Digital Collections. https://naldc.nal.usda.gov/catalog/CAT86200912

Conner, J. M., Paterson, A., & Piggott, J. R. (1999). Changes in wood extractives from oak cask staves through maturation of Scotch malt whisky. *Journal of the Science of Food and Agriculture, 79*(3), 287–294. https://doi.org/10.1002/(SICI)1097-0010(19990301)79:3<287::AID-JSFA235>3.0.CO;2-9

Fahrasmane, L., & Ganou-Parfait, B. (1998). Spirits and traditional beverages from sugarcane. *Food Research International, 31*(6–7), 365–371. https://doi.org/10.1016/S0963-9969(98)00094-0

Gutiérrez-Gamboa, G., & Moreno-Simunovic, Y. (2019). Volatile composition of tequila and mezcal. *Food Reviews International, 35*(3), 248–271. https://doi.org/10.1080/87559129.2018.1503380

Hughes, D., & Baxter, J. (2021). The science and sensory evaluation of gin. *Beverages, 7*(1), Article 11. https://doi.org/10.3390/beverages7010011

Jackson, R. S. (2020). *Wine science: Principles and applications* (5th ed.). Academic Press. https://doi.org/10.1016/C2018-0-03808-6

Piggott, J. R. (2011). *Alcoholic beverages: Sensory evaluation and consumer research*. Woodhead Publishing. https://doi.org/10.1533/9780857095176

Zheng, X.-W., & Han, B.-Z. (2016). Baijiu, Chinese liquor: History, classification and manufacture. *Journal of Ethnic Foods, 3*(1), 19–25. https://doi.org/10.1016/j.jef.2016.03.001

Observations on Identifying Spirits on First Sniff or Sip

Initial recognition depends on the presence of dominant volatiles, such as esters, phenols, or aldehydes, and how ethanol delivers them through nasal airflow and trigeminal tone. The brain matches these signals to stored olfactory templates: fruit–solvent for brandy, cereal–smoke for whisky, citrus–resin for gin. Such recognition is provisional. Overlapping congener profiles, glass shape, and strength often blur category lines—a high-ester rum may resemble fruit brandy, or a phenolic gin a light whisky.

First impressions are hypotheses that are refined by retronasal and palatal cues. Consistency comes from trained memory and recall. The first sniff suggests a direction; deliberate reevaluation confirms the identity. Regional character is often revealed before conscious recognition through subtle shifts in ester balance, oak extractives, or fermentation markers, which are shaped by climate and local practices. Sense them as familiarity rather than discrete notes. With experience, geography becomes a scent memory: Highland peat, Caribbean molasses, and Cognac's rancio each announce their origin.

Chapter 31: Recognizing Global Differences in Distillation

For the sensory evaluator, recognition of spirits is not limited to categories such as "whiskey" or "rum." Instead, evaluators must understand how regional production practices, from fermentation biology to still design and maturation climate, create unique sensory signatures. Many of these aroma and flavor cues are unfamiliar in American or Western European products; thus, their identification is essential for accurate, context-sensitive evaluation.

Global distillation practices are shaped by cultural tradition, raw material availability, microbial ecology, and regulatory frameworks. Sensory evaluators who learn to connect aroma markers with these practices can distinguish authenticity, prevent misattributing unusual notes to faults, and better appreciate the diversity of distilled beverages.

Regional Practices and Sensory Markers (Fig. 31.1)

Solid-State Fermentation (China: Baijiu, Maotai): Most Western spirits rely on liquid fermentation, but Chinese baijiu production uses solid-state fermentation. Grains such as sorghum are stacked into bricks or heaps, inoculated with *qu* starter cultures containing yeasts, molds, and bacteria. The result is a mixed microbial ecology producing hundreds of volatile compounds, including esters, pyrazines, and short-chain fatty acids (Zheng & Han, 2016).
Sensory markers: soy sauce, sesame oil, fermented bean paste, ripe tropical fruit, strong umami.
Evaluator cue: savory sauce-like volatiles rarely encountered in grain spirits outside East Asia.

Koji Fermentation (Japan: Shochu, Awamori): Japanese shochu and Okinawan awamori utilize koji (Aspergillus oryzae) fermentation, which converts starch into sugar while producing glutamates and umami-active molecules (Yoshizaki et al., 2010).
Sensory markers: toasted cereal, nutty, mushroom-like umami, earthy balance.
Evaluator cue: the presence of umami in spirits—a hallmark absent in most Western distillates.

Pit-Roasted Agave (Mexico: Mezcal): Unlike tequila, mezcal production involves roasting agave hearts in underground pits. This process caramelizes polysaccharides and introduces smoke phenols such as guaiacol and cresols (Márquez et al., 2020).
Sensory markers: roasted agave, charred citrus, vegetal smoke, mineral undertone.
Evaluator cue: deep smoke and char distinct from the lighter roasted agave of tequila.

Wild or Spontaneous Fermentations (Caribbean: High-Ester Rums, Brazil: Cachaça): In Jamaica and Guyana, extended, wild fermentations with mixed yeasts and bacteria create extremely high ester loads, sometimes exceeding 1,000 mg/L. These esters are key to "funk"

(Murtagh et al., 2021).

Sensory markers: isoamyl acetate (banana), ethyl butyrate (pineapple), ethyl acetate (solvent-like fruit).

Evaluator cue: "overripe fruit" or "funk" intensity not found in column-distilled American rums.

Use of Non-Traditional Woods (Brazil: Cachaça; India: Local Whiskies). While most global aging relies on oak, Brazilian cachaça producers often age their products in amburana, bálsamo, or jequitibá woods, each contributing unique volatiles (da Silva et al., 2019).

Sensory markers: cinnamon-like spice (cinnamic aldehydes), balsamic resin, herbal tones.

Evaluator cue: unexpected cinnamon/balsamic notes in cane spirits = Brazil.

Earthenware or Clay Distillation (China, Korea, Mezcal villages): Porous clay stills and vessels used in parts of Asia and Mexico influence both heat dynamics and subtle mineral character (Chen et al., 2014).

Sensory markers: soft mineral notes, earthy undertones, muted metallics.

Evaluator cue: mineral aroma not typical of copper or stainless-steel stills.

Climate-Driven Maturation (Tropics vs Temperate): Hot, humid regions accelerate maturation reactions. Tropical rum or Indian whisky may undergo oxidative and extractive changes in 2–3 years, comparable to those observed in Scotland over 10–12 years (Madrera & Suárez, 2018).

Sensory markers: intense vanillin, lactones (coconut), and dried fruit volatiles at a young age.

Evaluator cue: rapid maturity markers (deep oak, dried fruit) in spirits with low stated age.

Fig. 31.1	Regional Practices and Sensory Cues		
Region/Spirit	**Distinctive Practice**	**Dominant Aromas**	**Sensory Cues**
China, Baijiu	Solid state fermentation with qu	Esters, pyrazines, acids	Soy sauce, sesame, tropical fruit, umami
Japan, Shochu	Koji (A. oryzae) saccharification	Glutamates, esters	Umami, nutty, mild, mushroom
Mexico, Mezcal	Underground pit roasting	Smoky phenols (guaiacol, cresols)	Charred citrus, vegetal smoke, earthy tones
Jamaica, Rum	Long, wild fermentation	Isoamyl acetate, ethyl butyrate	Funk, banana, pineapple, solventy fruit
Brazil, Cachaça	Aging in native woods	Cinnamic aldehydes, terpenes	Cinnamon spice, balsamic resin, herbal edge
Mexico, Mezcal	Clay/earthenware distillation	Mineral traces, earthly volatiles	Soft mineral, muted metallic, earthy
Caribbean Rum, Whisky	Accelerated maturation	Vanillin, lactones, oxidized esters	Coconut, dried fruit, rapid maturity notes

Training Implications for Sensory Evaluators (Fig. 31.1)

Interpretation, not misattribution: A "soy sauce" note in baijiu or a "funky pineapple" in Jamaican rum is not a defect—it reflects authentic regional microbiology.

Comparative tasting: Training should involve side-by-side evaluations of American bourbon

with Jamaican rum or vodka with baijiu to highlight contrasts in fermentation and still design.
Cultural literacy: Recognizing global differences requires understanding not only chemical markers but the traditions that anchor them.

Case Study 31.1: Scotch Whisky Copper Contact and Sulfur Management: Traditional Scottish pot stills are engineered for extensive copper–vapor interaction, removing sulfur compounds such as DMT and methanethiol. Controlled trials at Glenmorangie and the Scotch Whisky Research Institute (SWRI, 2019) showed that shortened still necks or reduced reflux increased meaty, vegetal tones, while elongated necks and higher reflux produced cleaner, fruit-forward spirit. Evaluators recognize these differences by the presence or absence of metallic sharpness and by ester brightness.
Summary: Awareness of the expected low-sulfur, high-ester profile in Highland malt enables evaluators to detect departures in still design or operation, transforming minor deviations into interpretable sensory nuances rather than perceived flaws.

Case Study 31.2: Caribbean Rums—Column vs. Pot Still: Column stills in Trinidad and Puerto Rico yield lighter, high-purity rums with restrained congeners, while pot stills in Jamaica and Barbados retain dense esters and volatile acids responsible for fruity and oily weight. These distinctions persist even within the same raw material base. Evaluators note that column rums fade quickly in headspace, while pot still rums linger with layered fruit and phenolic warmth.
Summary: Recognizing whether a rum's volatility pattern conforms to its traditional still type allows evaluators to read *process intent*—and to appreciate subtle departures (a cleaner-than-expected Jamaican rum, for instance) as evidence of innovation rather than defect.

Case Study 31.3: Cognac vs. Armagnac— Temporal Heat Profiles: Cognac's double distillation in Charentais alembics promotes higher reflux and ethyl ester formation, producing polished fruit and vanilla tones. Armagnac's single-pass continuous distillation at lower strength preserves heavier acids and aldehydes, yielding notes of prune, cocoa, and spice after aging. Sensory training teaches the evaluator to distinguish elegance and lift from depth and rustic warmth—both valid expressions of grape distillate.

Summary

Global differences in distillation are not merely stylistic; they are sensory signposts. Solid-state fermentation, microbial diversity, pit roasting, unusual woods, and climate-driven maturation all leave chemical signatures in the glass. For the evaluator, mastery of these cues enables accurate identification, prevents misjudgment of authenticity, and enriches appreciation of the world's distilled traditions.

"It is not our differences that divide us. It is our inability to recognize, accept, and celebrate those differences."— Audre Lorde, feminist, writer, philosopher

"Alcohol consumption is rooted in culture and traditions."— Monaco (2020), reaffirming the cultural basis for beverage choices

References

Chen, S., et al. (2014). Influence of pottery and metal distillation vessels on aroma compounds in Chinese spirits. *Journal of Agricultural and Food Chemistry, 62*(47), 11336–11344. https://doi.org/10.1021/jf504011p

Da Silva, P. M., et al. (2019). Influence of Brazilian native woods on cachaça maturation. *Journal of the Institute of Brewing, 125*(2), 239–248. https://doi.org/10.1002/jib.563

Madrera, R. R., & Suárez, V. (2018). Influence of aging in tropical vs. temperate conditions on the composition of rum and whisky. *Food Chemistry, 266*, 382–389. https://doi.org/10.1016/j.foodchem.2018.06.006

Márquez, C., et al. (2020). Characterization of phenolic compounds in mezcal. *Food Research International, 131*, 109008. https://doi.org/10.1016/j.foodres.2020.109008

Murtagh, J., et al. (2021). Jamaican rum: High-ester production and sensory profile. *Applied Microbiology and Biotechnology, 105*(23), 8713–8726. https://doi.org/10.1007/s00253-021-11663-1

Yoshizaki, Y., et al. (2010). Characteristics of aroma compounds generated by *Aspergillus oryzae* in rice fermentation. *Journal of Bioscience and Bioengineering, 110*(2), 152–157. https://doi.org/10.1016/j.jbiosc.2010.02.013

Zheng, X. W., & Han, B. Z. (2016). Baijiu, Chinese liquor: History, classification and manufacture. *Journal of Ethnic Foods, 3*(1), 19–25. https://doi.org/10.1016/j.jef.2016.03.001

Observations on Global Differences in Distillation

Distillation diversity reflects culture as much as chemistry; sensory recognition requires contextual literacy, not just detection skill. Evaluators must distinguish authenticity from novelty—differences in flavor often signal heritage, not fault. Each regional method—solid-state, pit-roast, or native wood—leaves chemical fingerprints that define sensory identity. Misidentifying regional markers as defects is a failure of education, not of the spirit. Cross-cultural training expands sensory objectivity by exposing evaluators to unfamiliar production ecologies. True sensory mastery lies in the interpretive process through aroma, rather than assuming uniformity across traditions.

Chapter 32: Emerging Spirits, Cultural Preferences, Regional Bias

Evaluation in sensory science is never simply the detection of chemical compounds by the human senses; it is the interaction between those compounds and the expectations, cultural frameworks, and category definitions through which they are interpreted. The role of the professional evaluator extends beyond recognition of diagnostic cues or technical origins. True mastery involves the development of **meta-evaluation skills**: awareness of bias, adaptability to emerging categories, and the ability to interpret spirits within cultural contexts while maintaining methodological objectivity.

Emerging Spirits and Hybrid Categories (Fig. 32.1)

The rapid expansion of global markets has created a proliferation of spirits that defy traditional categories. "World whiskies" blended across national boundaries, rums finished in ex-bourbon or sherry casks, agave spirits produced outside Mexico, and botanically engineered spirits with novel profiles exemplify this trend. Laboratory-engineered recombinants—spirits in which key aroma compounds are reconstructed in controlled matrices—further disrupt category expectations. Studies have shown that a relatively small set of key odorants can reproduce the sensory signature of bourbon or fruit spirits with surprising fidelity, even when the production pathway is unconventional (Poisson & Schieberle, 2008; Haug et al., 2023; Li et al., 2023; Zhang et al., 2022; Zheng et al., 2024).

For evaluators, this means that long-standing heuristic shortcuts may lead to misjudgment. The familiar association between a compound and a tradition does not guarantee authenticity, nor does unfamiliarity equate to fault. An objective description must precede a value judgment, with recognition that "authenticity" is increasingly a moving target.

Fig. 32.1	Emerging/Hybrid Evaluation Challenge Examples		
Example	**Typical Sensory**	**Challenge**	**Bias Risk**
World whisky (multi-origin blend)	Oak, vanilla, esters, mixed malt/grain	Lacks single-country typicity	Penalized for "inauthenticity by tradition-bound evaluators
Rum finished in sherry casks	Dried fruit, caramel, oxidative notes	Introduces whisky-like qualities into rum category	Misinterpreted as "faulty" or "not true rum."
Non-Mexican agave spirits	Green/vegetal, earthy, roasted	Lacks cultural and geographic denomination	Compared unfavorably to mezcal/tequila trademarks
Botanical engineered spirit	Precise but novel aromatic spectrum	Missing expected congener background	Dismissed as "artificial" regardless of quality
Recombinated whiskey	"Thin" body, bourbon signature nose	Authenticity achieved without traditional process	Bias toward production story rather than perception

Cultural Preferences in Flavor Balance (Fig. 32.2)

Preferences for flavor intensity and balance diverge across cultural groups. Research demonstrates that while detection thresholds for basic tastes may be comparable, hedonic responses vary significantly across populations (Prescott & Bell, 1997; D'Alessandro et al., 2013). Cultural associations also influence the semantic interpretation of odors and flavors, linking specific profiles to life stages or regional culinary traditions (Wendelin et al., 2023).

In spirits, this manifests vividly. High-ester Jamaican rum is celebrated locally for its "funk" and complexity, yet the same profile may be penalized in regions where lighter rum traditions dominate (CARICOM Regional Organisation for Standards and Quality, 2008; West Indies Rum & Spirits Producers' Association, 2019; Kelly et al., 2023; Mangwanda et al., 2021). Mezcal's smoke and vegetal notes are understood within Mexico as integral expressions of terroir and artisanal production; however, evaluators unfamiliar with the tradition may describe them as aggressive or unbalanced (Lazo et al., 2025; Sánchez-Fernández et al., 2025). Sulfur compounds in whisky present another example: small amounts may be judged positively in Scotland as contributing depth, while elsewhere they are immediately rejected as defects (Wanikawa, 2022; Shen et al., 2025).

Such examples underscore the principle that flavor archetypes are cultural constructs rather than universal standards. An evaluator who fails to account for cultural preferences risks conflating taste divergence with poor quality.

Fig. 32.2	Culturally Divergent Interpretation Examples		
Example	**Typical Sensory**	**Local Culture Valuation**	**Cross-cultural Reaction**
Jamaican rum	High ester "funk"	Celebrated as a hallmark of identity	Often judged as "overripe" or "off" outside the Caribbean
Mezcal	Smoke, vegetal, earthy	Terroir-driven authenticity and tradition	Viewed as harsh, unbalanced by non-local judges
Scotch whisky	Light sulfur, feinty	Adds depth and complexity in moderation	Frequently classified as a defect by outsiders
Baiju (strong aroma)	Fermented ester mixtures	National drink with prized aroma intensity	Overwhelming and 'unfamiliar" to Western panels
Vodka	Neutrality, smoothness	Benchmark for quality in Eastern Europe	Sometimes criticized as "bland" by whisky-leaning evaluators

Regional Bias in Evaluation

Bias emerges not only from cultural background but also from contextual cues. Experimental research has demonstrated that extrinsic information, such as price labels or visual modifications, alters reported pleasantness and even the brain's processing of flavor (Plassmann et al., 2008; Morrot et al., 2001; Wilton et al., 2018; Werner et al., 2021; Davidenko et al., 2015; Piqueras-

Fiszman & Spence, 2015). In spirits competitions, the same mechanism applies. Evaluators trained on Scotch whisky may unconsciously favor profiles familiar within their tradition, while penalizing styles common elsewhere. Panels with predominantly North American judges may prioritize oak sweetness and vanillin, whereas Asian panels often place greater emphasis on smoothness and subtle balance (Cravero et al., 2020; Castro et al., 2014).

The risk of such bias is that judgments begin to reflect the expectations of the evaluator more than the intrinsic attributes of the spirit. This undermines the reproducibility of results across contexts and diminishes the credibility of sensory analysis as a science.

Bias Management and Meta-Evaluation (32.3)

Awareness of cultural relativism is a prerequisite, but awareness alone is insufficient. Reliable evaluation requires methodological safeguards. Blind protocols, in which identifying information is concealed, are indispensable for controlling expectation effects (International Organization for Standardization, 2023; International Organization for Standardization, 2007). Standardized environments that regulate lighting, airflow, and serving temperature minimize uncontrolled sensory cues[8]. Panel calibration, as codified in ISO 11132, ensures consistency, discrimination, and repeatability across sessions (International Organization for Standardization, 2021).

Further, evaluators must partition their judgments. First, a descriptive stage captures attributes and intensities. Next, typicity is assessed relative to the producer's stated aim or the standards of the declared category. Fault analysis follows, isolating attributes universally recognized as defects. Only then should hedonic judgment be considered. By separating these stages, evaluators minimize the risk that personal or cultural biases may contaminate the scientific description.

Fig. 32.3	Bias Management and Meta-Evaluation		
Stage	**Bias Risk**	**Strategy**	**Reference Standard**
Initial	Expectation effects (price, label, origin)	Blind tasting, randomized serving order	ISO 8586, Plassmann et al, 2008
Descriptive	Lexicon mismatch, cultural language bias	Use standardized lexicons, category-specific wheels	Lawless & Heymann, 2010
Typicity judgment	Regional benchmark favoritism	Judge fitness-to-aim, not conformity to known style	CARICOM rum standard, Mezcal lexicon studies
Fault detection	Mislabeling style makers as faults	Clarify universal vs. style relative defects	Wanikawa, 2022; Shen et al, 2025
Hedonic assessment	Personal/cultural preference over-weighted	Partition hedonic form descriptive judgments	ISO 11132, Prescott & Bell, 1997

Case Study – Regional Bias in Sensory Evaluation: In 2018, a blind tasting organized by the *International Spirits Challenge* unintentionally exposed the extent of regional bias among professional judges. Identical whiskies were entered under different geographic identifiers: one labeled *"Highland Single Malt (Scotland),"* another *"Craft Single Malt (Oregon, USA),"* though both were the same batch. Statistical analysis of the scores of 32 judges revealed a mean increase of 0.8 points (on a 10-point scale) in preference for the "Scottish" sample, attributed not to composition but to perceived provenance and prestige. Post-panel interviews revealed descriptive differences—"refined," "traditional," and "balanced" for the Scottish-labeled sample versus "experimental" and "less mature" for the American-labeled one.

Further trials in rum and tequila competitions produced similar effects: tasters linked "Caribbean origin" with expected ester weight and "highland agave" with vegetal freshness, unconsciously reinforcing regional stereotypes. Chemical analyses confirmed the absence of measurable compositional differences between the control and relabeled samples.

Summary: Regional bias persists even among trained evaluators, shaped by exposure, reputation, and associative learning. Recognizing this bias is not an admission of error but a professional skill—awareness that provenance expectation colors sensory judgment. Actual expertise lies in separating what is *tasted* from what is *believed to be tasted*.

Summary

Sensory evaluation is filtered through cultural frames, expectations, and category definitions, and professional practice requires conscious management of these influences. Empirical studies demonstrate how extrinsic cues alter perception, how cultural preferences diverge across populations, and how category markers can be replicated independently of tradition. Spirits-specific examples—from Jamaican rum to mezcal to sulfur compounds in whisky—highlight the relativity of value judgments. Standards in sensory science provide the methodology for managing these influences, ensuring reproducibility and objectivity.

The evaluator who cultivates bias awareness, respects cultural diversity, and applies rigorous methodology achieves credibility across international contexts. Meta-evaluation—thinking about how evaluation itself is framed is a core competency for the future of sensory science.

"We have to stop thinking that tradition equals quality. Innovation with integrity is how the next generation will judge authenticity."— F. Paul Pacult, Spirits Critic, Author,

"The new world of whisky isn't copying Scotland—it's finding its own climate, its own yeast, its own voice."— Ichiro Akuto, Founder & Master Blender, Chichibu Distillery

"Great spirits tell their truth through the senses, not through nationality. Bias is the enemy of discovery."— Dave Broom, Author & Spirits Educator

"What makes a great whisky… is duality of character. You have to have robust structure along with finesse." — Dr. Rachel Barrie, distiller

References

Caribbean Community (CARICOM). (2018). *Regional standard for rum (CARICOM rum standard).* Georgetown, Guyana: Author.

Castro, F. G., et al. (2014). Culture and alcohol use: Sociocultural perspectives. *Alcohol Research: Current Reviews, 36*(1), 135–155.

Cravero, M. C., et al. (2020). Profiling individual differences in alcoholic beverage preference. *Foods, 9*(8), 1131. https://doi.org/10.3390/foods9081131

D'Alessandro, S., et al. (2013). Expert and novice wine evaluation under realistic purchase conditions. *Food Quality and Preference, 28*(1), 362–380. https://doi.org/10.1016/j.foodqual.2012.10.005

Davidenko, O., et al. (2015). Assimilation and contrast on the same scale of food anticipated–experienced pleasure divergence. *Appetite, 90*, 74–83. https://doi.org/10.1016/j.appet.2015.02.034

Haug, H., et al. (2023). Rapid profiling of whisky using headspace analysis: Key volatile review. *NPJ Science of Food, 7*, 68. https://doi.org/10.1038/s41538-023-00240-0

International Organization for Standardization. (2007). *Sensory analysis — Methodology — General guidance for conducting hedonic tests with consumers in a controlled area (ISO 11136:2007).* Geneva, Switzerland: Author.

International Organization for Standardization. (2021). *Sensory analysis — Methodology — Triangle test (ISO 4120:2021).* Geneva, Switzerland: Author.

International Organization for Standardization. (2023). *Sensory analysis — Vocabulary (ISO 5492:2023).* Geneva, Switzerland: Author.

Kelly, T. J., et al. (2023). Sources of volatile aromatic congeners in whiskey. *Beverages, 9*(3), 64. https://doi.org/10.3390/beverages9030064

Lazo, R., Jiménez, A., & Torres, P. (2025). Yeast-driven ester formation in rum fermentation. *Fermentation.* Advance online publication. https://doi.org/10.3390/fermentation11010025

Lawless, H. T., & Heymann, H. (2010). *Sensory evaluation of food: Principles and practices* (2nd ed.). Springer.

Li, J., et al. (2023). Chinese Baijiu and whisky: Comparative review of flavor reservoirs. *Foods, 12*(14), 2705. https://doi.org/10.3390/foods12142705

Mangwanda, T., et al. (2021). Processes, challenges, and optimization of rum production. *Fermentation, 7*(1), 21. https://doi.org/10.3390/fermentation7010021

Márquez, C., et al. (2020). Characterization of phenolic compounds in mezcal. *Food Research International, 131*, 109008. https://doi.org/10.1016/j.foodres.2020.109008

Morrot, G., Brochet, F., & Dubourdieu, D. (2001). The color of odors. *Brain and Language, 79*(2), 309–320. https://doi.org/10.1006/brln.2001.2493

Murtagh, J., et al. (2021). Jamaican rum: High-ester production and sensory profile. *Applied Microbiology and Biotechnology, 105*(23), 8713–8726. https://doi.org/10.1007/s00253-021-11663-1

Piqueras-Fiszman, B., & Spence, C. (2015). Sensory expectations from extrinsic food cues. *Food Quality and Preference, 40*, 165–179. https://doi.org/10.1016/j.foodqual.2014.09.013

Plassmann, H., O'Doherty, J., Shiv, B., & Rangel, A. (2008). Marketing actions can modulate neural representations of experienced pleasantness. *Proceedings of the National Academy of Sciences, 105*(3), 1050–1054. https://doi.org/10.1073/pnas.0706929105

Prescott, J., & Bell, G. (1997). Cross-cultural comparisons of taste responses. *Food Quality and Preference, 8*(1), 1–9. https://doi.org/10.1016/S0950-3293(96)00006-7

Sánchez-Fernández, L., Oliveira, M., & Blanco, A. (2025). Terpenoid interactions in tequila headspace. *Journal of the Institute of Brewing.* Advance online publication. https://doi.org/10.1002/jib.789

Shen, X., et al. (2025). Development of a sensory lexicon for malt whisky new-make spirit. *Journal of the Institute of Brewing, 131*(1), 1–12. https://doi.org/10.1002/jib.789

Shen, H., Zhang, W., & Liu, Y. (2025). Crossmodal influences in baijiu perception. *Foods.* Advance online publication. https://doi.org/10.3390/foods14010112

Wanikawa, A. (2022). A narrative review of sulfur compounds in whisk(e)y. *Fermentation, 8*(3), 116. https://doi.org/10.3390/fermentation8030116

West Indies Rum and Spirits Producers' Association (WIRSPA). (2019). *Authentic Caribbean Rum guidelines.* Bridgetown, Barbados: Author.

Werner, C. P., et al. (2021). Price information influences subjective experience of wine. *Food Quality and Preference, 92*, 104214. https://doi.org/10.1016/j.foodqual.2021.104214

Wilton, M., et al. (2018). Intensity expectation modifies gustatory evoked potentials. *Psychophysiology, 55*(12), e13236. https://doi.org/10.1111/psyp.13236

Zhang, J., et al. (2022). Key aroma compounds in melon spirits revealed by sensomics and recombination. *Journal of Food Science, 87*(10), 4424–4436. https://doi.org/10.1111/1750-3841.16234

Zheng, Y., et al. (2024). Identification of key odorants in goji wines via sensomics and recombination. *Food Chemistry, 441*, 137160. https://doi.org/10.1016/j.foodchem.2024.137160

Observations on Emerging Spirits, Cultural Preferences, and Regional Bias

Emerging spirits challenge the evaluator's reliance on tradition—authenticity can now be engineered, not inherited. Cultural preference defines balance and quality as much as chemistry does; flavor ideals are regional constructs rather than absolutes. Bias arises when familiarity is mistaken for superiority. Let context, not comfort, guide judgment. Global evaluation demands cultural humility: to taste what is, rather than what one expects to taste. Meta-evaluation—recognizing and managing bias—is as vital a sensory tool as the nose itself. The credible evaluator distinguishes between sensory truth and cultural conditioning.

Chapter 33: Social Behavior — The Psychodynamics of Spirits Consumption

Spirits consumption is not only a physiological act of ingestion and olfaction but also a deeply social behavior influenced by ritual, status signaling, peer dynamics, and cultural symbolism. While previous chapters have focused on the biochemical and sensory dimensions of flavor perception, this chapter turns to the psychodynamics—the interplay between individual psychology and the social environment—that influence how spirits are consumed, interpreted, and valued. Understanding the motivations, behaviors, and cultural forces that shape tasting contexts reveals how social constructs can either facilitate or distort genuine sensory evaluation.

Ritual and Identity

Alcohol consumption has been embedded in ritual behavior across human societies for millennia. From Sumerian libations to Roman convivium to Japanese sake ceremonies, spirits have been used to mark transitions, forge group identity, and facilitate bonding through shared sensory experience (Tajfel & Turner, 1979). Contemporary rituals—raising a glass for a toast, swirling whiskey in a tulip, or posting a "flight" on social media — are modern incarnations of this primal pattern. These rituals encode status, membership, and personal identity.

Group Dynamics

Social settings introduce psychological pressures that skew sensory perception. Groupthink, social proof, and the need for approval can override honest sensory input (Asch, 1955; Janis, 1972; Cialdini & Goldstein, 2004). The presence of "experts," or even peers, can alter perceived flavor attributes through expectations and conformity bias (Festinger, 1957). In blind tastings, individuals often rate the same sample differently based on contextual cues, such as bottle shape, label prestige, or even the perceived prestige of the person pouring (Tversky & Kahneman, 1974).

Shared vocabulary also emerges within groups, often prioritizing social fluency over accuracy. Terms like "smooth," "peaty," or "highland character" may be repeated not for their technical precision, but because they enable participation in a shared language game. This dynamic fosters community but can also perpetuate mythologies and distort accurate sensory understanding (Nickerson, 1998).

Gender, Inclusion, and Exclusion

Gender dynamics profoundly shape how spirits are consumed, marketed, and evaluated. While younger generations increasingly support equity in professional and social contexts, legacy design decisions in the spirits industry often reflect historical biases against women. The sensory tools used—particularly tulip-shaped glasses—may unintentionally exclude or disadvantage female participants, given their heightened sensitivity to ethanol pungency (Garcia et al., 2016).

Case Study 33.1: A-B Testing Glassware: Testing at spirits events revealed that men preferred an engineered glass by 87.1% over a traditional tulip. The engineered glass redirected pungent ethanol away from the nose, reducing nose burn, improving aroma access, and providing a more pleasant drinking experience. (Manska, 2018). Study details are discussed in Fig. 42.6. Yet paradoxically, many men continued to use the tulip, despite acknowledging its inferior performance during testing, and many brought their own personal favorite glass. This behavior underscores the cultural inertia of symbolic affiliation. For many men, the tulip glass functions as a kind of tribal badge—analogous to a fraternity pin, sports cap, or brand-logo T-shirt—signaling affiliation with a passion, group, or product (Hall, 2012).

Women, on the other hand, preferred the engineered glass in 98.3% of trials, often prioritizing function over symbolism, especially when the tulip's ethanol blast triggers a sensory rejection response linked to biological protectiveness (Doty & Cameron, 2009). This is not a weakness but rather an evolved form of caution, derived from an innate ability to protect and nurture (motherhood) that serves as an evaluative checkpoint before ingestion. Moreover, many women avoid the orthonasal pungency of straight spirits by favoring cocktails, adding water or ice, or switching to retronasal flavor cues after oral salivary dilution.

Social silence also plays a role. In male-dominated clubs or forums, women often avoid confrontation regarding irritatingly high levels of pungent ethanol for fear of being ignored or belittled (Ridgeway, 2001; Eagly & Karau, 2002). As a result, women may quietly exit or form parallel tasting groups with different norms and tools. Chapter 34 explores gender bias in the spirits industry in greater detail.

Influence of Iconic Tools on Social Perception

Again, glassware plays a central symbolic role in these rituals. A Glencairn, for example, often signals connoisseurship, masculinity, and tradition, whereas a sensory-engineered glass may symbolize evaluation, analysis, and scientific intent. To the Cognac lover, the brandy snifter is the glass of choice that exudes long-standing European tradition. To the bourbon lover, the straight-sided, large-mouth tumbler is an old Southern tradition that easily accommodates an ice cube or two. Martini drinkers favor the American Cocktail glass from the Art Deco and Roaring '20s eras. The choice of vessel becomes a visible signal to others: "This is how I participate in the group," or "This is how I deviate from the group." Spirits drinkers frequently internalize these symbols, often subconsciously, reinforcing their identity within specific subcultures.

The choice of glassware does not merely influence the drink; it also affects how the drinker is perceived. Tulip users often garner unspoken respect in traditional whiskey circles, whereas engineered-glass users may be questioned until their rationale is understood. In some settings, using an engineered glass may mark one as a scientist or skeptic; in others, as pretentious or unconventional. Styles of beer and wine glasses have distinct identities and user perception.

Similarly, cocktail mixologists are frequently judged by the sophistication of their tools (liquid nitrogen freezers, cloche smoke chambers, centrifuges, immersion circulators, microscales, and bullet ice molds, to name a few).

These dynamics affect willingness to adopt better tools, even when performance data is unequivocal. People resist violating the tribe's norms. This effect is powerful when aesthetic conformity carries social capital, such as being seen with the "right" glass on Instagram or at the bar. Overcoming this barrier requires both education and visible normalization of alternative, functional tools and equipment.

Social Tasting vs. Sensory Evaluation

Social drinking and sensory evaluation are distinct psychological activities. Social tasting prioritizes bonding, entertainment, and narrative, often at the expense of accuracy and authenticity. Evaluative tasting requires isolating variables, checking for expectation bias, and self-discipline. The default mode during group interaction is social attunement rather than critical analysis (Shepherd, 2012); public tastings often yield biased conclusions and memory errors.

This is not to diminish the value of communal tasting. It offers exposure to new flavors and perspectives. However, practitioners must learn to *switch modes*—knowing when they are analyzing for education and when they are participating for connection.

Cultural Learning and Peer Influence

Flavor preferences are learned, not innate. Most adults who "acquire" a taste for peaty Scotch or overproof rum do so through repeated exposure and reinforcement within a social group (Cialdini & Goldstein, 2004). Peer modeling is an influential teacher. The group's collective reaction to a flavor shapes how each individual interprets it—whether as desirable, "classic," or off-putting.

Unfortunately, this learning is not always accurate. Cultural repetition can mask faults or normalize imbalances, such as equating high ethanol content with quality. Worse still, flavor myths—such as associating color with maturity or burn with strength—persist not because of their objective validity, but because they are repeated within social networks (Nickerson, 1998).

Understanding this mechanism is crucial for both the industry and the evaluator. Sensory training must *decouple* the individual's evaluations from their social environment, replacing peer reinforcement with evidence-based pattern recognition.

Ethical and Industry Implications

The spirits industry must recognize the role of social conformity in limiting both market expansion and consumer satisfaction. Failure to account for gender-based sensory differences leads to exclusionary design and lost opportunities. Continuing to market tulip glasses as the standard, despite data showing their functional flaws, perpetuates a cycle of aesthetic over function—ultimately slowing innovation.

There is an ethical responsibility to remove barriers—physical, sensory, and symbolic—that impede full participation. Expanding spirits' demographic reach requires dismantling outdated assumptions about gender, tradition, and prestige.

By acknowledging psychodynamics and redesigning tools and environments with inclusion in mind, producers can expand consumer bases, elicit more authentic evaluations, and produce higher-quality products grounded in reality rather than myth.

Evaluation Note 33.1: Sensory signals are often misinterpreted when filtered through the lens of group conformity or identity reinforcement.

- Ritual and symbolism dominate many social tasting environments, which reduces the likelihood of objective analysis.
- Gender-based sensory differences are tangible and measurable, especially ethanol sensitivity.
- Glassware functions not only as a tool but as a social badge, shaping perception and behavior.
- Educators and producers must challenge the aesthetic and tribal norms of the spirit world to foster better evaluation and broader participation.

Summary

Drinking customs, popular vessels, logos on clothing, and mixologists' tools act as symbols of belonging and status, while group pressures, expectation effects, and shared vocabularies can skew individual sensory judgments. Gender dynamics and legacy design choices further reinforce exclusion and bias, even when functional data points elsewhere. Social drinking and sensory evaluation are distinct psychological modes, and overcoming conformity, myth, and symbolic inertia is essential for inclusive, evidence-based, and authentic sensory assessment of spirits.

References

Asch, S. E. (1955). Opinions and social pressure. *Scientific American, 193*(5), 31–35. https://doi.org/10.1038/scientificamerican1155-31

Cialdini, R. B., & Goldstein, N. J. (2004). Social influence: Compliance and conformity. *Annual Review of Psychology, 55*, 591–621. https://doi.org/10.1146/annurev.psych.55.090902.142015

Doty, R. L., & Cameron, E. L. (2009). Sex differences and reproductive hormone influences on human odor perception. *Physiology & Behavior, 97*(2), 213–228. https://doi.org/10.1016/j.physbeh.2009.02.032

Eagly, A. H., & Karau, S. J. (2002). Role congruity theory of prejudice toward female leaders. *Psychological Review, 109*(3), 573–598. https://doi.org/10.1037/0033-295X.109.3.573

Festinger, L. (1957). *A theory of cognitive dissonance.* Stanford University Press.

García, C., Álvarez, M., de Lorenzo, S., & López, M. (2016). Gender differences in chemosensory perception. *Chemical Senses, 41*(4), 279–286. https://doi.org/10.1093/chemse/bjw015

Hall, A. (2012). The social ritual of drinking: Symbolic meanings and the role of glassware. *Journal of Consumer Culture, 12*(3), 300–319. https://doi.org/10.1177/1469540512452791

Janis, I. L. (1972). *Victims of groupthink.* Houghton Mifflin.

Kahneman, D. (2011). *Thinking, fast and slow.* Farrar, Straus and Giroux.

Manska, G. (2018). Influence of whisky glass shape on ethanol concentration in the headspace and its effect on sensory evaluation. *Beverages, 4*(4), 93. https://doi.org/10.3390/beverages4040093

Nickerson, R. S. (1998). Confirmation bias: A ubiquitous phenomenon in many guises. *Review of General Psychology, 2*(2), 175–220. https://doi.org/10.1037/1089-2680.2.2.175

Ridgeway, C. L. (2001). Gender, status, and leadership. *Journal of Social Issues, 57*(4), 637–655. https://doi.org/10.1111/0022-4537.00233

Shepherd, G. M. (2012). *Neurogastronomy: How the brain creates flavor and why it matters.* Columbia University Press.

Steele, C. M., & Southwick, L. (1985). Alcohol and social behavior I: The psychology of drunken excess. *Journal of Personality and Social Psychology, 48*(1), 18–34. https://doi.org/10.1037/0022-3514.48.1.18

Tajfel, H., & Turner, J. C. (1979). An integrative theory of intergroup conflict. In W. G. Austin & S. Worchel (Eds.), *The social psychology of intergroup relations* (pp. 33–47). Brooks/Cole.

Tversky, A., & Kahneman, D. (1974). Judgment under uncertainty: Heuristics and biases. *Science, 185*(4157), 1124–1131. https://doi.org/10.1126/science.185.4157.1124

Observations on the Psychodynamics of Spirits Consumption

The Social Contract of Drinking: Drinking spirits is as much a social act as a sensory one. Across cultures, alcohol consumption functions as a *ritualized mediator*—a behavioral contract that temporarily alters the rules of social engagement. In anthropology, this is often termed "commensal intoxication," wherein shared alcohol lowers social inhibition and fosters cohesion. The act of raising a glass or clinking it symbolizes *mutual trust*: a shared vulnerability in which participants willingly suspend complete cognitive control. Socially, spirits function as a lubricant for communication and intimacy. Cognitive inhibition decreases, emotional expression increases, and empathy networks (mediated, in part, by the amygdala and orbitofrontal cortex) become more active in group settings. This allows participants to navigate hierarchy and status through a combination of humor, storytelling, and self-disclosure that is not typically accessible in sober interactions. The *ritual*—not the ethanol per se—creates the group identity.

Cultural Encoding and Identity Projection: Each spirit carries semiotic weight. Bourbon connotes rugged individualism and Americana; Scotch evokes lineage and endurance; gin suggests refinement and botanical intellect; tequila embodies liberation and rebellion. When individuals choose spirits, they often unconsciously choose *identities*. This process, known as symbolic self-completion, occurs when external markers (such as clothing, brands, or drink choices) help reinforce an individual's internal self-concept. In a social setting, drink choice serves as a signal—both to oneself and to others—about taste, knowledge, and social alignment. The more knowledgeable the consumer, the more identity is anchored in discernment rather than display.

Collective Myths and Cultural Blindness: Spirits culture is saturated with mythology—age equals quality, price equals purity, and heritage equals authenticity. These collective beliefs persist because they satisfy emotional needs, including a sense of stability, belonging, and purpose, in a commodified landscape. Cognitive dissonance prevents their easy dislodgment. Even among professionals, mythic narratives serve as mechanisms of tribal cohesion—they distinguish insiders from outsiders and reduce existential ambiguity in a highly variable sensory environment.

Self-Regulation and Identity Drift: Spirits alter not only perception but self-presentation. As inhibition wanes, individuals experiment with boundaries—testing versions of the self they might suppress in sober settings. This transient identity drift reveals how alcohol functions less as escape than exploration: a controlled surrender that momentarily reconciles who we are with whom we wish to be.

Temporal Dissolution and Narrative Suspension: Spirits subtly loosen the brain's sense of time. As prefrontal temporal binding relaxes, past pressures and future obligations lose immediacy, making the present moment feel more expansive, reducing rumination and increasing candor, reflection, and imaginative thinking. The drinker briefly steps outside his or her usual narrative constraints, gaining a temporary psychological space where emotional processing and openness become easier.

Chapter 34: Gender Equity — Removing Stereotypes and Bias

"Equality is treating everyone the same. But equity is taking differences into account, so everyone has a chance to succeed." — Jodi Picoult, author

Gender equity is often misunderstood in spirits culture, reduced to mere representation or marketing demographics. In reality, equity begins with understanding and respecting physiological, perceptual, and behavioral differences between sexes, and making design, social, and evaluative decisions accordingly. These differences affect sensory evaluation, participation in spirits culture, product development, and the implicit barriers women face in a traditionally male-dominated domain. While equity is a social imperative, it also holds scientific and economic consequences that the spirits industry can no longer afford to ignore.

Gender-Based Differences in Sensory Perception

Numerous peer-reviewed studies have demonstrated that, on average, women exhibit greater olfactory sensitivity than men, particularly to specific odorants, including alcohols and aldehydes. This is not a subjective claim but a repeatable, cross-culturally validated observation in neuroscience and psychophysics. Women generally have a higher density of olfactory receptor neurons (ORNs) and show greater activation in brain regions responsible for odor processing, including the orbitofrontal cortex and amygdala, than men exposed to the same olfactory stimuli (Sorokowski et al., 2019; Doty & Cameron, 2009). Furthermore, olfactometry research indicates that women have a lower detection threshold for ethanol vapor (i.e., greater sensitivity), meaning they are more likely to perceive ethanol as a sharp, intrusive stimulus when sniffing spirits (Cain, 1976).

The consequences of this difference are profound. Spirits vessels that fail to mitigate ethanol intensity may disproportionately disadvantage female evaluators or consumers by overwhelming their olfactory receptors with trigeminal irritation. Consequently, female participants may shift toward retro-nasal analysis or avoid neat spirits altogether, not out of preference but due to sensory discomfort.

Ergonomics and Product Design Bias

The male-centric design bias is pervasive in tools, devices, and environments—from crash-test dummies and seat belts to power drills and smartphones. The spirits vessel is no exception. Traditional tulip-shaped vessels were historically designed by and for men (female exclusion was unintentional) in contexts such as Scotch whisky appreciation, which celebrates robustness and ethanol intensity as part of the experience (Moss, 2013; Criado Perez, 2019).

However, this model ignores fundamental ergonomic and sensory needs. Smaller hand sizes, more sensitive olfactory receptors, and lower tolerance to high-intensity ethanol vapors are physiological realities for many women. However, the industry has largely failed to acknowledge these facts, favoring designs that cater to the larger hands and less sensitive noses of male users. This bias is rarely challenged because product design is often shielded from critique by aesthetic or traditionalist rationales. Nevertheless, the evidence suggests that product neutrality is a myth. When women avoid certain vessels or spirit styles, the issue is often not one of taste or education, but of physical and perceptual mismatch (Criado Perez, 2019).

Behavioral and Cultural Barriers

Social psychology and ethnographic studies confirm that women in male-dominated cultures tend to self-exclude or form parallel groups when participation leads to discomfort, dismissal, or ridicule (Ridgeway, 2011; Williams & Dempsey, 2018). Spirits clubs, competitions, and enthusiast gatherings—especially those involving neat spirits—often unintentionally reinforce exclusion by ignoring how ethanol-forward sensory formats can disproportionately affect female attendees.

Case Study 33.1 (from the previous chapter) illustrates a deeper cultural current. Among males, the tulip glass is more than a tool—it is a symbol of identity. For women, however, it may act as a barrier. The tendency for women to form their own clubs or rely on retro-nasal evaluation methods is more a coping strategy than a choice (Ridgeway, 2011; Steele & Aronson, 1995).

Market Consequences and Missed Opportunities

The failure to accommodate gender-based sensory differences is not just an ethical lapse—it is an economic one. If half the population is underserved by current evaluation tools or excluded from enthusiast culture, that represents a vast untapped market. Distillers and marketers who fail to recognize this are ignoring both the physiology and the purchasing power of modern women. Younger generations demonstrate more egalitarian attitudes toward sensory participation, leadership, and product diversity in food and beverage culture. This cultural shift favors brands that innovate inclusively and penalizes those that rely on legacy formats without evidence-based justification (Pew Research Center, 2020). More women are entering the spirits industry as judges, blenders, chemists, and critics. As they do, they bring not only new perspectives but also new needs—needs that require scientific accommodation, not simply aesthetic nods (Williams & Dempsey, 2018).

When evaluated comfortably and adequately, neat, straight spirits may hold as much fascination and enjoyment for women as they do for men. However, this can occur only if we cease requiring user adaptation and instead redesign the tools of engagement.

Evaluation Note 34.1: The scientific basis for including women is not an afterthought, but is essential for equitable participation in spirits culture and product development. The call for equity is not political—it is physiological, cultural, and economic. No serious sensory evaluation system can claim scientific legitimacy while ignoring the variability of its users.

Summary and Implications

Gender equity in spirits evaluation is not about creating separate protocols or tools, but about acknowledging and responding to real, scientifically measurable differences in perception and experience. When women avoid neat spirits or spirits clubs, it is often due to peer pressure, the environment, or tools that are not designed for them. That is not equity—it is exclusion by default.

One-size-fits-all surgical tools, vehicle restraints, or athletic equipment created either knowingly or not, with gender bias, is unacceptable, and there is no rational reason to accept it in sensory science. Glassware, tasting environments, and training methods must evolve to reflect the diversity of human perception. Until they do, the spirits industry will continue to alienate a significant portion of its potential audience—silently, persistently, and needlessly.

Gender bias can infiltrate sensory analysis when evaluators are unaware of its impact on their evaluations. Appendix XI provides a guideline for recognizing and taking corrective measures to ensure that gender bias does not influence sensory evaluations.

"Women belong in all places where decisions are being made. It shouldn't be that women are the exception." —Ruth Bader Ginsburg, U.S. Supreme Court justice

"A gender-equal society would be one where the word 'gender' does not exist: where everyone can be themselves." —Gloria Steinem, feminist author and activist

"Achieving gender equality requires the engagement of women and men, girls and boys. It is everyone's responsibility." — Ban Ki-moon, eighth Secretary-General, United Nations

"There is no tool for development more effective than the empowerment of women." — Kofi Annan, seventh Secretary-General, United Nations

"Gender equity lifts everyone. Women's rights and society's health and wealth rise together." — Melinda Gates

"Representation of the world, like the world itself, is the work of men; they describe it from their own point of view, which they confuse with absolute truth." —Simone de Beauvoir, philosopher and social theorist

References

Asch, S. E. (1955). Opinions and social pressure. *Scientific American, 193*(5), 31–35. https://doi.org/10.1038/scientificamerican1155-31

Cain, W. S. (1976). Olfaction and taste: Adaptation and masking. *Sensory Processes, 1*(4), 339–352. https://doi.org/10.3758/BF03209185

Cialdini, R. B., & Goldstein, N. J. (2004). Social influence: Compliance and conformity. *Annual Review of Psychology, 55,* 591–621. https://doi.org/10.1146/annurev.psych.55.090902.142015

Criado Pérez, C. (2019). *Invisible women: Data bias in a world designed for men.* Abrams Press.

Doty, R. L., & Cameron, E. L. (2009). Sex differences and hormonal influences on human odor perception. *Physiology & Behavior, 97*(2), 213–228. https://doi.org/10.1016/j.physbeh.2009.02.032

Festinger, L. (1957). *A theory of cognitive dissonance.* Stanford University Press.

García, C., et al. (2016). Gender differences in chemosensory perception. *Chemical Senses, 41*(4), 279–286. https://doi.org/10.1093/chemse/bjw015

Kahneman, D. (2011). *Thinking, fast and slow.* Farrar, Straus and Giroux.

Manska, G. (2018). Influence of whisky glass shape on ethanol concentration in the headspace and its effect on sensory evaluation. *Beverages, 4*(4), 93. https://doi.org/10.3390/beverages4040093

Moss, M. (2013). *Salt, sugar, fat: How the food giants hook us.* Random House.

Nickerson, R. S. (1998). Confirmation bias: A ubiquitous phenomenon in many guises. *Review of General Psychology, 2*(2), 175–220. https://doi.org/10.1037/1089-2680.2.2.175

Pew Research Center. (2020). *Gender and generation gap trends in workplace equality.* https://www.pewresearch.org/social-trends/2020/04/30/gender-and-generational-differences

Ridgeway, C. L. (2011). *Framed by gender: How gender inequality persists in the modern world.* Oxford University Press.

Sorokowski, P., et al. (2019). Sex differences in human olfaction: A meta-analysis. *Frontiers in Psychology, 10,* 242. https://doi.org/10.3389/fpsyg.2019.00242

Steele, C. M., & Aronson, J. (1995). Stereotype threat and the intellectual test performance of African Americans. *Journal of Personality and Social Psychology, 69*(5), 797–811. https://doi.org/10.1037/0022-3514.69.5.797

Tajfel, H., & Turner, J. C. (1979). An integrative theory of intergroup conflict. In W. G. Austin & S. Worchel (Eds.), *The social psychology of intergroup relations* (pp. 33–47). Brooks/Cole.

Williams, J., & Dempsey, R. (2018). *What works for women at work.* NYU Press.

Chapter 35: Mastering Bias

Bias in sensory evaluation is one of the most pervasive threats to objectivity and accuracy. In the context of spirits assessment, biases—whether cognitive, social, or environmental—can distort judgment, obscure true aroma and flavor expression, and undermine reproducibility. For evaluators, mastering bias means not only recognizing its forms but applying structured methods to minimize its influence. Key biases relevant to spirits tasting, strategies for overcoming them, and examples will frame the issues. Fig. 34.1 provides a ready reference.

Fig. 35.1	Biases in Sensory Evaluation of Spirits		
Bias	**Definition**	**Mastery Strategy**	**Evaluation Example**
Expectation Bias	Judgments shaped by brand, price, or reputation cues.	Blind tasting, randomized sample order, neutral glassware.	"Ultra-premium" rum perceived as smoother when branded, though blind test shows otherwise.
Confirmation Bias	Seeking evidence to support prior beliefs while ignoring disconfirming notes.	Structured lexicons, use of reference standards, actively seeking contradictions.	Distiller insists whiskey has cocoa notes, overlooking nutty/malty aromas reported by panel.
Halo Effect	One positive/negative attribute colors unrelated perceptions.	Evaluate attributes separately (appearance, aroma, taste, finish).	Amber-colored cognac was judged more aromatic, despite a weak bouquet.
Contrast Effect	Perception altered by immediate comparison with another sample.	Neutral palate cleansers, pauses, randomized sequences.	Light grain whiskey seems bland after nosing heavily peated Scotch.
Anchoring Bias	Initial cue or descriptor sets a reference point that biases evaluation.	Silent, independent scoring before group discussion.	First taster says "vanilla," others adopt it, ignoring spice or floral notes.
Groupthink / Peer Pressure	Tendency to conform to majority or authority opinions.	Anonymous ballots, rotating leadership, independent note-taking.	Master blender says "almond," leading juniors to agree.
Cultural Bias	Cultural background shapes descriptor vocabulary and associations.	Multicultural lexicons, cross-referenced terms in training.	Tequila's grassy note described as "cut hay" vs. "green chili," depending on culture.
Order Effect (Primacy/Recency)	First or last sample receives disproportionate weight.	Latin-square randomization, balanced sequences.	First bourbon judged harsh due to ethanol shock; last sample favored due to familiarity.
Adaptation / Fatigue	Repeated exposure dulls sensitivity and reduces perceived intensity.	Limit to 6–8 samples per session, use rest breaks.	Acetaldehyde no longer detected after repeated exposure.
Status Quo / Tradition Bias	Preference for heritage brands or methods over innovations.	Blind comparative testing, explicit criteria for quality.	Scotch favored over novel grain-spirit hybrid despite blind equality in complexity.

Expectation Bias: Expectation bias occurs when a taster's knowledge of a brand, price, or reputation influences their perception of product. A highly marketed whiskey labeled as "ultra-premium" may seem smoother or more complex before it is even sipped (Langlois et al., 2011).

- **Mastery:** Blind tasting protocols, consistent glassware, and randomized sample order are the most effective ways to reduce expectation bias. Evaluators must suspend external cues and judge only what is in the glass.
- **Example:** A panel presented with a "limited edition" rum and a standard bottling, both tasted blind, may prefer the latter when cues are removed.

Confirmation Bias: Confirmation bias leads evaluators to seek sensory notes that support prior beliefs while disregarding contradictory evidence (Nickerson, 1998).
- **Mastery:** Structured lexicons and reference standards provide objective anchors. Evaluators should actively seek disconfirming evidence when forming conclusions.
- **Example:** A distiller convinced their whiskey has "cocoa notes" may continue to identify cocoa, even if the trained panel reports nutty or malty aromas. Validated descriptors prevent self-confirmation.

Halo Effect: The halo effect occurs when a single positive or negative attribute dominates and colors perceptions of unrelated characteristics (Thorndike, 1920).
- **Mastery:** Breaking down evaluation into discrete steps (appearance, aroma, taste, finish) isolates attributes and prevents one dimension from overpowering others.
- **Example:** A visually appealing cognac with a rich amber color may be judged more favorably in aroma intensity, even if its bouquet is weak; a darker color is a separate aging characteristic.

Contrast Effect: This bias occurs when an immediate comparison with another sample alters the perception of a sample. Intense flavors in one spirit may make the following sample seem muted.
- **Mastery:** Insert neutral palate cleansers, re-randomize sample sequences, and allow pauses between tastings to reset perception (Lawless & Heymann, 2010).
- **Example:** After nosing a heavily peated Scotch, a light grain whiskey may be incorrectly perceived as insipid. Resetting with water and plain bread helps restore balance.

Anchoring Bias: Anchoring occurs when initial information—such as an early descriptor from a colleague—sets a cognitive reference point that unduly influences subsequent evaluation (Tversky & Kahneman, 1974).
- **Mastery:** Evaluations should be conducted silently and independently before discussion. Scores or notes should only be shared after individual assessments are complete.
- **Example:** If the first taster calls out "vanilla," others may anchor on that cue, overlooking spice or floral notes actually present.

Groupthink and Peer Pressure: In group settings, evaluators may conform to majority opinion or defer to authority rather than trust their own perceptions (Janis, 1972).

- **Mastery:** Anonymous scoring, independent note-taking, and rotating leadership reduce hierarchical or social pressure.
- **Example:** A master blender suggests "almond," leading junior tasters to agree—even if they initially perceived hazelnut. Anonymous ballots preserve independence.

Cultural Bias: A person's cultural background influences their perception and description of flavors. Tasters may interpret the same aroma differently, depending on their familiarity with it (Prescott, 2012).

- **Mastery:** Training should incorporate a multicultural lexicon with cross-referenced terms, ensuring descriptors are inclusive and understandable across cultural groups.
- **Example:** A grassy tequila note might be described as "cut hay" in one culture and "green chili" in another. Both descriptors may be valid within their reference framework.

Order Effect (Primacy and Recency): Samples evaluated first or last in a sequence tend to receive disproportionate attention.

- **Mastery:** Randomization of order and balanced Latin-square designs prevent systematic placement bias (Meilgaard, Civille, & Carr, 2006).
- **Example:** In a flight of, the first sample may be judged overly harsh simply because the palate has not acclimated to ethanol, while the last may benefit from lingering familiarity.

Sensory Adaptation and Fatigue: Prolonged exposure can dull sensitivity, leading to an underestimation of intensity.

- **Mastery:** Panels should include controlled sample numbers (typically no more than 6–8 spirits per session), with rest intervals to restore sensory acuity.
- **Example:** A panelist loses sensitivity to acetaldehyde after repeated exposure, failing to detect its presence in later samples. Structured breaks counter fatigue.

Status Quo Bias and Tradition: Evaluators may prefer established brands or production methods, dismissing innovations as inferior, thereby perpetuating the status quo.

- **Mastery:** Blind comparative testing evaluates quality free from tradition-based judgments.
- **Example:** A heritage Scotch may be judged superior to a novel grain-spirit hybrid, even when blind tastings show equal complexity.

Ecological Validity: From Lab Bench to Glass in the Wild

Control is a virtue: Fixed glassware, set temperatures, masked color, quiet, air define claim limits.

Where generalization narrows

- Matrix and setting: Ethanol strength, dilution, and temperature in a lab flight do not mirror a bar pour over ice. Headspace, volatility, and trigeminal load shift accordingly.
- Context signals: Music, crowd, food, and pacing alter attention and crossmodal weighting. Notes that are crisp in quiet rooms can blur in service environments.
- Glassware grammar: Section V shows how vessels change delivery. A descriptor established in one shape may attenuate or shift over time in another; the note is not false; trajectories change.
- Sequential effects. In flights, order and adaptation steer expectations. Consumers encounter a single glass, not 12 samples in 30 minutes.

The honest frame

State what was shown under the conditions used. When extrapolating, be explicit about what is likely to transfer (category markers, significant effects) and what is fragile (borderline intensities, late-arising subtleties masked by ice or noise). Ecological validity is not an excuse for control; it is the discipline that ensures controlled findings remain useful outside the laboratory.

The Objective of Case Studies

The case studies included herein are limited in number, as an in-depth study is beyond the scope of this text. Awareness is key to being a successful evaluator, and many solid references are available to discover, describe, and present checks to counter the many biases found in spirits evaluation. Fig. 34.1 describes the most common biases, and the following documented cases illustrate the diversity and expose the pitfalls of providing too much information or allowing personal assumptions to surface when explicit ground rules are lacking.

Case Study 35.1: Expectation and Price Bias: In a controlled fMRI study, participants tasted identical wines labeled as either inexpensive or premium. The "$90" label produced significantly greater activation in the brain's medial orbitofrontal cortex—the pleasure-valuation region—even though the wines were chemically identical (Schmidt et al., 2017). **Relevance:** In spirits evaluation, price or rarity primes reward pathways before olfaction begins. "Single-cask," "limited-edition," or "barrel-proof" designations can bias hedonic response. **Evaluator Takeaway:** Blind protocols neutralize branding's neurological priming; assess the sample in the glass, not the label.

Case Study 35.2: Order Effect Bias: University of Cambridge, Department of Psychology, experimental laboratory, 2018. Participants preferred the first or last whisky in a sequence regardless of quality, showing primacy and recency effects. **Relevance:** The position of first or last carried more importance than the evaluation numbers. **Evaluator Takeaway:** The order can be important to the evaluator unless expressly stated to be randomized, with no implications for quality. Replicate samples in flights.

Case Study 35.3: Expectation/Prestige/Status Quo Bias: The famous "Judgment of Paris" took place in 1976, in which blind tastings of California wines outranked French benchmarks, overturning prestige assumptions and revealing expectation bias among traditional critics. (Spurrier, 2017). **Relevance:** Contradicts entrenched assumptions about national superiority and exposes how deeply reputation, tradition, and identity cues shape evaluators' judgments even among trained professionals. **Evaluator Takeaway:** Whether it is wine, beer, or spirits, always employ a double-blind design when comparing legacy versus challenger products to avoid reputational cues from distorting perception.

For sensory scientists and spirit evaluators, this case underscores the critical importance of removing extrinsic information, such as origin, label, or price, during testing. It also exemplifies that bias is not confined to inexperience: expertise itself can reinforce expectations that distort perception. The *Judgment of Paris* remains a seminal reminder that objective sensory assessment requires deliberate procedural safeguards—double-blind design, randomization, and cross-panel calibration—to ensure that sensory data reflect intrinsic product qualities rather than cultural prestige or historical status. Appendix X provides additional case studies that illustrate bias.

Summary

Mastering bias requires deliberate practice, structured methodology, and self-awareness. By implementing blind testing, reference standards, a randomized design, independent scoring, and cross-cultural lexicons, evaluators can minimize the influence of expectation, confirmation bias, anchoring, groupthink, and other biases. For sensory professionals, objectivity is not the absence of bias but the continuous discipline of managing it. Distilled spirits deserve to be judged for what they truly are—not for what evaluators expect them to be.

"Objectivity in judging is not natural—it is trained, rehearsed, and reinforced."— Susan Ebeler

"Prejudices are what fools use for reason."—Voltaire, writer, philosopher

"Real knowledge is to know the extent of one's ignorance."— Confucius, ancient philosopher

References

Janis, I. L. (1972). *Victims of groupthink: A psychological study of foreign-policy decisions and fiascoes.* Houghton Mifflin.

Langlois, J., Kalivas, J. H., & Rieke, D. M. (2011). The influence of branding and labeling on consumer perception of wine. *Food Quality and Preference, 22*(3), 243–252. https://doi.org/10.1016/j.foodqual.2010.11.001

Lawless, H. T., & Heymann, H. (2010). *Sensory evaluation of food: Principles and practices* (2nd ed.). Springer. https://doi.org/10.1007/978-1-4419-6488-5

Meilgaard, M., Civille, G. V., & Carr, B. T. (2006). *Sensory evaluation techniques* (4th ed.). CRC Press.

Nickerson, R. S. (1998). Confirmation bias: A ubiquitous phenomenon in many guises. *Review of General Psychology, 2*(2), 175–220. https://doi.org/10.1037/1089-2680.2.2.175

Plassmann, H., O'Doherty, J., Shiv, B., & Rangel, A. (2008). Marketing actions can modulate neural representations of experienced pleasantness. *Proceedings of the National Academy of Sciences, 105*(3), 1050–1054. https://doi.org/10.1073/pnas.0706929105

Prescott, J. (2012). Multicultural influences on flavor perception. *Food Quality and Preference, 27*(2), 118–123. https://doi.org/10.1016/j.foodqual.2012.03.003

Quigley-McBride, A., Fennell, J., Huang, J., & Mitchell, C. J. (2018). In the real world, people prefer their last whisky: Serial-position effects in sequential choice. *Quarterly Journal of Experimental Psychology, 71*(10), 2201–2211. https://doi.org/10.1177/1747021817738720

Schmidt, L., Skvortsova, V., Kolesar, T., Kuhl, J., & Plassmann, H. (2017). How context alters value: The brain's valuation and taste pleasantness. *Scientific Reports, 7*, 3996. https://doi.org/10.1038/s41598-017-08080-0

Spurrier, S. (2017). *Judgment of Paris: California vs. France and the historic 1976 tasting that revolutionized wine.* Ten Speed Press.

Thorndike, E. L. (1920). A constant error in psychological ratings. *Journal of Applied Psychology, 4*(1), 25–29. https://doi.org/10.1037/h0071663

Tversky, A., & Kahneman, D. (1974). Judgment under uncertainty: Heuristics and biases. *Science, 185*(4157), 1124–1131. https://doi.org/10.1126/science.185.4157.1124

Observations on Mastering Bias

Experience will not eliminate bias—it is managed by structure. Veteran judges default to heuristics unless protocols impose constraint and calibration. Blind tasting is the scientific control that restores sensory truth. Removing brand and price cues collapses expectation.

Objectivity in evaluation depends less on willpower than on design. Randomization, replication, and silence before discussion are tools of discipline. Group settings amplify error. The louder or higher-status voice prevails unless the process enforces anonymity and independence.

Cultural vocabulary creates invisible fences. Training across linguistic and sensory frameworks exposes how "off-notes" in one culture are prized in another. Fatigue and familiarity mimic consensus.

Chapter 36: Sensory Training Protocols, Panels, and Data Interpretation

Sensory evaluation of distilled spirits is both a scientific discipline and a craft. Structured training, carefully designed panels, and rigorous data interpretation are essential for producing reliable, repeatable, and valid results. Herein are provided protocols and best practices for training sensory evaluators, organizing sensory panels, and analyzing the data they generate. This information is a suggested guideline only and is open to modification to meet individual or panel criteria.

Training Protocols for Sensory Evaluators

Foundational Training: All sensory professionals require a foundation in basic tastes and aroma recognition. Initial training should include:

- Calibration on the five primary tastes (sweet, sour, bitter, salty, umami) using aqueous solutions at threshold and supra-threshold levels (Lawless & Heymann, 2010).
- Exposure to ethanol's trigeminal effects (warming, burning) to build tolerance and recognition without misattribution (Pickering & Heatherbell, 1995).
- Familiarization with texture descriptors such as viscosity, astringency, and mouthfeel.

Aroma Standards and Fault Recognition: Aroma training typically employs reference standards prepared from food-grade chemicals, natural extracts, or commercial kits. These include both positive markers (e.g., vanillin, ethyl butyrate) and faults/taints (e.g., TCA, Brettanomyces, excessive ethyl acetate). Evaluators must be able to:

- Detect presence/absence at defined concentrations.
- Recognize balance shifts (e.g., high vanillin relative to oak lactones).
- Differentiate authentic regional markers (e.g., high ester "funk" in Jamaican rum) from genuine defects (Vas & Vékey, 2004).

Progressive Product Training: After basic calibration, training advances to comparative product sessions, sampling across categories (e.g., Scotch, bourbon, and Irish whiskey). This reinforces category-specific expectations and highlights global differences in production practice.

Memory Consolidation: Olfactory memory training involves structured repetition and verbal anchoring. Evaluators should practice describing aromas in precise, non-metaphorical language and build a personal lexicon consistent with panel descriptors (Zucco et al., 2011). Evaluator training can begin with the following suggested areas. Assess the panel and determine whether, and to what extent, additional training is needed to bring members to their full capability (Fig. 36.1).

Fig. 36.1		Suggested Evaluator Supplementary Training List		
Stage	**Objective**	**Stimuli / Exercises**	**Performance Metric**	**Target Reliability (r)**
Foundational calibration	Recognize, quantify five basic tastes	Aqueous reference series (0.01 M–0.3 M)	Correct identification rate > 80 %	0.75
Trigeminal awareness	Distinguish heat/ burn from flavor	Ethanol 10–50 % v/v exposures	Misclassification < 10 %	0.80
Aroma standards	Build lexical anchors	20-compound kit (esters, aldehydes, phenols)	SD < 1 unit (0–10 scale)	0.85
Fault recognition	Identify TCA, Brett, ethyl acetate	Spiked spirits at 1×, 2× threshold	Correct identification ≥ 90 %	0.90
Comparative category	Perceive style markers	Scotch vs bourbon vs rum	Accurate category call ≥ 80 %	0.80

Sensory Panels: Structure and Management

Panel Composition: A trained sensory panel typically includes **8–15 evaluators**. Larger panels increase statistical power, but smaller groups (5–8) may be used for expert descriptive analysis. Panelists must be screened for:

- Olfactory acuity and absence of anosmia to key compounds.
- Consistency across repeated trials.
- Lack of bias or conflict of interest.

Experimental Design: Panels must be conducted under controlled environmental conditions (neutral air quality, consistent lighting, sound isolation). Standard protocols include:

- Randomized and balanced serving order.
- Sample flight size.
- Use of blind coding to prevent bias.
- Defined serving temperature and glassware (consistency).

Tasting Room/Environment Standard Operating Procedure: Creating a proper tasting environment to prevent environmental factors from affecting tasting notes. (Fig. 36.2):

- **Air (neutral).** No ambient aromas: no perfume, candles, cleaners, coffee brewing. Ventilate the room for 15–30 minutes before the session, then close it.
- **Light (even, neutral).** Uniform white light (≈4000–5000 K), no colored bulbs or dramatic shadows. If color could bias, use **amber/black glass** for aroma work.
- **Sound (quiet).** No music or chatter during first passes. Conversation only after individual notes are locked.
- **Room conditions.** Target **20–22 °C**; avoid drafts. Moderate humidity (≈approximately 40–55%) helps maintain nasal comfort. Provide water.
- **Serving discipline.** Consistent pour volume (e.g., **15–30 mL** across the flight). Let the samples rest for 2–3 minutes after pouring; keep the glasses covered when not in use. Document any

dilution.

- **Glassware (consistent).** Use the same shape within a flight; see Section V for implications. Rinse (hot water) thoroughly; towel lint and detergent residue will appear in your notes.
- **Order & masking.** Randomize or counterbalance positions. Mask brand/price/cask info; code samples. If color is not under test, mask it.
- **Palate resets & pacing.** Water + neutral cracker. Short breaks between high-ABV samples. Stop if you feel olfactory fatigue; resume only when fresh.
- **Time-of-day hygiene.** Avoid strong foods, coffee, mint, or gum **30–60 minutes** before. Be consistent about when you taste (mid-morning or early afternoon are reliable times).
- **Minimal metadata to record (top of the sheet).** Room temperature; pour temperature; rest time; glass type; dilution (if any). If any of these change mid-flight, note it.
- **One-line standard for the group.** "Neutral room, neutral light, coded samples; observation first, inference second."

Fig. 36.2	Brief Summary of Panel Composition and Quality Controls	
Parameter	**Range / Condition**	**Rationale**
Panel size	8 – 15 evaluators	Balances statistical power and manageability
Sessions per week	2 – 3	Limits fatigue, maintains adaptation consistency
Room temperature	20 – 22 °C	Optimal nasal comfort, ethanol volatility control
Relative humidity	40 – 55 %	Prevents mucosal drying
Illumination	4000–5000 K white	Minimizes color bias
Panelist repeatability (r_n)	≥ 0.8 within attribute	Acceptable consistency threshold
Panel agreement (r_p)	≥ 0.7 w/group mean	Ensures a cohesive dataset

Panel Leader Responsibilities: The panel leader is not an evaluator; instead, they ensure compliance with the protocol, moderate discussions, and monitor data quality. Data collection should be electronic when possible to minimize transcription errors.

Test types: (Fig. 36.3) See Appendix IX for more common tests and descriptions.

- Difference tests (triangle, duo–trio) for discrimination tasks (Lawless et al., 2010).
- Descriptive analysis for profiling intensity of defined attributes.
- Hedonic/acceptance testing (less common in spirits evaluation) (Stone & Sidel, 2004).
- Note: A/B testing is generally reserved for consumer preferences

Evaluator Personal Readiness

The evaluator's nose and palate are precision tools, sometimes un Addenda/Clarifications to a Prior 2018 MDPI Article prepared or up to the task.

30-Second Self-Check (Y/N) and fixes: If ≥2 "Yes": switch roles (scribe/steward), or postpone. Ensure that all relevant circumstances are recorded on the session log sheet.

- Congestion/allergy today: Postpone or switch roles
- Strong fragrance/hair product/aftershave: Wash hands, remove scented items, ventilate room

- Recent coffee, mint, gum, spicy food: Wait for 60 minutes
- Mouth dryness or fatigue: Water, neutral cracker, brief rest
- New meds/antihistamines/decongestants?: Best if not needed, discuss with panelists
- Headache, poor sleep, allergies active, or high stress: Retire or postpone (ethical pass)
- During evaluations: Ethanol adaptation (nose burn): 2–4 min breaks with tulip vessels; cover glasses between passes.

Fig. 36.3	Training and Evaluation Testing Protocols for Sensory Panels		
Protocol	**Purpose**	**Method**	**Evaluator notes**
Threshold testing	Establish detection limits	Dilution series of pure compounds	Essential for calibration and screening
Difference testing	Identify perceptible differences	Triangle, duo-trio	Requires statistical treatment for validity
Descriptive profiling	Map aroma/taste intensities	Attribute scaling (0–10, or 0–15)	Builds quantitative sensory fingerprints
Fault recognition	Detect spoilage and taints	Spiked samples at known levels	Must be contextualized (fault vs style marker)
Panel calibration	Ensure consistency	Reference samples before sessions	Reduces drift and lexicon mismatch

Ethical Pass Options (keep the session honest) Role to be recorded on datasheet

- **Observer/scribe:** record others' notes; do not vote.
- **Steward:** pour, code, time; no evaluation.
- **Calibration only:** smell references; skip samples.

Anosmia and Variability–Recovery and Pacing (specific anosmias are common with tulips) Do not force an evaluation you cannot verify; mark "descriptor boundary/unknown" and move on.

- Max 8–10 high-ABV samples per flight; break between flights.
- Two-pass habit: quick first pass; second pass after 60–90 s.
- Hydrate; keep room 20–22 °C; cover glasses between sniffs.

The operating standard is *"Check yourself, declare your state, and if you are off, change roles so the data stays clean and accurate."* The data sheet includes a section to record readiness, role (evaluator, scribe, steward), and notes/explanations from self-check and interferences.

Decision Hygiene: Keeping Bias Out of Your Notes

Fig. 36.4 provides a quick summary. Also refer to Chapter 34 and Appendix X for a quick review of the vast diversity and effect of various biases. Accurate, descriptive notes are only as good as the conditions that produced them. Keep the process clean, free of bias, and ensure accurate results:

- **Quiet first pass:** No talk until everyone locks their individual observations. Confident words from one taster can cue labels in others.
- **Blind what matters:** Use sample ID codes. Mask brand/price/cask info. When color could bias the results, use amber/opaque glass.

- **Order control:** Randomize or counterbalance serving order so no sample is always first or last. Split large sets into smaller flights.
- **Two-pass notes:** Fast pass (immediate), then a slow pass after 60–90 s. Keep the glass lightly covered between passes. Record any airing/dilution you use.
- **Palate resets and pacing:** Water + neutral cracker. Add short breaks, especially with high-ABV flights. Stop if sensing olfactory fatigue—resume only when refreshed.
- **Duplicates and anchors:** Quietly insert one duplicate sample per flight to check consistency. Keep a straightforward reference standard handy (e.g., vanillin) to monitor drift.
- **Observation vs. inference:** Write what was perceived first (descriptor + intensity + phase). Place process guesses second, accompanied by confidence tags (low/moderate/high).
- **Fault discipline:** Suspect a fault? Recheck after brief air or slight dilution. If it **persists and dominates**, mark "probable fault" and cross-reference **Appendix II** thresholds.
- **Micro-log the context:** Note glassware, pour temperature, rest time, and dilution. If any of these changed mid-flight, record it.
- **30-second self-check (before you start):** Fragrance/cologne today? Congestion or allergy? Just had coffee/mint/spicy food? If "yes" to any, switch roles (scribe/steward) or postpone.
- **One-line standard for the group:** Individual notes first, then discussion; observation first, inference second; uncertainty stated.

Fig. 36.4	Bias Sources and Countermeasures	
Bias Source	**Mechanism**	**Preventive Design Element**
Expectation (brand, age, price)	Top-down semantic priming	Double-blind codes; opaque glass
Order and contrast	Sensory adaptation, recency	Randomized or Latin-square order
Groupthink	Social conformity	Silent first-pass notes
Anchoring	Early sample sets scale	Insert mid-flight neutral reference
Fatigue/adaptation	Reduced receptor sensitivity	Max 10 samples per flight; rest breaks
Carryover	Residual retronasal volatiles	Water + neutral cracker reset
Instrumental drift	Environmental or data error	Control sample for each flight
Personal Anosmia	Receptor non-response	Screen via threshold pre-tests

As a general guide, Fig. 36.5 provides a quick checklist to ensure high-quality results throughout the process.

Data Interpretation

Quantitative Methods

- **Threshold determination**: geometric mean of detection levels across panelists (ASTM International, 2011).
- **Descriptive profiling**: mean intensity ratings for each attribute, analyzed via **ANOVA** to detect significant sample differences, identifying those with > 2 standard deviations.
- **Multivariate analysis** (PCA, cluster analysis): maps sample relationships in sensory space, reveals key drivers of perception (Lawless et al., 2010).

Reliability and Repeatability: Panel reliability must be checked with:

- **Panelist performance plots**: individual consistency across replicates.
- **Panel agreement indices**: correlation of individual data to group mean.
- **Control samples**: inserted randomly to monitor drift.

Linking Sensory to Chemistry: Interpretation should not isolate sensory data from chemical analysis. Correlations between descriptive attributes and GC–MS or HPLC data strengthen conclusions. Example: high ethyl butyrate correlating with panel descriptors of "pineapple" in rum (Piggott et al., 1993).

Fig. 36.5		Panel Workflow to Maintain Quality Assurance
Step	**Action**	**Purpose**
1	Pre-session self-check	Confirm evaluator readiness (no congestion, caffeine, fragrance, etc.)
2	Blind coding	Mask sample identity to remove brand or price expectation bias
3	First pass observation	Record individual notes silently; prevent social influence
4	Second pass (60–90 s later)	Rest, confirm perceptions; note any changes in intensity or character
5	Duplicate check sample	Compare repeat sample results; calculate $|\Delta| \leq 1$ unit for consistency
6	Panel discussion (post-lock)	Discuss only after notes are submitted; align lexicon boundaries
7	Electronic data entry & ANOVA check	Input data; identify outliers >2 Standard Deviations; verify reliability indices
8	Panel leader validation	Ensure protocol compliance and document deviations

Reporting, Plain and Simple

Once the data has been analyzed, it needs to be documented so that others can reproduce it. Use Appendix I for wording, give each term an intensity (0–10) and a phase (orthonasal, palate, retronasal, finish), and keep opinions out of the observation line. If you suspect a fault, point to Appendix II and note what actions were taken to check the fault (airing, slight dilution). Separate what you perceived from what you think it implies.

Write records in this order: sample code, glassware, temperature, dilution:

1. **Aroma (orthonasal)** with your top 3–5 lexicon terms and intensities.
2. **Palate (0–20 s)** with key attributes plus mouthfeel (heat, viscosity, astringency);
3. **Finish (duration in seconds)** with any late-arising notes; Fault check if relevant (threshold/context test)
4. **Inference (optional)** with a confidence tag (low/moderate/high)
5. Two-line **Summary** in neutral language, short, precise, testable.

Descriptor syntax: *Family – term (modifier) – intensity – phase.*

Use uncertainty honestly: "banana/pear-drop boundary; consensus check needed." Mark inferences: "likely ex-bourbon; confidence: moderate." **Example: single sample**

- **Aroma:** Fruity, apple (green skin), 6, orthonasal; Sweet/Confectionary, vanilla, 4, orthonasal.

- **Palate:** Woody, vanilla, 4, palate; Spicy, clove, 3, palate; ethanol heat, 5; astringency, 2.
- **Finish** *(38 s)*: Phenolic, medicinal, 2, finish (style, not fault).
- **Fault check:** Solvent (ethyl acetate-like), 2, orthonasal; dissipates after 2 min air → below concern (Appendix II).
- **Inference:** likely ex-bourbon influence; confidence: moderate.
- **Summary:** apple-led profile with moderate vanilla/clove; clean, medium finish.

Language Discipline

- No hedonic words ("lovely," "elegant") in observation lines.
- Metaphor only if it maps to a defined lexicon entry.
- Inference second, and tag **confidence**: *low/moderate/high*.
- If uncertain: write the **boundary** ("banana/pear-drop boundary, 4/10, palate").

Keep it readable and anchorable. If someone else cannot test it, rewrite it. With standardized reporting, observations can be transferred seamlessly from bench to panel to production without loss of meaning. Sheets should also record date, time, location, temperature, and humidity.

Data Integrity and Reproducibility (Beyond Significance)

Sensory results are persuasive only when repeatable by other evaluators under comparable conditions. Most failures in perception are not due to bad faith, but to ways of deceiving ourselves.

Sources of false certainty

- Multiplicity by drift. Every extra comparison—more attributes, more post-hoc subgroups, more "quick looks" inflates the chance of "significant" differences that are noise. Optional stopping (also known as peeking and deciding to stop when results appear promising) achieves the same effect.
- Garden of forking paths. Choices made *after* seeing data (which panelists to exclude, which attributes to emphasize, which transform to use) can manufacture effects without intent.
- Overinterpreting ordinals. Treating 0–10 intensity scales as if they were precise instruments invites spurious decimals and false precision.
- Modeling without stability. PCA, clustering, or any multivariate summary is brittle when n is small, and the attributes are noisy; patterns can be artifacts of the specific panel on a specific day.

What counts as robust

- Pre-specify the claim. State, in plain language, *what difference you expect and why* before you run the session. Record it. If you explore, mark it as exploratory.

- Report the design, not just the outcome. N (panel size), serving order logic, blinding/masking, glass/temperature/dilution, and any deviations. Omit these, and the number is context-free.
- Favor effect sizes over lone p-values. How big is the difference on a scale that matters to tasters, and in which direction? If the magnitude is trivial, "significant" is irrelevant.
- Replicate across days. A claim that appears once and disappears on re-testing is anecdotal. Across-day stability is the simplest filter against enthusiasm.
- Separate observation from inference. Observation lines use lexicon terms and intensities; inference lines carry the production hypothesis with an explicit confidence tag. Mixing them makes both fragile.
- Make the result portable. A finding is more convincing when another panel, using the same conditions, can attempt to reproduce it without guessing your hidden steps.

What the reader should expect

These methods and interpretations survive replication, resist multiplicity, and respect the ordinal nature of much sensory data. Where claims are exploratory, we say so. When the effect size is small, we do not overstate its significance. Reproducibility is not a burden added to sensory work; it is the reason anyone should trust it.

Note on Training Implications for the Future

With increasing global competition, distillers are investing in automation and AI-driven sensory analytics. Digital noses, machine learning applied to GC–MS data, and real-time performance tracking will supplement—but not replace—the trained human sense of smell. The sensory evaluator of the near future must be adept at both traditional descriptive protocols and data-driven interpretation to ensure relevance in technologically augmented quality control systems.

Sensory Thresholds, Weber's Law, and Panel Repeatability

Consistent sensory work depends on understanding how people perceive differences. Weber's Law (Chapter 2) states that the most minor perceivable change in stimulus intensity (the just-noticeable difference) is proportional to the baseline intensity. In practice, detecting a 5% difference in ethanol concentration is easier at 20% ABV than at 60% ABV. Repeatable panels rely on anchoring reference standards near these differential thresholds. While advanced statistics quantify such variation, awareness alone improves calibration: identical samples should yield consistent median ratings within each panelist's perceptual bandwidth. Teaching threshold awareness reinforces objectivity and prevents over-interpretation of noise as genuine difference.

Summary

Sensory training in distilled spirits is both a scientific discipline and a learned craft. Professional evaluators must undergo structured calibration in taste and aroma, fault recognition, and memory consolidation to ensure accuracy and repeatability of their assessments. Panels function as controlled environments where trained tasters collectively generate reliable data through difference testing, descriptive profiling, and calibration protocols. The interpretation of results relies not only on statistical methods such as ANOVA and PCA, but also on linking sensory perception to chemical analysis. True mastery extends beyond technique, requiring continual practice, awareness of bias, and humility. As technology evolves, AI and digital tools will supplement—but never replace—the human sensory evaluator, whose role remains central to the validity of spirits analysis. Appendix III provides a basic evaluation sheet format that should be modified at the panel's discretion to meet specific objectives, with the understanding that complete documentation is essential for achieving the best results.

"We are what we repeatedly do. Excellence, then, is not an act, but a habit."— Aristotle, philosopher

"What gets measured gets managed."— Peter Drucker, educator, author

References

ASTM International. (2011). *Standard practice for determination of odor and taste thresholds by a forced-choice ascending concentration series method of limits (ASTM E679-04R11)*. Author. https://doi.org/10.1520/E0679-04R11

Lawless, H. T., & Heymann, H. (2010). *Sensory evaluation of food: Principles and practices* (2nd ed.). Springer. https://doi.org/10.1007/978-1-4419-6488-5

Lawless, H. T., Popper, R., & Kroll, B. J. (2010). Discrimination testing in sensory evaluation. *Food Quality and Preference, 21*(8), 902–908. https://doi.org/10.1016/j.foodqual.2010.07.004

Pickering, G. J., & Heatherbell, D. A. (1995). The influence of ethanol concentration on the detection threshold levels of wine esters. *American Journal of Enology and Viticulture, 46*(4), 529–534.

Piggott, J. R., Conner, J. M., & Paterson, A. (1993). The contribution of selected congeners to Scotch whisky aroma. *Journal of the Science of Food and Agriculture, 61*(3), 235–242. https://doi.org/10.1002/jsfa.2740610305

Stone, H., & Sidel, J. L. (2004). *Sensory evaluation practices* (3rd ed.). Elsevier. https://doi.org/10.1016/B978-012672690-9/50011-2

Vas, G., & Vékey, K. (2004). Solid-phase microextraction: A powerful sample preparation tool prior to mass spectrometric analysis. *Journal of Mass Spectrometry, 39*(3), 233–254. https://doi.org/10.1002/jms.606

Zucco, G. M., Paolini, M., & Schaal, B. (2011). Olfactory memory: The role of verbalization in odor naming and recognition. *Memory & Cognition, 39*(3), 451–461. https://doi.org/10.3758/s13421-010-0043-5

Observations on Data Interpretation

Statistical significance is not sensory importance: A difference of 0.3 points on a 0–10 scale may be "significant" by $p < 0.05$ yet imperceptible to trained tasters. Panels must report both effect size and direction, not merely the p-value. Sensory significance is anchored to perceptibility, not arithmetic difference.

Replicability is the first test of truth: One-day patterns can arise from noise, fatigue, or sample drift. Any claim must survive re-testing under matched conditions. Reproducibility is the minimum criterion for knowledge rather than coincidence.

Error compounds silently: A small, uncontrolled variable—such as temperature, pour volume, or sample order—introduces bias that propagates through the dataset. Each recorded variable is one less hidden assumption.

Averaging can conceal structure: The panel means smoothing away minority perceptions that may indicate subgroup sensitivity or product heterogeneity. Examine variance and outliers before discarding them; they often reveal the sensory story that the mean obscures.

Lexicon drift precedes data drift: When panelists unconsciously redefine descriptors ("vanilla" vs. "oak-sweet"), numeric consistency becomes meaningless. Periodic lexicon audits preserve comparability across sessions and analysts.

Bias leaves a mathematical fingerprint: anchoring, order effects, and fatigue can be diagnosed statistically using compressed dynamic range, directional residuals, or autocorrelation along the serving sequence. Detecting bias quantitatively strengthens panel credibility.

Context defines precision: No metric has meaning outside the system that generated it. Intensity scores, detection rates, or hedonic responses are conditional on glassware, environment, and evaluator calibration. Without contextual annotation, precision becomes an illusion—numbers divorced from the conditions that give them validity.

Trends outweigh single points: Data taken as a sequence—across sessions, batches, or replicates—reveals underlying sensory stability far more reliably than isolated outcomes. A one-time anomaly teaches little; a consistent direction of change signals process control or drift. Sensory data gains power when interpreted as a trajectory rather than a snapshot.

Absence of difference is not equivalence: A non-significant result does not prove two samples are identical; it only indicates insufficient evidence of distinction under current conditions. Equivalence requires deliberate proof through sensitivity analysis and threshold framing. True sameness must be demonstrated, not assumed.

Chapter 37: Developing Sensory Proficiency — Skill Development

Orientation – Awakening to the Task

For many, the first meaningful encounter with spirits is not marked by science but by style—an ambient ritual of glass swirling, head tilting, and brand admiration. Yet behind this aesthetic lies a latent invitation: to move beyond casual drinking into the disciplined realm of sensory evaluation. This is the moment of calling, recognition that flavor is not a passive experience but a perceptual phenomenon governed by the sciences of chemistry, physics, and biology.

This transition requires a shift in purpose. Tasting becomes an act of observation, not an act of opinion. Evaluation becomes inquiry. What was once passive becomes deliberate. Sensory work begins with a shift in intention, not the development of technique. Spirit's evaluation is not a leisure activity cloaked in sophistication, but sensory science grounded in methodology, calibration, and repeatability. The task is to train the senses to yield reliable data.

The Tools Within

Before acquiring any external tools, the evaluator must assess the internal instruments, including the nose, palate, attentional faculty, and memory system. These are the core evaluative devices: fallible, dynamic, and prone to bias, but trainable.

Sensory mastery is less about innate sensitivity than it is about deliberate cultivation. It begins with self-awareness—knowing one's own internal sensory limitations and potential. The evaluator must become both subject and instrument, focusing and refining internal senses and reason.

The olfactory system comprises more than 400 functional receptor types, collectively responsible for discriminating among thousands of odors (Buck, 2005). Unless deliberately activated, this powerful system functions below conscious awareness.

The tongue's taste buds detect sweet, sour, salty, bitter, and umami, but these inputs are coarse compared to olfactory richness (Chaudhari & Roper, 2010). The evaluator's personal focus becomes critical, as what is not noticed is not perceived. Memory is the internal archive against which all stimuli are subsequently judged (Stevenson, 2001).

When the evaluator accepts that these internal tools must be studied, calibrated, and maintained, the shift occurs. The body becomes the laboratory. The mind becomes the interpretive engine. From this point on, the spirit is no longer a product for pleasure; it is a subject for evaluation. The drinker becomes a practitioner.

The Mindset of Mastery

Mastery is not a trophy but a path. There are no finish lines, only deepening awareness and evolving control. One does not become a master by accumulating tasting notes, but by learning to notice what others overlook, to describe the imperceptible with clarity, and to detect the erroneous within oneself. Humility is not a virtue; it is a requirement.

Those who succeed in sensory science are not the most innately gifted but the most methodical. The concept of *deliberate practice*, as articulated by Ericsson and colleagues, also applies to flavor perception: improvement requires structured, feedback-driven effort over time (Ericsson, Krampe, and Tesch-Römer, 1993). This differs sharply from notions of natural-born connoisseurs. While individual thresholds vary, the brain's capacity for perceptual learning is immense, even in adulthood (Li, Howard, Parrish, & Gottfried, 2008).

Equally essential is the evaluator's attitude. True mastery entails a willingness to be wrong, to revise one's judgments, and to seek critique. The novice fears error; the expert studies it. Evaluation is not a declaration of authority; it is a hypothesis, revised with each new sample. The moment one claims infallibility, learning stops. The task is not to demonstrate discernment but to cultivate it.

"The expert in anything was once a beginner."— Helen Hayes

"You don't learn to walk by following rules. You learn by doing, and by falling over."— Richard Branson

Exposure – Building the Sensory Archive

Diverse Encounters

Sensory memory is not conjured from a void. It is built through experience. The richness of one's perceptual database depends directly on the diversity and frequency of sensory inputs. To evaluate well, Exposure must be broad. It is not enough to taste the same whiskey every Friday. Evaluators must expose themselves to styles, categories, origins, cultural differences, and production methods. All spirit types have different signatures, and each encounter leaves a memory trace in the sensory archive (Prescott, 1999). Early-stage evaluators benefit from guided exposure—flights organized around production variables rather than brand or price. Systematic exposure accelerates learning. Sensory acuity is achieved through the accumulation of experiences rather than isolation.

Naming and Mapping

The act of assigning names to sensations is more than linguistic. It has a cognitive structure. Language shapes memory, recall, and comparison (Majid & Burenhult, 2014). Without a name, the brain struggles to store accessible sensations. The neophyte's challenge is a lack of vocabulary.

This is where lexicons come in—not as rigid authorities but as scaffolds. Professional reference wheels offer domain-specific vocabulary to jump-start labeling. The purpose is not to memorize terms but to experiment with mapping internal experiences to external words (Noble et al., 1987). The result is a structured associative network that enables faster recognition in future sessions. Eventually, evaluators develop their own lexicons. One taster's "overripe banana" may be another's "bruised plantain." The key is to communicate meaningfully and consistently.

Recording and Reflection
Documentation transforms fleeting experiences into permanent lessons. Recording impressions during and after each tasting consolidates memory, reinforces perceptual anchoring, and reveals personal patterns over time (Lawless & Heymann, 2010). The format may vary: some prefer scorecards with numerical ratings for each dimension, while others opt for open-form journaling or visual mapping of aroma clusters.

What matters is the act of reflection. Evaluation without reflection is merely consumption. Reflection begins with the question: *What did I perceive, and why?* The student begins to build an identity as an evaluator, an emerging identity with a mental filter for future encounters, guiding attention and noting subtlety over spectacle.

Eventually, the record becomes a tool for calibration—allowing the evaluator to compare past and present perceptions of the same spirit and identify whether the change lies in the sample, the taster, or the context.

"Sensory expertise can be defined as 'the ability to discriminate among different aromas, to recognize aromas when cued, and to describe aromas by free recall.'"— Winemakers Research Exchange

Focus: Training Perceptual Attention
Slowing the Process
Most beginners evaluate too quickly. They form conclusions within seconds, often based on first impressions or past preferences. Mastery requires slowing down. By extending the observation window—waiting longer before sniffing again, pausing before swallowing, and delaying conclusions—tasters give their senses time to process a broader range of cues. This deliberate pacing allows slower-arising aromas and textures to emerge, many of which would otherwise be missed. Studies in flavor perception confirm that temporal dynamics shape how stimuli are recognized and interpreted (Delwiche, 2004). Slower engagement means fuller evaluation.

Active versus Passive Perception

Passive perception happens automatically—odors and tastes are registered without focused attention. Active perception, by contrast, involves intentional control over how and what we evaluate. This includes techniques such as adjusting sniff strength or pausing to recheck retronasal signals. Research indicates that voluntary sniffing activates distinct neural pathways compared to passive odor exposure, thereby enhancing olfactory discrimination (Mainland & Sobel, 2006). Control also includes knowing when to isolate variables. A taster might warm a sample slightly to emphasize volatility or try the same sample after dilution. These small decisions can clarify what is being perceived and why. Over time, the taster learns to direct sensory input rather than receive it.

Cognitive Isolation

Learning to isolate variables is critical for building sensory precision. When evaluating spirits, multiple sensory inputs coincide—ethanol burn, texture, aroma, temperature, and finish. Without training, these merge into a general impression. Isolation helps separate them.

One method is to vary just one parameter at a time, such as dilution, glass shape, serving temperature, or tasting environment. Another is to focus on one sensory dimension per session— such as detecting sweetness across different spirits or tracking mouthfeel under controlled conditions. This kind of focus is supported by perceptual learning research, which shows that repeated exposure to single-variable distinctions enhances detection and categorization abilities (Goldstone, 1998). Cognitive isolation turns vague impressions into specific judgments.

Structure – Establishing Methodology

Repetition with Purpose

Repetition is often misunderstood in tasting. Casual drinkers may repeat experiences, but without structure, they learn little from them. In a professional evaluation, repetition must be intentional and purposeful. This involves tasting the same spirit multiple times under the same conditions, recording the results, and comparing them for consistency. Structured repetition reduces error and improves calibration. Over time, it builds internal reference points that remain stable across sessions. Scientific studies in sensory analysis support this approach. Evaluators improve accuracy through repeated exposure and by comparing structured sensory inputs under controlled conditions (Meilgaard, Civille, & Carr, 2007). This distinguishes passive experience from skill-building.

Evaluation Frameworks

A structured tasting protocol provides order and repeatability, ensuring consistency and accuracy.

Common professional sequences include the five-step method: appearance, aroma, palate, finish, and faults. This order reflects how sensory inputs are received and processed by the brain. Visual assessment prepares expectations. Aroma offers the richest information. The palate confirms and expands. The finish gives durability and complexity. Fault detection overlays all steps.

By using the same framework every time, the evaluator can track specific attributes across different spirits. This helps build pattern recognition and minimizes distractions. Without a straightforward method, attention shifts randomly, leading to data loss.

Frameworks also help with timing. Evaluators learn how long to observe each sensory stage before moving on. Aroma assessment might involve multiple sniffs spaced 15–30 seconds apart, with notes taken between each sniff. This timing reinforces attention and deepens evaluation.

Lexicon Development

A consistent vocabulary bridges the gap between internal sensation and external communication (Chambers & Koppel, 2013). Lexicons are not only for group calibration; they also help individuals sharpen their perceptions by linking sensations to language. Sensory scientists often rely on reference-based lexicons developed through consensus panels. These serve as tools for identifying, naming, and communicating aroma and taste.

For individual tasters, lexicon development begins with exposure but matures through repeated usage. Each time a taster labels a term such as "clove," "acetone," or "wet cardboard," that term anchors the associated neural pattern. Accuracy improves with feedback and practice.

While formal lexicons exist for wine, beer, and whiskey, personalization is allowed as long as terms remain consistent and meaningful. A shared vocabulary becomes especially important in judging panels, research groups, and educational contexts. Clarity in terminology enhances self-understanding and facilitates effective dialogue with others.

Consensus and Dissent: What Agreement Means (and Does Not)

Panels seek shared language, not unanimity for its own sake. Consensus is a communication outcome; it is not a truth oracle.

- **Agreement is evidence of clarity**, not infallibility. High consensus under stable conditions raises confidence that signals are strong.
- **Dissent is diagnostic.** Minority reports often mark **masking**, adaptation, or boundary cases ("banana/pear-drop border"). They also surface individual-specific anosmias.
- **Record both.** A clear consensus line, accompanied by a brief note of principled dissent, preserves information without derailing interpretation.

Aim for convergence on defined terms (*Appendix I*), allow documented disagreement where evidence is genuinely ambiguous, and resist the urge to smooth data into uniformity. That preserves reliability **and** intellectual honesty.

"Teachers can open the door, but you must enter it yourself."— Chinese proverb

Calibration – Internal Consistency and External Benchmarks

Blind Comparison Training

One of the most effective ways to eliminate bias in spirits evaluation is through blind tasting. By concealing labels, prices, and reputations, evaluators can focus solely on sensory information. This approach helps prevent expectation effects, where perceived quality is influenced by branding or cost (Siegrist & Cousin, 2009). Additionally, blind tasting can uncover personal inconsistencies that might otherwise remain unnoticed.

Blind tasting is particularly valuable when practiced repeatedly with known reference samples. Evaluators learn to recognize their own tendencies, such as favoring familiar styles or undervaluing subtle flavors. Over time, regular blind comparisons foster trust in the evaluator's perceptions, free from external cues.

In professional environments, blind tastings are usually randomized, coded, and scored according to preset criteria. In educational settings, simpler methods, such as three-glass triangle tests, can still yield valuable insights. The key is repeatability and honest scoring.

Reference Standards and Control Samples

Calibration improves when known benchmarks are used for comparison (Ferreira & López, 2019). This might include training with specific aroma compounds (e.g., isoamyl acetate for banana or ethyl hexanoate for apple) or by comparing commercial spirits known for their typicality. These reference points help the evaluator align their internal scale with measurable, external standards.

Control samples are also helpful. A neutral base spirit can be altered in a single variable—such as dilution, temperature, or ethanol content—to train perception of that specific element. This kind of calibration isolates cause and effect, strengthening sensory precision.

Industry organizations and academic labs often maintain libraries of such reference standards. While not always accessible to individuals, simplified at-home versions can be created using food-safe compounds and documented procedures. Even informal exposure to known aromatic markers reinforces calibration.

Tasting with Peers

Sensory training benefits from social comparison. When evaluators taste together and discuss results, differences in perception become apparent. These group sessions help identify blind spots, clarify language use, and establish boundaries of agreement. The process is not about enforcing uniformity but about expanding understanding.

Research in sensory science confirms that group tasting increases both reliability and learning (Guinard & Mazzucchelli, 1996). Repeated dialogue over time helps participants converge toward shared criteria, even when initial perceptions differ widely.

Peer calibration also introduces variability. Different tasters may detect different attributes based on prior exposure, olfactory acuity, or cultural associations. Sharing and debating these impressions increases awareness of the many ways a spirit can be experienced. Over time, this dialogue reinforces both self-confidence and humility—two essential traits for skilled evaluation.

"The Novice Mindset is the belief that we grow continuously through fearless, deliberate practice."— Jon Eckert

Interpretation – Moving Beyond Sensory Data

Pattern Recognition

Once basic perception is trained and calibrated, the next stage is recognizing patterns. This involves identifying recurring combinations of aromas, flavors, textures, or finishes and connecting them to likely causes—such as specific production methods, ingredients, or aging conditions. Pattern recognition is informed inference built on repeated observation.

For example, the consistent presence of dried fruit, rancio notes, and an oxidative tang in a sample may indicate sherry cask maturation. Heavy phenolic notes accompanied by iodine and seawater might indicate peated malt dried with coastal fuel sources. The more experience an evaluator gains, the more rapidly such patterns emerge.

This is the point at which the taster begins to link sensory input to knowledge processing. Neurological studies on expert tasting have shown that experienced evaluators engage more of their associative cortex, linking input not just to immediate sensation but also to memory, logic, and interpretation (Parr, Heatherbell, & White, 2002).

Probabilistic Thinking

Skilled tasters learn to resist the urge to be definitive. While some conclusions can be drawn with high confidence—such as identifying a dominant ester or sulfur compound—many require nuance and interpretation. Spirit's evaluation involves some uncertainty. The ability to say, "this likely indicates column distillation," or "possibly aged in American oak," reflects sound reasoning. This mindset aligns with probabilistic thinking, which involves weighing evidence, identifying degrees of certainty, and avoiding overstatement (Kahneman & Tversky, 1979). It is a critical skill for judges, educators, and researchers. Overconfidence not only misleads others, but it also dulls further observation. In contrast, probability language encourages rechecking, revision, and intellectual honesty.

Narrative Building

Interpretation culminates in building a narrative—a logical explanation for what the spirit is, how it was made, and what its quality or flaws suggest. This narrative should be grounded in the data, but it also reflects the evaluator's reasoning process. In competition settings, such narratives inform scoring rationales. In education, they serve as demonstrations of skill. In the industry, they are called tasting notes or product profiles.

Compelling narratives combine sensory data, contextual knowledge, and logical coherence to create a cohesive, engaging story. They avoid speculation, unfounded metaphors, or storytelling that obscures rather than clarifies. The goal is to communicate how the spirit's sensory properties reflect its production and style.

Narrative construction enhances memory retention. When evaluators create a reasoned account of what they tasted and why, they are more likely to recall and apply the experience (Ericsson et al., 1993). This transforms isolated tastings into building blocks for long-term mastery.

Self-Mastery – Transcending Technique

Bias Awareness and Neutralization

By this stage, the evaluator has developed a reliable understanding of the data. However, one of the most persistent challenges is cognitive bias. Even seasoned tasters are vulnerable to preference, familiarity, and emotional association. Bias can be cultural (favoring certain regions), aesthetic (favoring clarity or intensity), or personal (favoring styles similar to one's training background).

The only defense is constant vigilance. Self-monitoring during tastings—questioning whether impressions are driven by sensation or expectation—is an advanced skill. Blind tasting helps, but so does intentional humility: assuming nothing and allowing the spirit to speak for itself. Bias awareness should not be confused with self-doubt. It is a disciplined mental filter that checks assumptions and preserves objectivity. Evaluators must learn to ask, "Is that aroma truly present, or am I imposing it from memory?" Bias Awareness is critical (Wansink & van Ittersum, 2005).

Recovery and Adaptation

Olfactory fatigue, contextual distraction, and health-related anosmia are part of every evaluator's experience. Professional tasters must learn how to manage and recover from these disruptions. Overexposure to ethanol vapor, for example, can cause a loss of receptor sensitivity within minutes (Cain, 1974). High-concentration sampling without breaks can impair judgment even in trained noses.

Best practices include spacing out high-ABV spirits, alternating with water or neutral aromas, and maintaining overall health. Some evaluators rotate different spirit categories (e.g., mezcal to rum) to prevent burnout. Others track sensory peaks and dips throughout the day or week.

Adaptation also applies to changing conditions, such as new judging environments, unfamiliar glassware, or varying room temperatures. Consistent evaluators adapt without compromising their protocols. They understand their own sensory profile well enough to adjust, recalibrate, and continue accurately.

Teaching and Translation

Teaching is the final test of mastery. Explaining a complex aroma or guiding someone through a tasting forces the evaluator to clarify their own process. Teaching reveals gaps in understanding, improves terminology, and refines logical structure.

This act of translation—from sensation to words, and then to another person—creates the highest level of internal calibration. Studies in skill acquisition show that teaching strengthens memory and consolidates expertise (Fiorella & Mayer, 2013).

Teaching does not require a formal classroom. It may happen during mentorship, judging panels, or casual tastings. What matters is the intent: using one's own skill to develop another's. At this stage, the evaluator becomes not just a practitioner but a steward of sensory knowledge.

"A person who never made a mistake never tried anything new."— Albert Einstein

Lifelong Refinement – The Master's Path

Re-exposure and Reinvention

True mastery in sensory evaluation is never final. Over time, the risk of stagnation increases—not from lack of skill, but from routine. Evaluators may develop dependence on ruts in vocabulary, perception, or judgment. To counteract this, periodic return to the fundamentals is essential.

Re-exposing oneself to unfamiliar spirits, emerging categories, or overlooked profiles restores a "beginner's mind" (Herz, R.S., & Engen, T., 1996). This is not regression—it is a deliberate act of humility and curiosity. It broadens the evaluator's archive and sharpens critical faculties by disrupting automatic responses. Some professionals schedule regular exercises designed to reawaken forgotten sensory modes, such as retronasal focus or orthonasal isolation.

Reinvention can also involve updating methods—such as testing new glassware, revisiting dilution ratios, or incorporating insights from other sensory domains, like wine, coffee, or perfumery. Continuous learning prevents ossification and maintains a flexible perception.

"If you have a willingness to carve out time to work on a skill every day, you'll slowly make that transition from novice to expert!"— S. J. Scott

Specialization and Deep Study

As general competence stabilizes, many evaluators develop a specific focus. This could be based on spirit category (e.g., tequila, rum, Armagnac), chemical class (e.g., esters, phenols), or production method (e.g., pot vs. column stills). Specialization enables deeper understanding, greater precision in comparison, and a greater contribution to the field.

Deep study often involves structured learning—consulting scientific literature, conducting personal trials, or working alongside producers. It may also involve repeated exposure to narrow sensory profiles until patterns and outliers become clear.

Expertise in a focused area also reinforces broader skills and sharpens perception (Lawless & Heymann, 2010). It teaches the limits of generalization and trains the evaluator to identify rare or subtle traits with higher accuracy.

Contribution to the Field

At the highest level, the evaluator becomes a contributor. This might involve publishing research, developing new training methods, organizing competitions, or mentoring the next generation. The act of codifying what has been learned solidifies one's legacy.

Contribution can also mean advancing standards. Skilled evaluators often challenge flawed assumptions, correct sensory myths, and advocate for better practices. Their value lies not just in what they perceive, but in how they shape the framework through which others perceive.

In this way, the arc from neophyte to master closes and begins again. Lifelong refinement is not about perfection—it is about persistent, humble progress. The journey never ends, but each stage elevates both the individual and the field of study. These stages are summarized in Fig. 37.1.

"Never forget the beginner's mind." (Shōshin)— Zen concept

"Lumps of coal turn into diamonds." — Henry Kissinger, statesman

Fig. 37.1	Stages of Sensory Proficiency		
Stage	**Cognitive Focus**	**Training Emphasis**	**Metric of Progress**
Orientation	Realization of the task	Awareness of perception as data	Conceptual shift from enjoyment to evaluation
Exposure	Expanding sensory archive	Diverse encounters, naming, mapping	Descriptor diversity and recall accuracy
Structure	Systemization of the method	Repetition, framework, and lexicon development	Consistency across controlled trials
Calibration	Objectivity and standardization	Blind tasting, reference standards, peer calibration	Agreement indices (r ≥ 0.8) and reduced variance
Interpretation	Meaning-building and reasoning	Pattern recognition, probabilistic reasoning	Logical coherence between data and inference
Self-Mastery	Metacognition and teaching	Bias control, mentorship, and translation of skill	Ability to teach or replicate outcomes
Refinement	Renewal and continuous learning	Re-exposure, specialization, methodological reinvention	Sustained curiosity, adaptability, and publication of findings

Case Study 37.1: Scotch Whisky Panel Re-Calibration: University of Strathclyde (1985 – 1990)

Context: In the mid-1980s, Professor A. P. Pigott and colleagues at the University of Strathclyde operated one of the first academic whisky-sensory laboratories in the world, running weekly descriptive sessions for Scotch distillers. Over several years, the group observed a drift: panelists' intensity scores for phenolic and smoky attributes gradually increased, even though GC-MS data showed no change in cresol or guaiacol levels (Pigott et al., 1993). The "inflation" is traced to the conditioning of expectations. Many program samples were Islay malts; panelists subconsciously expected smoke, sought it out, and over-reported it. Panelists were retrained using neutral control non-Islay whiskies and non-Islay styles; mandatory blind duplicates and cross-day replications were introduced. The variance decreased by half, and the correlation returned to its baseline level.

Expectation/anchoring bias: — panelists' prior knowledge skewed perception despite objective data. **Evaluator Takeaway:** Mastery is not a straight ascent; it is a loop of *drift, discovery, and discipline.* Even seasoned experts can be caught up in context. Pigott's solution was structured: stop, reset, and rebuild confidence through verified reference anchors. The professional evaluator learns humility not by doubt but by testing themselves against data. Regular calibration is not a clerical routine; it is intellectual hygiene. Skill without feedback becomes self-confirmation. Treat every panel cycle as an opportunity to rediscover objectivity.

Summary

The evaluator's journey from casual drinker to sensory master requires expertise, cultivated through structured training, reflection, and humility rather than innate talent. The stages include accepting the scientific nature of spirit evaluation, building a sensory archive, developing a

280

lexicon, and practicing deliberate focus to isolate variables. Structured frameworks, blind comparisons, and peer calibration provide reliability and reduce bias. Interpretation requires probabilistic thinking and narrative construction. Evaluators must confront and manage bias, teach, and continually refine skills through re-exposure and reinvention. Mastery is not a final destination but a lifelong process of disciplined practice, self-awareness, and contribution. Periodically pausing to review the basic elements of sensory proficiency is a valuable aid to maintaining professionalism and accuracy. Fig. 37.2 is useful to check skill development.

Fig. 37.2	Basic Elements of Sensory Proficiency		
Domain	Skill	Training Method	Evaluator Metric
Perceptual	Detection & discrimination	Threshold & difference testing	Correct % at threshold
Cognitive	Categorization & naming	Lexicon drills, aroma mapping	Descriptor accuracy
Reflective	Bias monitoring	Blind duplicates, peer review	Self-consistency,($\Delta \leq 1$ unit)
Interpretive	Pattern recognition	Causal reasoning exercises	Logical coherence score
Communicative	Teaching & reporting	Verbalization & narrative practice	Clarity rating by peers

References

Buck, L. B. (2005). Unraveling the senses of smell and taste. *Cell, 121*(1), 9–12. https://doi.org/10.1016/j.cell.2005.02.009

Cain, W. S. (1974). Perception of odor intensity and the time-course of olfactory adaptation. *Perception & Psychophysics, 16*(2), 302–312. https://doi.org/10.3758/BF03203961

Chambers, E., & Koppel, K. (2013). Associations of volatile compounds with sensory aroma and flavor: The complex nature of flavor. *Molecules, 18*(5), 4887–4905. https://doi.org/10.3390/molecules18054887

Chaudhari, N., & Roper, S. D. (2010). The cell biology of taste. *The Journal of Cell Biology, 190*(3), 285–296. https://doi.org/10.1083/jcb.201003144

Delwiche, J. (2004). The impact of perceptual interactions on perceived flavor. *Food Quality and Preference, 15*(2), 137–146. https://doi.org/10.1016/S0950-3293(03)00041-7

Ericsson, K. A., Krampe, R. T., & Tesch-Römer, C. (1993). The role of deliberate practice in the acquisition of expert performance. *Psychological Review, 100*(3), 363–406. https://doi.org/10.1037/0033-295X.100.3.363

Ferreira, V., & López, R. (2019). The actual and potential aroma of winemaking grapes. *Biomolecules, 9*(12), 818. https://doi.org/10.3390/biom9120818

Fiorella, L., & Mayer, R. E. (2013). The relative benefits of learning by teaching and teaching expectancy. *Contemporary Educational Psychology, 38*(4), 281–288. https://doi.org/10.1016/j.cedpsych.2013.06.001

Goldstone, R. L. (1998). Perceptual learning. *Annual Review of Psychology, 49*, 585–612. https://doi.org/10.1146/annurev.psych.49.1.585

Guinard, J. X., & Mazzucchelli, R. (1996). Panelist training and performance in descriptive analysis of espresso coffee. *Food Quality and Preference, 7*(3-4), 217–229. https://doi.org/10.1016/0950-3293(96)00016-7

Herz, R. S., & Engen, T. (1996). Odor memory: Review and analysis. *Psychonomic Bulletin & Review, 3*(3), 300–313. https://doi.org/10.3758/BF03210754

Kahneman, D., & Tversky, A. (1979). Prospect theory: An analysis of decision under risk. *Econometrica, 47*(2), 263–291. https://doi.org/10.2307/1914185

Lawless, H. T., & Heymann, H. (2010). *Sensory evaluation of food: Principles and practices* (2nd ed.). Springer. https://doi.org/10.1007/978-1-4419-6488-5

Li, W., Howard, J. D., Parrish, T. B., & Gottfried, J. A. (2008). Aversive learning enhances perceptual and cortical discrimination of indiscriminable odor cues. *Science, 319*(5871), 1842–1845. https://doi.org/10.1126/science.1152837

Majid, A., & Burenhult, N. (2014). Odors are expressible in language, as long as you speak the right language. *Cognition, 130*(2), 266–270. https://doi.org/10.1016/j.cognition.2013.11.004

Mainland, J. D., & Sobel, N. (2006). The sniff is part of the olfactory percept. *Chemical Senses, 31*(2), 181–196. https://doi.org/10.1093/chemse/bjj010

Meilgaard, M., Civille, V. G., & Carr, B. T. (2007). *Sensory evaluation techniques* (4th ed.). CRC Press. https://doi.org/10.1201/9781420004334

Noble, A. C., Arnold, R. A., Buechsenstein, J., Leach, E. J., Schmidt, J. O., & Stern, P. M. (1987). Modification of a standardized system of wine aroma terminology. *American Journal of Enology and Viticulture, 38*(2), 143–146.

Parr, W. V., Heatherbell, D., & White, K. G. (2002). Demystifying wine expertise: Olfactory threshold, perceptual skill and semantic memory in expert and novice wine judges. *Chemical Senses, 27*(8), 747–755. https://doi.org/10.1093/chemse/27.8.747

Pigott, A. P., Conner, J. M., Paterson, A., & Piggott, J. R. (1993). Flavour development in Scotch whisky: Phenolic components and sensory interactions. *Journal of the Science of Food and Agriculture, 61*(3), 329–338. https://doi.org/10.1002/jsfa.2740610313

Prescott, J. (1999). Flavour as a psychological construct: Implications for perceiving and measuring the sensory qualities of foods. *Food Quality and Preference, 10*(4-5), 349–356. https://doi.org/10.1016/S0950-3293(99)00048-9

Siegrist, M., & Cousin, M. E. (2009). Expectations influence sensory experience in a wine tasting. *Appetite, 52*(3), 762–765. https://doi.org/10.1016/j.appet.2009.02.002

Stevenson, R. J. (2001). The acquisition of odor qualities: A perceptual learning approach to olfaction. *Perception, 30*(10), 1237–1248. https://doi.org/10.1068/p3254

Wansink, B., & van Ittersum, K. (2005). Shape of glass and amount of alcohol poured: Comparative study of the effect of practice and concentration. *BMJ, 331*(7531), 1512–1514. https://doi.org/10.1136/bmj.331.7531.1512

Observations on Developing Sensory Proficiency

Mastery is process fidelity, not perception sharpness. The evaluator who repeats the proper process consistently will outperform the one who "feels" more.

Exposure is data acquisition. Every aroma encountered is a data point in a lifelong sensory database. The larger the archive, the faster the recognition speed.

Reflection converts experience into knowledge. Recording and revisiting notes is the cognitive equivalent of statistical replication.

Bias is an ever-present variable. Awareness does not eliminate bias; only method does. Each blind test is a recalibration of humility.

Teaching reveals comprehension gaps. When an evaluator can guide another to the same conclusion under blind conditions, reliability has become transferable.

Pattern recognition precedes interpretation. Before language or analysis, the brain learns aroma through repetition. Recognition is not discovery; it is recall in disguise.

Consensus without calibration is a coincidence. Agreement among evaluators proves nothing until their methods align. Replication of process, not opinion, defines credibility.

Precision grows from constraint: Limiting variables—such as glass, temperature, and pour size—reduces noise and amplifies learning. Control is not rigidity; it is clarity.

Consistency is memory reinforcement. Repeated exposure under controlled conditions strengthens neural pathways, reducing perceptual noise and stabilizing recall across sessions.

Confidence is a hypothesis, not a conclusion. Every sensory judgment should be treated as provisional until verified by repetition or peer replication.

Discomfort signals growth. When aromas defy categorization, the evaluator stands at the boundary of current competence—an opportunity for expansion, not failure.

Objectivity begins with discipline. The moment a taster favors expectation over evidence, perception bends toward preference. The professional's task is to remain the instrument, not the audience.

Calibration precedes insight. Before perception can deepen, the evaluator must anchor their senses in stable references; without fixed points, refinement has nothing to which to attach.

Ambiguity is data, not failure. When an aroma sits between categories, it reveals interaction effects or borderline compounds—signals that sharpen discrimination more than clear-cut notes ever can.

Precision grows from subtraction. Eliminating distractions clarifies signal strength and accelerates the evaluator's ability to isolate the actual driver of the perception.

Chapter 38: The Language of Sensory Evaluation in Spirits

Sensory evaluation is not merely a private encounter with aroma and taste; it is a collaborative enterprise that depends on language to make individual perception legible to others. In spirits, where the volatile composition is complex and ethanol modifies perception, descriptions must be precise enough to guide technical decisions—such as fermentation control, cut points, and cask choice—yet accessible enough to support panel consensus and to communicate clearly to students and professionals. Language, then, is both a measuring instrument and a social contract; build a rigorous vocabulary, use it responsibly in group settings, and know its limits.

From Metaphor to Method

Historically, beverage description leaned on metaphor and cultural shorthand: "masculine," "feminine," "elegant," "harsh," or "old leather armchair." Such phrasing can be evocative, but it is unstable across cultures and experience levels. Modern sensory science replaces private metaphor with shared terms anchored to reference stimuli and operational definitions (Suwonsichon, 2019). The shift does not eliminate poetry; it quarantines it from the analytic record. In professional practice, metaphor may appear in consumer-facing prose, but the panel table, lab notebook, and technical report demand language that supports reproducibility.

What is an Acceptable Descriptor

Acceptable terminology meets three criteria: it names a perceivable quality, is understandable to trained peers, and can be stabilized by reference. Describing a whisky as "solvent-like (nail-polish remover)" is acceptable when it points to a recognizable sensation most panelists can reproduce upon training. Calling the same whisky "immature" or "careless" fails the test: those are judgments, not sensory attributes. "Smooth" fails for a different reason—it bundles multiple sensations (reduced trigeminal burn, lower bitterness, perceived viscosity) without specifying which qualities are being perceived. Acceptable language favors specificity, names one attribute at a time, and leaves hedonic verdicts for a separate channel. Fig. 38.1 is a simplified representation of subjective versus standardized descriptors, illustrating the differences and providing conceptual examples to help readers create a lexicon that carries the same descriptors for everyone.

The Architecture of a Professional Lexicon

A professional lexicon is more than a list; it is a living system whose terms are defined, trained, and periodically audited. A robust lexicon distinguishes at least four categories. First, basic gustatory terms—sweet, sour, bitter, salty, umami—are reserved for taste proper. Second, olfactory terms describe volatile character in families the field fully recognizes—fruity, floral, spicy, smoky, woody, cereal, nutty, lactic, sulfurous—each with finer subdivisions (e.g., fruity → apple/pear/banana/stone-fruit/citrus, and tropical). Third, trigeminal and tactile terms capture chemesthetic sensations such as burn, astringency, cooling, prickle, viscosity, oiliness, and dryness. Fourth, temporal terms specify onset, evolution, finish, and aftereffects. A lexicon assigns an operational definition to each term. It pairs it with one or more reference standards so that "banana" does not float as a vague idea but is anchored to a training stimulus presented at a known, replicable level in aqueous ethanol appropriate for spirits work (Suwonsichon, 2019, and Su et al., 2022).

Fig. 38.1 Examples of Subjective vs Standardized Descriptors in Scotch Whisky			
Subjective Descriptor	**Potential Problem**	**Standardized Lexicon Term**	**Notes for Evaluators**
Smells like a smoky fireplace	Too personal, lacks reproducibility	Peaty / Phenolic	Links aroma to chemical origin (peat smoke, phenols)
Smells like Band-Aids	Cultural/context-dependent	Medicinal	Recognized in standard whisky lexicons; consistent meaning
Candy shop sweet	Overly broad, not chemically anchored	Vanilla / Caramel	Oak lactones (vanillin) and cask sugars are specific sources
Like a wet dog in the rain	Negative, imprecise	Sulfurous / Reductive	May indicate sulfides or faults; more precise terminology
Cheap perfumy	Subjective and generational	Floral (rose, lavender)	Anchors aroma to floral volatiles; avoids personal bias
Hot and burning	Sensation is confused with flavor	Pungent / Ethanol Trigeminal	Distinguishes chemesthetic burn from authentic flavor

Cognitive Foundations and the Problem of Odor Words

Language for describing smell is sparse in many languages compared to the language for describing color or shape. Individuals, therefore, tend to reach for source-based guesses ("fermenty," "oak," "distillery floor") or hedonic shortcuts ("nice," "harsh") when their vocabulary is thin. Training works because odor memory is malleable when tied to stable verbal anchors and repeated exposure. Two cognitive hazards are persistent: suggestion and expectation. Suggestion occurs when one panelist's confident term elicits the same term in others; expectation occurs when cask type, age statement, or brand primes the lexicon before the glass is near the nose. Both distort word choice. The practical remedy is to use a blinding, randomized order, independent note-taking before discussion, and a disciplined use of references to keep words tethered to sensation rather than reputation (Olofsson & Gottfried, 2015; Majid, 2021).

Semantics Is Not Neutral: Odor Language Across Cultures

Language does not merely report perception; it also shapes it. Odor terms vary in granularity (the level of fine distinctions), boundaries (what is included within a term and what is excluded), and translatability (how well a word maps across languages and training backgrounds). The result is that two honest evaluators can mean different things by the same word—and not know it.

Three recurring hazards:

- **Polysemy disguised as precision.** Words like *"smoky" cover distinct clusters—peat smoke, wood char, tar, and incense—*each with different causes and implications. Without containment, the term drifts, and so do conclusions.

- **Metaphor creep.** Vivid metaphors ("grandfather's library") feel accurate but do not travel across panels or languages. They bias memory while eroding reproducibility.

- **Translation loss.** Some languages carve odor space differently (some finer, some coarser). Back-translated terms rarely return to their original scope; they tend to expand or shrink.

What to carry forward in practice is not more words but tighter words. A lexicon term must have: (1) a narrow core (what it certainly is), (2) a nearest-neighbor set (what it is often confused with), and (3) a brief differentiator ("charred wood, not peat smoke"). That is semantic containment—keeping a term small enough that panelists can use it the same way, semester after semester, without chasing metaphor or fashion. *Appendix I* provides the anchor points; Chapter 34 supplies the discipline that keeps those anchors from dragging.

Clarity here is not pedantry. It is what allows evaluation notes written in one program to be read and acted upon in another—across borders, markets, and time.

Building the Lexicon: From Exposure to Definition (Fig. 38.2)

Lexicons grow by deliberate exposure followed by disciplined naming. The novice begins with broad families and converges toward finer distinctions as exemplars accumulate. For "fruity," early training separates ester-driven banana/pear/apple from citrus or stone-fruit clusters. Later, separate the citrus into lemon, lime, orange, and grapefruit, each with a characteristic sensory profile. Advancement depends on repeated triangulation: revisit the reference, compare it against new spirits, and assess the term's stability over time. A term earns permanence in the lexicon when it proves discriminative (it actually helps tell products apart), reliable (different panelists use it the same way after training), and communicative (the term's definition and reference are unambiguous to others) (Suwonsichon, 2019; Su et al., 2022).

The Role of Chemical Families—With Caution

Mapping descriptors to chemical families is pedagogically useful—esters are often perceived as fruity, certain phenols as smoky or medicinal, particular lactones as coconut/woody, and some sulfur compounds as cabbage/onion/rubber. This mapping provides structure to the lexicon and helps panels anticipate co-occurrence patterns. The caution is simple: do not present causal attributions as facts unless confirmed analytically. In the record, descriptors should be sensory ("smoky, medicinal") rather than chemical claims ("guaiacol present"), unless instrument data justifies the latter. Panels are designed to produce a disciplined perception, not chemical identification.

Interpersonal Communication on Panels

Panels are social systems. Language quality depends on how people interact, not just what they smell. Effective panels are calibrated before evaluation begins: each attribute in the session has a definition, a reference, and a brief instruction on scale use. During the session, panelists record their observations individually and silently before engaging in any group discussion. When reconciling terms afterward, the facilitator's task is to surface convergences, not to impose vocabulary by authority. Disagreements are resolved by returning to the glass and, if necessary, to the reference. Dominance, deference to reputation, and rhetorical flourish are managed through a process that includes equal speaking time, evidence-based discussion, and a clear separation between sensory description and interpretation.

Preferred Phrasing and What to Avoid

Good phrasing identifies the sensory channel, the attribute, and, when relevant, the temporal location. "Orthonasal: ripe pear; palate: moderate sweetness; retronasal: light vanilla; trigeminal: medium-low ethanol burn; finish: medium length with toasted coconut." This style conveys more than "smooth and elegant" while remaining readable. Avoid hedonic judgments ("tasty," "disgusting") in analytic notes; they belong in consumer preference studies, not descriptive analysis. Avoid prestige and process claims ("well-made," "poor cut") unless separate evidence supports them. Avoid empty intensifiers and mood words ("bold," "brooding," "feminine"). When an everyday metaphor helps a novice, fence it with a primary term: "solvent-like (nail-polish remover)" is more explicit than "nail-polish remover" standing alone.

Capturing Intensity and Dynamics Without Losing Meaning

Words must work with numbers. Panels that record only terms quickly lose the ability to compare samples rigorously; panels that record only numbers lose the richness of description. A disciplined approach pairs each attribute with a clearly explained scale, used consistently across sessions. Temporal language should be more than "long" or "short": it should say where an attribute appears (attack, mid-palate, finish) and whether it grows, plateaus, or fades. Precision in temporal phrasing is essential for chemesthetic sensations (burn, prickle) and astringency, which often build with repeated sips even as olfactory notes adapt.

Dealing With Synonyms, Dialects, and Translation

Groups bring different food and drink histories to the table, which translates into different anchors for aroma words. One panelist's "plantain" may be another's "banana," or "tonka" may not be part of the culinary memory of some regions. A lexicon anticipates this by listing primary terms and accepted synonyms, then choosing one preferred label for the official record. The choice is pragmatic: select the term most likely to be recognized and trained across the broadest audience, and document alternatives so no information is lost. When translation is required, include a short note describing the reference and its sensory boundaries rather than assuming a one-to-one equivalent exists in another language (Lippi et al., 2023).

Fault Language: Necessary, but Disciplined

Spirits carry their own vocabulary of defects: sulfurous/rubbery notes, cooked cabbage/onion, excessive solvent, plastic/phenolic, rancid/soapy, musty/moldy, metallic, and more. Fault terms serve as a quality control measure, but they also carry reputational weight and can smuggle in untested causal narratives. Panels should separate the sensory naming of a fault from its origin story. "Rubbery, sulfurous" is a description; "dimethyl trisulfide from inadequate copper contact" is a hypothesis that belongs in a technical discussion, not the primary sensory record, unless corroborated elsewhere.

Documentation That Survives Scrutiny

A record that will be trusted later is written as if the future reader will not remember the session, and often, they will not. Each entry should identify the sample, context (glassware, pour volume, temperature, rest time), sensory channels covered, attributes with intensities, and temporal notes, and any uncertainties or disagreements recorded without embarrassment. A good record is an instrument for future decisions. The more the language is tied to defined attributes and references, the more useful the record is months later when the cask must be chosen or the blend repeated.

Training for Fluency

Fluency develops through a cycle of perception, naming, and feedback. Students should maintain a running, dated lexicon, in which each term is accompanied by a definition in their own words, a reference sample used in training, notes on any confusion resolved ("pear vs. apple"), and a brief list of instances where the term has been applied confidently. Re-calibration sessions prevent drift;

blind proficiency checks reveal which terms are stable and which collapse under pressure. Over time, the lexicon contracts (fewer, cleaner terms used consistently) even as the evaluator's discrimination increases. That contraction is a sign of maturity, not lingual poverty.

Limits of Language and the Duty of Restraint

Language cannot fully capture the complexity of interactions between volatile mixtures and human physiology. Individual differences in receptor genetics, learned associations, adaptation rates, and even nasal airflow alter what is perceived and how easily it is named. Context also shapes expression: glass geometry, ethanol concentration, and tasting order all modulate the signal available to the nose and palate. Recognizing these limits does not weaken description; it protects it from overconfidence. The discipline is not to say everything, but to say only what can be supported by sensation and shared training (Olofsson & Gottfried, 2015; Majid, 2021).

Putting It All Together in Practice

Professional notes in spirits evaluation should read as disciplined observation, not advocacy. They partition channels, use agreed terms, and respect temporal structure. They avoid hedonic shortcuts and process speculation unless the session's purpose is to do so. In panels, they are produced independently before consensus is attempted, and consensus is mediated by the lexicon and its references, not by authority. Over time, the shared vocabulary becomes a tool for education, as students inherit a language they can trust, one built to withstand rigorous testing.

Fig. 38.2	Building the Lexicon: Examples of Precision and Practical Limits			
Narrow Core Definition	**Nearest Neighbor**	**Differentiator**	**Reference Standard**	**Cross-Cultural Notes**
Phenolic smoke derived from peat combustion; earthy, tar-like aroma distinct from wood char.	Woody smoke	Charred wood ≠ peat smoke; peat imparts earthy and marine notes absent in char.	Guaiacol / Creosol reference (peat-smoke distillate)	Term not universal in Asian lexicons; 'smoky' preferred.
Phenolic, antiseptic aroma typical of 4-ethylphenol and related volatiles; evokes disinfectant or iodine.	Smoky	No tar or wood tone; lacks the sweetness of peated smoke.	4-ethylphenol / phenol solution	Stable across English-language whisky lexicons; limited use in non-phenolic spirits.
Woody, creamy sweetness from lactones (γ-nonalactone); characteristic of American oak cask maturation.	Vanilla	Less sweet, oilier; lacks vanillin sharpness of vanilla note.	γ-nonalactone in ethanol solution (10 ppm)	Common in bourbon and rum panels; less frequent in Asian and gin vocabularies.

Building the Lexicon with New Terms

The *lexicon* is a controlled vocabulary for aromas. A *descriptor* is a specific, referenceable term within the lexicon (e.g., *apple*), and a *modifier* narrows a descriptor (e.g., *green skin*). Any new entry must be *falsifiable*: specific enough that a competent evaluator could confirm or disconfirm it under agreed conditions. If you cannot test it, it is not helpful for professional communication. A sensory claim is falsifiable when it is:

- **Observable:** uses a concrete, referenceable descriptor (from the lexicon), not a vibe.
- **Operationalized:** states **phase** (orthonasal / palate / retronasal / finish) and **intensity** (0–10).
- **Reproducible:** another trained panel, under the same setup, could replicate or refute it.
- **Bounded by diagnostics:** if it is a **fault** claim, cites thresholds or context tests (Appendix III).
- **Separated from opinion:** keep hedonic words out of the observation line.

Rule of thumb: if a note names a defined attribute, places it in time, gives it a magnitude, and distinguishes observation from interpretation, it is falsifiable. If it judges ("nice," "balanced," "elegant"), metaphorizes without anchors, or asserts certainty without evidence, it is not. Always pair each descriptor with an intensity (0–10) and the phase in which it appears: orthonasal (before sip), palate (during), retronasal (after you swallow/exhale), or finish (lingering). First check Appendix I; if no existing term fits, propose an addition.

Evaluation Note 38.1: Steps to add "dill" to the spirit aroma lexicon

1. Capture. Use house syntax: *Green/Herbaceous, dill, 4, palate.*

2. Anchor. Prepare a simple reference standard by placing sprigs of fresh dill in ~100 mL of room-temperature water, capping it, incubating for 10 minutes, and then smelling the headspace.

3. Replicate. On a different day, have trained peers assess the same spirit and the same reference, blind to your term. If ~**70%** independently recognize the same note when cued by the reference, it is reproducible.

4. Define. Write a simple definition (how to recognize), list common confusions (e.g., not *fennel/anise*), and record observed intensity/phase ranges.

5. Map & submit. Map to the nearest **Appendix I** family (here, *Green/Herbaceous*) and add it to the lexicon. If the panel adopts it, please submit the following information for inclusion in the shared lexicon: name, family, reference preparation, distinction from the nearest term, and typical intensity range. If not adopted, you may still use it to cross-reference the closest approved term to ensure clear communication.

Bottom line: Verify the gap, anchor the term with a simple reference, replicate the findings with others, and document so that other evaluators can assess. Lexicons grow as evidence warrants.

Predictive Processing and Expectation in Olfaction

The brain predicts first, then tests.

Perception is not a camera; it is a prediction machine. The brain brings priors—probabilistic expectations from memory, language, and learning—and compares them with incoming sensory evidence. The outcome is a posterior: what you experience as the aroma and flavor of a spirit at a given moment.

Precision weighting: when priors dominate.

The brain weighs whichever source it judges more reliable: the prediction or the incoming signal. In olfaction, the raw signal is often ambiguous, noisy, and volatile-dependent—especially in ethanol-rich matrices. When the likelihood (sensory evidence) is weak or masked (e.g., high ABV, fatigue, ambient odors), the system relies more heavily on priors, including category knowledge, labels, price, cask type, age statements, regional stereotypes, and even the words someone else has just used. When evidence is clear and stable, prediction becomes more relaxed and sensation carries more weight.

Language as a prior.

A lexicon labels, but it also constrains hypotheses. Offer the term "clove," and panelists search that region of perceptual space; offer "medicinal," and they search for phenolics instead. This is helpful when terms are tight and shared (as in Appendix I), and hazardous when they are loose or loaded. Suggestion ("he said banana") and expectation ("this was sherry cask") are not bad manners; they are computational shortcuts. They become errors when they are allowed to outrun the evidence.

First-pass recognition and its risks.

"First sniff" and "first sip" judgments are predictive triumphs: the brain compresses years of learning into a fast classification. They are also where over-confident priors most easily substitute for perception. If you are told "Islay," smoke will be "found" even in its absence—because the model is powerful. Experts are not immune; expertise often strengthens priors, which helps under noise and hurts when priors are wrong.

Panel dynamics.

Groups create shared priors in real time. A confident term spoken early narrows the hypothesis space for everyone else. That can accelerate convergence when correct or lock in an error when premature. Silence is not neutrality; it simply leaves private priors unobserved.

Active sampling changes the evidence.

Sniffing is not passive. Sniff strength, duration, and timing alter the likelihood by changing airflow and volatility; sip size and dilution change the trigeminal load and retronasal access. Predictive systems depend on sampling; changing the sampling changes the evidence.

Implications for interpretation (not instructions).

- Treat labels, prices, and origin stories as priors, not proofs.
- Expect language to shape perception; your lexicon must be stable and narrow to be useful.
- Recognize that ethanol, fatigue, and context reduce sensory reliability and tilt experience toward expectation.
- Value speed and pattern recognition, but reserve certainty when first-pass models and evidence disagree.
- In written reports, separate observation (what the evidence supports) from inference (what the model proposes) because they are different computational steps.

Summary

The language of sensory evaluation is not ornamental; it is the infrastructure that allows a field to grow. When words are chosen carefully, trained against references, and used with restraint, they transform personal experience into sharable knowledge. In spirits, where perception is complex, and ethanol is both subject and confounder, the duty to be precise is higher still. Mastery of sensory language does not promise universal agreement. However, it delivers something more valuable: a reliable way to disagree, test, and converge—glass by glass—on descriptions that others can understand and reproduce.

Predictive processing does not undermine expertise; it **explains** it. Expertise is the training of priors—what to expect and when to distrust expectation. The disciplined evaluator learns to sense when prediction is carrying too much of the load and to let the evidence speak, even when it is slower or less familiar. That is not etiquette; it is a matter of cognition.

"The meanings of words are not in the words; they are in us."—S.I. Hayakawa, linguist, author

"But if thought corrupts language, language can also corrupt thought."—George Orwell (writer)

"A language is not just words. It's a culture, a tradition, a unification of a community, a whole history that creates what a community is."—Noam Chomsky (linguist)

References

Lippi, N., Bianchi, E., Martínez, A., Zhao, H., & Rossi, P. (2023). Development and validation of a multilingual sensory lexicon. *Foods, 12*(5), 1123. https://doi.org/10.3390/foods12051123

Majid, A. (2021). Human olfaction at the intersection of language, culture, and biology. *Trends in Cognitive Sciences, 25*(2), 111–123. https://doi.org/10.1016/j.tics.2020.11.005

Olofsson, J. K., & Gottfried, J. A. (2015). The muted sense: Neurocognitive limitations of olfactory language. *Trends in Cognitive Sciences, 19*(6), 314–321. https://doi.org/10.1016/j.tics.2015.04.007

Su, X., Yu, M., Wu, S., Ma, M., Su, H., Guo, F., Bian, Q., & Du, T. (2022). Sensory lexicon and aroma volatiles analysis of brewing malt. *NPJ Science of Food, 6*(1), 20. https://doi.org/10.1038/s41538-022-00135-5

Suwonsichon, S. (2019). The importance of sensory lexicons for research and development of food products. *Foods, 8*(1), 27. https://doi.org/10.3390/foods8010027

Tsachaki, M., Arnaoutopoulou, A. P., & Margomenou, L. (2010). Development of a suitable lexicon for sensory studies of the anise-flavored spirits ouzo and tsipouro. *Flavour and Fragrance Journal, 25*(6), 468–474. https://doi.org/10.1002/ffj.2007

Observations on the Language of Sensory Evaluation

Language is the most powerful calibration tool in sensory science, yet also its most significant source of drift. Every word carries both information and assumptions. Precision in vocabulary turns perception into data; imprecision turns it back into opinion—evaluators who discipline their language discipline their thought.

The lexicon is not a dictionary of flavors but a framework of operational agreements. Each descriptor must be testable, repeatable, and teachable. A term that cannot be trained or falsified has no standing in the analytic record. The strength of a professional language lies not in its eloquence but in its replicability.

Training for linguistic fluency is inseparable from perceptual training. When a taster learns to accurately distinguish among various sensations, the sensory cortex itself becomes more selective. Naming reinforces noticing; stable words carve stable percepts. The evaluator's lexicon is a mirror of neural adaptation refined through exposure, verified through dialogue, sustained through recalibration.

Cultural and cognitive biases infiltrate language more easily than they do chemistry. Metaphor, prestige terms, and unexamined synonyms transmit collective prejudice. The only remedy is conscious linguistic hygiene: define, document, and defend every word you use. Shared vocabulary is a social contract built on mutual restraint and understanding.

Ultimately, the professional evaluator's task is not to invent words but to earn them. When a descriptor survives replication across tasters, time, and translation, it becomes part of the field's shared truth. When it fails, it becomes an artifact of personal memory. Knowing the difference is the essence of sensory literacy.

Ambiguity in language reveals ambiguity in perception: when tasters reach for uncertain words, they are signaling not confusion but unresolved sensory structure that needs further investigation, not better poetry.

Chapter 39: The Professional Pathway — Formal Sensory Science Integration

Navigational Compass

Function: Provide the reader—especially undergraduate or novice researchers—with a basic foundation and guidance to achieve full professional competence, including a curated set of exercises and reference tools for further professional development.

Beyond What This Book Provides (Fig. 39.1)

Fig. 39.1	The Professional Pathway: Text Scope
What is Provided	**Not Provided – The Next Step**
Mechanistic explanations of anatomy, aroma release, ethanol–water interactions, matrix effects, psychophysics, and cross-modal interactions (Ch. 1–17)	Complete theories of experimental design, power analysis, generalized linear mixed models, and advanced panel architectures
Detection and fault vocabulary, lexicon tables, and rejection thresholds (Appendices I–II)	Full-scale consumer panel methods, regulatory sensory certification, or proprietary industry panel practice manuals
Basic test procedures (Appendix IX) with guiding principles for bias, temperature, rest, and coding	All versions of ASTM/ISO method variants, custom method development, or complete automation
Evidence-based analysis of vessel design (Ch. 36–38; Appendices VI–VII)	Complete comparative vessel trials across all ethanol levels, aging conditions, and headspace geometries

What Lies Ahead: Mapping Topics to Professional Sources (Fig. 39.2)

Fig. 39.2	Mapping Topics to Professional Sources	
Text Topics	**Recommended Source(s)**	**Purpose**
Basic discrimination tests (triangle, duo–trio, paired)	Meilgaard, Carr, & Civille, Sensory Evaluation Techniques	Shows complete test design, validation, replicates, and more variants beyond basics
Odor or taste threshold methods	ASTM E679 Standard Practice for Odor and Taste Thresholds	Gives a stepwise method for ascending series, forced-choice designs, and replicate validation
Descriptor lexicon development	ISO 11035; ASTM E253-20 (Quantitative Descriptive Analysis Guideline)	Standardizes vocabulary across labs and disciplines
Bias, repeatability, and panel training	Advanced chapters in Meilgaard et al.; ASTM E18 standards	Provides procedures for assessor screening, calibration, repeatability/reproducibility measures
Distillation, matrix effects, volatile interaction	Buglass, *Handbook of Alcoholic Beverages*; Kuno, *Introductory Science of Alcoholic Beverages*	Lays out deeper chemical pathways, production constraints, and compound interactions
Vessel design and headspace physics	This text, chapters 37–39; appendix VI	Basis for evaluating and quantifying how vessel geometry modifies aroma release

Suggested Exercises & Protocol Sources (Fig. 39.3)

Fig. 39.3	Suggested Exercises and Protocol Sources	
Suggested Exercise	**Goal and Learning Outcome**	**Protocol Source or Starting Point**
Triangle test on dilution effect	Learn detectability limits when small volumetric changes occur	Appendix IX, Chapter 11, and Meilgaard et al.
Duo–Trio on vessel shape	Compare aromatic differentiation caused by vessel geometry	Appendix IX, ISO 4120, and ASTM E1879 variants
Threshold determination of a key ester	Practice individual sensitivity, concentration series	Appendix II and ASTM E679
Time–Intensity or TCATA profiling of a spirit sample	Capture temporal perceptions (rise, decay, dominance)	Appendix IX, ISO 11036, and recent sensory journal articles
Descriptor–Chemistry crosswalk	Map common sensory descriptors (e.g., banana, clove) to volatile families and sources	Appendices I & II, and supplementary literature reviews
Assessor repeatability check	Use replicate sessions to compute correlation, identify misfitting assessors	Appendix IX and advanced chapters in Meilgaard et al.

Professional Pathways & Resource Guide

Transitioning into rigorous sensory work typically involves engaging with:

Professional societies and divisions: The Sensory & Consumer Science Division of the Institute of Food Technologists (IFT) and the Sensory Science Society offer workshops, webinars, and access to their journals.

Standards bodies, such as ASTM International (especially Committee E18 – Sensory Evaluation), publish numerous standard methods. ISO maintains relevant standards (e.g., ISO 8589—Sensory rooms; ISO 5492—vocabulary; ISO 11035—descriptor list development; ISO 11036—time–intensity).

Contract sensory labs and university centers: Many universities host sensory laboratories; industry sometimes outsources panel testing to specialized labs.

Statistical and software resources: Open-source (e.g., R packages such as sensR, PanelANOVA, agricolae) and GUI tools (e.g., JASP) offer entry points.

The Future of Sensory in Spirits

The preceding chapters have laid out the first structured framework for evaluating spirits within the discipline of sensory science. What began as a fragmented set of practices—often steeped in tradition, repetition, and commercial bias—has here been reassembled into a coherent discipline grounded in physiology, chemistry, and defensible methodology. This text has emphasized the distinctiveness of spirits, highlighting their ethanol volatility, chemesthetic effects, and the profound influence of raw materials, distillation, and cask aging. These attributes set spirits apart from the sensory science of food and wine, necessitating a dedicated framework. That framework is now complete.

It is important, however, to mark the limits of the present scope. Food sensory science encompasses a much broader range of flavor systems, textural attributes, and cultural contexts. It employs larger consumer panels, more diverse statistical tools, and a broader set of temporal and cross-modal methods. To pursue those directions fully would be to dilute the focus of this text. Here, the boundaries are drawn deliberately. This book does not attempt to replace foundational works in general sensory science, nor does it provide a statistical manual or a laboratory course in experimental design. It equips evaluators of spirits to practice with clarity, consistency, and confidence—and no further mission is required to justify that achievement.

The path forward in spirits sensory science, however, is open. Emerging questions will continue to drive inquiry. The molecular dynamics of the ethanol–aroma interaction at the glass interface remains incompletely understood and would benefit from new tools in physical chemistry. The growing availability of gas chromatography–mass spectrometry (GC–MS) and liquid chromatography–mass spectrometry (LC–MS) in distilleries and universities will enable a more systematic correlation of sensory descriptors with measurable compounds. Neuroimaging techniques such as fMRI and EEG hold promise for exploring how complex stimuli stimulate the olfactory, gustatory, and trigeminal systems in concert. Cross-disciplinary work combining data science and psychometrics integrates panel outputs with consumer behavior, clarifying the relationship between sensory fidelity and market response.

These prospects mark the beginning of further study, not the continuation of this text. For students and professionals, the essential foundation lies here; the more advanced pursuits belong in specialized research, broader courses in sensory science, or laboratory-based training. By clarifying this boundary, the book both affirms the scope of its accomplishment and signals a clear entry point for those who wish to push the frontier.

Summary and Outlook

This book accomplishes what the spirits industry has long lacked: a structured, scientific, and reproducible method for evaluating aroma, flavor, and mouthfeel in distilled spirits. It does not claim to close the field—sensory science is dynamic and future developments will expand its tools and insights—but it provides, for the first time, a reliable foundation for students, professionals, and enthusiasts alike. Instructors may adopt it as a core text in sensory courses. Distillers may use it to bring science into a domain long shaped by anecdote and marketing. Enthusiasts may employ it to sharpen their appreciation and understanding. Above all, the book demonstrates that serious sensory analysis of spirits is not only possible but also necessary and transformative. The scope of this text is complete; the next frontier belongs to those who will extend to reach the science.

References

ASTM International. (2019). *ASTM E679-19: Standard practice for determination of odor and taste thresholds by a forced-choice ascending concentration series method of limits.* ASTM International. https://www.astm.org/e0679-19.html ASTM International | ASTM+1

ASTM International. (2020). *ASTM E253-20: Standard terminology relating to sensory evaluation of materials and products.* ASTM International. https://doi.org/10.1520/E0253-20 ANSI Webstore+1

Buglass, A. J. (Ed.). (2011). *Handbook of alcoholic beverages: Technical, analytical and nutritional aspects* (Vols. 1–2). Wiley. https://doi.org/10.1002/9780470976524

International Organization for Standardization. (1993). *ISO 8589: Sensory analysis — General guidance for the design of test rooms.* Geneva, Switzerland: ISO. https://www.iso.org/standard/36385.html

International Organization for Standardization. (1996). *ISO 11035: Sensory analysis — Identification and selection of descriptors for establishing a sensory profile by a multidimensional approach.* Geneva, Switzerland:

ISO. https://www.iso.org/standard/19270.html

International Organization for Standardization. (2020). *ISO 11036: Sensory analysis — Methodology — Texture profile and time-intensity.* Geneva, Switzerland: ISO. https://www.iso.org/standard/80365.html

Kuno, M. (2015). *Introductory science of alcoholic beverages.* NTS Publishing.

Meilgaard, M. C., Civille, G. V., Carr, B. T., & Osdoba, K. E. (2024). *Sensory evaluation techniques* (6th ed.). CRC Press. https://doi.org/10.1201/9781003352082

Observations on Professional Sensory Science

The move from evaluator to professional scientist is marked not by sharper senses but by accountability. Perception becomes professional when it produces data that others can verify. Formal sensory science replaces intuition with structure—controlled design, transparent records, and defensible results—the evaluator's role shifts from opinion to evidence.

Authority rests on repeatability, not reputation. Each variable—temperature, glass, order—must be documented because credibility depends on control.

Scientific literacy anchors sensory skills. Knowing how perception behaves under test conditions lends validity to the tasting process.

Professionalism in sensory work means rigor, humility, and truth before pride. The expert no longer seeks affirmation but evidence.

Section V:

Presenting the Flavors – The Role of The Evaluation Vessel

If Science had Intervened Sooner

History's most enduring errors are not born of ignorance, but of early success mistaken for truth. Once a convenient explanation gains traction, it hardens into doctrine—accepted, defended, and passed down until no one remembers to ask if it was ever true or correct. From Aristotle's geocentric universe to centuries of bloodletting as a medical practice, progress has often stalled not for lack of intelligence, but for lack of doubt.

The same pattern echoes wherever observation is replaced by tradition: an idea once useful becomes an anchor that prevents movement. As Einstein warned, *"Concepts that have proved useful in ordering things easily achieve such authority over us that we forget their earthly origins,"* and, as Confucius cautioned long before, *"Real knowledge is to know the extent of one's ignorance."*

Scientific advancement depends not on defending what we believe, but on testing what we assume to be true. The tragedy of accepted error is not its initial mistake—it is the generations who build on it, unaware that the foundation was never sound. Nowhere is this lesson more visible than in the way the spirit world has accepted vessel design as a settled truth.

Perspective on Glassware and Sensory Integrity

The discussion of glassware in spirits evaluation is not a matter of preference or brand advocacy, but of sensory accuracy. Vessel geometry governs how volatile compounds—especially ethanol—disperse and reach the olfactory system, shaping perception before a single sip is taken. The chapters that follow emphasize this relationship because the glass is not neutral; it is an instrument whose design directly influences the data we perceive.

The flared-rim sensory glass emerged as a scientifically informed example of how vessel geometry can improve accuracy by mitigating ethanol dominance, thereby revealing hidden aroma profiles. Nevertheless, the principle extends beyond any single design: any wide-mouth, low-concentration vessel that allows natural vapor dispersion can improve accuracy.

What matters is not allegiance to a shape but adherence to physics and physiology. As research advances and more independent laboratories and distillers contribute experimental data, improved methods may emerge. The essential point is to remain open to evidence, to question assumptions, and to recognize that every advance in sensory tools—however small—moves evaluation closer to measurable truth.

Chapter 40: Historical Glassware Sensory Design and Development

Mesopotamia (2000 BCE)

The earliest known evidence of human-made glass dates to beads found in Mesopotamia and dated at 2000 BCE, in India around 1700 BCE, and in Persia around 1600 BCE. Lead oxide (chemical symbol PbO) found in the Nippur blue glass fragment from 1400 BCE is the oldest known example of leaded glass, often misnamed lead crystal. Note that glass is amorphous and has no crystalline structure (Tait, 1991). Design Intent: Slow cooling of lead oxide glass allowed longer forming times, without sensory intent, but with artistic license. The process was lost until the mid-1700s, when it was reintroduced by George Ravenscroft (Godfrey, 1975).

Greece (8th to 1st century BCE)

Wine was the beverage of choice, and several vessel designs became ritualistically important to Greek society. Common shapes were the Krater for mixing wine with water, the Skyphos, and the Kylix for individual drinking cups. Design Intent: Ritualistic symbolism, no sensory intent, cups designed to move the beverage from the Krater to the mouth (Tait, 1991).

Roman and Early Medieval Era (~1st–7th century AD)

Drinking vessels were primarily hand-blown soda-ash glass beakers with thick walls and straight sides, occasionally adorned with wheel-cut relief (Whitehouse, 1997). Design intent: Primarily to improve production and decorating capability, without sensory or flavor enhancement considerations.

Medieval Islamic and European Glass (10th–13th century)

Hedwig–style beakers from Norman Sicily or Islamic workshops featured squat, thick-walled forms with decorative motifs (Carboni, 2001). Design Intent: Visual prestige over sensory, minimal impact on aroma or mouthfeel.

Renaissance Venetian Glass (15th–17th century)

Murano glassmakers in Venice perfected clear "cristallo," made with crushed quartz river pebbles that have a high refractive index, yielding rainbows and sparkling prisms at sharp edges, and elaborately shaped goblets with elegant bowls and ornate stems (e.g., wings, fruited prunts). Design Intent: Refined tactile and *visual* sensory appeal. Thin walls, clarity, refined shapes, enhanced visual appearance, and drinking experience: no sensory intent, visual only (Tait, 1991).

Renaissance to Early Modern Glass (also 15th–17th century)

Across Europe, Berkemeyer and Römertyp stemware emerged, with prunted stems and flared bowls (Eales, 1985). Design Intent: Larger stems were intended to improve grip, but any sensory impact was incidental.

Leaded Crystal Glassware (17th century onward)

Schooled in Venetian glass techniques, George Ravenscroft (1618-1681) pioneered English leaded crystal by reintroducing lead oxide in 1674 (Godfrey, 1975). The English "crystal" became prized for its brilliance, clarity, and thin rims—by creating glassware similar to the elegant Murano table settings of European and Russian royalty, yet affordable to the middle class. Longer forming times produced longer stems, thereby enhancing the visual appearance of the glass and its contents in candlelit areas. Design Intent: Easier forming and longer stems improve appearance and ergonomics (Klein & Lloyd, 1989). This opened the door for sensory design. However, 20th-century health concerns over the harmful leaching of lead into beverages and subsequent legal bans made its usage less common (Landrigan & Todd, 1994).

Georgian & Excise-era Britain Glass (18th century)

Georgian glass created baluster stems (bulbous "knops" added for grip enhancement). In 1746, the British Government imposed an excise tax based on weight, targeting lead oxide glass, resulting in smaller, thinner, and lighter glasses and hollow stems **(Charleston, 1984)**. Design Intent: Lighter forms with longer stems preserve temperature; an early attempt at sensory enhancement, with visuals improved by capturing more candlelight, enhancing color, and adding sparkle.

Champagne and Cocktail Stemware (19th–early 20th century)

The introduction of the flute and coupe (MacNeil, 2015). Design Intent for Flutes: Control CO2 release to prolong effervescence and enhance visual appeal. Design Intent for Coupes: Visual display of bubbles at the expense of aroma. Today, the surviving flute is in question for optimal sensory delivery, as many opt for larger rim/bowl white wine glasses for improved nosing space.

Late 20th–21st Century: Functionally-Driven Sensory Glassware

Major manufacturers employ marketing attempts to sell more glassware by introducing varietal-specific wine glasses designed by carefully selected "experts" who perform in-house empirical studies to arrive at marketable glass shapes, with no consideration for the dynamics of sensory science and relying on the long-proven erroneous tongue-map as a marketing tool (Guinard & Pangborn, 1987). A single effort by the ISO (International Organization for Standardization) resulted in the release of a wine-glass standard for wine evaluation, rather than for scientific sensory analysis (ISO, 1977). Riedel changed the wine industry with a single sensory achievement, creating a large headspace in the wine glass (Burgundy Grand Cru), reducing the numbing effect of ethanol in low ABV wines, and allowing better aroma detection (Delwiche, 2004), marking the days of popularity for the ISO glass and copita-shaped wine glasses.

Case Study 40.1: The first sensory alcoholic beverage glass: In the 1950s, Claus Josef Riedel transformed wine glass design by placing aroma performance above decorative tradition. Rejecting heavy cut crystal, he spent sixteen years studying how bowl curvature, volume, and rim taper affect vapor concentration and palate delivery. His breakthrough came in 1958 with the Grand Bourgogne (also known as the Burgundy Grand Cru) glass, specifically designed for Pinot Noir and Burgundian red wines. The design featured a vast bowl with a gently inward-tapered rim, maximizing surface area for volatilization while concentrating aromas toward the nose. This geometry enhanced the wine's complex bouquet and softened acidity through balanced air exposure, making it the first scientifically tuned, varietal-specific wineglass. Exhibited at the Brussels World's Fair, the glass won a gold medal and was added to the Museum of Modern Art's collection as a model of functional minimalism. By prioritizing olfactory optimization over aesthetic flourish, Riedel redefined glassware as a sensory instrument rather than an ornament. The Burgundy glass became the foundation of a new era linking glass engineering with the science of wine perception.

Sensory, Experimental, and Neuromorphic Designs (21st century)

In 2012, Arsilica, Inc. introduced a sensory-designed spirits glass applying Graham's law of diffusion to divert pungent, anesthetic ethanol for better olfactory detection, identification, and evaluation of spirits. Each dimension was examined for its contribution to controlling the sensory experience. In full disclosure, this book's author is a co-inventor of the glass.

In the near future, contemporary concepts like cymatics cups—glasses whose liquid surfaces are dynamically shaped by sound—represent a multimodal sensory frontier, supporting the growing interest in integrating sensory stimulation with tasting (Spence & Gallace, 2011). What does the future of artificial intelligence (AI) hold for vessel product design? As with other products, it will be incredible as the world wakes up to science.

Summary

The history of glassware design reveals a gradual evolution from utility and ritual to visual prestige and, eventually, to sensory purpose. Ancient Mesopotamian and Greek vessels were functional or symbolic, with no sensory intent, whereas Roman and medieval glass reflected the production capabilities and cultural aesthetics of their respective times. Venetian cristallo and English leaded crystal emphasized clarity, brilliance, and elegance, elevating glassware as a status symbol. The 18th–19th centuries saw shifts driven by taxation, ergonomics, and sparkling wine traditions, although they were still not truly sensory-focused. In the late 20th century, varietal-specific wine glasses emerged, marketed as scientifically justified but often built on flawed tongue-map theories. The first true sensory breakthrough came with large-headspace wine glasses (Riedel) and later with the NEAT spirits vessel in 2012. Looking ahead, experimental designs such as cymatics-inspired vessels represent the frontier of multisensory glassware, merging science and experience.

"Glass is the most magical of all materials. It transmits light in a special way, unlike any other substance."— Dale Chihuly, renowned glass artist

"The Venetians did not merely make glass, they made it sing with light."— Hugh Tait, author

References

Carboni, S. (2001). *Glass from Islamic lands.* Thames & Hudson / The al-Sabah Collection, Kuwait National Museum.

Charleston, R. J. (1984). *English glass and the glass used in England, c. 1500–c. 1800.* HMSO.

Delwiche, J. (2004). The impact of perceptual interactions on perceived flavor. *Food Quality and Preference, 15*(2), 137–146. https://doi.org/10.1016/S0950-3293(03)00041-7

Eales, J. (1985). *Glass through the ages.* Ward Lock Educational.

Godfrey, E. (1975). *English glassware of the seventeenth and eighteenth centuries.* Arco Publishing.

Guinard, J. X., & Pangborn, R. M. (1987). The influence of glass shape on wine flavor. *Journal of Sensory Studies, 2*(2), 133–145. https://doi.org/10.1111/j.1745-459X.1987.tb00401.x

International Organization for Standardization. (1977). *ISO 3591: Sensory analysis — Apparatus — Wine-tasting glass.* International Organization for Standardization. https://www.iso.org/standard/9004.html ISO+2Iteh Standards+2

Klein, D., & Lloyd, W. (1989). *Glass: A world history.* Abrams.

Landrigan, P. J., & Todd, A. C. (1994). Lead in drinking glassware: A hidden hazard. *New England Journal of Medicine, 331*(25), 1669. https://doi.org/10.1056/NEJM199412223312520

MacNeil, K. (2015). *The wine bible* (2nd ed.). Workman Publishing.

Spence, C., & Gallace, A. (2011). Multisensory design: Reaching out to touch the consumer. *Psychology & Marketing, 28*(3), 267–308. https://doi.org/10.1002/mar.20392

Tait, H. (1991). *Five thousand years of glass.* University of Pennsylvania Press.

Whitehouse, D. (1997). *Roman glass in the Corning Museum of Glass* (Vol. 1). Corning Museum of Glass.

Chapter 41: Tulip Glass Evolution — The Myth of Concentrating Aromas

Vessel shapes for spirits have traditionally been designed to satisfy one of two objectives: (1) to appeal to one's sense of artistic style or (2) strictly as an intermediary vessel to transport a beverage from container to mouth. Sensory science is a relatively new field, and little has been published regarding the potential of shape to display or manipulate aroma for identification and detection. The tulip shape has long been the traditional shape for spirits glasses. However, its origin did not begin with spirits, but with fortified wines. It evolved through centuries of cultural habits, colonial trade, post-war industry shifts, and ultimately, commercial marketing identity. Today, it stands as the most recognized vessel for whiskey tasting, yet its basic design limits its use to tradition, as its inherent shape works against diagnostics and evaluation. The standard justification for the tulip is that its tiny rim size collects all aromas for easy detection. However, that justification overlooks the significant presence of anesthetic, pungent ethanol, implying that ethanol's pungency is also a flavor when, in fact, it is a trigeminal irritant.

Definition of a Tulip Vessel

Tulip-shaped vessels have smaller cross-sectional areas at the rim than at the maximum bowl diameter, forcing vapor molecules closer together at the rim plane. They also have a high aspect ratio (height/width > 2.0), so fewer molecules with masses greater than ethanol's reach the rim for detection and evaluation.

1700s – The Copita and the Sherry Trade

In Jerez, Spain, the copita emerged as the workhorse of the sherry industry. This small, stemmed glass was used at the wharf to verify sherry quality and proof before trade deals were sealed— earning it the nickname "dock glass." Over time, the copita's compact tulip shape was adopted throughout Spain, Portugal, and their colonies for fortified wines like sherry, port, and *Madeira*. It became the standard drinking vessel for high-strength wines (18–22 % ABV), admired for its ability to concentrate aromas and support small servings (Kissack, n.d.).

1800s – The Copita Crosses into British Households

As fortified wines gained popularity in the U.K., the copita's tulip shape followed. It was adopted in upper- and middle-class British households as the preferred glass for wine and port, particularly suited to the standard single-serving size of fortified wines. This widespread domestic familiarity would later inspire the shape chosen for the ISO wine-tasting glass standard (Unwin, 1991).

Early 1900s – Tulip Shape Becomes the Scotch Lover's Glass

The crossover into the spirit world began subtly. With sherry and port firmly entrenched in the

UK's drinking culture, the same households targeted by the scotch industry already had tulip-shaped glasses. For Scotch whisky, also a strong, aromatic spirit, the copita's shape proved convenient—small, elegant, and already present in homes. Rather than invent a new glass, distillers and drinkers gravitated toward what was already available, avoiding the costs of development and consumer re-education (MacLean, 2003).

1950s – Post-War Spirits and International Expansion

After World War II, malted barley spirits, such as Scotch whisky, gained global traction—not only in Europe but also in the United States, Japan, and India. The tulip-shaped glass came along, piggybacking on the growing prestige of Scotch and the legacy of sherry glassware. It became an international symbol of "correct" drinking, despite its wine-based origin (Milne, 2020).

1960s – American Resistance and Marketing the Tulip

Selling Scotch in the United States proved challenging. The country was still deeply entrenched in a cocktail culture born of Prohibition. Between 1920 and 1933, illegal and often dangerous spirits (e.g., "bathtub gin") drove drinkers to mix alcohol with sugar, citrus, and ice to mask harsh or poisonous flavors. Even after Prohibition was repealed, cocktails remained popular throughout the decades and continue to be popular today (Wondrich, 2007). Unlike Europeans who drank spirits neat, Americans were reluctant to embrace the pungency of Scotch.

Scotch whisky marketers needed a solution—not just to change drinking habits, but to elevate perception of straight whisky at drinking room temperature without water, mix, or ice. The tulip-shaped glass was promoted as the "proper" way to enjoy straight Scotch, suggesting sophistication, aroma control, and European tradition. However, the shape increased ethanol's harshness by directing vapors into the nose, while giving the appearance of a refined ritual (Williams, 2024). The sharp pungency was mitigated through several acclimation techniques, but these "fixes" did not resolve the trigeminal irritation associated with ethanol, which detracts from enjoyment and aroma evaluation. Drinking Scotch on the rocks in a large tumbler was later encouraged, as was adding water to dilute Scotch in tulips, and instructing drinkers to breathe through both the mouth and nose simultaneously to reduce the ethanol's impact on the olfactory system. Pre-tulip in America, whiskey and Bourbon drinkers were satisfied with a large-mouth tumbler for straight spirits, and the same glass could hold an ice cube or two to cool the spirit, thereby reducing its pungency. No special glassware requirement needed, and they were better off without tulips.

1977 – ISO 3591:1977 and the Confusion of Standards

In 1977, the International Organization for Standardization (ISO) introduced ISO 3591: *Sensory*

Analysis—Apparatus—Wine-Tasting Glass. This specification codified a glass with a tulip-like shape designed for standardized wine evaluation (International Organization for Standardization, 1977). Seizing the opportunity, glassware companies began branding lookalike glasses as "ISO whisky glasses" even though ISO had never authorized or defined a version for spirits. Educational organizations, such as the Court of Master Sommeliers and the WSET, began using these glasses, reinforcing a false sense of legitimacy. The "ISO spirits glass" was never officially sanctioned, never existed, yet became widely accepted through clever marketing and similarity to a copita.

2001 – Glencairn: The Tulip Rebranded

In 2001, the Glencairn Glass was introduced by Raymond Davidson and endorsed by five prominent Scotch whisky master blenders. Based on a copita (therefore tulip) shape, the Glencairn has become the most successful spirits glass in history. It received the Queen's Award for Innovation in 2006 and the Queen's Award for International Enterprise in 2012 (MacLean, 2003). Glencairn solidified the tulip shape as the default and spirits industry icon for whiskey tasting, primarily through branding and strategic industry adoption rather than sensory validation.

2000s–Present: Criticism and Resistance to Change

Despite its popularity, scientists and consumers have raised questions about the tulip glass's performance. Ethanol vapor is known to concentrate near the rim in the headspace above the liquid surface due to the bowl's sides converging to a tiny rim diameter. The ethanol content in headspace aromas from over 40% ABV spirits has been estimated to range from 65–75% due to ethanol's volatility. This creates a sensory flaw: ethanol's volatility and anesthetic properties suppress the sensitivity of aroma receptors, thereby dominating the nose (MacLean, 2003).

A recent peer-reviewed study confirmed that increasing ethanol concentrations in the headspace measurably and significantly reduces the detectability of a spirit's characteristic aroma compounds (Lee, Paterson & Piggott, 2001).

Additionally, a master's thesis from Edinburgh Napier University reported that high-strength spirits have a greater olfactory impact when ethanol overwhelms the aromatic profile—a problem exacerbated by glassware that directs volatiles directly upward (Williams, 2024). Enthusiast reviews on spirits forums have criticized the tulip, calling it the "ethanol bomb" of spirits.

Still, most modern glassware designs retain the tulip's narrow rim and tall bowl (Fig. 41.1). Driven by tradition, branding, and caution, the industry resists change—even at the expense of olfactory accuracy. New product design, development, and marketing introduction is costly, and the fear of rejection caused by any departure from accepted tulip size, shape, and capacity is too big a risk for product manufacturers who have relied on style for centuries and are unaware of the science of controlling pungency and displaying aromas through technical and science-based means.

The common statement that tulips concentrate all aromas for detectability ignores the fact that most of the concentrated aroma molecules are pungent, anesthetic ethanol, which significantly affects aroma detection and identification.

Fig. 41.1   Evolution of the Tulip as a Spirits Vessel

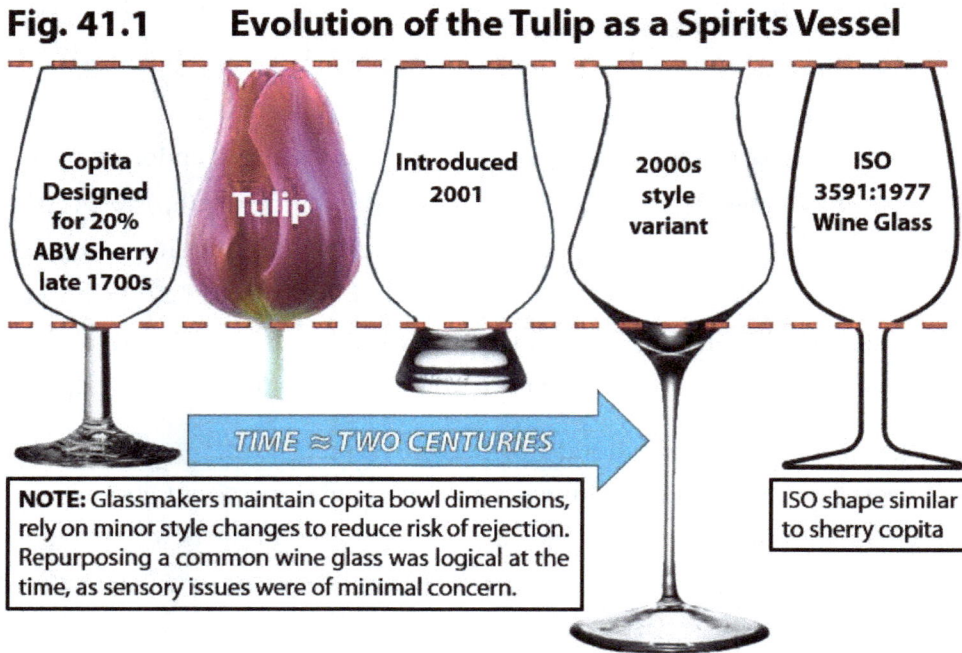

Copita Designed for 20% ABV Sherry late 1700s

Tulip

Introduced 2001

2000s style variant

ISO 3591:1977 Wine Glass

TIME ≈ TWO CENTURIES

NOTE: Glassmakers maintain copita bowl dimensions, rely on minor style changes to reduce risk of rejection. Repurposing a common wine glass was logical at the time, as sensory issues were of minimal concern.

ISO shape similar to sherry copita

Summary

The tulip-shaped spirits glass—originally a practical trade confirmation vessel for sherry—rose through history on the coattails of colonial commerce, household familiarity, and mass marketing. Its current status as the "official" whiskey glass was not born of scientific necessity, but from path dependency and strategic positioning. Its functional limitations are key to its future acceptability in evaluation and diagnostics, emphasizing the need for better, purpose-built alternatives.

"Man prefers to believe what he prefers to be true." — Francis Bacon, philosopher

"Design is not just what it looks like and feels like. Design is how it works." — Steve Jobs

"Form follows function." — *Louis H. Sullivan*, American architect, 1896.

"Design is intelligence made visible." — *Alina Wheeler*, branding and design strategist.

References

Chilled Magazine. (n.d.). *Inside the tulip: Evaluating traditional whiskey glassware. Chilled Magazine.* Retrieved from https://chilledmagazine.com [Trade source, not peer-reviewed]

International Organization for Standardization. (1977). *ISO 3591: Sensory analysis — Apparatus — Wine-tasting glass.* Geneva, Switzerland: Author. https://www.iso.org/standard/9004.html

Kissack, C. (n.d.). Sherry glassware: The copita. *The Wine Doctor.* Retrieved from https://www.thewinedoctor.com [Trade/gray literature, not peer-reviewed]

Lee, K., Paterson, A., & Piggott, J. R. (2001). Origins of volatile compounds in Scotch malt whisky. *Trends in Food Science & Technology, 12*(10), 391–401. https://doi.org/10.1016/S0924-2244(01)00127-5

MacLean, C. (2003). *Scotch whisky: A liquid history.* Cassell Illustrated.

Milne, D. (2020). *Whisky and the global market.* Edinburgh University Press.

Poisson, L., & Schieberle, P. (2008). Characterization of the most odor-active compounds in an American bourbon whisky by AEDA. *Journal of Agricultural and Food Chemistry, 56*(14), 5813–5819. https://doi.org/10.1021/jf800382m

Unwin, T. (1991). *Wine and the vine: An historical geography of viticulture and the wine trade.* Routledge.

Williams, T. (2024, Month Day). Inside the tulip: Evaluating traditional whiskey glassware. *Chilled Magazine.* Retrieved from https://chilledmagazine.com

Wondrich, D. (2007). *Imbibe!* Penguin Books.

Observations on the Tulip Vessel for Drinking Spirits

The tulip glass—more a product of tradition than of science—has long symbolized refinement while quietly undermining sensory accuracy. Its narrow rim and inward taper compress vapor flow, forcing ethanol-rich air directly into the nose. In spirits at or above 40% ABV, this creates a harsh, anesthetic effect that suppresses olfactory receptor sensitivity and masks delicate esters, aldehydes, and phenols that define character.

Rather than concentrating on aroma, the tulip concentrates on ethanol, producing a false sense of intensity and a shorter sniff duration. The small bowl restricts surface area and prevents controlled volatilization. What results is distortion—a pungent burn mistaken for complexity. Evaluators compensate by adding water, altering breathing, or shortening exposure, all unsatisfactory workarounds to justify a flawed geometry.

Its survival rests on ritual, not performance. Adopted from wine and fortified-wine traditions, the tulip persists due to its aesthetic appeal and marketing inertia. In truth, a proper diagnostic vessel must disperse ethanol and separate volatiles by weight, allowing meaningful access to aroma. The tulip remains an icon of elegance—but also a poignant reminder of sensory misunderstanding and dysfunction.

Perspective: The Evaluator as a Measuring Instrument

Every sensory evaluation is a measurement, and the evaluator functions *as* the measuring instrument. The human sensory system, like analytical equipment, requires calibration and maintenance to ensure reliability. Olfactory receptor neurons fatigue with repeated stimulation, and interpretation shifts in response to expectation, context, and memory. Vivid aromas can dull within seconds—a physiological adaptation that reduces neural response intensity.

Consistency in evaluation, therefore, depends on deliberate human calibration. Trained assessors re-establish baselines by inhaling neutral air, referencing known standards, and maintaining hydration and rest cycles. Panel protocols that fix sample order, exposure duration, and sniff spacing are not rituals; they are statistical controls that limit perceptual variance. The professional evaluator treats sensory acuity as a measurable capability—tracking personal thresholds over time, noting adaptation trends, and re-aligning perception using validated references. Control and calibration transcend opinion and approach metrology.

Controlling the Enemy of Accurate Evaluation: Signal-to-Noise Ratio

Instruments are designed to maximize signal-to-noise ratio by reducing noise; therefore, the evaluator must do the same—reduce sensory, environmental, and cognitive noise to achieve higher perceptual resolution." "Signal" is the set of volatile cues that communicate origin, material, and process; "noise" is everything that interferes—ethanol pungency, temperature imbalance, ambient odor, expectation, or fatigue. High noise masks subtle signals; low noise clarifies them.

Glass geometry, sample temperature, and tasting protocol all serve as filters that control this ratio. A wide bowl with a small rim diameter amplifies aroma density but can also trap ethanol, increasing chemical noise. Poor ventilation, perfume contamination, or inconsistent sniff timing can all contribute to environmental noise. Evaluator bias and language priming add cognitive noise, distorting perception before the stimulus even reaches conscious awareness.

Improving signal-to-noise ratio requires mechanical, chemical, and behavioral control: Stable ambient air, neutral surroundings, measured glass geometry, correct pour volume, rest intervals, and calibrated expectation. When these conditions align, the evaluator perceives a sharper image of the spirit's actual aromatic structure—the sensory equivalent of increasing resolution in a microscope. Signal-to-noise management, not poetic description, defines the difference between hobbyist tasting and sensory science.

Chapter 42: Diverting Ethanol from the Olfactory—Vessel Design

Ethanol is both the defining psychoactive molecule of spirits and the greatest obstacle to accurate sensory evaluation. While its volatility transports aroma to the nose, ethanol vapor overwhelms odorant receptors through trigeminal irritation, anesthetic masking, and intensity, thereby hindering aroma detection and identification. Conventional tulip glasses with tiny rims and tall profiles exacerbate the problem by concentrating ethanol vapors at the rim plane. Peer-reviewed studies have shown that headspace ethanol can exceed 65–75% for a 40% ABV spirit, creating a sensory blockade to aroma detection. Fig. 42.1 compares ethanol vapors in an ordinary tulip glass versus an existing sensory glass engineered to diffuse ethanol. The importance of ethanol diffusion in accurate evaluation is discussed in depth in Chapters 10 and 11.

Fig. 42.1 Atmosphere Composition by Vessel Design (40% ABV Spirit)

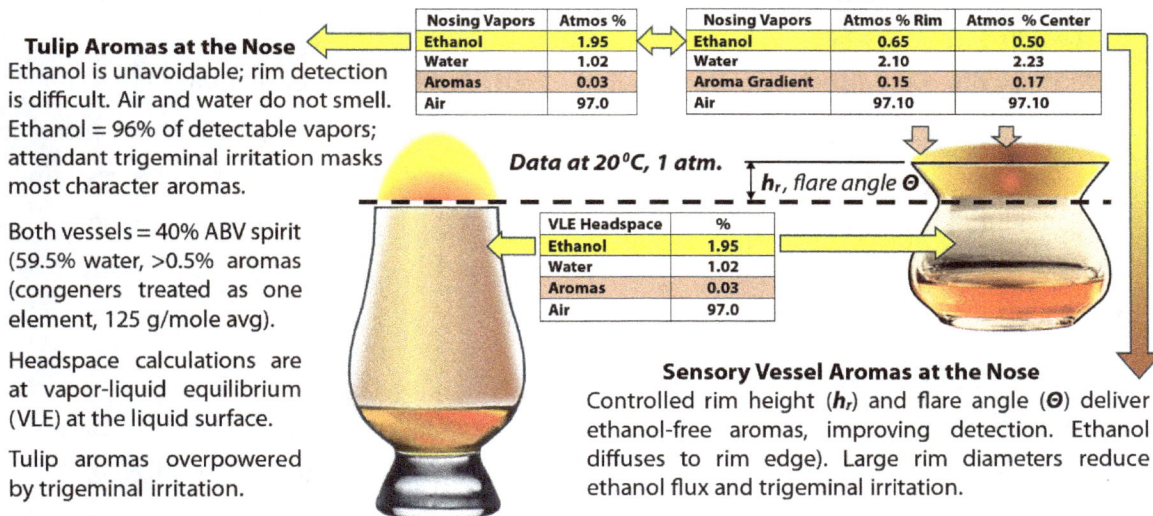

Tulip Aromas at the Nose
Ethanol is unavoidable; rim detection is difficult. Air and water do not smell. Ethanol = 96% of detectable vapors; attendant trigeminal irritation masks most character aromas.

Both vessels = 40% ABV spirit (59.5% water, >0.5% aromas (congeners treated as one element, 125 g/mole avg).

Headspace calculations are at vapor-liquid equilibrium (VLE) at the liquid surface.

Tulip aromas overpowered by trigeminal irritation.

Nosing Vapors	Atmos %
Ethanol	1.95
Water	1.02
Aromas	0.03
Air	97.0

Nosing Vapors	Atmos % Rim	Atmos % Center
Ethanol	0.65	0.50
Water	2.10	2.23
Aroma Gradient	0.15	0.17
Air	97.10	97.10

Data at 20°C, 1 atm.

h_r, flare angle Θ

VLE Headspace	%
Ethanol	1.95
Water	1.02
Aromas	0.03
Air	97.0

Sensory Vessel Aromas at the Nose
Controlled rim height (h_r) and flare angle (Θ) deliver ethanol-free aromas, improving detection. Ethanol diffuses to rim edge). Large rim diameters reduce ethanol flux and trigeminal irritation.

Sensory Ethanol Management

This specific sensory glass was designed and patented in 2011 to mitigate excessive olfactory ethanol by leveraging the laws of physics (gaseous diffusion). The vessel brand name is NEAT, an acronym for (Naturally Engineered Aroma Technology). Sensory glass refers to any vessel engineered to minimize ethanol interference through passive diffusion.

- **Headspace volume** – Located below the neck, headspace is a "holding tank" for aromas as the vapor pressure pushes molecules upward through the neck.
- **Small neck area** – Increases molecular impingement (collisions), imparting higher velocities to lower mass compounds (Graham's Law of Diffusion). Ethanol, among the lowest-mass

compounds and the most abundant, diffuses faster than higher-mass aromas (conservation of momentum). Some aromas may *appear* less intense due to the absence of ethanol pungency.

- **Controlled Diffusion Chamber** – Flare angle and rim height above the neck control diffusion rates, displays all aromas by molecular mass, with ethanol diffusing to the rim edge. Large rim area lowers ethanol flux by 67% compared to standard tiny-rim tulips.

- **The "Sweet Spot"** –At the center of the rim plane, ethanol concentration is undetectable, and at the rim edge, much less than one-half of 1% (Fig. 42.1); creating a functional nosing space sans ethanol, and improving the evaluation process by:
  1. Removing ethanol trigeminal irritation improves focus and general tasting comfort.
  2. Exposing subtle aromas masked by trigeminal irritation.
  3. Permitting the evaluator to focus on detection and identification.
  4. Eliminating the error-prone process of stepped 5% dilutions commonly used in laboratory tulip glasses to attempt to uncover compounds obscured by concentrated ethanol.
  5. Avoiding olfactory fatigue due to ethanol anesthesia.
  6. Shifting the ethanol away from orthonasal to retronasal "finish" evaluation.

- **Ergonomics** – With lips placed on the rim, most noses tilt into the rim plane center, ensuring consistent sampling placement for all panel evaluators. Holding a glass by the neck restrains unwanted hand heat. However, the sample can be quickly hand-warmed to release more aromas, replacing profile-altering stepped dilution. Larger bowl diameters significantly improve aroma release by increasing swirling velocity.

The Importance of Headspace and Expansion Volume

Relating headspace physics to laboratory data bridges the gap between chemical quantification and sensory reality, ensuring that measurements support perception rather than replace it. Analytical instruments quantify volatiles under idealized conditions—controlled temperature, carrier gas flow, and equilibrium sampling—whereas human evaluation occurs in a turbulent, ethanol-rich micro-climate above liquid surfaces. Gas-chromatographic data indicate what is *present*; the glass headspace reveals what is *perceived*.

Engineering vessel dimensions improves spirit perception by controlling molecular diffusion and displaying aromas within a functional nosing space in the rim plane. The headspace collection of random aroma molecules is refined and displayed in order of molecular mass for easy detection and identification. In vessels that do not control aroma expansion and display, the aroma is vented and mixed with the pungent ethanol molecules, leaving many aromas undetectable.

Estimating Aroma Position in Rim Plane (Fig.42.2)

The mathematical calculation (App. XII) of the actual aroma concentration at any given position in the rim plane relies on many proven laws of gas behavior. Using Graham's Law to calculate aroma position reduces evaluators' uncertainty and provides greater confidence in detection, especially when verifying data from available GC–MS studies. **Graham's Law of Diffusion is** expressed as the ratio of diffusion velocities, equal to the square root of the inverse ratio of their molecular masses. A simple derivation provides a valuable tool for organizing aromas.

$$\frac{\text{Rate}_1}{\text{Rate}_2} = \sqrt{\frac{M_2}{M_1}}$$

Where $\text{Rate}_2 = 1$, the assumed rate for ethanol, and $M_1 = 46.07$
M_2 = molecular mass of the aroma. Solve for Rate_1

Order the rates in a table, estimate rim plane location and sniff at the estimated location to verify.

- **High mass aromas** appear near the center point of the rim plane
- **Low mass aromas** appear outward, or closer to the rim edge
- **Orderly dis**play of aromas by molecular mass (Fig. 42.3)

Tulips force the evaluator to seek specific aromas while inhaling pungent, nose-numbing ethanol, which hinders accurate evaluation and masks subtler aroma detection, as every sniff contains anesthetic ethanol. Sniff sampling occurs at the rim, not the constriction, as in a tulip glass.

Fig. 42.2　　　　　**Aroma Detection in Sensory Glass**

Ethanol	Furfurol	Gualacol	Vanillin	Syringaldehyde
Low pungency	Almond, grain	Smoky	Vanilla, cream	Woody, spicy

Close mouth, move nose slowly, across the rim plane, from center to edge.
Placing the nose at or below the neck yields ethanol pungency and burn similar to tulip.

Aroma cloud at the rim plane is the "sweet spot." All aromas clearly displayed Trigeminal irritation is gone, and pungent, anesthetic ethanol is diffused. Swirl prior to sniff, cover to reach headspace equilibrium.

Aroma Separation and Display

Fig. 42.3 illustrates how the aromas are displayed at the rim plane, unencumbered by ethanol for sampling. Higher mass aromas appear near the center, while lighter mass aromas appear further away from the center. Note that relative diffusion rates increase as molar mass decreases.

Compound	Aroma	MM (g/mole)	Diffusion Rate	Rim Plane	Type
β-Damascenone	rose, grape, earthy, green floral	190.29	0.49	center	ketone
Syringaldehyde	spicy woody	182.17	0.50	center	aldehyde
Gamma-Decalactone	peach, apricot	170.25	0.52	center	lactone
Eugenol	medicinal, spicy cloves	164.20	0.53	center	phenolic
cis-3-Methyl-4-octanolide	spicy coconut	156.23	0.54	center	lactone
trans-3-Methyl-4-octanolide	wood, coconut, honey	156.23	0.54	center	lactone
Vanillin	sweet vanilla, creamy	152.15	0.55	mid	aldehyde
Ethyl Hexanoate	sweet apple, pineapple, waxy	144.21	0.57	mid	acid ethyl ester
2-methyl-3-(methyl-disulfanyl) furan	meat, sulfur	140.14	0.57	mid	heteroaromatic
Isoamyl Acetate	banana, pear, ripe	130.18	0.59	mid	acid ethyl ester
Sotolon (caramel furanone)	caramel, maple, fenugreek	128.13	0.60	mid	lactone
Guaiacol	peat smoke, spice, leather	124.14	0.61	mid	phenolic
Phenethyl alcohol	floral, rose	122.16	0.61	mid	alcohol
Ethyl isobutyrate	butterscotch	116.16	0.63	mid	acid ethyl ester
Meta-Cresol	band-aid	108.14	0.65	mid	phenolic
Ortho-Cresol	medicinal, sweet phenolic	108.14	0.65	mid	phenolic
Para-Cresol	medicinal, sweaty, pig-sty	108.14	0.65	mid	phenolic
Ethyl propanoate	fruity, rum, molasses, eggnog	102.13	0.67	mid	acid ethyl ester
Hexanal	green grass, vegetal, leafy	100.16	0.68	mid	aldehyde
Furfural	almond, grain	96.08	0.69	mid	aldehyde
Diacetyl	butter	86.09	0.73	rim	vicinal diketone
2-methyl propanol	malty, dark chocolate	74.12	0.79	rim	aldehyde
Ethyl Alcohol	ethanol	46.07	1.00	rim	alcohol

Fig. 42.3 Sensory Glass Rim Plane Aroma Location (Whiskey Sample)*

*Illustrative sample whiskey using identified GC-MS compounds.

The Importance of Swirling

Swirling plays a significant role in releasing aromas, as wine aficionados have known for centuries. Swirling is difficult at best with tall, narrow, high-profile glasses (Fig. 42.4).

Fig. 42.4   Aroma Release: Swirling and Shape Dependent

Shear releases aromas, aided by Marangoni effect and vertical shear as liquid descends from sides. Compounds are remixed.

- Small bowl diameters limit swirling. Less liquid surface area + less wetted surface = less aromas

- Large bowl diameters = large liquid surface area + large wetted surface = more aromas

h_s = swirl height   Surface Area $A = \pi r^2$ Wetted Area $A = \pi d h_s$

Swirling a wide vessel releases more aromas than a narrow vessel

highest swirl level

h_s

liquid level

Small vessel diameters result in slower swirling velocities, leading to suboptimal aroma release.

The double-edged sword of "do not swirl" advice given to tulip users to justify the higher concentration of ethanol ignores the important methodical step (swirling) that ensures as many aromas as possible reach the nose for detection and identification. Tall aspect ratios and convergent rims swirl poorly. Appendix XIII describes the physical science of swirling in depth.

Aroma Compound Molecular Behavior

Each aroma compound has its own unique geometry, chain length, polarity, and other attributes that influence its motion and determine its velocity, momentum, and kinetic energy. All of these factors play an important role in aroma location and in how to deliver them to the nose for detection and identification (Fig. 42.5). Once ethanol diffuses to the rim edge of the glass (as in the sensory glass), aromas are released. The simple view that vessel function is to hold the spirit for delivery to the mouth is challenged by the concept that vessel shape and dimensions control the availability of aroma—the result: ethanol < 4% at the rim.

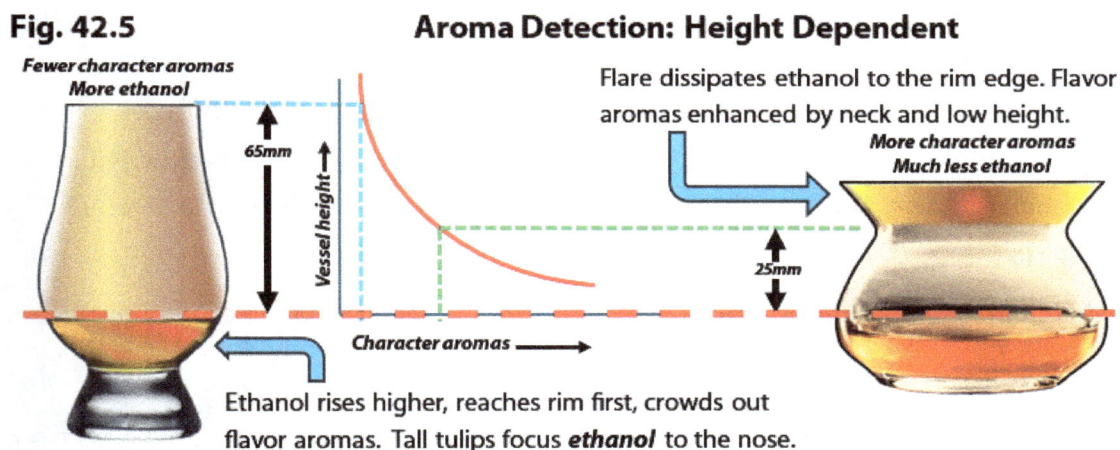

Fig. 42.5

Aroma Detection: Height Dependent

Fewer character aromas
More ethanol

65mm

Vessel height

Flare dissipates ethanol to the rim edge. Flavor aromas enhanced by neck and low height.

More character aromas
Much less ethanol

25mm

Character aromas

Ethanol rises higher, reaches rim first, crowds out flavor aromas. Tall tulips focus *ethanol* to the nose.

Gender Bias Testing of the Sensory Vessel

Fig. 42.6	Vessel Preference by Gender			
Category	**Male Preferences**		**Female Preferences**	
Preferences	**Tulip**	**Sensory**	**Tulip**	**Sensory**
Individual trials Percent (%)	272 (12.9 %)	1830 (87.1 %)	14 (1.7%)	798 (98.3 %)
Std. Deviation	2.23		1.37	
Confidence Level	2σ (97.5 %)		2σ (97.5 %)	
17 Locations, Miami to Portland, LA to NYC, 2012 — 2015, spirits festivals & spirits judging competitions				

Clearly, the more sensitive female noses preferred the sensory-engineered glass over the tulip. Although many males preferred the sensory glass, most stated they would continue to use the tulip. When asked why, the most common replies were "It is my favorite,' or "everyone else uses it," or "It represents tasting whiskey." The male resistance to the sensory-engineered glass suggests that the social group aspect of tasting spirits is more important to male drinkers than to females.

Summary

Removing ethanol from the olfactory is not about eliminating alcohol, but about managing its disproportionate volatility and trigeminal effects. Every sensory glass should focus on sensory accuracy by enabling evaluators to perceive the full spectrum of aromas, advancing spirits analysis from the cultural ritual of a predetermined, casually adopted wine glass to a scientific discipline, powered by applying the laws of simple physics, vapor pressure, and gaseous diffusion. In the example, the sensory glass reduced ethanol to $1/650^{th}$ at the rim edge and 66% at the centerline. This reduction in trigeminal effects is crucial to accurate aroma detection and evaluation.

Any well-designed vessel that is an efficient method for diffusing ethanol is a boon to human evaluation, as it will (1) eliminate the tedious and inaccurate practice of diluting spirits in arbitrary 5% v/v increments, often necessary in tulip glasses by dispersing the ethanol prior to nosing the sample, (2) enable more efficient swirling which releases more aromas and improves intensity.

Among male drinkers, social conformity aspects of tasting and enjoying spirits outweigh considerations of improved flavor detection and identification as the primary motives for choosing a personal spirits-tasting vessel.

"The details are not the details. They make the design."— Charles Eames, Designer and architect

"We shape our tools and thereafter our tools shape us." — Marshall McLuhan, Theorist

"Perfection is achieved not when there is nothing more to add, but when there is nothing left to take away." — Antoine de Saint-Exupéry, Author and philosopher

References

Delwiche, J. (2004). The impact of perceptual interactions on perceived flavor. *Food Quality and Preference, 15*(2), 137–146. https://doi.org/10.1016/S0950-3293(03)00041-7

Graham, T. (1833). On the law of diffusion of gases. *Philosophical Transactions of the Royal Society of London, 123*, 573–631. https://doi.org/10.1098/rstl.1833.0029

Lee, K., Paterson, A., & Piggott, J. R. (2001). Origins of volatile compounds in Scotch malt whisky. *Trends in Food Science & Technology, 12*(10), 391–401. https://doi.org/10.1016/S0924-2244(01)00127-5

Manska, G. (2021). Diagnostic beverage tasting vessel for ethanol-based beverages. *Beverages, 7*(3), 57. https://doi.org/10.3390/beverages7030057

Appendices

Evaluators' Guide to the Dual Role of Trigeminal Irritants				
Compound	**Source**	**Nasal Cavity Sensation**	**Oral Cavity Sensation**	**Notes**
Ethanol	All spirits	Burning, pungency, anesthesia at high concentrations	Warming burn, diffuse sting	Dominant trigeminal driver
Methanol	All distillates (trace)	Mild burn at high mg/L	Light burn	Usually below sensory significance
n-Propanol	Whiskey, rum, brandy	Sharp prickling burn	Warming, solvent burn	Increases in tail-heavy cuts
Isobutanol	All spirits	Solvent-like sting	Hot, numbing heat	More potent irritant per mole than ethanol
Isoamyl alcohol	All spirits	Heavy, solvent pungency	Lingering burn, numbing	Marks higher alcohols
1-Butanol	Rum, grain spirits	Acrid, stinging	Hot, bitter burn	Potent even at low ppm
Acetaldehyde	All, high in new whiskey, brandy	Sharp, piercing irritation	Sour-bite, burning	Strongest irritant in beverages
Furfural	Barrel-aged spirits	Dry-smoke irritation	Astringent burn	Increases with toasted/charred oak
5-Methylfurfural	Barrel-aged spirits	Sweet-smoke irritant	Woody burn	Toast/char effect
Acrolein	Peated whisky, mezcal smoke	Intense nasal sting	Harsh burning	Powerful irritant unsaturated aldehyde
Crotonaldehyde	Charred barrels	Acrid irritation	Harsh burn	Rums, charred spirits
Acetone	Grain whiskey, rum	Solvent sting	Sharp burn	Poor cuts
Acetic acid	All spirits	Vinegar-like sting	Sharp sour burn	Volatile acid faults elevate impact
Propionic acid	Rum, brandy	Sour, acrid irritation	Harsh sour burn	Trace but potent
Isovaleric acid	Whiskey, Scotch	Pungent cheesy/fusel sting	Acrid burn	Strong oral–nasal bite
Hydrogen sulfide	All spirits	Sharp, rotten-egg sting	Metallic bite	Potent at ppb volatile
Sulfur dioxide	Brandy (rare)	Strong nasal irritation	Harsh chemical burn	Often absent in well-made spirits
Thiols	Whiskey, rum	Acrid, piercing irritation	Sharp chemical bite	Potent even at low concentration
Guaiacol	Peated Scotch, mezcal	Smoky burn	Medicinal burn	Defining irritant in peated whisky
Cresols	Heavily peated whisky	Sharp tar-like irritation	Spicy medicinal burn	Very potent
Phenol	Brandy, Scotch	Sharp chemical burn	Medicinal burn	Benchmark phenolic irritant
Syringol	Peated whisky	Sweet-smoke irritation	Lingering burn	High-affinity irritant
α-Pinene	Gin, rum, agave	Pine-sharp tingle	Cooling bite	Trigeminal & olfactory
β-Pinene	Gin	Sharper pine-burn	Astringent burn	More irritant than α-pinene
Oxidized limonene	Gin, citrus spirits	Stinging irritant (oxidized form only)	Bitter burn	Fresh limonene is not irritant
Eucalyptol (1,8-cineole)	Gin, agave, rum	Cooling–pungent	Cooling burn	Dual activation
Menthol	Botanical spirits	Cooling tingle	Cooling burn	Classic oral–nasal trigeminal
Camphor	Botanical spirits	Cooling-stinging	Cooling burn	Strong somatosensory

Appendix I: Creating the Aroma Lexicon Table

The Aroma Lexicon Table provides a structured reference for describing the sensory attributes of spirits. Each entry lists a descriptor family (e.g., fruity, floral, smoky), specific terms, reference standards for calibration, associated chemical compounds, and usage guidelines. The purpose of the table is not to prescribe what an evaluator must perceive, but to ensure that when a term is used, it carries the same meaning across different individuals and tasting panels.

A lexicon becomes valuable only when evaluators converge on terminology. "Apple" should refer to the same aroma profile whether noted in a Scotch whisky, a rum, or a brandy. Without shared definitions, communication collapses into metaphor and ambiguity. Training sessions anchored to reference standards—such as pure compounds, food analogs, or prepared solutions—are critical for aligning vocabulary.

- The table is designed as a living document. New terms may be added as novel aroma references are agreed upon, while redundant or confusing terms can be retired. Expansion should follow three principles:
- **Clarity:** Every term must point to a distinct, recognizable sensory experience.
- **Calibration:** Each descriptor must be linked to an accessible reference standard.
- **Consensus:** Additions should be adopted only when evaluators can apply them consistently.

Used correctly, the Aroma Lexicon Table allows evaluators to document spirits with precision, compare findings across panels, and build a cumulative, reproducible language of sensory science.

Panels must agree on the terminology prior to evaluation. The table serves as a suggested starting point for building a comprehensive lexicon as new aroma descriptors become necessary.

App. I	Aroma Lexicon (Common Aromas)			
Descriptor	**Term**	**Reference Standard**	**Compound**	**Notes/Guidelines**
Fruity (Esters)	Apple/ Pear	Ethyl hexanoate (apple), pear drops	Ethyl hexanoate	Use only for fresh fruit aromas; avoid generic 'fruit'.
Fruity (Esters)	Banana	Isoamyl acetate (banana candy)	Isoamyl acetate	High ester yeast strains.
Fruity (Esters)	Citrus	Lemon peel, lime juice, orange zest	Ethyl butyrate, limonene, citral	Distinguish lemon vs. lime vs. orange.
Fruity (Esters)	Stone Fruit	Apricot preserves, peach nectar	γ-Decalactone, β-Ionone	Often linked to yeast metabolism.
Fruity (Esters)	Tropical	Pineapple juice, mango flesh	Ethyl butyrate, ethyl hexanoate	Strong in some rum and brandy.
Floral	Rose	Rosewater	2-Phenylethanol	Associated with fermentation bouquet.

App. I   (continued)

Descriptor	Term	Reference Standard	Compound	Descriptor
Floral	Violet	Violet candies	Ionones	Low-intensity, easily masked.
Floral	Lavender	Lavender oil	Linalool, linalyl acetate	Common in gin botanicals.
Spicy	Clove	Clove bud	Eugenol	Oak and maturation derived.
Spicy	Vanilla	Vanilla extract	Vanillin	Strong marker of cask aging.
Spicy	Cinnamon	Cinnamon bark	Cinnamaldehyde	Oak influence and additives.
Woody/Oaky	Fresh Oak	New oak shavings	Oak lactones (cis/trans-β-methyl-γ-octalactone)	Bourbon and first-fill casks.
Woody/Oaky	Toasted Wood	Toasted oak staves	Furfural	Caramelized, toasty notes.
Woody/Oaky	Smoky Oak	Charred oak	Guaiacol, syringol	Heavily charred barrels.
Nutty	Almond	Almond extract	Benzaldehyde	It can overlap with marzipan.
Nutty	Hazelnut	Roasted hazelnuts	2,5-Dimethyl-pyrazine	Toasty cask influence.
Lactic/ Dairy	Butter	Margarine/butter	Diacetyl	By-product of lactic bacteria.
Lactic/ Dairy	Creamy	Fresh dairy cream	Lactones	Positive only when soft, not rancid.
Lactic/ Dairy	Cheesy	Parmesan cheese	Isovaleric acid	Unpleasant if dominant.
Sulfurous	Cooked Cabbage	Boiled cabbage water	Dimethyl trisulfide (DMTS)	Fault in the new make.
Sulfurous	Rubber	New rubber tubing	Mercaptans	Generally considered a flaw.
Sulfurous	Onion/ Garlic	Chopped onion	Diallyl sulfides	Sign of stressed fermentation.
Smoky/ Phenolic	Medicinal	Iodine solution	Phenol, cresols	Typical of peated whisky.
Smoky/ Phenolic	Tar	Coal tar	Cresols	Heavy phenolics.
Smoky/ Phenolic	Smoked Meat	Smoked bacon	Guaiacol derivatives	Overlap with BBQ/meatiness.
Sweet/Confectionery	Caramel	Caramelized sugar	Furfural, maltol	From cask toasting.
Sweet/Confectionery	Honey	Clover honey	Phenylacetaldehyde	Associated with floral carryover.
Sweet/Confectionery	Chocolate	Dark chocolate	Theobromine, pyrazines	Toasty/malty notes.
Green/ Herbaceous	Fresh Grass	Cut grass	Hexanal, (Z)-3-hexenol	Linked to immaturity.
Green/ Herbaceous	Mint	Peppermint oil	Menthol	Cooling trigeminal overlap.
Green/ Herbaceous	Eucalyptus	Eucalyptus oil	1,8-Cineole	Common in Australian spirits.
Earthy	Damp Soil	Fresh potting soil	Geosmin	Considered a fault if dominant.
Earthy	Mushroom	Dried mushrooms	Octenol	Cask-derived or storage influence.
Chemical/ Solvent	Nail Polish Remover	Acetone, ethyl acetate	Ethyl acetate, acetone	Common in faulty distillation.
Chemical/ Solvent	Paint Thinner	Turpentine	Higher alcohols	Avoid vague usage.
Chemical/ Solvent	Plastic	PVC tubing	Styrene derivatives	Usually fault-related.
Fruity (Esters)	Citrus	Lemon peel, lime juice, orange zest	Ethyl butyrate, limonene, citral	Distinguish lemon vs. lime vs. orange.
Fruity (Esters)	Stone Fruit	Apricot preserves, peach nectar	γ-Decalactone, β-Ionone	Often linked to yeast metabolism.

App. I (continued)				
Descriptor	**Term**	**Reference Standard**	**Compound**	**Descriptor**
Fruity (Esters)	Tropical	Pineapple juice, mango flesh	Ethyl butyrate, ethyl hexanoate	Strong in some rum and brandy.
Floral	Rose	Rosewater	2-Phenylethanol	Associated with fermentation bouquet.
Floral	Violet	Violet candies	Ionones	Low-intensity, easily masked.
Floral	Lavender	Lavender oil	Linalool, linalyl acetate	Common in gin botanicals.
Spicy	Clove	Clove bud	Eugenol	Oak and maturation derived.

Strengthening the Discipline to Establish a Practical Professional Lexicon

1. Anchor Every Descriptor to Perceptual Reality, Not Vocabulary Tradition: A term is useless unless it corresponds to a reproducible sensory event. Many inherited terms (e.g., *"clean," "smooth," "balanced"*) carry emotional or hedonic weight but no measurable referent. **Advice:** Reject any descriptor that cannot be replicated through a standard or sample. A lexicon should represent *observables*, not opinions.

2. Quantify Intensity Levels with Scale Anchors: Words alone do not convey strength or dominance. A shared scale—typically 0–5 or 0–10 anchored with sensory standards—ensures that "moderate vanilla" means the same magnitude to every evaluator. **Advice:** Attach an intensity scale to each descriptor with calibration points (e.g., "vanillin 0.5 ppm = moderate"). This allows statistical treatment and inter-panel comparison.

3. Define Descriptor Scope and Boundaries: Overlap between terms like *"buttery," "creamy," "lactic,"* or *"dairy"* creates confusion. **Advice:** Each descriptor family should include inclusion/exclusion rules—for instance, "use *lactic* for fermented milk note; do not use *buttery* unless diacetyl reference confirmed." Clear boundaries prevent redundancy and semantic drift.

4. Classify by Chemical or Biosynthetic Pathway When Known: Organizing lexicons only by sensory family (fruity, floral, etc.) obscures valuable production insights. **Advice:** Include an auxiliary column or metadata grouping by chemical class (esters, aldehydes, phenols, terpenes, sulfur compounds) to strengthen causal understanding and support traceability to process origin.

5. Document Matrix and Ethanol Interference Effects: Descriptors shift with ethanol concentration and matrix interactions—"banana" (isoamyl acetate) in high-ABV rum differs perceptually from its expression in liqueurs or beers. **Advice:** Annotate reference standards with effective concentration and solvent system, and specify whether the note is perceptible in aqueous, hydroalcoholic, or neat conditions.

6. Mandate Controlled Environmental Conditions for Lexicon Validation: Ambient odor, humidity, temperature, and lighting alter descriptor consensus. **Advice:** Require all lexicon calibration sessions to be conducted under clean air, neutral lighting (D65 or 5000 K), and a stable temperature (21–23 °C). Always include fresh-air breaks to mitigate adaptation bias.

7. Distinguish Olfactory-Only from Trigeminal or Cross-Modal Descriptors: Terms like *"peppery," "burning,"* or *"cooling"* reflect chemesthetic sensations, not aromas. **Advice:** Maintain a dual lexicon or explicitly flag such terms. Mixing olfactory and trigeminal descriptors into a single list confuses dimensional analysis.

8. Use Panel Statistics to Confirm Descriptor Validity: A descriptor is only valid if multiple evaluators can identify it consistently above chance. **Advice:** Apply binomial probability or frequency thresholds (e.g., ≥ 70 % agreement across ≥ 10 panelists) before admitting new terms. This ensures stability and prevents individual bias from contaminating the lexicon.

9. Maintain Version Control and Traceability: Lexicons evolve—without documentation, reproducibility collapses. **Advice:** Assign a date, editor, and revision number to each version. Archive versions to track term evolution and justify decisions during audits or competitions.

10. Use Cross-Modal Verification for Ambiguous Descriptors: Some aromas co-occur with mouthfeel or taste cues—e.g., *"honeyed"* often implies both sweet aroma and viscous mouthfeel. **Advice:** Tag such descriptors as cross-modal and discuss whether they belong in olfactory, gustatory, or integrated sensory categories before inclusion.

11. Integrate Training Frequency and Memory Reinforcement: Olfactory memory decays rapidly without repetition. **Advice:** Schedule quarterly or semiannual calibration refreshes in which panelists re-evaluate reference standards to maintain term stability, prevent descriptor drift.

12. Culturally Normalize Vocabulary: Cultural context influences analogy—"molasses" may be unfamiliar in some regions. **Advice:** Choose references that are accessible globally, or provide alternative analogs ("molasses" ↔ "dark syrup"). Include a column for regional synonyms to aid international panels.

A lexicon without calibration is vocabulary; with calibration, it becomes data. Calibration transforms words into measurements; it binds language to perception through shared reference, converting description into evidence.

Appendix II: Spirits, Faults, and Rejection Thresholds

Spirit's evaluation requires clear criteria for identifying when a fault is severe enough to reject a sample. Faults can arise from the use of poor raw materials, fermentation errors, improper distillation cuts, inadequate cask hygiene, or contamination. This table summarizes common faults, their typical descriptors, chemical associations, likely causes, and estimated sensory rejection thresholds.

Evaluators should rely on consensus training samples for calibration (e.g., pre-prepared spiked solutions, reference kits, or known faulty spirits). Not every perception of a fault warrants rejection; thresholds guide when a compound becomes perceptually dominant and disrupts balance. Fault detection must be paired with documentation of intensity, context (including glassware, dilution, and serving temperature), and evaluator agreement. This table should be updated as new research refines thresholds or identifies emerging fault compounds.

App. II	Spirits Faults and Rejection Thresholds			
Fault Descriptor	Typical Sensory Note	Probable Compound(s)	Likely Cause	Rejection Threshold
Sulfurous/ Cooked Cabbage	Rotten cabbage, onion, sewer gas	Hydrogen sulfide, dimethyl trisulfide (DMTS), mercaptans	Yeast stress, inadequate copper contact	~1–5 µg/L for DMTS
Rubber/ Burnt	Burnt rubber, plastic	Mercaptans, thiophenes	Contaminated fermentation, bacterial spoilage	~10–20 µg/L (mercaptans)
Solvent/ Nail Polish Remover	Acetone, nail varnish, glue	Ethyl acetate, acetone	Faulty fermentation, excessive head cut	>200 mg/L ethyl acetate
Metallic	Copper, iron, blood-like	Iron, copper ions	Contaminated water, still corrosion	>1–2 mg/L (Cu/Fe)
Musty/ Moldy	Damp cellar, cardboard	Geosmin, TCA (2,4,6-trichloroanisole)	Contaminated casks, storage issues	ng/L range for TCA
Soapy/ Fatty	Soap, rancid oil	Long-chain fatty acid esters, sodium salts	Excessive fermentation lipids, poor cut	Variable; noticeable >1 mg/L
Phenolic/ Medicinal	Band-aid, tar, creosote	4-ethylphenol, 4-ethylguaiacol, cresols	Poor cask hygiene, bacterial spoilage	~400–600 µg/L (4-EP)
Earthy	Damp soil, beetroot	Geosmin	Contaminated raw material or water	~10 ng/L
Excessive Acidity	Vinegar, sour milk	Acetic acid, lactic acid	Bacterial spoilage, poor hygiene	>1 g/L (acetic acid)
Oxidized/ Stale	Stale nut, wet cardboard	Acetaldehyde	Poor storage, excessive oxygen exposure	>100 mg/L

Fault recognition and rejection thresholds are essential to professional evaluation standards. Spirits that exceed threshold levels of sulfurous, solvent, metallic, moldy, phenolic, or oxidized notes may be considered unsuitable for competition judging, blending, or commercial bottling. Enhance abilities to distinguish between stylistic differences and genuine sensory faults.

Appendix III: Sensory Evaluation Data Sheet (Master Form)

A) Session Cover Sheet — Room, Panel, and Protocol

Session ID: _____ Date: ____ / ____ / _____ Start–End: _____

Location: _____ Panel Leader: _____

Panel Size (n): ____ Evaluators: _____

Test Type (check):

[ ] Descriptive [ ] Triangle [ ] Duo–Trio [ ] A–Not–A [ ] Screening [ ] Other: _____

Blinding / Masking (check):

[ ] Sample codes only [ ] Brand/price masked [ ] Cask info masked [ ] Color masked (amber/opaque glass)

Order Control:

[ ] Randomized [ ] Counterbalanced / Latin square Notes: _____

Flight Setup:

- **# Samples:** ____ **Pour volume:** ____ mL **Rest after pour:** ____ min
- **Glassware:** _____ [ ] Cover used (watch glass / lid)
- **Water:** [ ] Yes **Palate reset:** [ ] Neutral cracker [ ] Water only [ ] Other: _____

Room Conditions:

- **Temp:** ____ °C **Humidity:** ____ % **Air prep (ventilate):** ____ min before session
- **Lighting:** Type _____ CCT _____ K [ ] Uniform / no color casts
- **Ambient odors present:** [ ] None [ ] Notes: _____
- **Noise:** [ ] Quiet [ ] Notes: _____

Reference Standards on hand (if any): _____

Duplicates included? [ ] Yes Code(s): _____ [ ] No

Anchors included? [ ] Yes Descriptor(s): _____ [ ] No

Deviations from SOP (if any): _____

B) Evaluator Readiness (complete once per evaluator)

Evaluator ID/Name: _____ **Role today:** [ ] Evaluator [ ] Scribe [ ] Steward

30-Second Self-Check (Y/N): Fragrance/cologne _____ Coffee/mint/pepper <60 min _____ Congestion/allergy _____

New meds (antihistamine/decongestant) _____ Poor sleep/high stress _____ Mouth dryness _____

Readiness status: [ ] OK [ ] CAUTION — action: [ ] switch role [ ] postpone [ ] note & proceed

Interferences / Notes: _____

C) Per-Sample Note Sheet (duplicate for each sample code)

Session ID: _____ **Sample Code:** _____ **Position in flight:** ____ / ____

Pour time: _____ **Time to first sniff:** ____ min **Eval time:** _____

Glassware: _____ **Pour temp:** ____ °C **Room temp:** ____ °C

Dilution: [ ] None [ ] Water ___ mL → final ABV approx. ____ % (note exact)

Color masked: [ ] Yes [ ] No

Onset (0–5 s, orthonasal)

1. __ Family — Term (modifier) __ Intensity (0–10): __
2. __ Family — Term (modifier) __ Intensity (0–10): __
3. __ Family — Term (modifier) __ Intensity (0–10): __

Mid-Palate (5–20 s, palate / retronasal)

1. __ Family — Term (modifier) __ Intensity (0–10): __
2. __ Family — Term (modifier) __ Intensity (0–10): __
3. **Mouthfeel:** Ethanol heat __ /10 Viscosity __ /10 Astringency __ /10

Finish (20–60+ s, lingering)

- **Duration:** ____ seconds
- **Late-arising notes:** __ Family — Term (modifier) __ Intensity (0–10): __

10

- Additional: _____

Fault Check (Appendix III discipline)
- **Suspected fault?** [ ] No   [ ] Yes → Descriptor _____
- **Context test:** [ ] 2 min air [ ] small dilution [ ] both   **Outcome:** _____
- **Threshold cue (Appendix II):** compound/level (if known): _____
- **Assessment:** [ ] Style marker at this level   [ ] Probable fault (dominant/persistent)

Observation vs. Inference
- **Observations (lexicon terms only):** _____
- **Inference (optional):** likely process/cask/style _____
  Confidence: [ ] Low   [ ] Moderate   [ ] High

Summary (≤ 2 lines, neutral language)_____
Duplicate present for this code? [ ] Yes   [ ] No   **Congruent with duplicate?** [ ] Yes [ ] No
Notes: _____
Evaluator signature/initials: _____ **Timestamp:** _____

Legend & Reminders

- **Intensity scale (0–10):** 0 = not perceptible; 5 = moderate; 10 = extremely strong.
- **Phases:** *Onset* (orthonasal), *Mid-palate* (palate/retronasal), *Finish* (lingering).
- **Syntax: Family –Term (modifier) – Intensity –Phase** (phase implied by section).
- **Language discipline:** Use **Appendix I** terms; keep hedonic words out of observation lines; mark uncertainty honestly (e.g., "banana/pear-drop boundary").
- **Environment:** If room/glass/temperature/dilution changed mid-flight, note on sample sheet.

Note that this sample format serves as a suggested starting point. Each evaluator and each lab panel should modify the format to reflect their respective objectives. Conduct repeat evaluations on different days and at different times, and compare the results to identify sources of repeatability discrepancies.

Observations on Using the Master Sensory Evaluation Form

- Regular, disciplined use of the master form converts tasting into measurable science.
- Evaluator readiness is non-negotiable; fatigue, fragrance, or medication distorts results.
- The cover sheet anchors every variable, making comparisons meaningful.
- Blinding and randomized order prevent expectation bias and false learning.
- The separation of *observation* from *inference* forces objectivity and exposes Type III errors, where perception is correct but reasoning is not.
- Duplicate samples verify repeatability, and intensity scoring turns impression into useful data.
- Fault checks require patience—confirmation, not speculation.
- Completed forms are data, not paperwork; archived over time, they become the lab's cumulative sensory intelligence repository and benchmark for evaluator precision.

Appendix IV: Digital Resources and Continuing Education

The study of sensory science and spirits evaluation does not end with this text. The study begins here, at the foot of the bridge, and crosses over to the other side, rich in technical and personal sensory development. The rapid pace of new research in chemistry, neuroscience, psychophysics, and spirits production requires continually refreshed knowledge and skills. Digital resources, online databases, and continuing education opportunities support professional development.

Reference Databases: Use PubChem, PubMed, and ISO/AOAC standards when validating chemical associations, testing protocols, and threshold references.

Standards Bodies: Regularly consult ASTM and ISO updates to ensure compliance with evolving international sensory testing guidelines.

Formal Courses: Pursue structured qualifications through WSET, IBD, or equivalent to maintain credibility in spirits evaluation.

Professional Societies: Join the SSP and IFT to stay connected with the global sensory community and access the latest methodological research.

Open Learning: Supplement expertise with online courses in neuroscience, statistics, and psychophysics to strengthen interdisciplinary knowledge.

Professional sensory science is a lifelong pursuit. Evaluators who use digital resources and engage in structured continuing education are best positioned to adapt to new scientific discoveries, rigorously evaluate spirits, and contribute to the evolution of global tasting standards.

App. IV-A	Digital Resources	
Resource	**Description**	**Access / Use**
PubChem (NIH)	Open database of chemical structures, properties, and safety data. Useful for confirming compound identities linked to aroma descriptors.	https://pubchem.ncbi.nlm.nih.gov
PubMed	Primary biomedical literature database indexing neuroscience, physiology, and sensory science articles.	https://pubmed.ncbi.nlm.nih.gov
ScienceDirect	Repository of peer-reviewed journals in food chemistry, neuroscience, and sensory evaluation.	https://www.sciencedirect.com
SpringerLink	Access to sensory science and psychophysics texts, including food and beverage studies.	https://link.springer.com
MDPI Foods / Beverages	Open-access journals publishing sensory evaluation and food chemistry research.	https://www.mdpi.com
ISO Standards Database	The International Organization for Standardization (ISO) publishes documents relevant to sensory evaluation, including ISO 8586 and ISO 3591.	https://www.iso.org
AOAC INTERNATIONAL	Official methods of analysis, including alcohol and beverage testing standards.	https://www.aoac.org
ASTM International	Standards for sensory testing methods (triangle, duo-trio, descriptive analysis).	https://www.astm.org

App. IV-B	Continuing Education Platforms	
Platform	**Offerings**	**Notes**
WSET (Wine & Spirit Education Trust)	Globally recognized structured qualifications in wine and spirits, including tasting methodology.	https://www.wsetglobal.com
Institute of Brewing & Distilling (IBD)	Courses and diplomas in brewing, distilling, and sensory evaluation.	https://www.ibd.org.uk
ASTM Sensory Division	Workshops and technical symposia on sensory evaluation methods.	https://www.astm.org
Society of Sensory Professionals (SSP)	Annual conferences, webinars, and a professional community for sensory scientists.	https://www.sensorysociety.org
Institute of Food Technologists (IFT)	Sensory and consumer sciences division, offering continuing education modules.	https://www.ift.org
EdX / Coursera	University-level online courses on neuroscience, perception, food chemistry, and statistics for sensory science.	https://www.edx.org/ https://www.coursera.org

Digital Sensory Analytics and Open Data Integrity

Digital sensory analysis is redefining how evaluators interpret aroma and flavor. Advanced data systems now quantify what once depended solely on human judgment—linking gas chromatography–mass spectrometry outputs, molecular modeling, and machine learning algorithms to human descriptive data. Properly used, these tools expose relationships between compound structure, volatility, and perceptual intensity that manual observation alone cannot resolve. Their value lies not in replacing sensory panels but in verifying and calibrating them, thereby improving repeatability across laboratories and over time. The emerging standard of practice pairs trained evaluators with digital pattern recognition to establish reproducible benchmarks for aroma families, thresholds, and interaction effects.

This digital expansion also demands scientific accountability. The credibility of sensory science relies on open data, transparent methodology, and the ability to replicate results. Repositories such as Zenodo, Figshare, and the Open Science Framework now host datasets, lexicons, and experimental protocols that can be independently tested or reanalyzed. As sensory evaluation becomes increasingly data-driven, public access to well-curated results will determine which findings endure and which are discarded. Future professionals must regard open access not as optional generosity but as a professional obligation central to the integrity of sensory science.

Appendix V: Health & Ethics: Alcohol Management in Sensory Sessions

Spit-first policy: Evaluation ≠ consumption. Provide spit cups for all tasters. Swallowing is not required and should be discouraged during training.

Exposure budget (track ethanol, not just samples): Log **pure ethanol per sample**: mL poured × ABV (%) ÷ 100. If spitting, assume a residual of ~0.3–0.7 mL ingested per sample (contact and micro-swallow); note your lab's assumption on the sheet. Cap total **ingested ethanol** per session conservatively (program policy), and design flight sizes accordingly.

Flight design to minimize impairment: Limit high-ABV flights to **8–10** samples with large rim glassware, 4-6 with tulips; split large sets.

- Standard metered pour **15–30 mL**, covered, 2–3 min rest; dilution allowed but recorded.
- Enforce **breaks** between flights, hydrate.

Impairment checks & pass rights: Any evaluator may take an **ethical pass** at any time (scribe/steward roles available). If coordination, speech, or judgment are degraded, end participation and annotate the dataset (do not use the affected data).

Medication & health disclosures (private, minimal): Remind participants that certain medications (sedatives, antihistamines, decongestants) and conditions (pregnancy, liver disease) may contraindicate alcohol consumption. Provide a no-questions-asked pass.

Allergen & ingredient notice. List potential allergens in samples or references (e.g., nuts, botanicals). Provide alternatives.

Transport & legal compliance. No driving after participation if alcohol was swallowed beyond your program's zero-ingestion policy. Follow institutional age, service, and insurance guidelines; secure samples in a locked storage facility.

Sample handling & disposal. Use clean glassware (no detergent odor). Discard leftovers; do not re-bottle. Rinse to waste; keep ethanol away from heat/flame sources.

Documentation (minimum): On the **Session Cover Sheet** and **Per-Sample Note Sheet** record: pour volume, ABV, dilution, spitting policy, assumed residual per sample, total session ethanol exposure (estimated), and any role changes or passes.

One-line standard. "Spit by default, track exposure, respect pass rights, and document anything that could affect safety or data quality."

Managing Exposure per Sample, per Session

Liability considerations are crucial for those hosting spirits tastings. Many courts of law consider the management of ethanol intake the responsibility of the tasting panel administration. Adopt and enforce the Spit First Rule. Suggested print on session cover sheets:

"Spit by default; declare time (start/end and per-sample times); report grams and g/hour; any BAC math is illustrative only and never a substitute for the program's zero-ingestion/transport rules." (Or similar)

Ensure space is provided on the evaluator's log to record and calculate intake. Using a fixed residual model, print policy declaration on the session cover sheet. Decide on the average ml of pure ethanol ingested per sample (e.g., 1.0 ml per taste). Multiply by the number of samples, and add swallow event estimates (deliberate or accidental) to calculate total ethanol consumed. The final total yields estimated ethanol consumed. Given the session duration, one can readily calculate the amount of pure ethanol consumed per hour. Establish a cap guideline. Several procedures can be enforced to minimize ethanol exposure:

- Adopt the Widmark-style rough check to calculate BAC (blood-alcohol level). This approximation requires the evaluator's body weight, an assumed ethanol elimination rate, and session duration. The exact formula for calculation may be found at: https://alcohol.indianapolis.iu.edu/calculators/bac.html
- Evaluator consents to a mandatory breathalyzer test (maintenance and calibration of apparatus is the responsibility of the panel administrator or hosting group) at session end.
- Sign a consent form that acknowledges alcohol will be consumed, with clear safety guidelines and acceptance of liability for consumption.
- Eligibility screening to exclude underage, health risks (pregnancy, medication contraindications, liver conditions), and history of alcohol abuse, and alternatives for transportation post-session.
- Provide spit cups and dump-buckets. Neutral crackers and readily available water can slow absorption. Cap the maximum number of samples per session and per flight (6-8 per flight). Further limiting metered pours to 10-15ml per sample, provided there is still a high enough quantity to support the two-pass method. Keep sessions under 90 minutes to reduce fatigue and cumulative BAC. Timed breaks.
- Behavior observation for signs of impairment. Provide transportation, follow regulatory guidelines (ASTM E253–16, ISO sensory protocols), and integrate safety language into SOPs

Ethics Beyond Safety: Conflicts, Independence, and Public Trust

Safety and exposure management protect people. **Ethics** protects the credibility of the work. Spirit's evaluation intersects with commerce, which requires clear boundaries.

Conflict of interest (COI) and disclosure

- **Financial ties.** Paid consulting, product development, travel, hospitality, or gifts connected to a producer constitute a COI when evaluating that producer's spirits.
- **Institutional ties.** Employment, funding, or supervisory relationships that could benefit from a particular outcome are COIs.
- **Disclosure standard.** When any COI exists, disclose it **before** evaluation and recuse from decision-making where appropriate. Blind codes do not erase conflicts; they only reduce expectancy.

Separation of roles

- **Evaluation vs promotion.** The same person should not serve as evaluator and marketer for the same product in the same time window. If public comments are later used in advertising, they must be quoted **accurately**, with consent, and not imply institutional endorsement.
- **Data ownership and confidentiality.** Panel data collected under academic or internal terms are not press kits. Do not release raw notes, panelist identities, or partial results without prior agreement.

Sampling and fairness

- **Provenance.** Evaluate only legally obtained samples with a documented chain of custody. Avoid hand-selected "show bottles" unless the claim is explicitly about that selection.
- **Access and equal treatment.** Apply the same protocols and timing to all entries in a set. Subtle schedule advantages are still advantages.

Integrity of the record

- **No fabrication, no back-filling.** If a data point is missing, state it; do not invent one. Do not retro-fit lexicon terms to match a later narrative.
- **Version clarity.** If a sample is reformulated or proof-adjusted, treat it as a new sample. Keep version identifiers in the record.

Cultural respect - style without stereotyping. Discuss regional typicality and production grammars without importing myths or denigrating practices. Curiosity is compatible with rigor. These are not legal codes; they are principles of independence that keep sensory language meaningful to producers, consumers, and the academy alike. Ethics, like safety, is not a performative concept. It is the quiet framework that lets the rest of the work stand.

Appendix VI: Sensory Comparison (Tulip vs Snifter vs Engineered Vessels)

In September 2019, an independent glassware comparison by Sensation Research evaluated three different spirits in three different shapes. Tulip shape glass is a Glencairn, the snifter is the Libbey small snifter, and the Sensory engineered glass is NEAT: See https://sensationresearch.com/wp-content/uploads/2020/01/Glass-Vessel-Aroma-DA-Study-Sept-2019.pdf, including aroma rose plots, bar charts, and test circumstances. See the complete study at the above URL

App. VI-A   Scotch Sampling Vessel Comparison

Glass Vessel Evaluation; Scotch: n=21 responses	Sensory Engineered Glass	Tulip Shaped Whisky Glass	Snifter Glass	LSD	Prob
Nasal Burn	27.3c	63.0b	72.8a	2.6	0.0001
Overall Aroma	83.8a	80.4b	82.0ab	2.2	0.0200
Fruity Aromas Complex	69.2a	60.2b	51.1c	2.4	0.0001
Banana Aroma	21.5a	13.6b	6.8c	5.1	0.0002
Apple Aroma	63.5a	55.1b	46.9c	2.0	0.0001
Pear Aroma	41.5a	29.9b	17.7c	2.9	0.0001
Raisin Aroma	30.0a	27.4b	22.7c	2.2	0.0001
Rose Floral Aroma	29.9b	31.6ab	33.8a	2.5	0.0158
Sweet Aromas Complex	67.1a	51.7b	40.9c	2.1	0.0001
Caramel Aroma	52.4a	44.5b	37.4c	2.5	0.0001
Vanilla Aroma	59.9a	45.5b	29.7c	4.3	0.0001
Brown Sugar/Light Molasses Aroma	36.1a	26.5b	18.3c	2.6	0.0001
Coumarin Aroma	26.9a	22.3b	17.8c	2.8	0.0001
Cinnamon Aroma	15.7a	5.0b	0.0c	2.8	0.0001
Black Pepper Aroma	18.5b	21.0b	27.5a	3.4	0.0003
Coffee Aroma	5.9a	2.9ab	0.3b	4.1	0.0392
Creamy/Nutty Aroma	12.9a	4.5b	0.0b	4.7	0.0002
Wood Aromas Complex	68.7b	71.0b	75.6a	3.2	0.0015
Fresh Oak Aroma	40.3c	45.4b	51.6a	3.4	0.0001
Charred Oak Aroma	63.8c	66.9b	70.5a	2.9	0.0013
Smoky Aroma	26.0c	33.3b	37.2a	2.6	0.0001
Phenolic Aroma	1.4b	5.2b	15.8a	3.9	0.0001
Leather Aroma	29.0b	39.0a	39.1a	3.7	0.0001
Tobacco Aroma	26.2b	39.0a	37.9a	3.8	0.0001
Fresh Hay	22.9a	12.1b	2.2c	4.3	0.0001
Grain Aroma	16.9b	24.8a	23.6a	3.5	0.0007
Ethanol Aroma	18.1c	38.7b	51.3a	4.7	0.0001

*Within an attribute, products labeled with different letters are significantly different at the 95% confidence as determined by Anova followed by Fisher's LSD.

✳ **Sensation Research**

Highest in attribute
Lowest in attribute

For the evaluation of Scotch, the Sensory Engineered Glass was significantly highest in aromas of fruity, banana, apple, pear, raisin, sweet aromas complex, caramel, vanilla, brown sugar/light molasses, coumarin, cinnamon, creamy/nutty and fresh hay but lowest in nasal burn, fresh oak, charred oak, smoky, leather, tobacco, grain and ethanol. The Tulip Shaped Whisky Glass is in the middle of the aroma profiles, and significantly lower than the Sensory Engineered Glass in overall aroma. It is high in leather, tobacco, and grain, along with the Snifter Glass. The Snifter Glass is significantly higher in notes of nasal burn, black pepper, wood complexity, fresh oak, charred oak, smokiness, phenolic notes, and ethanol. The Snifter Glass is characterized by a fruity complexity, including notes of banana, apple, pear, raisin, sweet notes such as caramel, vanilla, and brown sugar/light molasses, as well as hints of coumarin, cinnamon, and fresh hay. It is higher than the Sensory Engineered Glass in rose floral and lower than the Sensory Engineered Glass in coffee.

App. VI-B   Cognac Sampling Vessel Comparison

Glass Vessel Evaluation; Cognac: n=21 responses	Sensory Engineered Glass	Tulip Shaped Whisky Glass	Snifter Glass	LSD	Prob
Nasal Burn	31.4[b]	62.3[a]	65.9[a]	4.3	0.0001
Overall Aroma	86.8[a]	75.8[b]	68.2[c]	1.9	0.0001
Sweet Aromas Complex	68.2[a]	62.3[b]	58.5[c]	2.0	0.0001
Vanilla Aroma	48.4[a]	42.9[b]	37.8[c]	2.1	0.0001
Caramelized/Maple Aroma	64.0[a]	58.6[b]	54.6[c]	1.4	0.0001
Cinnamon Aroma	37.3[a]	30.9[b]	24.3[c]	2.2	0.0001
Fruit Aromas Complex	76.9[a]	60.7[b]	49.8[c]	3.0	0.0001
Fresh Pressed Cider Aroma (red/green apples)	72.6[a]	56.0[b]	44.1[c]	2.6	0.0001
Baked Apple Aroma	62.3[a]	53.0[b]	43.6[c]	2.5	0.0001
Dried Fruit Aroma	40.3[a]	30.0[b]	24.8[c]	2.3	0.0001
Pineapple Esters Aroma	22.2[a]	18.1[b]	19.1[b]	2.1	0.0027
Cooked Cherries/Berries Aroma	45.1[a]	33.4[b]	25.7[c]	2.9	0.0001
Rose Floral Aroma	33.2[a]	32.6[a]	27.0[b]	1.6	0.0001
Wood Aromas Complex	70.1[a]	69.8[a]	66.2[b]	2.5	0.0088
Wood Shavings Aroma	36.5[a]	34.0[a]	29.3[b]	2.8	0.0003
Toasted Aroma	67.4[a]	67.0[a]	62.9[b]	2.4	0.0030
Smoky Aroma	35.4[a]	27.7[b]	26.0[b]	2.7	0.0001
Tar Aroma	16.1	13.0	14.5	3.7	0.2243
Crayon Aroma	13.0[a]	3.7[b]	2.8[b]	3.6	0.0001
Ethanol Aroma	14.0[c]	34.7[b]	48.2[a]	3.4	0.0001

*Within an attribute, products labeled with different letters are significantly different at the 95% confidence as determined by Anova followed by Fisher's LSD.

✳ Sensation Research

| Highest in attribute |
| Lowest in attribute |

For the evaluation of Cognac, there was no difference in tar aroma among the glass vessels. The Sensory Engineered Glass was significantly highest in overall aromas, complex sweet aromas, vanilla, caramelized/maple, cinnamon, fruit aromas, complex, fresh pressed cider, baked apple, dried fruit, pineapple esters, cooked cherries/berries, smoky, and crayon, and lowest in nasal burn and ethanol aroma. The tulip glass is positioned in the middle of the aroma profiles, but it scores high in rose floral, wood aromas, and complex notes, including wood shavings and toasted notes, with the Sensory Engineered Glass. It is also high in nasal burn along with the Snifter Glass. The Snifter Glass is significantly highest in ethanol, lowest in overall, sweet aromas complex, vanilla, caramelized/maple, cinnamon, fruit aromas complex, fresh pressed cider, baked apple, dried fruit, cooked cherries/berries, rose floral, wood aromas complex, wood shavings, and toasted.

App. VI-C   Tequila Sampling Vessel Comparison

Glass Vessel Evaluation; Tequila: n=21 responses	Sensory Engineered Glass	Tulip Shaped Whisky Glass	Snifter Glass	LSD	Prob
Nasal Burn	25.3[b]	73.1[a]	70.0[a]	3.7	0.0001
Overall Aroma	80.5[a]	68.4[b]	61.9[c]	2.1	0.0001
Salty Aroma	22.4[a]	21.6[a]	18.9[b]	2.6	0.0297
Marshmallow Aroma	23.0[a]	11.4[b]	0.0[c]	2.9	0.0001
Smoky Aroma (ashtray)	53.1[b]	59.8[a]	54.2[b]	2.8	0.0004
Black Pepper Aroma	43.1[a]	38.1[b]	35.2[c]	2.1	0.0001
Overall Fruit Aromas Complex	68.7[a]	54.0[b]	33.3[c]	2.5	0.0001
Grapefruit Citrus Aroma	44.0[b]	47.4[a]	29.7[c]	3.3	0.0001
Lime Citrus Aroma	62.7[a]	44.1[b]	28.4[c]	2.5	0.0001
Banana Aroma	21.8[a]	1.0[b]	0.0[b]	2.8	0.0001
Floral Aroma	33.8[a]	31.8[a]	22.3[b]	3.1	0.0001
Grass/Bell Pepper Aroma	37.8[b]	46.7[a]	41.3[b]	3.6	0.0006
Cactus Aroma	48.8[a]	34.4[b]	16.6[c]	4.7	0.0001
Mint Aroma	42.4[a]	32.6[b]	19.4[c]	2.4	0.0001
Yellow Aromas (Squash/Sweet Potato)	32.4[a]	12.8[b]	5.2[c]	4.4	0.0001
Ethanol Aroma	21.0[c]	47.7[b]	59.0[a]	3.2	0.0001

*Within an attribute, products labeled with different letters are significantly different at the 95% confidence as determined by Anova followed by Fisher's LSD.

✳Sensation Research

| Highest in attribute |
| Lowest in attribute |

For the evaluation of tequila, the Sensory Engineered Glass was highest in aromas of overall fruit complexity, marshmallow, black pepper, lime citrus, banana, cactus, mint, yellow squash/sweet potato, and lowest in nasal burn and ethanol aroma. The tulip is in the middle of most of the aroma profiles; however, it scores highest in smoky (ashtray), grapefruit, and grass/bell pepper. It also scores high in salty and floral, along with the Sensory Engineered Glass. The Snifter Glass is highest in ethanol and lowest in overall, salty, marshmallow, black pepper, overall fruit complex, grapefruit citrus, lime citrus, floral, cactus, mint, and yellow squash/sweet potato.

Dissipation of ethanol in the sensory glass significantly reduced trigeminal irritation, unmasking most subtle flavors in spirits and eliminating the need for stepped dilution to evaluate.

Appendix VII: Sensory Effects of Different Still Constructions

Sensory nuances in sensory are typical and characteristic of different still types. This reference serves as a general guide for solving distillation problems or identifying the source of product flaws or flavor imbalances. See Appendix VII Fig.1.

App. VII-A	Sensory Contribution by Still Type			
Still Type	**Process Mode**	**Key Features**	**Typical Spirits**	**Sensory Contribution**
Column (Coffey / Patent / Multi-Column)	Continuous	Tall analyzer + rectifier columns with plates/ packing; high reflux ratios; copper or copper-lined sections	Vodka, light rum, neutral grain spirits, grain whisky	Clean, light, neutral spirit with high efficiency; minimal congeners; consistent batch-to-batch profile
Pot Still	Batch	Copper body, swan neck, limited internal reflux; manual heads/tails cuts; worm or shell-and-tube condenser	Scotch malt whisky, Irish pot still, brandy, mezcal	Rich, complex, heavy congeners; full body and texture; high aroma intensity, but more operator-dependent
Hybrid Still (Pot + Plates)	Semi-batch	Pot base with removable or fixed plates; variable dephlegmator and reflux control; mixed copper/stainless	Craft whiskey, new-style gins, brandies	Adjustable output: can emulate pot-like heaviness or lighter column spirit depending on settings
Charentais (Cognac) Still	Double-batch	Copper onion head, swan neck, chauffe-vin pre-heater, worm condenser; legal max 72.4 % ABV	Cognac	Fruity, floral, elegant balance; copper suppresses sulfur while retaining aromatic esters
Armagnac Continuous Still	Continuous (single pass)	Short low-rectification column (~15 trays); pre-heater; all copper; ~52–60 % ABV spirit	Armagnac	Deeper dried-fruit, nut, prune notes; heavier oils than Cognac; rustic and expressive
Alembic (Historic / Eaux-de-vie)	Batch	Traditional flame-heated copper or brass; onion or pear head; long vapor path; small capacity	Fruit brandies, small-scale eaux-de-vie	Rustic, aromatic, robust; retains many congeners but is sensitive to heat management
Double Retort Pot Still (Jamaican Rum)	Semi-batch	Pot still feeding one or two retorts charged with low wines/high wines/dunder; copper vapor path	High-ester Jamaican rum	Extremely high ester content, tropical fruit "funk;" layered complexity
Thumper Keg (U.S. Tradition)	Semi-batch	Simple intermediate vessel between pot and condenser charged with weak, low wines; usually copper	Traditional American moonshine, heritage ryes	Slight ABV boost; retains grain character, but risk of heads/tails smearing
Three-Chamber Still (Leopold Bros Revival)	Semi-continuous steam-percolation	Three stacked chambers; live steam strips alcohol and oils sequentially; copper construction	Old-style American rye whiskey	Dense, oily, aromatic spirit with nutty/cocoa/maple notes; intense grain oil contribution
Carter Head / Gin Vapor-Infusion Still	Batch or Semi-batch	Pot or short column, botanical basket above vapor path; controlled vapor speed; mixed material	London dry gin, vapor-infused gins	Bright, high floral and citrus top notes; less heavy oil; clean extraction of botanicals

App. VII-A	(continued)			
Still Type	**Process**	**Key Features**	**Spirits**	**Sensory Contribution**
Vacuum (Low-Pressure) Still	Continuous or batch under vacuum	Lower boiling temperatures preserve fragile volatiles; often stainless or glass with copper inserts	Experimental gins, fruit distillates, and liqueurs	Preserves delicate aromatics; fresh, bright "raw fruit" notes; minimal thermal degradation
Glass / Lab-Scale Pilot Still	Batch (analytical)	Borosilicate or quartz; Vigreux column; inert (no copper catalysis)	Research, sensory trials	Reveals unmasked fermentation volatiles, including sulfur; used for R&D, not production
Industrial Multi-Effect Continuous Still	Continuous	Multiple columns (analyzer, rectifier, fusel oil, aldehyde, demethylizer); extreme rectification	Industrial neutral ethanol, vodka bases	Ultra-neutral ethanol with virtually no congeners; smooth but characterless; blending base
Lomond / Modified Plate Still	Semi-batch	Pot base, vertical column, adjustable plates, and cooling coils; all copper	Specialty malt whisky styles (Inverleven, Scapa)	Allows modulation between heavy and light malt spirit in the same still; flexible congener retention

Distillers often tweak the process or design to solve recurring issues. Appendix VII, Fig. 2, provides insight into the types of adjustments distillers must make to improve product quality. This table can also serve as a valuable reference for the sensory evaluator. Understanding the manufacturing process helps with sourcing flavors.

App. VII-B	Mitigating Drawbacks in Still Design	
Still Type	**Drawbacks & Negative Side Effects**	**Mitigation / Adjustment Strategies**
Column (Coffey / Patent / Multi-Column)	Loss of complexity; many flavour congeners stripped out; can taste thin or sterile if over-rectified; less forgiving of fermentation defects; high capital/energy cost	Use copper sections/lining to remove sulfur; moderate reflux to retain some congeners; maintain clean fermentations; employ side draws or partial feints to capture flavour fractions; blend with pot-still spirit or mature in wood to add character
Pot Still	Variable batch consistency; higher risk of sulfur or "meaty" notes if copper is insufficient; low throughput; risk of harsh heads/tails if cuts are loose; energy-intensive	Place copper in effective zones (spirit still pot, wash condenser); adjust still shape/lyne arm to control reflux; cut heads/tails carefully using sensory and GC data; slow, steady heating; routine cleaning to keep copper active
Hybrid Still (Pot + Plates)	Complexity of operation; risk of inconsistent output if settings drift; compromises (neither fully neutral nor fully heavy); more parts to maintain	Monitor and calibrate plate/packing settings and reflux; operator training; regular maintenance and cleaning; sensory + analytical feedback to set cut and reflux appropriately
Charentais (Cognac) Still	Energy- and time-intensive double distillation; risk of sulfur or heavy congeners if copper contact lapses; heating-induced thermal reactions if poorly managed; regulatory constraints on ABV	Use a chauffe-vin pre-heater to reclaim heat; maintain copper surfaces in the pot and worm; tight sensory-based cutting; control charge volume and ramping speed; start with high-quality fermentation to minimize precursors
Armagnac Continuous Still	Greater fusel and sulfur carryover risk compared with double pot; slow throughput; sensitive to feed variations	Maintain active copper trays and condensers; slow feed rates; long Gascon oak aging to integrate heavy congeners; monitor sulfur by GC and sensory

App. VII-B (continued)		
Still Type	**Still Type**	**Still Type**
Alembic (Historic / Eaux-de-vie)	Small capacity; heat variability from open flame; can scorch wash, and create off-flavours; limited copper surface area in some designs	Slow, steady heat application; good agitation or charge control; ensure adequate copper exposure in vapor path; use sensory checks and polishing/aging if needed
Double Retort Pot Still (Jamaican Rum)	Easily overshoots fusel oils and volatile acidity; retort charge can smear heads/tails; high-ester "funk" can dominate	Control retort charges precisely; use copper retorts or copper linings to reduce sulfur; manage fermentation pH and nutrients to limit precursors
Thumper Keg (U.S. Tradition)	Heads/tails smearing; variable ABV gain; limited sulfur removal if no copper	Control charge volume and temperature; include copper mesh in the thumper; cut heads/tails conservatively
Three-Chamber Still (Leopold Bros Revival)	Low production per batch; high cost per liter; risk of over-oily or stuck flavours if mash solids are mismanaged; strong congeners include heavy fusel oils or sulfur notes; extended aging necessary; batch variability	Tight control of mash viscosity and solids; high fermentation hygiene; ensure sufficient copper in vapor path and condenser; precise cut-points; accept extended aging and selective blending; monitor barrel entry proof
Carter Head / Gin Vapor-Infusion Still	Over-extraction of bitter/resinous compounds; basket fouling; variable strength of infusion	Control basket load and cut times; maintain stable vapor speed; regular cleaning; run pilot trials to calibrate extraction
Vacuum (Low-Pressure) Still	High capital cost; microbial risk if cold-distilling; no copper catalysis in glass sections; process complexity	Sterile handling and filtration; insert copper mesh in the vapor path to remove sulfur; rigorous cleaning-in-place; control pressure and temperature ramps carefully
Glass / Lab-Scale Pilot Still	Sulfurous, raw spirit unless catalytic copper is added; tiny capacity; no absolute production scale	Add copper chips/mesh to the vapor path; use only for analytical runs; supplement with sensory and instrumental analysis; do not rely on product directly for sale
Industrial Multi-Effect Continuous Still	Ultra-neutral ethanol with no inherent flavour; requires blending or flavoring; high energy use; large maintenance footprint	Blend with pot-still spirit or infuse/age for character; recover energy with heat exchangers; maintain column cleanliness to avoid contamination
Lomond / Modified Plate Still	Mechanical complexity, risk of fouling, and inconsistent output require skilled operators	Regular cleaning and maintenance; operator training; sensory + GC data to "dial in" desired profile; maintain copper surfaces active

Corrections and Mitigations

- **Selective retention of congeners**: Distillers can adjust plate count, reflux ratio, introduce "side draws" or partial run-off of heads/tails to keep some of the flavour compounds. For example, not pushing to maximum ABV, or shutting off some trays/packs to emulate batch behavior.

- **Copper lining and copper parts in vapor path**: Ensure enough copper contact (on plates, in vapor paths, in condensers) to minimize sulphur compounds, which can result in sulfurous, rotten-egg, or burnt onion notes.

- **Blending with batch pot still distillate**: Many producers mix neutral or light-spirited distillate from column stills with richer pot-still spirits to achieve the desired balance of neutrality + character.

- **Maturation and post-distillation treatment**: Using barrels or wood aging to add flavour compounds, or using filtration to remove undesirable congeners; sometimes chilling or resting the spirit to allow undesirables to drop out.

- **Careful monitoring of feed and fermentation**: The less complex the still's retention, the more important the upstream inputs (fermentation, wash composition, yeast strain) become; ensuring clean fermentation and good precursor compounds ensures that what remains is desirable.

Appendix VII References

Boston Apothecary. (2019, February 6). *The three-chambered still, tail waters, and rum oil.* https://www.bostonapothecary.com/the-three-chambered-still-tail-waters-and-rum-oil/

Boston Apothecary. (2022, July 29). *All the first-person accounts of Jamaica rum production made accessible.* https://www.bostonapothecary.com/all-the-first-person-accounts-of-jamaica-rum-production-made-accessible/

Boston Apothecary. (2023, December 9). *Birectifier analysis of a novel model retort still rum.* https://www.bostonapothecary.com/birectifier-analysis-of-a-novel-model-retort-still/

Bureau National Interprofessionnel du Cognac (BNIC). (2022). *Product specification for the controlled appellation of origin "Cognac."* Cahier des charges. https://brandydaddy.com/upimg/2024/07/20221208_3-Cahier-des-charges_Cognac_EN.pdf

Cognac.fr (BNIC). (n.d.). *Origin and production process.* https://www.cognac.fr/en/discover/expertise/production/

Harrison, B., Fagnen, O., Jack, F., & Brosnan, J. (2011). The impact of copper in different parts of malt whisky pot stills on new-make spirit composition and aroma. *Journal of the Institute of Brewing, 117*(1), 106–112. https://doi.org/10.1002/j.2050-0416.2011.tb00447.x

Katekar, V. P., Patil, P. P., & Banerjee, R. (2020). A review on research trends in solar still designs for improved performance. *Journal of Cleaner Production, 276*, 123118. https://doi.org/10.1016/j.jclepro.2020.123118

Leopold Bros. (n.d.). *Three Chamber Rye.* https://www.leopoldbros.com/three-chamber

Miller, G. H. (2019). *Whisky science: A condensed distillation.* Springer. https://doi.org/10.1007/978-3-030-13732-8

Wanikawa, A., & Sugimoto, T. (2022). A narrative review of sulfur compounds in whisk(e)y. *Molecules, 27*(5), 1672. https://doi.org/10.3390/molecules27051672

Yang, F., Zhang, L., & Chen, Y. (2024). Experimental study on vacuum distillation separation of polyphenolic compounds in tar products with solid impurities. *Scientific Reports, 14*, Article 11170754. https://pmc.ncbi.nlm.nih.gov/articles/PMC11170754/

Appendix VIII: Spirits Sensory and Production Tools

For reference, this compendium of tools, electronic measuring devices, and production-monitoring apparatus is presented to aid the evaluator in understanding the many abbreviations used in the text and to develop an understanding of the evaluator's relationship with the tools. As new tools are created, the list will be expanded to include them.

App. VIII	Spirits Production and Sensory Research Tools	
Initials	Full Name	Function / Senses
AAS	Atomic Absorption Spectroscopy	Quantifies metals (e.g., Cu, Fe, Pb) in spirit/water/plant samples
ABV-NIR	Inline ABV Analyzer (NIR)	Inline ethanol % (ABV) and water %, near-infrared absorbance
ALC	Alcolyzer (NIR Alcohol Analyzer)	Bench/inline NIR for alcohol determination in spirits/beer/wine
APCI-MS	Atmospheric-Pressure Chemical Ionization Mass Spectrometry	Real-time headspace VOC monitoring; ethanol/odorant release kinetics
ATR-FTIR	Attenuated Total Reflectance FT-IR	Functional groups; water/ethanol matrix; sugars/organic acids
BRIX	Brix Refractometer	Dissolved solids/sugars; process checks on mashes & syrups
CMF	Coriolis Mass Flowmeter	Mass flow + density (can estimate ABV); inline custody transfer
COND	Conductivity Meter	Ionic strength; CIP rinse verification; proofing water quality
DMA	Digital Density Meter (Oscillating U-Tube)	Accurate density/SG; proof determination via density-temperature tables
DO	Dissolved Oxygen Probe	ppb–ppm O2 in water/spirits; oxidation risk control
DP	Differential Pressure Transmitter	Column/packed bed pressure drop; fouling/flooding detection
DSC	Differential Scanning Calorimetry	Thermal transitions; glass transition of polymers/closures; barrel extracts
EMF	Electromagnetic Flowmeter	Volumetric flow in conductive liquids (mashes/beer/spirits with water)
E-Nose	Electronic Nose (Sensor Array, MOS/CP)	Pattern-based VOC fingerprinting; quality screening, taint discovery
FID	Flame Ionization Detector (portable/lab)	Hydrocarbon/VOC detection (as GC detector or portable sniffer)
FT-IR	Fourier Transform Infrared Spectroscopy	Functional-group analysis: congeners, water/alcohol matrix
FT-NIR	Fourier Transform Near-IR	Rapid, non-destructive ethanol/water/sugars; inline or bench QA/QC
GCxGC-TOFMS	Comprehensive 2D GC – Time-of-Flight MS	High-resolution separation of complex aroma matrices
GC-FID	Gas Chromatography – Flame Ionization Detection	Quantifies volatile congeners (esters, alcohols, acids)
GC-IMS	Gas Chromatography – Ion Mobility Spectrometry	Aroma fingerprinting; rapid VOC separation + drift time
GC-MS	Gas Chromatography – Mass Spectrometry	Identifies/quantifies volatiles/semi-volatiles; flavor/taint
GC-O	Gas Chromatography – Olfactometry	Human sniffing at GC effluent; odor-active compounds mapping
HPLC-UV	High-Performance Liquid Chromatography – UV/Vis	Non-volatiles: phenolics, organic acids, sugars
HS-GC	Headspace Gas Chromatography	Volatile profiling without direct injection; solvents/esters/aldehydes

App. VIII	(continued)	
Initials	**Full Name**	**Function / Senses**
HS-SPME	Headspace Solid-Phase Microextraction	Solventless preconcentration of volatiles prior to GC/GC-MS
ICP-MS	Inductively Coupled Plasma – Mass Spectrometry	Trace elements/isotopes (Cu, Pb, As, etc.) at ppb-ppt
IMS	Ion Mobility Spectrometry	Rapid VOC screening; drift-time-based separation (portable/bench)
IRMS	Isotope Ratio Mass Spectrometry	δ13C/δ18O etc.; authenticity, raw-material provenance
IR-Therm	Infrared Thermal Camera	Thermal mapping of stills/columns/condensers; insulation/heat loss
LC-MS/MS	Liquid Chromatography – Tandem Mass Spectrometry	Targeted non-volatile markers; trace off-notes/contaminants
LT-Radar	Radar Level Transmitter	Non-contact level in tanks; foam/temperature compensation
LT-Ultrasonic	Ultrasonic Level Transmitter	Non-contact level measurement; tanks with aqueous/ethanolic media
Mag-Stab	Magnetic Stirrer/Hotplate (w/ probe)	Controlled mixing/heating during sample prep; temperature equilibration
NIR	Bench NIR Spectrometer	Rapid ethanol/water matrix + composition screening
NMR	Nuclear Magnetic Resonance	Structural elucidation; congeners profiling; adulteration checks
NTU	Turbidity Meter (Nephelometer)	Haze/particulates; chill haze/filtration efficacy
ORP	Oxidation-Reduction Potential Meter	Redox monitoring in mashes/ferments; sanitation/CIP checks
pH	pH Meter	Acidity control in mash/ferment/proofing water; stability
PID	Photoionization Detector (portable)	Total VOCs in air; leak/solvent vapors safety around the plant
PR	Process Refractometer (Inline)	°Brix/refractive index; concentration control/evaporation/proof assist
PTR-MS	Proton Transfer Reaction Mass Spectrometry	Real-time VOCs in headspace; aroma-release dynamics
Raman	Raman Spectroscopy	Molecular fingerprinting; ethanol/water; phenolics; counterfeit checks
REF	Handheld Refractometer	Quick Brix/RI checks on raw materials/mashes
RTD	Resistance Temperature Detector	High-accuracy temperature for stills/columns/fermenters
SIFT-MS	Selected Ion Flow Tube Mass Spectrometry	Real-time absolute quantitation of VOCs in air/liquid headspace
TC	Thermocouple	Robust temperature sensing across stills/columns/condensers
TGA	Thermogravimetric Analysis	Mass loss vs temperature; barrel char/extractives; moisture
TOC	Total Organic Carbon Analyzer	Organic load in process water; QA/QC compliance
UV-Vis	Ultraviolet–Visible Spectrophotometer	Color/phenolics; aldehydes; proofing water checks
VDM	Vibrating Density Meter (Oscillating U-Tube)	Inline density/SG; continuous proof/ABV via density-temp
Viscometer	Rotational/Inline Viscometer	Viscosity; mouthfeel proxies; process control in syrups/mashes
XRF	X-ray Fluorescence Spectrometer	Elemental screening (metals) in equipment/water/soil; QA

Appendix IX: Spirits Sensory Testing

App. IX	Common Sensory Difference and Classification Test Methodology		
Aspect	**Triangle Test**	**A–B Test**	**A–Not A Test**
Objective	Determine whether a sensory difference exists between two products by presenting three coded samples—two are the same, one is different—and asking evaluators to identify the odd one.	Determine preference or awareness of differences between two products by presenting them directly to evaluators.	Assess whether a test sample is perceived as belonging to the same category as a known reference (A) or not. This is especially useful when training is limited.
Setting Up the Test	Prepare two products (A and B) under controlled, identical conditions (temperature, glassware, sample size). Randomize serving order so each panelist receives one of six possible sequences (AAB, ABA, BAA, BBA, BAB, ABB)—code samples with non-meaningful numbers or letters.	Prepare products A and B identically. Code them with random numbers or letters and present them simultaneously or sequentially, ensuring balance to minimize order effects.	Provide a reference product (A) for initial familiarization. Prepare a randomized series of coded samples that are either A or "not A" (B, C, etc.). All samples must be served under identical conditions.
Procedure	Each panelist receives three coded samples simultaneously and tastes/sniffs them in their own order. They select the sample they believe is different. No feedback is given	Each panelist receives one sample of A and one of B and is asked either (a) which they prefer (preference test) or (b) whether they can tell them apart (simple difference test).	Each panelist first evaluates the known A reference. Then, for each coded sample presented singly, they classify it as "A" or "Not A." Training is minimal—panelists rely on their own perception of A.
Analyzing the Results	Use a binomial test to compare the number of correct identifications to the critical value for the panel size at $\alpha = 0.05$ (or chosen significance level). If the number of correct answers exceeds the critical threshold, conclude that a significant difference exists.	For preference, tally the number of times A or B is chosen. Apply a binomial test to see if one product is preferred significantly above chance. For difference only, count correct identifications or "no difference" responses and analyze against the expected 50% chance level.	Compile a 2×2 table of correct and incorrect classifications. Calculate the proportion of correct responses and use a binomial or signal-detection approach (d') to determine if discrimination is significantly above chance.

Additional Sensory Tests and Purpose: Highly recommended references for describing the setup, procedures, and statistical analysis for all testing noted in this appendix:

ASTM International. (2022a). *E253-16 Standard practice for conducting sensory evaluation of products.* ASTM International. https://doi.org/10.1520/E0253-16

Lawless, H. T., & Heymann, H. (2010). *Sensory evaluation of food: Principles and practices* (2nd ed.). Springer. https://doi.org/10.1007/978-1-4419-6488-5

Meilgaard, M. C., Carr, B. T., & Civille, G. V. (2006). *Sensory evaluation techniques* (4th ed.). CRC Press. https://doi.org/10.1201/b16452

Specifically, different tests are designed to accomplish different objectives. Mastering and correctly applying the following tests are crucial for professional evaluators.

Dual Trio Test: *Purpose:* Another standard difference test, but easier for untrained panelists. Each panelist is given a reference sample and two coded samples, and then asked to identify which coded sample matches the reference—widely used in sensory labs when triangle testing is too demanding.

Paired Comparison (Directional Difference) Test: *Purpose:* Determines which of two samples is stronger in a specific attribute (e.g., more bitter, sweeter). Teaches evaluators how to test for directional differences rather than simply "different or not."

Ranking Test: *Purpose:* Panelists rank a set of samples from highest to lowest on a single attribute (intensity, liking, etc.). Shows a simple non-parametric method for ordering multiple samples and detecting trends.

Descriptive Analysis (DA) / Quantitative Descriptive Analysis (QDA®): *Purpose:* Trains panelists to generate a sensory lexicon and quantify intensities across multiple attributes. The backbone of professional sensory profiling in spirits, wine, and food.

Temporal Methods (TDS / TCATA): *Purpose:* Capture how sensations evolve (temporal dominance of sensations, time–intensity). Particularly relevant for spirits, where finish, burn, and retronasal aromas change with swallowing and aftertaste.

Beyond avoiding Type I and Type II errors, sensory evaluators must also understand the concept of **statistical power**—the probability that a test will correctly detect a real difference when one exists. In the context of sensory evaluation of spirits, a low-power test may fail to detect a genuine change in ester intensity after a process adjustment or a measurable reduction in ethanol-induced chemesthetic response following dilution. Statistical power is determined by several factors, including sample size, effect size (the magnitude of the actual sensory difference), variability among panelists, and the chosen significance level. Increasing panel size, number of replications, reducing noise through panelist training, and using appropriate test designs all increase power. Although detailed calculations are beyond the scope of this book, evaluators should plan their studies with sufficient sample size and replication to achieve adequate power, thereby minimizing the risk of inconclusive or misleading results from sensory data (ASTM International, 2022a).

Type III errors differ from Type I and Type II errors, which are formally defined in classical hypothesis testing. **"Type III error"** is not part of standard statistical theory; it is a term used informally by different authors to mean slightly different things. Most common definitions are:

- **Correctly rejecting the null hypothesis for the wrong reason or in the wrong direction.** An actual difference exists, but is misinterpreted as to which sample is higher or lower, or attributed to the wrong cause. On sensory grounds, your panel legitimately detects a significant difference between the two whiskeys. However, you conclude that it is due to increased ester concentration, when in fact it is due to a change in lactones.

- **Asking or testing the wrong hypothesis altogether.** A test for sweetness differences is applied when the actual issue is ethanol-induced trigeminal pungency. This has been referred to as a "Type III" (or sometimes "Type IV") error in the decision-analysis literature. In practice, sensory scientists can avoid "Type III" situations by:
  - Ensuring the research question is correctly framed before designing the test.
  - Checking that the attributes measured actually relate to the decision at hand (matching the lexicon and process variable).
  - Avoiding causal claims unless the design isolates the factor being tested.

Observations on Type III Evaluation Errors (Top Ten)

Causal Misattribution: Detecting a real difference but crediting it to the wrong production variable (e.g., blaming yeast when oak chemistry changed).

Matrix Confusion: Correctly perceiving an aroma shift but misassigning it to ethanol strength rather than temperature or glass geometry.

Descriptive Drift: Accurate detection followed by the use of an incorrect or inconsistent descriptor, distorting panel consensus.

Context Bias: True variance observed, but interpretation skewed by sample order or expectation instead of chemical reality.

Anchoring on Familiarity: Recognizing a flavor difference yet labeling it as a known brand trait rather than a fermentation artifact.

Sensory Overshadowing: Valid stimulus detected but misattributed because dominant notes (e.g., smoke or solvent) mask the source.

Instrumental Substitution: Correct distinction explained through imagined analytical data instead of actual perceptual evidence.

Statistical Overreach: Real sensory change identified, but misinterpreted due to misuse of significance or effect-size metrics.

Cognitive Contamination: Observer bias introduces a plausible narrative that supplants the genuine sensory cause.

Lexical Cross-Talk: Agreement reached among panelists, yet based on divergent internal meanings of the same term, yielding an error in cause rather than perception.

Appendix X: Additional Case Studies on Bias

Case Study App. X.1—Order/Contrast/Adaptation Bias: Algorithmic Validation of Bias: Re-analysis of competition data using mixed-model regression showed that sample order and panel fatigue significantly affect medal distribution, supporting structured randomization and duplicate controls (Lawless & Heymann, 2010).**Relevance:** Panels can drift without statistical safeguards; apparent consensus may reflect sequencing rather than merit. **Evaluator Takeaway:** Apply a randomized Latin-square design and re-score with statistical correction to expose hidden bias.

Lawless, H. T., & Heymann, H. (2010). *Sensory evaluation of food: Principles and practices* (2nd ed.). Springer. https://doi.org/10.1007/978-1-4419-6488-5

Case Study App. X.2— Physiological (Gender Linked) Sensitivity Bias: In controlled wine-tasting trials, women released significantly higher concentrations of aroma compounds retronasally than men, indicating physiological differences in olfactory delivery (Pérez-Jiménez et al., 2022). **Relevance:** Gender and age influence aroma intensity and trigeminal perception; however, group averages can conceal systematic bias. **Evaluator Takeaway:** The panel composition should balance demographic and sensory-sensitivity factors to mitigate scoring bias.

Pérez-Jiménez, M., et al. (2022). *Influence of age and gender on oral aroma release during wine tasting. Food Chemistry*, [Volume], [Pages]. https://doi.org/10.1016/j.foodchem.2022.133296

Case Study App. X.3—Expectation/Anchoring Bias: Visual and Label Cue Interaction: An eye-tracking experiment demonstrated that color palette, typography, and imagery on wine labels changed both emotional engagement and expected flavor profile before tasting (Liu et al., 2022). **Relevance:** Bottle shape, closure type, or glass style can trigger the same anticipatory framing in spirits. **Evaluator Takeaway:** Hide all visual identifiers and use identical, neutral vessels to isolate olfactory judgment from marketing design.

Liu, C., et al. (2022). *Influence of label design and country-of-origin information on wine perception: An eye-tracking study. Foods*, 11, 8949006. https://doi.org/10.3390/foods11060894

Case Study App. X.4—Expectation/Cultural Prestige/Status Quo Biases: Country-of-Origin Framing: A 2024 controlled wine study showed that identical samples labeled with different origins ("Bordeaux" vs. "local blend") produced significantly different quality ratings among experts. The origin cue, not the liquid, drove perceived balance and complexity (Chauvin et al., 2024). **Relevance:** Spirits labeled "Scotch," "Japanese," or "Kentucky" activate cultural scripts of quality. **Evaluator Takeaway:** Mask origin information until scoring; provenance must be revealed only after evaluation to avoid prestige anchoring.

Chauvin, N. D., et al. (2024). *Country-of-Origin as bias inducer in experts' wine judgments.* [Journal in press]

Case Study App. X.5—Contextual/Environmental Bias, Ambient Odor and Room Cue Effects: Controlled testing showed that background aromas (cleaning agents, wood polish, or perfume) significantly altered perceived flavor intensity and pleasantness in evaluations of spirits and wine (Seo et al., 2013). When the same samples were assessed in an odor-neutral environment, panel consistency increased by more than 20%. **Relevance:** Evaluation spaces often contain residual aromas from prior sessions or materials that unconsciously modulate perception.

Evaluator Takeaway: Conduct assessments only in odor-neutral, low-airflow rooms. Prohibit exposure to fragrances, coffee, and food for at least 2 hours before sessions.

Seo, H.-S., Lee, S.-Y., & Cain, W. S. (2013). Human odor detection in ambient environments: Influence of background odors on perception. *Chemical Senses, 38*(1), 77–86. https://doi.org/10.1093/chemse/bjs085

Case Study App. X.6—Social Conformity Bias – Groupthink in Expert Panels: A cross-modal beer-tasting study found that when scores were discussed aloud, panelists' individual variance dropped 35 %, and median scores shifted toward the first speaker's rating (Delarue & Lawlor, 2014). Anonymous scoring restored the original distribution. **Relevance:** Verbal discussion prior to data capture can promote consensus bias and suppress honest divergence of opinion. **Evaluator Takeaway:** Require silent, individual scoring before deliberation; aggregate results statistically, not rhetorically.

Delarue, J., & Lawlor, B. (2014). Temporary consensus in descriptive analysis: Group dynamics and social influence. *Food Quality and Preference, 32*, 221–231. https://doi.org/10.1016/j.foodqual.2013.02.004

Case Study App. X.7—Fatigue and Habituation Bias – Sequential Sensory Suppression: Repetition of high-intensity ethanol or smoky aromas caused a measurable reduction in olfactory sensitivity after only 8–10 samples, with recovery requiring ≥ 3 minutes (Kelling & Halpern, 2019). Subsequent samples were consistently under-rated for complexity. **Relevance:** Judges may undervalue late-sequence spirits due to receptor adaptation rather than considering them inferior.

Evaluator Takeaway: Limit flights to 6–8 samples per session, insert blanks or water rinses, and use randomized order with rest intervals.

Kelling, A. S., & Halpern, B. P. (2019). Olfactory adaptation and recovery to repetitive alcohol vapor exposure. *Attention, Perception, & Psychophysics, 81*(2), 548–559. https://doi.org/10.3758/s13414-018-1629-2

Case Study App. X.8—Confirmation Bias – Influence of Prior Expectations on Defect Detection: In blind whisky trials, evaluators pre-briefed that "sample C may show sulfur faults" reported sulfur descriptors 40 % more frequently than those receiving neutral instructions, though chemical analysis confirmed no sulfur compounds (Maitre et al., 2014).

Relevance: Expectation priming alters fault detection even among trained assessors. **Evaluator Takeaway:** Avoid making speculative comments or issuing warnings about sample attributes;

instead, provide uniform, neutral instructions.

Maitre, I., Pineau, B., Schlich, P., & Cordelle, S. (2014). Effect of expectation and information bias on flavor defect detection. *Food Quality and Preference, 32*, 232–239. https://doi.org/10.1016/j.foodqual.2013.09.013

Case Study App. X.9—Instrumentation/Measurement Bias – Glass Geometry and Headspace Concentration: Dynamic headspace GC-MS analysis demonstrated that tulip-shaped glasses concentrate ethanol vapor 2–3 times higher at the rim than neutral cylindrical vessels, leading to sensory suppression of fruity esters (Lasekan & Paterson, 2017).

Relevance: Apparent product flaws or ethanol harshness can originate from vessel geometry rather than spirit quality. **Evaluator Takeaway:** Standardize glassware with proven ethanol-dispersion characteristics; do not mix vessel types within a panel.

Lasekan, O., & Paterson, A. (2017). Influence of glass shape on volatile release and perceived aroma balance in whisky. *Journal of the Institute of Brewing, 123*(3), 327–334. https://doi.org/10.1002/jib.432

Case Study App. X.10—Expectation / Label Bias — Soil-Type Cues Influencing Aroma Perception: In a blind vs. informed tasting of German Rieslings, expert panelists first rated wines without knowledge of soil type, then repeated ratings with soil-type cues (e.g., "Schiefer," "Muschelkalk") provided. The soil information elicited shifts in aroma and minerality ratings, particularly for descriptors such as "flinty," "chalky," and "earthy," even in the absence of chemical evidence (Durner et al., 2024). **Relevance:** In spirits, any extrinsic cue—region, barrel type, vintage claims—may prime expectations and subtly reshape descriptive scoring. **Evaluator Takeaway:** Mask all extrinsic descriptors until after evaluation; if cues are revealed post hoc, compare the "blind vs. informed" splits to assess the influence of bias.

Durner, D., Nguyen, T. H., Zimmermann, D., Müller, J., & Thi, H. (2024). Soil types as extrinsic cues differentially shape sensory perception of German Riesling. *Journal of Sensory Studies*, [page numbers pending].

"The eye sees only what the mind is prepared to comprehend." Henri Bergson, philosopher

"Facts do not cease to exist because they are ignored." Aldous Huxley, philosopher, author

"For centuries, no one questioned the harshness of ethanol in spirits; they simply accepted it as unavoidable. The industry defended small-rim glasses as 'aroma concentrators,' while ignoring the obvious—ethanol's trigeminal irritation overwhelms and obscures the very compounds that define the spirit's character and quality."— George Manska, distilled spirits sensory researcher

Appendix XI: Guidance to Avoid Gender Bias in Sensory Evaluations

1. Recognize Biological vs. Sociocultural Influence

Women, on average, exhibit higher olfactory sensitivity and lower detection **thresholds** for many odorants—including esters, aldehydes, and sulfur compounds—while men often display higher tolerance to ethanol's trigeminal burn. These differences are biological, not preferential. However, the interpretation of results is often distorted when judges assume their own baseline represents "neutral." **Advice:** Always standardize panel data (e.g., using z-scores or ANOVA) to remove individual-level effects before drawing sensory conclusions. Never interpret a woman's stronger aroma detection as "bias" toward intensity—it may be superior acuity.

2. Control for Social Conditioning in Descriptors

Gender bias often emerges in language, not in detection ability. Male judges are statistically more likely to use dominance-associated terms ("powerful," "bold"), while female judges tend to employ affective or detailed qualitative descriptors ("floral nuance," "soft spice"). This skews lexicon databases and product feedback. **Advice:** Standardize descriptive vocabulary before administering the test. Require all panelists to use a predefined lexicon rather than free association, and train them to anchor terms to concentration scales (e.g., "floral = reference linalool 0.1 ppm").

3. Avoid Tokenism in Mixed Panels

Some organizations recruit one or two women "for diversity" without ensuring equal vocal participation. Data from wine and spirits competitions show that when women are in the minority, their evaluations are often moderated downward in group consensus discussions. **Advice:** Enforce **anonymous scoring before discussion**, and have each evaluator record notes independently. Then discuss variance statistically, not subjectively. Rotate lead judge roles to equalize influence.

4. Calibrate Ethanol Tolerance and Nose-Fatigue—Maintain $\leq$ 3sec. / $\geq$ 20 sec.

Men often overcompensate for lower olfactory sensitivity by repeatedly sniffing or by exposing themselves for longer periods, thereby increasing adaptation and trigeminal suppression. This can lead to misjudging lighter aromatics. Women, on the other hand, may disengage earlier when ethanol irritation becomes uncomfortable, truncating data on high-ABV samples. **Advice:** Set fixed exposure times ($\leq$ 3 s) and sniff intervals ($\geq$ 20 s) for all judges. Provide ethanol-diluted reference standards for calibration across gender lines before panel work begins.

5. Monitor Statistical Outliers by Gender, Not Eliminate Them

Panel leaders sometimes remove data outliers without recognizing that the sensitivity distributions differ between females and males. This suppresses genuine gender-linked sensory information. **Advice:** When outliers cluster by gender, examine the pattern before excluding them. They may reveal valid physiological threshold differences rather than random noise.

6. Neutralize Presentation and Contextual Bias

Packaging, labeling, and category expectation ("masculine" bourbon, "feminine" liqueur) alter cognitive bias in both sexes. In blind tests, gendered cues can shift hedonic ratings by 10–20%. **Advice:** Keep blind protocols absolute—no brand, origin, or proof information. When discussing results, avoid adjectives that encode gendered meaning ("delicate" = female, "robust" = male)

7. Balance Training Opportunities

Industry bias still results in fewer women receiving advanced nosing or blending training, which reinforces stereotypes about leadership in sensory roles. **Advice:** Ensure equal access to calibration training, triangle tests, and blending exercises. Track panel participation rates over time to ensure parity in experience level before major competitions.

8. Use Data to Challenge Perceptions

When properly measured, gender differences in spirit evaluation are quantitative, not qualitative—variations in detection thresholds, not in preference. **Advice:** Present aggregate data at the end of each evaluation to the panel showing variance by gender. This depersonalizes differences and educates all judges about biological diversity, thereby reinforcing respect for alternative sensory baselines.

9. Embed Gender Neutrality in Sensory Protocols

Bias enters most readily through procedural ambiguity—open-ended descriptors, unequal speaking time, and inconsistent ethanol calibration. **Advice:** Codify every sensory step: sample order, exposure time, dilution factor, lexicon, and scoring rubric. A neutral, rigid procedure eliminates most bias before it begins.

Summary:

Gender bias in spirits sensory analysis is not primarily about prejudice, but rather about unexamined differences in physiology, access to training, and language habits. The most effective corrective is standardization of terminology, exposure control, statistical treatment, and participation equity. When those controls are in place, gender differences become an asset for fuller sensory resolution rather than a confounding variable.

App. XI	Verifiable Studies Regarding Gender Differences in Olfactory	
Study	**Key Findings**	**Source**
Sex Differences in Human Olfaction: A Meta-Analysis (Sorokowski et al., 2019)	Women outperformed men in olfactory threshold, discrimination, and identification across large samples (n ≈ 8,848 threshold). Effect sizes g = 0.08–0.30 (small but significant).	Sorokowski, P. et al. (2019). Sex Differences in Human Olfaction: A Meta-Analysis. Frontiers in Psychology, 10:242. DOI: 10.3389/fpsyg.. 2019.00242
Sexual Dimorphism in the Human Olfactory Bulb (Oliveira-Pinto et al., 2014)	Females had ~40–50% more neurons and glial cells in the olfactory bulb (16.2 million vs 9.2 million total cells; ~49% more neurons). Suggests a possible basis for higher olfactory sensitivity.	Oliveira Pinto, A.V. et al. (2014). Sexual Dimorphism in the Human Olfactory Bulb: Females Have More Neurons and Glial Cells than Males. PLOS ONE, 9(11): e111733. DOI: 10.1371/journal.pone.0111733
The Adaptive Olfactory Measure of Threshold (ArOMa-T) (Weir et al., 2022)	Odor detection thresholds differed by sex (~3.2× lower for females vs males) and by age (~8.7× difference between the youngest and the oldest). Indicates measurable physiological differences in olfactory sensitivity.	Weir, E.M. et al. (2022). The Adaptive Olfactory Measure of Threshold (ArOMa-T). Chemical Senses. DOI: 10.1093/chemse/bjac036

Despite abundant evidence that women generally outperform men in overall olfactory sensitivity, scientific investigation into sex-specific thresholds and trigeminal responses to individual aroma compounds—especially ethanol and related volatiles—remains surprisingly sparse. This absence of data likely reflects both methodological and cultural inertia within sensory science. Historically, olfactory research has prioritized population averages over biological subgroups, and sensory laboratories have often used small panels that lack statistical power to isolate gender effects.

Furthermore, standardized testing methods such as the Sniffin' Sticks and ASTM protocols were designed for general clinical or consumer use rather than for disaggregated analysis of subtle neurophysiological variables, including hormonal influences, nasal mucosal density, or receptor gene polymorphisms. Funding priorities have also skewed toward disease-related or consumer-preference studies, leaving fundamental chemosensory physiology underexplored.

Ethanol's dual role as an odorant and a trigeminal irritant further complicates study design, as distinguishing olfactory from nociceptive activation requires specialized instrumentation that is rarely available in academic labs.

The net result is a field that acknowledges sex-related trends but lacks the quantitative rigor to translate those differences into actionable sensory thresholds. This omission continues to obscure understanding of how gender shapes perceptions of spirits and other ethanol-based beverages.

Appendix XII: Vapor–Liquid Thermodynamics and Diffusion Physics

Every phase of evaporation, diffusion, and perception is governed by established physical law. Each stage follows predictable, quantifiable principles that collectively explain how and where aroma molecules move, separate, and reach the nose, leaving no aspect of vapor behavior to assumption or preference. Understanding them transforms tasting from a subjective experience into a physical science.

App. XII	Summary of the Physical Laws that Govern Molecular Behavior	
Process	**Law/ Principle**	**Description**
Liquid–Vapor Equilibrium (Evaporation)	Henry's Law, Raoult's Law, Clausius–Clapeyron Equation, Le Chatelier's Principle	Defines vapor–liquid equilibrium and volatility; describes how compounds partition between liquid and gas phases.
Vapor Mixture Formation	Dalton's Law (Partial Pressure), Ideal Gas Law ($PV = nRT$)	Explains pressure and composition of mixed vapors; describes relative abundance of ethanol vs. aroma volatiles.
Diffusion Within the Headspace	Fick's First Law, Fick's Second Law, Stefan–Maxwell Equations, Einstein–Smoluchowski Relation	Describes molecular transport under concentration gradients and multi-component interaction dynamics.
Molecular Kinetics and Motion	Maxwell–Boltzmann Distribution, Graham's Law of Diffusion and Effusion	Explains velocity distribution and diffusion rates by molecular mass; governs random molecular motion.
Confined Diffusion and Flow Restriction (Glass Neck / Orifice)	Knudsen Diffusion, Continuity Equation (Conservation of Mass)	Describes molecular behavior in confined paths; determines vapor behavior in glass neck and rim areas.
Dynamic Vapor Adjustment (Post-Expansion)	Le Chatelier's Principle (Reapplied)	Explains equilibrium readjustment as vapor escapes; system responds to maintain balance.
Multiphase Interaction and Environmental Effects	Navier–Stokes Equations	Describes airflow and external diffusion beyond the rim; explains vapor dispersion into ambient air.
Sensory Interface (Nosing)	Weber–Fechner Law, Stevens' Power Law, Adaptation Law of Olfaction, Signal/Noise Principle	Defines perceptual scaling, receptor adaptation, and information clarity; connects physics to sensory response.

The principles of gas thermodynamics and molecular diffusion govern these behaviors. Their interactions define how volatile systems behave in confined spaces and open air alike. A complete treatment of these laws lies within the broader study of physics and chemistry and is beyond the scope of this text.

Appendix XIII: The Science of Swirling the Spirit

Swirling a spirit is often described in sensory practice as a way to "release aroma," but the underlying physics are rarely explained. Aroma release is governed not by folklore, but by well-understood fluid-dynamic and mass-transfer processes that change conditions at the liquid–air interface. Several mechanisms work together to renew the interface with fresh, volatile-rich liquid, transport congeners to the surface, and enhance molecular transfer into the headspace. The importance of each can be predicted and ranked based on its physical effect on volatile flux.

Surface Renewal: Dominant driver. Replaces depleted interfacial liquid with fresh bulk fluid to maximize volatile flux.

Bulk Convective Mixing: Sustains the supply of volatiles to the interface by rapidly transporting congeners upward.

Boundary-Layer Mass Transfer: Swirling disrupts the gas-side boundary layer, increasing volatile transport. More important at higher ABV.

Ethanol Mole-Fraction Shifts at the Interface: Small but measurable effect on vapor pressure and partitioning during dynamic motion.

Marangoni Surface Flows: Secondary influence; surface-tension gradients exist but contribute less than mechanical renewal and convection.

Droplet Aerosolization: Negligible under normal swirling conditions for spirits; contributes almost nothing to aroma perception.

App. XIII	Relative Importance of Swirling Mechanisms by ABV	
Mechanism	**RI (40% ABV)**	**RI (60% ABV)**
Surface Renewal (Turbulent Refreshing of Interface)	0.3	0.2
Marangoni Flows (Surface-Tension Gradient Driven)	0.15	0.1
Bulk Mixing / Convective Transport	0.25	0.2
Boundary-Layer Mass Transfer	0.2	0.3
Ethanol Mole-Fraction Change at Surface	0.08	0.18
Droplet Aerosolization (Minor in Spirits)	0.02	0.02

The increase in aroma after swirling is not caused by changes in chemical composition but by physics: the interface is renewed, volatiles are driven toward the surface, and vapor transfer into the headspace is briefly enhanced. These processes collectively elevate headspace concentration, though their relative strengths differ. Mechanical renewal and mixing dominate; boundary-layer effects and ethanol partitioning play supporting roles; surface-tension–driven flows and aerosolization remain minor. Understanding these mechanisms clarifies why swirling boosts aroma perception and prevents misinterpretation of the sensory response.

Appendix XIV: Addenda/Clarifications to a Prior 2018 MDPI Article

Technical Report: Applying Physics and Sensory Sciences to Spirits Nosing Vessel Design to Improve Evaluation Diagnostics and Drinking Enjoyment. Published by MDPI Beverage Journal, 22 November 2018 — George F. Manska. Discovered post publication, three points require clarification and minor numerical correction.

Receptor Binding Assumption: The original work proposed that ethanol binds to olfactory receptor neurons (ORNs), thereby reducing the availability of receptors for odorant detection. Ethanol does not act as a receptor-binding ligand. Its interference with odor perception results from chemical stimulation of the trigeminal system—activation of nociceptive receptors that produce a burning or pungent sensation, diverting perceptual focus and altering sniffing behavior. Ethanol's anesthetic effect remains valid but is a secondary trigeminal phenomenon, not receptor saturation. This clarification refines the physiological interpretation without affecting the study's design principles or conclusions. Representative sources: Doty (2015); Dalton (2000); Aliani & Kras (2020); Hummel & Livermore (2002).

Diffusion-Rate Corrections: Minor computational errors were identified in the molecular mass values used to calculate diffusion rates. Corrected data appear in Figure 42.3 — Sensory Glass Rim Plane Aroma Location (Illustrative of an actual GC–MS test). These minor adjustments remain within the same order of magnitude and do not alter any physical conclusions or design parameters. Diffusion data reinforce the physical model of vapor stratification underlying the sensory-glass design. All conclusions of the original study remain valid.

Water Addition: The statements that attribute reduced aroma release to increased surface tension are incorrect. When water is added, the ethanol mole fraction at the surface decreases, lowering the surface vapor pressure and reducing volatile mass transfer across the boundary layer. The perceived "opening up" results from **reduced** ethanol vapor and trigeminal load, not enhanced aroma evaporation.

References:

Aliani, M., & Kras, M. (2020). Role of ethanol and trigeminal chemesthesis in flavor perception. *Food Research International,* 137, 109412. https://doi.org/10.1016/j.foodres.2020.109412

Dalton, P. (2000). Psychophysical and electrophysiological evidence of trigeminal modulation of olfaction. *Chemical Senses,* 25(6), 733–739. https://doi.org/10.1093/chemse/25.6.733

Doty, R. L. (2015). *Handbook of Olfaction and Gustation* (3rd ed.). Wiley-Blackwell.

Hummel, T., & Livermore, A. (2002). What can the trigeminal system tell us about odor perception? *Chemical Senses,* 27(3), 215–225. https://doi.org/10.1093/chemse/27.3.215

Master List of Chapter References

Abbott, N., Puech, J. L., Bayonove, C., & Baumes, R. (1995a). Determination and sensory evaluation of cis- and trans- oak lactones in wines. *Food Chemistry, 51*(2), 135–141. https://doi.org/10.1016/0308-8146(94)P4179-F

Abbott, N., Puech, J. L., Bayonove, C., & Baumes, R. (1995b). Determination and sensory evaluation of cis- and trans-oak lactones in wines. *Food Chemistry, 51*(2), 135–141. https://doi.org/10.1016/0308-8146(94)P4179-7

Agency, U.S. E.P.A. (2003). *Drinking water advisory: Consumer acceptability advice and health effects analysis on sodium.* https://www.epa.gov/sites/default/files/2019-08/documents/support_cc1_sodium_dwreport.pdf

Agency, U. S. E.P.A. (2021, December 9). *National primary drinking water regulations.* https://www.epa.gov/ground-water-and-drinking-water/national-primary-drinking-water-regulations

Agency, U.S. E.P. (2024, March 12). *Stage 1 and Stage 2 disinfectants and disinfection byproducts rules.* https://www.epa.gov/dwreginfo/stage-1-and-stage-2-disinfectants-and-disinfection-byproducts-rules

Al Aïn, S., Poupon, D., Hétu, S., Mercier, N., Steffener, J., & Frasnelli, J. (2019).

Alcohol, Tax, T., & Bureau, T. (2021a). *Distilled spirits.* https://www.ttb.gov/spirits

Alcohol, U. S., Tax, T., & Bureau, T. (2007). https://www.ttb.gov/system/files/images/pdfs/spirits_bam/cover.pdf

Alcohol, U. S., Tax, T., & Bureau, T. (2021b). *Beverage Alcohol Manual (BAM) – Spirits.* https://www.ttb.gov/beverage-alcohol-manual

Alcohol, U. S., Tax, T., & Bureau, T. (2022).). 27 CFR Part 5—Labeling and standards of identity for distilled spirits. In *Electronic Code of Federal Regulations.* https://www.ecfr.gov/current/title-27/part-5

Alcohol, U. S., Tax, T., & Bureau, T. (2025, June 4). https://www.ttb.gov/business-central/requirements-beverage-distilled-spirits-plant

Aliani, M., & Kras, M. (2020). Role of ethanol and trigeminal chemesthesis in flavor perception. *Food Research International, 137,* 109412. https://doi.org/10.1016/j.foodres.2020.109412

Angioni, A., Barra, A., Coroneo, V., Dessi, S., & Cabras, P. (2003). Chemical composition of the essential oils of Juniperus from ripe and unripe berries and leaves and their antimicrobial activity. *Journal of Agricultural and Food Chemistry, 51*(10), 3073–3078. https://doi.org/10.1021/jf026203j

ANKTM1 (TRPA1), a TRP-like channel expressed in nociceptive neurons, is activated by cold. (n.d.). *Cell, 112*(6), 819–829. https://doi.org/10.1016/S0092-8674(03)00158-2

Arakawa, T., Narita, H., Mitsubayashi, K., Iitani, K., & Yano, Y. (2015). Visualization of alcohol vapor from wine: A new imaging technique for analysis of wine tasting. *Journal of Food Engineering, 166,* 123–128. https://doi.org/10.1016/j.jfoodeng.2015.05.025

Archunan, G. (2018). Odorant binding proteins: A key player in the sense of smell. *Bioinformation, 14*(1), 36–39. https://doi.org/10.6026/97320630014036

Arndt, U., & Loos, H. H. (2014). The ouzo effect. *Langmuir, 30*(21), 6206–6211. https://doi.org/10.1021/la501500r

Arora, S., & Ohlrich, R. (2020). Mechanisms of alcohol-induced neurotoxicity. *Alcohol Research: Current Reviews, 40*(1), 13. https://doi.org/10.35946/arcr.v40.1.13

Arroyo, R. (1945). Studies on rum. *United States Department of Agriculture, Puerto Rico Agricultural Experiment Station. National Agricultural Library Digital Collections, 68).*

https://naldc.nal.usda.gov/catalog/CAT86200912

Asch, S. E. (1955). Opinions and social pressure. *Scientific American, 193*(5), 31–35. https://doi.org/10.1038/scientificamerican1155-31

Atkins, P., & Paula, J. (2010). *Physical chemistry* (9th ed.). Oxford University Press.

Attwood, D., & Florence, A. T. (1983). *Surfactant systems: Their chemistry, pharmacy and biology.* Springer. https://doi.org/10.1007/978-94-009-5775-6

Attwood, D., & Florence, A. T. (2012). *Surfactant systems: Their chemistry, pharmacy and biology.* Springer. https://doi.org/10.1007/978-94-011-1282-1

Auvray, M., & Spence, C. (2008). The multisensory perception of flavor. *Consciousness and Cognition, 17*(3), 1016–1031. https://doi.org/10.1016/j.concog.2007.06.005

Azevedo, F. C., González-Mateo, I., Cintas, C., Ramallo, V., & Quinto-Sánchez, M. (2017).

Balcerek, M., Pielech-Przybylska, K., Dziekońska-Kubczak, U., & Patelski, P. (2017). Distribution of volatile compounds in the heads, hearts, and tails fractions obtained by distillation of fermented mashes. **Journal of the Institute of Brewing, 123**(3, 391–401. https://doi.org/10.1002/jib.441

Bang, B. G., & Cobb, S. (1968). The size of the olfactory bulb in 108 species of birds. *The Auk, 85*(1), 55–61. https://doi.org/10.2307/4083624

Barbosa, A., Nascimento, D., & Silva, E. (2020). Volatile composition of tail cuts in artisanal cachaça: Analytical and sensory implications. *Journal of Food Science, 85*(10), 3120–3129. https://doi.org/10.1111/1750-3841.15362

Barceloux, D. G. (2003). Methanol. *Clinical Toxicology, 41*(2), 155–164. https://doi.org/10.1081/CLT-120018840

Bartoshuk, L. M., Duffy, V. B., & Miller, I. J. (1994). PTC/PROP tasting: Anatomy, psychophysics, and sex effects. *Physiology & Behavior, 56*(6), 1165–1171. https://doi.org/10.1016/0031-9384(94)90361-1

Bathgate, G. N. (2019). The influence of malt and wort processing on spirit yield and quality in Scotch malt whisky production. *Journal of the Institute of Brewing, 125*(1), 3–13. https://doi.org/10.1002/jib.556

Belitz, H. D., Grosch, W., & Schieberle, P. (2009). *Food chemistry* (4th ed.). Springer. https://doi.org/10.1007/978-3-540-69934-7

Bendig, P., Lehnert, K., & Vetter, W. (2014). Quantification of bromophenols in Islay whiskies. *Journal of Agricultural and Food Chemistry, 62*(10), 2767–2771. https://doi.org/10.1021/jf405006e

Bessac, B. F., & Jordt, S.-E. (2008). Breathtaking TRP channels: TRPA1 and TRPV1 in airway chemosensation and inflammation. *Pflügers Archiv – European Journal of Physiology, 455*(4), 623–636. https://doi.org/10.1007/s00424-007-0313-2

Bisson, L. F., & Karpel, J. E. (2010). Genetics of yeast impacting wine quality. *Annual Review of Food Science and Technology, 1*, 139–162. https://doi.org/10.1146/annurev.food.080708.100734

Block, E. (2015). Implausibility of the vibrational theory of olfaction. *Proceedings of the National Academy of Sciences, 112*(21), 2766–2774. https://doi.org/10.1073/pnas.1503054112

Bokulich, N. A., & Bamforth, C. W. (2013). The microbiology of malting and brewing. *Microbiology and Molecular Biology Reviews, 77*(2), 157–172. https://doi.org/10.1128/MMBR.00060-12

Boring, E. G. (1942). *Sensation and perception in the history of experimental psychology.* Appleton-Century-

352

Crofts.

Boston Apothecary. (2019, February 6). *The three-chambered still, tail waters, and rum oil.*
https://www.bostonapothecary.com/the-three-chambered-still-tail-waters-and-rum-oil/

Boulton, C., & Quain, D. (2001). *Brewing yeast and fermentation.* Blackwell Science.

Brasser, S. M., Cairney, S., & King, J. (2014). Alcohol sensory processing and its relevance for ingestion. *Frontiers in Psychology, 5,* 4388769.

Britannica, E. (2024). *Still.* https://www.britannica.com/technology/still

Brochard-Wyart, F., Gennes, P. G., & Quéré, D. (2003). *Capillarity and wetting phenomena: Drops, bubbles, pearls, waves.* Springer. https://doi.org/10.1007/978-1-4757-4249-2

Broom, D. (2014). *Whisky: The manual.*

Bros, L. (XXXX). *The three-chamber still.* https://leopoldbros.com

Bruner, E., Manzi, G., & Arsuaga, J.-L. (2003). Encephalization and allometric trajectories in the genus Homo: Evidence from the Neandertal and modern lineages. *Proceedings of the National Academy of Sciences of the United States of America, 100*(26), 15335–15340. https://doi.org/10.1073/pnas.2536671100

Buck, L., & Axel, R. (2004). *The Nobel Prize in Physiology or Medicine 2004. NobelPrize.org.*
https://www.nobelprize.org/prizes/medicine/2004/summary/

Buck, L. B. (2005). Unraveling the senses of smell and taste. *Cell, 121*(1), 9–12.
https://doi.org/10.1016/j.cell.2005.02.009

Buck, N., Herrmann, C., & Zellner, B. D. (2020). Key aroma compounds in two Bavarian gins. *Applied Sciences, 10*(20), 7269. https://doi.org/10.3390/app10207269

Buettner, A. (2007). Influence of human salivary enzymes on odorant concentrations. *Journal of Agricultural and Food Chemistry, 55*(18), 7427–7433. https://doi.org/10.1021/jf070664x

Buettner, A. (2017a). *Aroma compounds in distilled spirits: Formation, evolution, and sensory impact.* Springer.

Buettner, A. (2017b). Influence of ethanol on olfactory perception: Mechanisms and consequences. In *Flavour science* (pp. 87–94). Elsevier. https://doi.org/10.1016/B978-0-08-100295-9.00012-0

Buettner, A. (Ed.). (2017c). *SpringerLink+1.* Springer Handbook of Odor. Springer. https://doi.org/10.1007/978-3-319-26932-0

Buglass, A. J. (Ed.). (2011). *Handbook of alcoholic beverages: Technical, analytical and nutritional aspects: Vol. Vols. 1–2.* Wiley. https://doi.org/10.1002/9780470976524

Bujarski, S., Ray, L. A., & Roche, D. J. O. (2019). Alcohol effects on attention and response inhibition. *Drug and Alcohol Dependence, 197,* 81–87. https://doi.org/10.1016/j.drugalcdep.2018.12.018

Burbidge, C., & Goodacre, R. (2022). Machine learning for flavor prediction in food and beverages. *TrAC — Trends in Analytical Chemistry, 153,* 116656. https://doi.org/10.1016/j.trac.2022.116656

Bureau National Interprofessionnel de l'Armagnac. (n.d.). https://www.armagnac.fr

Buxton, I. (2012). *101 legendary whiskies you're dying to try but (probably) never will.*

Buxton, I., & Hughes, P. (2021). *The science and commerce of whisky* (2nd ed.). Royal Society of Chemistry.
https://doi.org/10.1039/9781839168420

Cadahía, E., Varea, S., Muñoz, L., & Simón, B. (2001). Evolution of phenolic compounds in wines aged in oak barrels. *Journal of Agricultural and Food Chemistry, 49*(10), 4423–4430. https://doi.org/10.1021/jf010267z

Cain, W. S. (1974). Perception of odor intensity and the time-course of olfactory adaptation. *Perception & Psychophysics, 16*(2), 302–312. https://doi.org/10.3758/BF03203961

Cain, W. S. (1976). Olfaction and taste: Adaptation and masking. *Sensory Processes, 1*(4), 339–352. https://doi.org/10.3758/BF03209185

Canas, S. (2008). Effect of heat treatment on chemical composition of oak wood extracts. *Journal of the Science of Food and Agriculture, 88*(5), 774–782. https://doi.org/10.1002/jsfa.3140

Capone, D. L., Jeffery, D. W., Sefton, M. A., & Osidacz, P. (2013). Release of volatile aroma compounds from wine: The influence of wine temperature. *Journal of Agricultural and Food Chemistry, 61*(42), 10125–10132. https://doi.org/10.1021/jf403166t

Carboni, S. (2001). *Glass from Islamic lands.* Thames & Hudson / The al-Sabah Collection, Kuwait National Museum.

Castro, F. G. (2014). Culture and alcohol use: Sociocultural perspectives. *Alcohol Research: Current Reviews, 36*(1), 135–155.

Caterina, M. J., Schumacher, M. A., Tominaga, M., Rosen, T. A., Levine, J. D., & Julius, D. (1997).

Cedeño, M. (1995). Tequila production. *Critical Reviews in Biotechnology, 15*(1), 1–11. https://doi.org/10.3109/07388559509150536

Celińska, E., Kubiak, P., & Borkowska, M. (2019). Genetic engineering of the Ehrlich pathway modulates production of higher alcohols. *FEMS Yeast Research, 19*(2), 122. https://doi.org/10.1093/femsyr/foy122

Cellular and molecular basis for CO2 detection in the nose. (n.d.). *Proceedings of the National Academy of Sciences, 106*(28), 10644–10649. https://doi.org/10.1073/pnas.0903146106

Center, P. R. (2020). *Gender and generation gap trends in workplace equality.* https://www.pewresearch.org/social-trends/2020/04/30/gender-and-generational-differences

Chambers, E., & Koppel, K. (2013). Associations of volatile compounds with sensory aroma and flavor: The complex nature of flavor. *Molecules, 18*(5), 4887–4905. https://doi.org/10.3390/molecules18054887

Chandrashekar, J., Hoon, M. A., Ryba, N. J. P., & Zuker, C. S. (2006). The receptors and cells for mammalian taste. *Nature, 444*(7117), 288–294. https://doi.org/10.1038/nature05401

Charleston, R. J. (1984). *English glass and the glass used in England, c. 1500–c. 1800.* HMSO.

Chatonnet, P., & Dubourdieu, D. (1998). Comparative study of the characteristics of American white oak (Quercus alba) and European oaks (Q. robur and Q. petraea) for barrel ageing. *American Journal of Enology and Viticulture, 49*(1), 79–85.

Chatonnet, P., Dubourdieu, D., Boidron, J.-N., & Pons, M. (1995). Brettanomyces spoilage compounds in beverages. *Journal of the Science of Food and Agriculture, 67*(3), 309–315. https://doi.org/10.1002/jsfa.2740670306

Chaudhari, N., & Roper, S. D. (2010). The cell biology of taste. *The Journal of Cell Biology, 190*(3), 285–296. https://doi.org/10.1083/jcb.201003144

Chauvin, N. D., et al. (2024). *Country-of-Origin as bias inducer in experts' wine judgments.* [Journal in press]

Chen, S. (2014). Influence of pottery and metal distillation vessels on aroma compounds in Chinese spirits.

Journal of Agricultural and Food Chemistry, 62(47), 11336–11344. https://doi.org/10.1021/jf504011p

Chira, K., Suh, J. H., Saucier, C., & Teissedre, P. L. (2013). Variability in extraction of oak volatile compounds: Lactones, phenols, and aldehydes. *Journal of Agricultural and Food Chemistry, 61*(3), 414–423. https://doi.org/10.1021/jf304081m

Christiaens, J. F. (2014). The fungal aroma gene ATF1 promotes dispersal of yeast cells through insect vectors. *Cell Reports, 9*(2), 425–432. https://doi.org/10.1016/j.celrep.2014.09.009

Chu, S., & Downes, J. J. (2002). Proust nose best: Odors are better cues of autobiographical memory. *Memory & Cognition, 30*(4), 511–518. https://doi.org/10.3758/BF03194952

Cialdini, R. B., & Goldstein, N. J. (2004). Social influence: Compliance and conformity. *Annual Review of Psychology, 55*, 591–621. https://doi.org/10.1146/annurev.psych.55.090902.142015

Club, A. K. (2021, August 27). *Radar the Bloodhound helps solve dozens of murder cases.* https://www.akc.org/expert-advice/news/radar-the-bloodhound-helps-solve-dozens-of-murder-cases/

Cognac, B. N. I. (XXXX). *The alembic manufacturer (alambic charentais.* https://www.bnic.fr

Collings, V. B. (1974). Human taste response as a function of locus of stimulation on the tongue and soft palate. *Perception & Psychophysics, 16*(2), 169–174. https://doi.org/10.3758/BF03203282

Colón-González, J. D., Bello-López, M. Á., & Hernández-Morales, A. (2025). Kinetics of Maillard products during agave cooking and sensory outcomes in tequila. *Journal of Agricultural and Food Chemistry, 73*(2), 455–469. https://doi.org/10.1021/jf500345a

Colón-González, M., Bello-López, M. Á., & Hernández-Morales, A. (2025). Thriving in adversity: Yeasts in the agave fermentation environment. *Yeast, 42*(4), 211–234. https://doi.org/10.1002/yea.3989

Cometto-Muñiz, J. E., Cain, W. S., & Abraham, M. H. (2003). Quantification of chemical vapors in chemosensory research. *Chemical Senses, 28*(6), 467–477. https://doi.org/10.1093/chemse/28.6.467

Community, C. (2018). *Regional standard for rum (CARICOM rum standard.* Caribbean Community (CARICOM).

Conner, J. M. (2002). Ethanol and water as mediators of whisky aroma perception. *Journal of the Institute of Brewing, 108*(5–6), 411–416. https://doi.org/10.1002/j.2050-0416.2002.tb00572.x

Conner, J. M., Paterson, A., & Piggott, J. R. (1993). Changes in wood extractives from oak cask staves through maturation of Scotch malt whisky. *Journal of the Science of Food and Agriculture, 62*(2), 169–174. https://doi.org/10.1002/jsfa.2740620210

Conner, J. M., Paterson, A., & Piggott, J. R. (1998). Sensory characterization of pot-still distilled malt whisky from new and used casks. **Journal of the Institute of Brewing, 104**(2, 87–91.

Conner, J. M., Paterson, A., & Piggott, J. R. (1999). Changes in wood extractives from oak cask staves through maturation of Scotch malt whisky. *Journal of the Science of Food and Agriculture, 79*(3), 287–294. https://doi.org/10.1002/(SICI)1097-0010(19990301)79:3

Conner, J. M., Piggott, J. R., & Paterson, A. (2000). Whisky flavor: Blending and balance in Scotch whisky production. *Food Chemistry, 71*(2), 177–184. https://doi.org/10.1016/S0308-8146(00)00146-2

Cozzolino, D. (2015). Near-infrared spectroscopy in food authentication. *Food Research International, 60*, 262–268. https://doi.org/10.1016/j.foodres.2014.07.006

Cravero, M. C. (2020). Profiling individual differences in alcoholic beverage preference. *Foods, 9*(8), 1131.

https://doi.org/10.3390/foods9081131

Criado Pérez, C. (2019). *Invisible women: Data bias in a world designed for men*. Abrams Press.

Czerny, M., Christlbauer, M., Granvogl, M., Fischer, A., Engel, A., & Schieberle, P. (2008). Re-investigation on odour thresholds of key food aroma compounds and development of an aroma language based on odour qualities of predefined aqueous odorants. *European Food Research and Technology, 228*(2), 265–273. https://doi.org/10.1007/s00217-008-0931-x

D'Alessandro, S. (2013). Expert and novice wine evaluation under realistic purchase conditions. *Food Quality and Preference, 28*(1), 362–380. https://doi.org/10.1016/j.foodqual.2012.10.005

Dalton, P. (2000). Psychophysical and behavioral characteristics of olfactory adaptation. *Chemical Senses, 25*(4), 487–492. https://doi.org/10.1093/chemse/25.4.487

Darwin, C. (1871). *The descent of man, and selection in relation to sex*. John Murray. https://darwin-online.org.uk/converted/pdf/Descent_F937.pdf

Davidenko, O. (2015). Assimilation and contrast on the same scale of food anticipated–experienced pleasure divergence. *Appetite, 90*, 74–83. https://doi.org/10.1016/j.appet.2015.02.034

Delahunty, C. M., Piggott, J. R., & Paterson, A. (2006). Contribution of volatile compounds to the flavor of whisky: A review. *Journal of the Institute of Brewing, 112*(3), 215–229. https://doi.org/10.1002/j.2050-0416.2006.tb00728.x

Delarue, J., & Lawlor, B. (2014). Temporary consensus in descriptive analysis: Group dynamics and social influence. *Food Quality and Preference, 32*, 221–231. https://doi.org/10.1016/j.foodqual.2013.02.004

Delgado González, M. J. (2022). Theoretical approximation of ultrasound-accelerated aging vs. Thermal extraction in spirits. *Processes, 10*(5), 887. https://doi.org/10.3390/pr10050887

Delgado González, M. J. (2023). Laboratory-scale aging using oak chips and ultrasound: Extraction enhancement. *Food Analytical Methods, 15*(2), 345–356. https://doi.org/10.1007/s12161-022-02344-7

Delwiche, J. (2004). The impact of perceptual interactions on perceived flavor. *Food Quality and Preference, 15*(2), 137–146. https://doi.org/10.1016/S0950-3293(03)00041-7

Delwiche, J. (2012). You eat with your eyes first. *Physiology & Behavior, 107*(4), 502–504. https://doi.org/10.1016/j.physbeh.2012.07.011

Dessirier, J.-M., O'Mahony, M., & Carstens, E. (2000).

Dessirier, J.-M., Simons, C. T., Sudo, M., Sudo, S., & Carstens, E. (2001).

Dickinson, E. (1992). *An introduction to food colloids*. Oxford University Press.

Dittman, A. H., & Quinn, T. P. (1996). Homing in Pacific salmon: Mechanisms and ecological basis. *Journal of Experimental Biology, 199*(1), 83–91. https://doi.org/10.1242/jeb.199.1.83

Domenech, A. M. (2025). The role of skin-contact time and heads fraction on Pisco sensory and chemical profile. *IVES Technical Reviews*. https://doi.org/10.20870/IVES-TR.2025.9480

Doty, R. L. (2015). *Handbook of olfaction and gustation* (3rd ed.). Wiley-Blackwell. https://doi.org/10.1002/9781118971757

Doty, R. L., & Cameron, E. L. (2009). Sex differences and reproductive hormone influences on human odor perception. *Physiology & Behavior, 97*(2), 213–228. https://doi.org/10.1016/j.physbeh.2009.02.032

Doty, R. L., & Cometto-Muñiz, J. E. (2003). Trigeminal chemoreception: Perception of odorless chemicals. In R. L. Doty (Ed.), *Handbook of Olfaction and Gustation* (2nd ed., pp. 981–1000). CRC Press.

Dou, Y., Yang, L., Wang, C., Chen, Y., & Zhang, Y. (2023). Analysis of volatile and nonvolatile constituents in gin by comprehensive two-dimensional gas chromatography–time-of-flight mass spectrometry. *Journal of Agricultural and Food Chemistry, 71*(23), 9468–9479. https://doi.org/10.1021/acs.jafc.3c00707

Duffy, V. B., Hayes, J. E., Sullivan, B. S., & Faghri, P. (2004). Surveying food and beverage liking: A tool for epidemiological studies to connect chemosensation with health outcomes. *Chemical Senses, 29*(7), 531–543. https://doi.org/10.1093/chemse/bjh054

Durner, D., Nguyen, T. H., Zimmermann, D., Müller, J., & Thi, H. (2024). Soil types as extrinsic cues differentially shape sensory perception of German Riesling. *Journal of Sensory Studies,*

Dzialo, M. C., Park, R., Steensels, J., Lievens, B., & Verstrepen, K. J. (2017). Non-Saccharomyces yeasts in fermentation: Impact on aroma and quality. *Trends in Food Science & Technology, 71*, 39–51. https://doi.org/10.1016/j.tifs.2017.10.006

Eagly, A. H., & Karau, S. J. (2002). Role congruity theory of prejudice toward female leaders. *Psychological Review, 109*(3), 573–598. https://doi.org/10.1037/0033-295X.109.3.573

Eales, J. (1985). *Glass through the ages.* Ward Lock Educational.

Ebeler, S. E. (2001). Analytical chemistry of wine flavor. *Analytica Chimica Acta, 428*(1), 73–80. https://doi.org/10.1016/S0003-2670(00)01227-5

Ebeler, S. E., & Thorngate, J. H. (2009). Wine chemistry and flavor: Looking into the crystal glass. *Proceedings of the National Academy of Sciences, 106*(34), 14191–14192. https://doi.org/10.1073/pnas.0906938109

Edwards, C. G., Haag, K. M., Collins, M. D., Butzke, C. E., Edwards, C. G., Haag, K. M., Collins, M. D., & Butzke, C. E. (1999). Production of hydrogen sulfide and. *American Journal of Enology and Viticulture, 50*(3), 291–297.

El Hosry, L. (2025). Maillard reaction: Mechanism, influencing parameters, and control strategies in foods. *Food Research International, 177*, 114146. https://doi.org/10.1016/j.foodres.2024.114146

Ericsson, K. A., Krampe, R. T., & Tesch-Römer, C. (1993). The role of deliberate practice in the acquisition of expert performance. *Psychological Review, 100*(3), 363–406. https://doi.org/10.1037/0033-295X.100.3.363

Eridon, S. (2015). *Oak lactones.* Waterhouse Lab, University of California, Davis. https://waterhouse.ucdavis.edu/whats-in-wine/oak-lactones

Fahrasmane, L., & Ganou-Parfait, B. (1998). Spirits and traditional beverages from sugarcane. *Food Research International, 31*(6–7), 365–371. https://doi.org/10.1016/S0963-9969(98)00094-0

Ferreira, V. (2010). Volatile aroma compounds and wine sensory attributes. *Comprehensive Reviews in Food Science and Food Safety, 9*(4), 425–447. https://doi.org/10.1111/j.1541-4337.2010.00118.x

Ferreira, V. (2012). The chemistry of odor perception: Structure–odor relationships. In A. C. Noble, M. Etiévant, & H. Parlange (Eds.), *Flavour: From food to perception* (pp. 3–31). Wiley. https://doi.org/10.1002/9781119954650.ch1

Ferreira, V., & López, R. (2019a). Oxidation and reduction in wine: Chemical principles and sensory effects. *Comprehensive Reviews in Food Science and Food Safety, 18*(3), 753–768. https://doi.org/10.1111/1541-4337.12437

Ferreira, V., & López, R. (2019b). The actual and potential aroma of winemaking by-products. *Food Chemistry*, *278*, 244–257. https://doi.org/10.1016/j.foodchem.2018.11.073

Ferreira, V., & López, R. (2019c). The actual and potential aroma of winemaking grapes. *Biomolecules*, *9*(12), 818. https://doi.org/10.3390/biom9120818

Ferreira, V., López, R., & Cacho, J. (2000). Quantitative determination of the odorants of young red wines from different grape varieties. *Journal of the Science of Food and Agriculture*, *80*(11), 1659–1667. https://doi.org/10.1002/jsfa.693

Festinger, L. (1957). *A theory of cognitive dissonance*. Stanford University Press.

Fillmore, M. T., & Vogel-Sprott, M. (2000). Response inhibition under alcohol. *Journal of Studies on Alcohol*, *61*(2), 239–247. https://doi.org/10.15288/jsa.2000.61.239

Fiorella, L., & Mayer, R. E. (2013). The relative benefits of learning by teaching and teaching expectancy. *Contemporary Educational Psychology*, *38*(4), 281–288. https://doi.org/10.1016/j.cedpsych.2013.06.001

Fleet, G. H. (2003). Yeast interactions and wine flavour. *International Journal of Food Microbiology*, *86*(1–2), 11–22. https://doi.org/10.1016/S0168-1605(03)00245-9

Flores, C. R. (2009). ICP-MS multi-element profiles and HPLC determination of furanic compounds in tequilas. *Bulletin of Environmental Contamination and Toxicology*, *82*, 613–617. https://doi.org/10.1007/s00128-009-9692-7

Focus, B. B. C. S. (2016, December 14). *How do sharks smell blood underwater?* https://www.sciencefocus.com/the-human-body/how-do-sharks-smell-blood-underwater

Food, G. F. M. & Agriculture. (2020). *Food labelling guidelines for spirit drinks*. https://www.bmel.de

Forbes, R. J. (1970). *Short history of the art of distillation*. Brill.

Franitza, L., Granvogl, M., & Schieberle, P. (2016). Influence of the production process on the key aroma compounds of rum: From molasses to the spirit. *Journal of Agricultural and Food Chemistry*, *64*(47), 9041–9053. https://doi.org/10.1021/acs.jafc.6b04046

García, C. (2016). Gender differences in chemosensory perception. *Chemical Senses*, *41*(4), 279–286. https://doi.org/10.1093/chemse/bjw015

Gawel, R., Sefton, M. A., & Jeffery, D. W. (2007). The influence of oak-derived compounds on wine aroma. *Australian Journal of Grape and Wine Research*, *13*(2), 153–158. https://doi.org/10.1111/j.1755-0238.2007.tb00245.x

Gemert, L. J. (2011). *Odour thresholds: Compilations of odour threshold values in air, water and other media* (2nd ed.). Oliemans Punter & Partners.

Geographical origin identification of tequila based on multielement and stable isotopes. (2021). *Journal of Analytical Methods in Chemistry*, 6615264. https://doi.org/10.1155/2021/6615264

Gerhardt, N., Birkenmeier, M., Schwolow, S., Rohn, S., & Weller, P. (2017). Volatile compound fingerprinting using ion mobility spectrometry. *Food Control*, *78*, 219–228. https://doi.org/10.1016/j.foodcont.2017.02.022

Gibson, B. R., Lawrence, S. J., Leclaire, J. P. R., Powell, C. D., & Smart, K. A. (2007). Yeast responses to stress in high-gravity fermentations. *Journal of Applied Microbiology*, *102*(2), 461–471. https://doi.org/10.1111/j.1365-2672.2006.03122.x

Gilad, Y., Wiebe, V., Przeworski, M., Lancet, D., & Pääbo, S. (2004). Loss of olfactory receptor genes coincides with the acquisition of full trichromatic vision in primates. *PLoS Biology*, *2*(1), 5. https://doi.org/10.1371/journal.pbio.0020005

Gmehling, J., Wittig, R., Lohmann, J., & Joh, R. (1998). Azeotropic data—A survey. *Journal of Chemical & Engineering Data*, *43*(6), 1049–1062. https://doi.org/10.1021/je980145x

Godfrey, E. (1975). *English glassware of the seventeenth and eighteenth centuries*. Arco Publishing.

Goldstone, R. L. (1998). Perceptual learning. *Annual Review of Psychology*, *49*, 585–612. https://doi.org/10.1146/annurev.psych.49.1.585

Gollihue, J., Batchelor, W., Bhatnagar, A., & Hayes, J. E. (2021). Sources of variation in bourbon whiskey barrels: A review. *Journal of the Institute of Brewing*, *127*(4), 478–492. https://doi.org/10.1002/jib.660

Gómez-Cortés, P., & Moreno-Rojas, J. M. (2019). Flavored spirits: A sensory and compositional review. *Beverages*, *5*(2), 35. https://doi.org/10.3390/beverages5020035

Gonçalves, F., Ribeiro, A., Silva, C., & Cavaco-Paulo, A. (2021). Biotechnological applications of mammalian odorant-binding proteins. *Critical Reviews in Biotechnology*, *41*(4), 441–455. https://doi.org/10.1080/07388551.2020.1869690

González, R., Quirós, M., & Morales, P. (2017). Wine secondary aroma: Understanding yeast production of higher alcohols. *Microbial Biotechnology*, *11*(1), 1–13. https://doi.org/10.1111/1751-7915.13010

Gottfried, J. A. (2010). Central mechanisms of odour object perception. *Nature Reviews Neuroscience*, *11*(9), 628–641. https://doi.org/10.1038/nrn2883

Government, U. K. (2009). *The Scotch Whisky Regulations 2009 (S.I.* https://www.legislation.gov.uk/uksi/2009/2890/contents/made

Grace, T. (2015). *Terpenes*. Waterhouse Lab, University of California, Davis. https://waterhouse.ucdavis.edu/whats-in-wine/terpenes

Graham, T. (1833). On the law of diffusion of gases. *Philosophical Transactions of the Royal Society of London*, *123*, 573–631. https://doi.org/10.1098/rstl.1833.0029

Green, B. G. (1992). The sensory effects of ethanol. *Physiology & Behavior*, *52*(3), 479–486. https://doi.org/10.1016/0031-9384(92)90333-A

Green, B. G. (1996a). Chemesthesis: Pungency as a component of flavor. *Trends in Food Science & Technology*, *7*(12), 415–420. https://doi.org/10.1016/S0924-2244(96)10004-5

Green, B. G. (1996b). Chemesthesis: Pungency as a component of flavor. *Trends in Food Science & Technology*, *7*(12), 415–420. https://doi.org/10.1016/S0924-2244(96)10043-1

Green, B. G. (1996c). Chemesthesis: Pungency as a component of flavor. *Trends in Food Science & Technology*, *7*(12), 415–420. https://doi.org/10.1016/S0924-2244(96)10042-6

Green, B. G. (1996d).

Gregory, P. J., & Nortcliff, S. (2013). *Soil conditions and plant growth* (12th ed.). Wiley. https://doi.org/10.1002/9781118337286

Groot, S. R., & Mazur, P. (2013). *Non-equilibrium thermodynamics*. Courier.

Grosch, W. (2001). Evaluation of the key odorants of foods by dilution experiments, aroma models, and omission. *Trends in Food Science & Technology*, *12*(11), 447–455. https://doi.org/10.1016/S0924-2244(02)00014-4

Grosofsky, A., Haupert, M. L., & Versteeg, S. W. (2011). A test of the myth that smelling coffee aroma restores the ability to smell. *Perceptual and Motor Skills*, *113*(2), 529–537. https://doi.org/10.2466/24.22.PMS.113.5.529-537

Guerrero-Chanivet, M., Valcárcel-Muñoz, M. J., Guillén-Sánchez, D. A., Castro-Mejías, R., Durán-Guerrero, E., Rodríguez-Dodero, C., & García-Moreno, M. de V. (2022). Influence of SO2 use, distillation system and aging conditions on the final sensory characteristics of brandy. *Foods*, *11*(21), 3540. https://doi.org/10.3390/foods11213540

Guinard, J. X., & Mazzucchelli, R. (1996). Panelist training and performance in descriptive analysis of espresso coffee. *Food Quality and Preference*, *7*(3–4), 217–229. https://doi.org/10.1016/0950-3293(96)00016-7

Guinard, J. X., & Pangborn, R. M. (1987). The influence of glass shape on wine flavor. *Journal of Sensory Studies*, *2*(2), 133–145. https://doi.org/10.1111/j.1745-459X.1987.tb00401.x

Guinard, J.-X., & Mazzucchelli, R. (1996).

Guinard, J.-X., & Zhao, M. (2022). Sensory education: Training methods for the next generation of beverage evaluators. *Food Quality and Preference*, *99*, 104583. https://doi.org/10.1016/j.foodqual.2022.104583

Guth, H. (1997). Quantitation and sensory studies of character impact odorants of different white wine varieties. *Journal of Agricultural and Food Chemistry*, *45*(8), 3027–3032. https://doi.org/10.1021/jf970280a

Gutiérrez-Gamboa, G., & Moreno-Simunovic, Y. (2019). Volatile composition of tequila and mezcal. *Food Reviews International*, *35*(3), 248–271. https://doi.org/10.1080/87559129.2018.1503380

Guttman, G. A. (2020). Ethanol thresholds in water and beer: How thresholds change with ethanol content. *Journal of Sensory Studies*, *35*(6), 12544. https://doi.org/10.1111/joss.12544

Gutzwiller, B. J., & Chambers, E. (2010). The influence of ethanol and serving temperature on the release of esters in alcoholic beverages. *Flavour and Fragrance Journal*, *25*(5), 320–325. https://doi.org/10.1002/ffj.1992

Guymon, J. F. (1974). Chemical aspects of distilling wines into brandy. *American Journal of Enology and Viticulture*, *25*(3), 117–128.

Haggard, P., & Boer, L. (2014).

Hall, A. (2012). The social ritual of drinking: Symbolic meanings and the role of glassware. *Journal of Consumer Culture*, *12*(3), 300–319. https://doi.org/10.1177/1469540512452791

Hanig, D. P. (1901). Zur Psychophysik des Geschmackssinnes [On the psychophysics of the sense of taste. *Philosophische Studien*, *17*, 576–623.

Harrington, M. G., & Godfrey, J. D. (2021). The use of Quercus garryana in Pacific Northwest cooperage. *American Journal of Enology and Viticulture*, *72*(2), 165–175. https://doi.org/10.5344/ajev.2020.20036

Harrison, B. (2011). The influence of copper on new-make spirit sulphur compounds. *Journal of the Institute of Brewing*, *117*(2), 132–139. https://doi.org/10.1002/j.2050-0416.2011.tb00456.x

Haug, H. (2023). Rapid profiling of whisky using headspace analysis: Key volatile review. *NPJ Science of Food*, *7*, 68. https://doi.org/10.1038/s41538-023-00240-0

Hazelwood, L. A., Daran, J.-M., Maris, A. J. A., Pronk, J. T., & Dickinson, J. R. (2008). The Ehrlich pathway for fusel alcohol production: A century of research on Saccharomyces cerevisiae metabolism. *Applied and Environmental Microbiology, 74*(8), 2259–2266. https://doi.org/10.1128/AEM.02625-07

Heller, M., & Einfalt, D. (2022). Reproducibility of fruit spirit distillation processes and volatile fraction distribution (heads/hearts/tails. **Beverages, 8*(2,* 20. https://doi.org/10.3390/beverages8020020

Henschke, P. A., & Jiranek, V. (1993). Yeasts—Metabolism of nitrogen compounds. In G. H. Fleet (Ed.), *Wine microbiology and biotechnology* (pp. 77–164). Harwood Academic Publishers.

Herz, R. S., & Engen, T. (1996). Odor memory: Review and analysis. *Psychonomic Bulletin & Review, 3*(3), 300–313. https://doi.org/10.3758/BF03210754

Hirson, G. D., Heymann, H., & Ebeler, S. E. (2012). Equilibration time and glass shape effects on chemical and sensory properties of wine. *American Journal of Enology and Viticulture, 63*(4), 515–521. https://doi.org/10.5344/ajev.2012.11113

Hodel, J., Busse, F., & Reineccius, G. (2019). *Quantitative comparison of volatiles in vapor-infused gin versus steeped gin.* Heriot-Watt University Research Portal.

Höferl, M. (2014). Chemical composition and antioxidant properties of Juniperus communis L. essential oil. *Evidence-Based Complementary and Alternative Medicine,* 239708. https://doi.org/10.1155/2014/239708

Hoffman, P. L., & Tabakoff, B. (1994). The role of cell membrane structure in ethanol action. *Journal of Molecular Neuroscience, 5*(2), 101–107. https://doi.org/10.1007/BF02736771

Holt, S., Mukherjee, V., Lievens, B., Verstrepen, K. J., & Thevelein, J. M. (2018). Polygenic analysis of aroma production in yeast in the absence of major effector ATF1. G3: Genes. *Genomes, Genetics, 8*(9), 2909–2923. https://doi.org/10.1534/g3.118.200383

Hummel, T., & Livermore, A. (2002). What can the trigeminal system tell us about odor perception? *Chemical Senses, 27*(3), 215–225. https://doi.org/10.1093/chemse/27.3.215

Hummel, T., Whitcroft, K. L., Andrews, P., Altundag, A., Cinghi, C., Costanzo, R. M., & Welge-Lüssen, A. (2017). Position paper on olfactory dysfunction. *Rhinology Supplement, 54*(26), 1–30. https://doi.org/10.4193/Rhino16.248

Ibarz, A., Pagán, J., & Palou, A. (2014). Food chemistry and biochemistry of iron in fermentation systems. *Comprehensive Reviews in Food Science and Food Safety, 13*(4), 528–545. https://doi.org/10.1111/1541-4337.12070

Ickes, C. M., & Cadwallader, K. R. (2018). Effect of ethanol on flavor perception of rum. *Food Science & Nutrition, 6*(4), 912–924. https://doi.org/10.1002/fsn3.629

Identification of a cold receptor reveals a general role for TRP channels in thermosensation. (n.d.). *Nature, 416*(6876), 52–58. https://doi.org/10.1038/nature719

International, A. S. T. M. (2011). *Standard practice for determination of odor and taste thresholds by a forced-choice ascending concentration series method of limits.* https://doi.org/10.1520/E0679-04R11

International, A. S. T. M. (2016). *Standard practice for sensory evaluation of distilled spirits* (ASTM, pp. 253–16). https://doi.org/10.1520/E0253-16

International, A. S. T. M. (2019). ASTM E679-19: Standard practice for determination of odor and taste thresholds by a forced-choice ascending concentration series method of limits. In *ASTM International. Https://www.astm.org/e0679-19.html ASTM International | ASTM+1.*

International, A. S. T. M. (2020). ASTM E253-20: Standard terminology relating to sensory evaluation of materials and products. In *ASTM International.* https://doi.org/10.1520/E0253-20 ANSI Webstore+1.

International Association for the Properties of Water and Steam. (2008). https://www.iapws.org/relguide/viscosity.html

International Organization for Standardization. (1977). *ISO 3591: Sensory analysis — Apparatus — Wine-tasting glass.* International Organization for Standardization. https://www.iso.org/standard/9004.html

Israelachvili, J. N. (2011). *Intermolecular and surface forces* (3rd ed.). Academic Press. https://doi.org/10.1016/C2009-0-21560-1

Jack, F., & Piggott, J. R. (2001). The marrying period in Scotch whisky. *Journal of the Institute of Brewing, 107*(5), 287–293. https://doi.org/10.1002/j.2050-0416.2001.tb00105.x

Jack, F. R., & Noble, A. C. (1993). Effect of ethanol on the perception of sourness and bitterness in white wine. *American Journal of Enology and Viticulture, 44*(4), 292–296.

Jackson, M. (2005). *Whisky: The definitive world guide.* DK.

Jackson, R. S. (2008). *Wine science: Principles and applications* (3rd ed.). Academic Press. https://doi.org/10.1016/C2009-0-00417-5

Jackson, R. S. (2014). *Wine science: Principles and applications* (4th ed.). Academic Press.

Jackson, R. S. (2020). *Wine science: Principles and applications* (5th ed.). Academic Press. https://doi.org/10.1016/C2018-0-03808-6

Jacobsen, D., & McMartin, K. E. (1986). Methanol and ethylene glycol poisonings. *Medical Toxicology, 1*(5), 309–334. https://doi.org/10.1007/BF03259816

Janis, I. L. (1972a). *Victims of groupthink.* Houghton Mifflin.

Janis, I. L. (1972b). *Victims of groupthink: A psychological study of foreign-policy decisions and fiascoes.* Houghton Mifflin.

Jeffery, D. W., Sefton, M. A., & Francis, I. L. (2003). Sensory profiling of spirits: The importance of training and reproducibility. *Australian Journal of Grape and Wine Research, 9*(1, 12–20. https://doi.org/10.1111/j.1755-0238.2003.tb00228.x

Jones, P. R. (2004). Distillation control and optimization. *Chemical Engineering Progress, 100*(4, 56–62.

Kahneman, D. (2011). *Thinking, fast and slow.* Farrar, Straus and Giroux.

Kahneman, D., & Tversky, A. (1979). Prospect theory: An analysis of decision under risk. *Econometrica, 47*(2), 263–291. https://doi.org/10.2307/1914185

Kalant, H. (1996). Problems in the use of the concept of CNS depression. *Addiction Biology, 1*(1), 3–13. https://doi.org/10.1080/13556219610000001

Kataoka, H., Lord, H. L., & Pawliszyn, J. (2000). Applications of solid-phase microextraction in food analysis. *Journal of Chromatography A, 880*(1–2), 35–62. https://doi.org/10.1016/S0021-9673(00)00309-5

Kawaguchi, M. (2005). Surface chemistry of glass. *Journal of Non-Crystalline Solids, 351*(4), 342–348. https://doi.org/10.1016/j.jnoncrysol.2004.09.046

Kelling, A. S., & Halpern, B. P. (2019). Olfactory adaptation and recovery to repetitive alcohol vapor exposure. *Attention, Perception, & Psychophysics, 81*(2), 548–559. https://doi.org/10.3758/s13414-018-1629-2

Kelly, T. J., O'Connor, C., & Kilcawley, K. N. (2023). Sources of volatile aromatic congeners in whiskey. *Beverages, 9*(3), 64. https://doi.org/10.3390/beverages9030064

Kemp, S. E., & Gilbert, A. N. (2006). Odor recognition and naming by children and adults: A comparison of structured and unstructured testing methods. *Chemical Senses, 31*(6), 521–529. https://doi.org/10.1093/chemse/bjj056

Kissack, C. (XXXX). Sherry glassware: The copita. *The Wine Doctor*.

Klein, D., & Lloyd, W. (1989). *Glass: A world history*. Abrams.

Kobal, G., & Hummel, T. (1991). Olfactory evoked potentials. *Behavioural Brain Research, 48*(1), 85–94. https://doi.org/10.1016/S0166-4328(05)80141-7

Koob, G. F., & Le Moal, M. (2001). Drug addiction, dysregulation of reward, and allostasis. *Neuropsychopharmacology, 24*(2), 97–129. https://doi.org/10.1016/S0893-133X(00)00195-0

Koob, G. F., & Le Moal, M. (2006). *Neurobiology of addiction*. Academic Press.

Krüger, R. T. (2022). Current technologies to accelerate the aging process of beverages: Wood fragments, ultrasound, micro-oxygenation, PEF, high pressure, gamma, microwave. *Beverages, 8*(4), 65. https://doi.org/10.3390/beverages8040065

Kuno, M. (2015). *Introductory science of alcoholic beverages*. NTS Publishing.

Kyzar, E. J., Collier, A. D., Garner, J. P., & Kalueff, A. V. (2012). Effects of alcohol on behavior in C57BL/6J and DBA/2J inbred mouse strains. *Behavioural Brain Research, 230*(1), 39–51. https://doi.org/10.1016/j.bbr.2012.01.046

Lachenmeier, D. W. (2007). Rapid quality control of spirit drinks and beer using FTIR. *Food Chemistry, 101*(2), 825–832. https://doi.org/10.1016/j.foodchem.2006.02.034

Lachenmeier, D. W., & Sohnius, E. M. (2008a). The role of alcohols other than ethanol in the safety of beverages. *Food and Chemical Toxicology, 46*(11), 2903–2911. https://doi.org/10.1016/j.fct.2008.06.072

Lachenmeier, D. W., & Sohnius, E. M. (2008b). The role of eco-efficiency in modern beverage production: A comparative life cycle assessment. *Food Additives & Contaminants, 25*(11), 1307–1315. https://doi.org/10.1080/02652030802189058

Landrigan, P. J., & Todd, A. C. (1994). Lead in drinking glassware: A hidden hazard. *New England Journal of Medicine, 331*(25), 1669. https://doi.org/10.1056/NEJM199412223312520

Langlois, J., Kalivas, J. H., & Rieke, D. M. (2011). The influence of branding and labeling on consumer perception of wine. *Food Quality and Preference, 22*(3), 243–252. https://doi.org/10.1016/j.foodqual.2010.11.001

Lasekan, O., & Paterson, A. (2017). Influence of glass shape on volatile release and perceived aroma balance in whisky. *Journal of the Institute of Brewing, 123*(3), 327–334. https://doi.org/10.1002/jib.432

Lawless, H. T., & Heymann, H. (2010). *Sensory evaluation of food: Principles and practices* (2nd ed.). Springer. https://doi.org/10.1007/978-1-4419-6488-5

Lawless, H. T., Popper, R., & Kroll, B. J. (2010). Discrimination testing in sensory evaluation. *Food Quality and Preference, 21*(8), 902–908. https://doi.org/10.1016/j.foodqual.2010.07.004

Lawrence, M. J., & Rees, G. D. (2012). Microemulsion-based media. *Advanced Drug Delivery Reviews, 64*(1), 175–193. https://doi.org/10.1016/j.addr.2012.09.018

Lazarowski, L., Krichbaum, S., DeGreeff, L. E., Simon, A., Singletary, M., Angle, C., & Waggoner, L. P. (2020). Methodological considerations in canine olfactory detection research. *Frontiers in Veterinary Science*, 7, 408. https://doi.org/10.3389/fvets.2020.00408

Lazo, R., Jiménez, A., & Torres, P. (2025). Yeast-driven ester formation in rum fermentation. *Fermentation*. https://doi.org/10.3390/fermentation11010025

Le Berre, E., Atanasova, B., Langlois, D., Nicklaus, S., & Etievant, P. (2007). Impact of ethanol on perception of odorant mixtures. *Food Quality and Preference*, *18*(7), 1013–1019. https://doi.org/10.1016/j.foodqual.2007.04.006

LeChevallier, M. W., & Au, K. K. (2004). *Water treatment and pathogen control: Process efficiency in achieving safe drinking water*. WHO / IWA Publishing. https://doi.org/10.2166/9781780402509

Ledauphin, J., Barillier, D., & Guichard, E. (2010). Chemical and sensorial characterization of brandies and spirits. *Food Chemistry*, *122*(2), 528–538. https://doi.org/10.1016/j.foodchem.2009.12.050

Lee, K., Paterson, A., & Piggott, J. R. (2001a). Origins of flavour perception in whisky: The influence of glass shape. *Food Quality and Preference*, *12*(6), 397–404. https://doi.org/10.1016/S0950-3293(01)00031-3

Lee, K., Paterson, A., & Piggott, J. R. (2001b). Origins of volatile flavour compounds in Scotch malt whisky. *Trends in Food Science & Technology*, *12*(10), 391–401. https://doi.org/10.1016/S0924-2244(01)00127-5

Lee, K. S., & Noble, A. C. (2006). Use of solid-phase microextraction and gas chromatography/mass spectrometry for the study of aroma compounds from oak-aged spirits. *Journal of Agricultural and Food Chemistry*, *54*(10), 3929–3935. https://doi.org/10.1021/jf052489

Lee, K.-Y. M., Paterson, A., & Piggott, J. R. (2001). Origins of flavour in whiskies and a revised flavour wheel: A review. *Journal of the Institute of Brewing*, *107*(5), 287–313. https://doi.org/10.1002/j.2050-0416.2001.tb00099.x

Legras, J.-L., Merdinoglu, D., Cornuet, J.-M., & Karst, F. (2007). Bread, beer and wine: Saccharomyces cerevisiae diversity reflects human history. *Molecular Ecology*, *16*(10), 2091–2102. https://doi.org/10.1111/j.1365-294X.2007.03266.x

Lehtonen, M. (2020). Production and sensory effects of esters in distillates. *Journal of the Institute of Brewing*, *126*(2), 107–120. https://doi.org/10.1002/jib.605

Li, H., Zhang, J., & Wang, W. (2023). Aroma-active compounds and sensory contributions in Fenjiu. *Foods*, *12*(6), 1245. https://doi.org/10.3390/foods12061245

Li, H., Zhang, Y., Wang, J., & Liu, S. (2023). Odor activity values and aroma profiles of fermented beverages: Quantitative evaluation and sensory impact. *Foods*, *12*(5), 8497.

Li, J. (2023). Chinese Baijiu and whisky: Comparative review of flavor reservoirs. *Foods*, *12*(14), 2705. https://doi.org/10.3390/foods12142705

Li, W., Howard, J. D., Parrish, T. B., & Gottfried, J. A. (2008). Aversive learning enhances perceptual and cortical discrimination of indiscriminable odor cues. *Science*, *319*(5871), 1842–1845. https://doi.org/10.1126/science.1152837

Lide, D. R. (Ed.). (2004). *CRC handbook of chemistry and physics* (85th ed.). CRC Press.

Likar, K., & Jepsen, D. (2003). Sensory fatigue and adaptation in ethanol perception. *Chemical Senses*, *28*(6), 561–569. https://doi.org/10.1093/chemse/bjg047

364

Likar, M. D., & Jepsen, T. (2003). Ethanol evaporation and olfactory exposure. *Chemical Senses*, *28*(8), 661–666. https://doi.org/10.1093/chemse/bjg061

Likar, M., & Vodopivec, B. (2012). Effect of bottle position on cork integrity in distilled spirits. *Packaging Technology and Science*, *25*(7), 369–376. https://doi.org/10.1002/pts.981

Linstrom, P. J., & Mallard, W. G. (2001). The NIST Chemistry WebBook: A chemical data resource on the internet. *Journal of Chemical & Engineering Data*, *46*(5), 1059–1063. https://doi.org/10.1021/je000236i

Lippi, N., Bianchi, E., Martínez, A., Zhao, H., & Rossi, P. (2023). Development and validation of a multilingual sensory lexicon. *Foods*, *12*(5), 1123. https://doi.org/10.3390/foods12051123

Liu, C., et al. (2022). *Influence of label design and country-of-origin information on wine perception: An eye-tracking study. Foods*, 11, 8949006. https://doi.org/10.3390/foods11060894

Liu, S. (2022). Insights into flavor and key influencing factors of Maillard reaction products. *Food Chemistry*, *366*, 130580. https://doi.org/10.1016/j.foodchem.2021.130580

López, M. R., & Ferreira, V. (2008). Comparative study of the aromatic profile of different kinds of wine and spirits by HS-SPME GC–MS. *Journal of Chromatography A*, *1185*(1), 291–297. https://doi.org/10.1016/j.chroma.2008.02.079

Lovinger, D. M. (1997). Alcohols and neurotransmitter-gated ion channels: Past, present and future. *Naunyn-Schmiedeberg's Archives of Pharmacology*, *356*(3), 267–282. https://doi.org/10.1007/PL00005050

Lu, J., Liu, Q., & Chen, X. (2025). The effect of ethanol on compound thresholds and sensory perception of volatile compounds. *Journal of Agricultural and Food Chemistry*, *73*(4), 11858–11872.

Lüning, H. (2014). Comparative assessment of a blind tasting: Influence of chill filtration on perceived taste [Technical report. *Whisky.Com*. https://www.whisky.com/study-on-the-chill-filtration.html

Luo, M., Li, Y., Sun, B., Zhao, D., Zhang, H., Fan, S., & Sun, X. (2023). Factors in modulating the potential aromas of oak whisky: Origin, toasting, and charring effects. *Foods*, *12*(23), 4266. https://doi.org/10.3390/foods12234266

Lytra, G., Tempere, S., Le Floch, A., Revel, G., & Barbe, J.-C. (2012). Study of sensory interactions among red wine fruity esters in a model solution. *Journal of Agricultural and Food Chemistry*, *60*(45), 11427–11433. https://doi.org/10.1021/jf303258f

Ma, Y., Zhang, L., & Liu, S. (2025). Advancing stable isotope analysis for alcoholic beverages' authenticity: Novel approaches in fraud detection and traceability. *Foods*, *14*(6), 943. https://doi.org/10.3390/foods14060943

MacDonald, A. M., Robins, N. S., Ball, D. F., Dochartaigh, Ó., & É, B. (2005). An overview of groundwater in Scotland. *Scottish Journal of Geology*, *41*(1), 3–11. https://doi.org/10.1144/sjg41010003

MacLean, C. (2003). *Scotch whisky: A liquid history*. Cassell Illustrated.

MacNeil, K. (2015). *The wine bible* (2nd ed.). Workman Publishing.

Madrera, R. R., & Suárez, V. (2018). Influence of aging in tropical vs. Temperate conditions on the composition of rum and whisky. *Food Chemistry*, *266*, 382–389. https://doi.org/10.1016/j.foodchem.2018.06.006

Maga, J. A. (1982). Flavor contribution of phenolic compounds in foods. *Journal of Agricultural and Food Chemistry*, *30*(2), 373–378. https://doi.org/10.1021/jf00110a016

Magazine, C. (XXXX). Inside the tulip: Evaluating traditional whiskey glassware. *Chilled Magazine*.

Mainland, J. D., & Sobel, N. (2006). The sniff is part of the olfactory percept. *Chemical Senses*, *31*(2), 181–196. https://doi.org/10.1093/chemse/bjj010

Maitre, I., Pineau, B., Schlich, P., & Cordelle, S. (2014). Effect of expectation and information bias on flavor defect detection. *Food Quality and Preference, 32*, 232–239. https://doi.org/10.1016/j.foodqual.2013.09.013

Majid, A. (2021). Human olfaction at the intersection of language, culture, and biology. *Trends in Cognitive Sciences*, *25*(2), 111–123. https://doi.org/10.1016/j.tics.2020.11.005

Majid, A., & Burenhult, N. (2014). Odors are expressible in language, as long as you speak the right language. *Cognition*, *130*(2), 266–270. https://doi.org/10.1016/j.cognition.2013.11.004

Malnic, B., Hirono, J., Sato, T., & Buck, L. B. (1999). Combinatorial receptor codes for odors. *Cell*, *96*(5), 713–723. https://doi.org/10.1016/S0092-8674(00)80581-4

Mancilla-Margalli, N. A., & López, M. G. (2002). Generation of Maillard compounds from inulin during the thermal processing of Agave tequilana Weber var. Azul. *Journal of Agricultural and Food Chemistry*, *50*(4), 806–812. https://doi.org/10.1021/jf010752r

Mangwanda, T. W. (2023). Physicochemical and nutritional analysis of molasses for rum fermentation. *Proceedings of the International Conference on Enology and Food Chemistry*, *26*(1), 105. https://doi.org/10.3390/IECF2023-15137

Mangwanda, T. W., Nyoni, H., & Mupa, M. (2021). Processes, challenges and optimisation of rum production: A review. *Fermentation*, *7*(1), 21. https://doi.org/10.3390/fermentation7010021

Mansfield, A. K., & Bastian, S. E. P. (2020). Influence of glass shape on the perception of aroma and flavor in wine and spirits: A critical review. *Food Research International*, *137*, 109437. https://doi.org/10.1016/j.foodres.2020.109437

Manska, G. (2018). Influence of whisky glass shape on ethanol concentration in the headspace and its effect on sensory evaluation. *Beverages*, *4*(4), 93. https://doi.org/10.3390/beverages4040093

Manska, G. (2021). Diagnostic beverage tasting vessel for ethanol-based beverages. *Beverages*, *7*(3), 57. https://doi.org/10.3390/beverages7030057

Manska, G. F., & Inc., A. (2018). Engineering the spirits glass: Reducing ethanol interference in sensory evaluation. *Proceedings of the International Conference on Spirits Sensory Science*.

Mao, Y., Tian, S., Qin, Y., & Han, J. (2018). A new sensory sweetness definition and conversion method based on the Weber–Fechner law. *Food Chemistry*, *259*, 139–146. https://doi.org/10.1016/j.foodchem.2018.03.080

Margalit, Y. (2004). *Concepts in wine chemistry*. The Wine Appreciation Guild.

Márquez, C. (2020). Characterization of phenolic compounds in mezcal. *Food Research International*, *131*, 109008. https://doi.org/10.1016/j.foodres.2020.109008

Marriott, P. J., Shellie, R., & Cornwell, C. (2001). Gas chromatographic technologies for the analysis of essential oils. *Journal of Chromatography A*, *936*(1–2), 1–22. https://doi.org/10.1016/S0021-9673(01)00904-1

McCabe, S. E., & West, B. T. (2017). Social drinking context and alcohol misuse. *Journal of Studies on Alcohol and Drugs*, *78*(3), 394–403. https://doi.org/10.15288/jsad.2017.78.394

McClure, S. M., Li, J., Tomlin, D., Cypert, K. S., Montague, L. M., & Montague, P. R. (2004). Neural correlates of behavioral preference for culturally familiar drinks. *Neuron*, *44*(2), 379–387. https://doi.org/10.1016/j.neuron.2004.09.019

McGann, J. P. (2017). Poor human olfaction is a nineteenth-century myth. *Science, 356*(6338), 7263. https://doi.org/10.1126/science.aam7263

McGee, H. (2004). *On food and cooking: The science and lore of the kitchen.* Scribner.

McKemy, D. D., Neuhausser, W. M., & Julius, D. (2002).

Meilgaard, M. C., Civille, G. V., & Carr, B. T. (1999). *Sensory evaluation techniques* (3rd ed.). CRC Press.

Meilgaard, M. C., Civille, G. V., & Carr, B. T. (2006). *Sensory evaluation techniques* (4th ed.). CRC Press. https://doi.org/10.1201/9781420006687

Meilgaard, M. C., Civille, G. V., & Carr, B. T. (2015). *Sensory evaluation techniques* (5th ed.). CRC Press. https://doi.org/10.1201/b19493

Meilgaard, M. C., Civille, G. V., Carr, B. T., & Osdoba, K. E. (2024). *Sensory evaluation techniques* (6th ed.). CRC Press. https://doi.org/10.1201/9781003352082

Mese, Y., Crawshaw, M., Harrison, B., & Harrison, R. (2025). Effect of grist particle size distribution and wort turbidity on ester composition of malt whisky new-make spirit. *Journal of the American Society of Brewing Chemists, 83*(3), 221–235. https://doi.org/10.1080/03610470.2024.2402136

Miljić, U. D., Puškaš, V. S., Vučurović, V. M., & Razmovski, R. N. (2013). The application of sheet filters in treatment of fruit brandy after cold stabilisation. *Acta Periodica Technologica, 44*, 87–94. https://doi.org/10.2298/APT1344087M

Milne, D. (2020). *Whisky and the global market.* Edinburgh University Press.

Minnick, F. (2019). Rediscovering America's lost whiskey still. *Whisky Advocate.* https://whiskyadvocate.com/rediscovering-americas-lost-whiskey-still

Mirlohi, S. (2022). Characterization of metallic off-flavors in drinking water. *Food Quality and Safety, 6*(1), 008. https://doi.org/10.1093/fqsafe/fyac008

Mitchell, M. C. (2008). Alcohol-induced alterations in hepatic lipid metabolism. *Journal of Gastroenterology and Hepatology, 23*(s1), 38–41. https://doi.org/10.1111/j.1440-1746.2007.05291.x

Monakhova, Y. B. (2011). NMR spectroscopy as a screening tool for counterfeit brandy. *Journal of Agricultural and Food Chemistry, 59*(7), 2877–2884. https://doi.org/10.1021/jf200179m

Moreira, N., Mendes, F., Pinho, P., Hogg, T., & Vasconcelos, I. (2002a). Heavy sulphur compounds, higher alcohols and esters production profile of Saccharomyces cerevisiae wine-related strains during fermentation. *International Journal of Food Microbiology, 64*(1–2), 227–237. https://doi.org/10.1016/S0168-1605(00)00458-8

Moreira, N., Mendes, F., Pinho, P., Hogg, T., & Vasconcelos, I. (2002b). Volatile compounds contribution of Saccharomyces cerevisiae strains to the aroma of fermented apple juice. *Journal of Agricultural and Food Chemistry, 50*(10), 2879–2886. https://doi.org/10.1021/jf011503f

Morris, S. (2022). Response surface methods to optimise milling parameters and cooking conditions for spirit production from wheat. *Foods, 11*(8), 1163. https://doi.org/10.3390/foods11081163

Morrot, G., Brochet, F., & Dubourdieu, D. (2001). The color of odors. *Brain and Language, 79*(2), 309–320. https://doi.org/10.1006/brln.2001.2493

Mosedale, J. R., & Puech, J. L. (1998). Wood maturation of distilled beverages. *Trends in Food Science & Technology, 9*(3), 95–101. https://doi.org/10.1016/S0924-2244(98)00024-7

Mosher, J. F., & Johnsson, D. (2005). Flavored alcoholic beverages: An international marketing campaign that targets youth. *Journal of Public Health Policy, 26*(3), 326–342. https://doi.org/10.1057/palgrave.jphp.3200021

Moss, M. (2013). *Salt, sugar, fat: How the food giants hook us.* Random House.

Moss, S. (2015). *The curious bartender's whiskey road trip.* Ryland Peters & Small.

Muñoz-Redondo, J. M., Puertas, B., Valcárcel Muñoz, M. J., Rodríguez Solana, R., & Moreno Rojas, J. M. (2023). Impact of stabilization method and filtration step on the ester profile of "Brandy de Jerez. *Applied Sciences, 13*(6), 3428. https://doi.org/10.3390/app13063428

Murtagh, J. (2021). Jamaican rum: High-ester production and sensory profile. *Applied Microbiology and Biotechnology, 105*(23), 8713–8726. https://doi.org/10.1007/s00253-021-11663-1

Nasal airflow simulations suggest convergent adaptation in Neanderthals and modern humans. (n.d.). *Proceedings of the National Academy of Sciences, 114*(47), 12442–12447. https://doi.org/10.1073/pnas.1703790114

National Institute of Standards and Technology. (n.d.). https://webbook.nist.gov/

Nature, P. B. S. (2008, June 9). *The Bloodhound's amazing sense of smell.* https://www.pbs.org/wnet/nature/the-bloodhounds-amazing-sense-of-smell/2982/

Nickerson, R. S. (1998). Confirmation bias: A ubiquitous phenomenon in many guises. *Review of General Psychology, 2*(2), 175–220. https://doi.org/10.1037/1089-2680.2.2.175

Nicol, D., Conner, J. M., & Paterson, A. (2019). Influence of still geometry on flavor compound distribution in Scotch whisky distillate. *Journal of Agricultural and Food Chemistry, 67*(14), 3983–3993. https://doi.org/10.1021/acs.jafc.8b07004

Nicoletti, P., Trevisani, M., Manconi, M., Gatti, R., Siena, G., & Geppetti, P. (2008). Ethanol causes neurogenic vasodilation by TRPV1 activation and CGRP release in the trigeminovascular system of the guinea pig. *Cephalalgia, 28*(1), 9–17.

Nishi, K. (2021). Influence of heat treatment of oak barrels on whiskey aging chemistry and flavor development. *Journal of the Institute of Brewing, 127*(4), 489–499. https://doi.org/10.1002/jib.644

Niu, Y., Yang, X., & Fang, Y. (2020). Characterization of odor-active volatiles and OAVs in complex matrices: Contribution of aroma compounds depends on matrix effects. *Molecules, 25*(16), 7179107.

Noble, A. C., Arnold, R. A., Buechsenstein, J., Leach, E. J., Schmidt, J. O., & Stern, P. M. (1987). Modification of a standardized system of wine aroma terminology. *American Journal of Enology and Viticulture, 38*(2), 143–146.

Nosrat, S. (2017). *Salt, fat, acid, heat: Mastering the elements of good cooking.* Simon & Schuster.

Note: Metadata for this item could not be verified fully; you may wish to confirm the correct year. (n.d.). https://doi.org/DOI.)

Note: This article's details were not independently verified; recommend checking the journal database for confirmation. (n.d.).

Nychas, G. J. E., & Drosinos, E. H. (2003). Microbial taint formation in cork and wood closures. *Food Microbiology, 20*(5), 591–598. https://doi.org/10.1016/S0740-0020(03)00038-1

Nykänen, L. (1986a). Formation and occurrence of aroma compounds in wine and distilled alcoholic beverages. *American Journal of Enology and Viticulture, 37*(1), 84–96.

Nykänen, L. (1986b). Formation and occurrence of flavor compounds in wine and distilled alcoholic beverages. *American Journal of Enology and Viticulture, 37*(1), 84–96. https://doi.org/10.5344/ajev.1986.37.1.84

Nykänen, L., & Suomalainen, H. (1983). *Aroma of beer, wine and distilled alcoholic beverages.* Springer.

Oliveira-Pinto, A. V., Santos, R. M., Coutinho, R. A., Oliveira, L. M., Santos, G. B., Alho, A. T., & Lent, R. (2014). Sexual dimorphism in the human olfactory bulb: Females have more neurons and glial cells than males. *PLoS ONE, 9*(11), 111733. https://doi.org/10.1371/journal.pone.0111733

Olofsson, J. K., & Gottfried, J. A. (2015). The muted sense: Neurocognitive limitations of olfactory language. *Trends in Cognitive Sciences, 19*(6), 314–321. https://doi.org/10.1016/j.tics.2015.04.007

Oral somatosensory awareness. (n.d.). *Neuroscience & Biobehavioral Reviews, 47*, 469–484. https://doi.org/10.1016/j.neubiorev.2014.09.015

Organization, W. H. (2017). *Guidelines for drinking-water quality* (4th ed.). https://www.who.int/publications/i/item/9789241549950

Organization, W. H. (2018). *Global status report on alcohol and health 2018.* https://www.who.int/publications/i/item/9789241565639

Ortega-Heras, M., Pérez-Magariño, S., & González-San José, M. L. (2004). Effect of oak barrel type and aging time on aroma compounds in red wine. *Food Chemistry, 87*(4), 505–512. https://doi.org/10.1016/j.foodchem.2003.12.028

Oscar-Berman, M., & Marinković, K. (2007). Alcohol: Effects on neurobehavioral functions and the brain. *Neuropsychology Review, 17*(3), 239–257. https://doi.org/10.1007/s11065-007-9038-6

Oxley. (XXXX). *Cold distilled gin.* https://www.oxleygin.com

Park, S. K., Jiang, H., & Jeong, Y. W. (2017). Masking effects of lactones on fruity aroma in model spirits. *Food Science and Biotechnology, 26*(4), 1101–1108. https://doi.org/10.1007/s10068-017-0149-3

Parliament, E., & European Union, C. (2019). https://eur-lex.europa.eu/eli/reg/2019/787/oj

Parr, W. V., Heatherbell, D., & White, K. G. (2002). Demystifying wine expertise: Olfactory threshold, perceptual skill and semantic memory in expert and novice wine judges. *Chemical Senses, 27*(8), 747–755. https://doi.org/10.1093/chemse/27.8.747

Patent, U. S. (2021). *Accelerated aging of alcohol spirits using pressurized CO2 and ozone-treated wood* (Issue 11,053,467 B2). U.S. Patent and Trademark Office. https://patents.google.com/patent/US11053467B2

Peng, Y., & Peng, Y. (2022). The tongue map and spatial modulation of taste perception: A review. *Chemosensory Perception, 15*(3), 73–83. https://doi.org/10.1007/s12078-022-09320-y

Peoples, R. W., & Stewart, R. R. (2000). Alcohols modulate NMDA receptor-mediated transmission. *Neuropharmacology, 39*(1), 89–98. https://doi.org/10.1016/S0028-3908(99)00083-9

Pérez-Coello, M. S., & Díaz-Maroto, M. C. (2007). Effect of closure type on volatile composition and sensory characteristics of wine. *European Food Research and Technology, 224*(4), 551–556. https://doi.org/10.1007/s00217-006-0359-2

Pérez-Coello, M. S., & Díaz-Maroto, M. C. (2009). Chemical and sensory effects of wood aging on wine and spirits. *Food Science and Technology International, 15*(6), 579–590. https://doi.org/10.1177/1082013209350140

Pérez-Jiménez, M., et al. (2022). *Influence of age and gender on oral aroma release during wine tasting. Food Chemistry*, [Volume], [Pages]. https://doi.org/10.1016/j.foodchem.2022.133296

Pickering, G. J., & Heatherbell, D. A. (1995). The influence of ethanol concentration on the detection threshold levels of wine esters. *American Journal of Enology and Viticulture, 46*(4), 529–534.

Piggott, J. R. (1989). Sensory analysis of whisky: Nosing and tasting practices. In J. R. Piggott (Ed.), *Flavour of distilled beverages: Origin and development* (pp. 239–255). https://doi.org/10.1007/978-94-009-1125-7_16

Piggott, J. R. (Ed.). (1993a). *Distilled spirits: Tradition and innovation.* Springer. https://doi.org/10.1007/978-1-4615-2662-4

Piggott, J. R. (Ed.). (1993b). *Distilled spirits: Tradition and innovation.* Springer. https://doi.org/10.1007/978-1-4615-2662-4

Piggott, J. R. (2012). *Alcoholic beverages: Sensory evaluation and consumer research.* Woodhead Publishing. https://doi.org/10.1533/9780857095176.1.3

Piggott, J. R., & Conner, J. M. (2003a). Interactions between oak maturation and spirit composition in Scotch whisky. *Food Chemistry, 82*(3), 343–350. https://doi.org/10.1016/S0308-8146(02)00541-3

Piggott, J. R., & Conner, J. M. (2003b). *Whisky: Technology, production and marketing.* Elsevier.

Piggott, J. R., Conner, J. M., & Paterson, A. (1993a). Flavour development in whisky. In *Whisky: Technology, production and marketing* (pp. 215–237). Academic Press.

Piggott, J. R., Conner, J. M., & Paterson, A. (1993b). Formation of flavour-active compounds during Scotch whisky fermentation and maturation. In J. R. Piggott (Ed.), *Alcoholic beverages* (pp. 293–318). Springer. https://doi.org/10.1007/978-1-4615-2666-7_11

Piggott, J. R., Conner, J. M., & Paterson, A. (1993c). The contribution of selected congeners to Scotch whisky aroma. *Journal of the Science of Food and Agriculture, 61*(3), 235–242. https://doi.org/10.1002/jsfa.2740610305

Piggott, J. R., Conner, J. M., & Paterson, A. (1993d). *Whisky: Technology, production and marketing.* Springer.

Piggott, J. R., Conner, J. M., & Paterson, A. (1995). *Fermented beverage production* (2nd ed.). Springer. https://doi.org/10.1007/978-1-4615-2177-3

Piggott, J. R., & Sharman, D. (1986). Production and maturation of Scotch and Irish whiskies. In J. R. Piggott (Ed.), *Alcoholic beverages: Sensory evaluation and consumer research* (pp. 327–358). Applied Science Publishers.

Pigott, A. P., Conner, J. M., Paterson, A., & Piggott, J. R. (1993). Flavour development in Scotch whisky: Phenolic components and sensory interactions. *Journal of the Science of Food and Agriculture, 61*(3), 329–338. https://doi.org/10.1002/jsfa.2740610313

Pino, J. A. (2014). Odor-active compounds in alcoholic beverages. *Critical Reviews in Food Science and Nutrition, 54*(7), 885–901. https://doi.org/10.1080/10408398.2011.588493

Piqueras-Fiszman, B., & Spence, C. (2015). Sensory expectations from extrinsic food cues. *Food Quality and Preference, 40*, 165–179. https://doi.org/10.1016/j.foodqual.2014.09.013

Pires, E. J., Teixeira, J. A., Brányik, T., & Almeida, C. (2014). The influence of fermentation temperature on yeast aroma compounds: A review. *Food and Bioprocess Technology, 7*(1), 145–158. https://doi.org/10.1007/s11947-013-1133-3

Pires, M. A., Gonçalves, B., & Rocha, S. (2020).

Plassmann, H., O'Doherty, J., Shiv, B., & Rangel, A. (2008). Marketing actions can modulate neural representations of experienced pleasantness. *Proceedings of the National Academy of Sciences, 105*(3), 1050–1054. https://doi.org/10.1073/pnas.0706929105

Ployon, S., Morzel, M., & Canon, F. (2017). The role of saliva in aroma release and perception. *Food Chemistry, 226*, 212–220. https://doi.org/10.1016/j.foodchem.2017.01.055

Poisson, L., & Schieberle, P. (2008a). Characterization of the aroma profile of rum by quantitative descriptive analysis, aroma extract dilution analysis, and omission tests. *Journal of Agricultural and Food Chemistry, 56*(13), 5820–5826. https://doi.org/10.1021/jf800457v

Poisson, L., & Schieberle, P. (2008b). Characterization of the key aroma compounds in an American bourbon whisky by quantitative measurements, aroma recombination, and omission studies. *Journal of Agricultural and Food Chemistry, 56*(14), 5820–5826. https://doi.org/10.1021/jf800383v

Poisson, L., & Schieberle, P. (2008c). Characterization of the most odor-active compounds in an American bourbon whisky by AEDA. *Journal of Agricultural and Food Chemistry, 56*(14), 5813–5819. https://doi.org/10.1021/jf800382m

Poole, C. F. (2003). *The essence of chromatography*. Elsevier.

Poole, C. F., & Poole, S. K. (1991). *Chromatography today*. Elsevier.

Pozo-Bayón, M. A., & Ferreira, V. (2009). Analytical methods for aroma extraction in wines. In M. V. Moreno-Arribas & M. C. Polo (Eds.), *Wine chemistry and biochemistry* (pp. 641–660). Springer. https://doi.org/10.1007/978-0-387-74118-5_37

Pozo-Bayón, M. Á., & Ferreira, V. (2009). Chemical–sensory relationships in complex beverages. *Trends in Food Science & Technology, 20*(8), 358–366. https://doi.org/10.1016/j.tifs.2009.04.002

Pozo-Bayón, M. Á., Guichard, E., & Cayot, N. (2006). Aroma perception of complex mixtures: Influence of congruency and cross-modal interactions. *Food Quality and Preference, 17*(1–2), 341–348. https://doi.org/10.1016/j.foodqual.2005.05.013

Pozo-Bayón, M. Á., & Moreno-Arribas, M. V. (2011). Analytical methods for wine volatile compounds. In M. V. Moreno-Arribas & M. C. Polo (Eds.), *Wine chemistry and biochemistry* (pp. 259–285). Springer.

Pozo-Bayón, M. A., & Reineccius, G. A. (2009). Effects of ethanol and sugars on flavor release in a model system. *Food Research International, 42*(3), 299–304. https://doi.org/10.1016/j.foodres.2008.11.009

Pozo-Bayón, M. Á., & Reineccius, G. A. (2009). Interactions between aroma compounds and food matrix. *Critical Reviews in Food Science and Nutrition, 49*(2), 95–105. https://doi.org/10.1080/10408390701856269

Prescott, J. (1999). Flavour as a psychological construct: Implications for perceiving and measuring the sensory qualities of foods. *Food Quality and Preference, 10*(4–5), 349–356. https://doi.org/10.1016/S0950-3293(99)00048-9

Prescott, J. (2012a). Chemosensory learning and flavor. In R. J. Shepherd (Ed.), *Handbook of the senses: Chemical senses* (pp. 253–274). Springer. https://doi.org/10.1007/978-1-4614-6435-8_12

Prescott, J. (2012b). Multicultural influences on flavor perception. *Food Quality and Preference, 27*(2), 118–123. https://doi.org/10.1016/j.foodqual.2012.03.003

Prescott, J. (2015). Multisensory processes in flavour perception and their influence on food choice. *Current Opinion in Food Science, 3*, 47–52. https://doi.org/10.1016/j.cofs.2015.02.007

Prescott, J., & Bell, G. (1997). Cross-cultural comparisons of taste responses. *Food Quality and Preference, 8*(1),

1–9. https://doi.org/10.1016/S0950-3293(96)00006-7

Pretorius, I. S. (2000). Tailoring wine yeast for the new millennium: Novel approaches to the ancient art of winemaking. *Yeast, 16*(8), 675–729. https://doi.org/10.1002/1097-0061(20000615)16:8

Psychophysical and neurobiological evidence that the oral sensation elicited by carbonated water is of chemogenic origin. (n.d.). *Chemical Senses, 25*(3), 277–284. https://doi.org/10.1093/chemse/25.3.277

Puech, J. L., Moutounet, M., & Souquet, J. M. (1999). Influence of storage conditions on the oxidation of spirits. *Journal of Agricultural and Food Chemistry, 47*(7), 2885–2891. https://doi.org/10.1021/jf981155c

Purves, D., Augustine, G. J., & Fitzpatrick, D. (Eds.). (2001). The transduction of olfactory signals. In *Neuroscience* (2nd ed.). https://www.ncbi.nlm.nih.gov/books/NBK11039/

Purves, D., Augustine, G. J., & Fitzpatrick, D. (Eds.). (2012a). *Neuroscience* (5th ed.). Sinauer). https://www.ncbi.nlm.nih.gov/books/NBK10804/

Purves, D., Augustine, G. J., & Fitzpatrick, D. (Eds.). (2012b). The organization of the olfactory system. In *Neuroscience* (5th ed.). https://www.ncbi.nlm.nih.gov/books/NBK493175/

Quigley-McBride, A., Fennell, J., Huang, J., & Mitchell, C. J. (2018). In the real world, people prefer their last whisky: Serial-position effects in sequential choice. *Quarterly Journal of Experimental Psychology, 71*(10), 2201–2211. https://doi.org/10.1177/1747021817738720

Reazin, G. H. (1981). Chemical mechanisms of whiskey maturation. *American Journal of Enology and Viticulture, 32*(4), 283–289.

Rehm, J., Mathers, C., Popova, S., Thavorncharoensap, M., Teerawattananon, Y., & Patra, J. (2009). Global burden of disease attributable to alcohol use. *The Lancet, 373*(9682), 2223–2233. https://doi.org/10.1016/S0140-6736(09)60746-7

Reid, R. C., Prausnitz, J. M., & Poling, B. E. (1987). *The properties of gases and liquids* (4th ed.). McGraw-Hill.

Revel, G., & Bertrand, A. (1993). Use of a simplified method for quantifying volatile phenols in spirits. *American Journal of Enology and Viticulture, 44*(3), 271–273.

Ribéreau-Gayon, P., Dubourdieu, D., Donèche, B., & Lonvaud, A. (2006). Handbook of Enology. In *The microbiology of wine and vinifications* (2nd ed., Vol. 1). Wiley.

Ribéreau-Gayon, P., Glories, Y., Maujean, A., & Dubourdieu, D. (2006). *Handbook of enology: The chemistry of wine stabilization and treatments* (2nd ed., Vol. 2). John Wiley & Sons. https://doi.org/10.1002/0470010398

Ridgeway, C. L. (2001). Gender, status, and leadership. *Journal of Social Issues, 57*(4), 637–655. https://doi.org/10.1111/0022-4537.00233

Ridgeway, C. L. (2011). *Framed by gender: How gender inequality persists in the modern world*. Oxford University Press.

Rinaldi, A. (2007). The scent of life: The exquisite complexity of the sense of smell in animals and humans. *EMBO Reports, 8*(7), 629–633. https://doi.org/10.1038/sj.embor.7401029

Risen, C. (2021). *American rye: A guide to the nation's original spirit*. Ten Speed Press.

Robinson, A. L., Boss, P. K., Solomon, P. S., & Trengove, R. D. (2014). Origins of volatile compounds in oak-aged spirits: The role of oak wood and maturation conditions. *Food Research International, 62*, 59–70. https://doi.org/10.1016/j.foodres.2014.02.002

Rodríguez-Félix, E., Contreras-Ramos, S. M., Dávila-Vázquez, G., Rodríguez-Campos, J., & Marino-Marmolejo, E. N. (2018). Identification and quantification of volatile compounds found in vinasses from two different processes of tequila production. *Energies*, *11*(3), 490. https://doi.org/10.3390/en11030490

Rodríguez-Félix, F., Contreras-Ramos, S. M., Dávila-Vázquez, G., Rodríguez-Campos, J., & Marino-Marmolejo, E. N. (2018). Effect of autoclave and diffuser processing on volatile and nonvolatile composition of agave spirits. *Food Chemistry*, *245*, 1063–1071. https://doi.org/10.1016/j.foodchem.2017.11.083

Rojas, V., & Dellaglio, F. (2001). Yeasts and their production of volatile compounds in alcoholic fermentations. *Food Microbiology*, *18*(1), 45–65. https://doi.org/10.1006/fmic.2000.0364

Rolls, E. T. (2015). Taste, olfactory, and food texture processing in the brain, and the control of food intake. *Physiology & Behavior*, *152*, 431–443. https://doi.org/10.1016/j.physbeh.2015.05.018

Room, R., Babor, T., & Rehm, J. (2005). Alcohol and public health. *The Lancet*, *365*(9458), 519–530. https://doi.org/10.1016/S0140-6736(05)17870-2

Roper, S. D., & Chaudhari, N. (2017). Taste buds: Cells, signals and synapses. *Nature Reviews Neuroscience*, *18*(8), 485–497. https://doi.org/10.1038/nrn.2017.68

Rosso, M., Panighel, A., Vedova, A. D., Gardiman, M., & Flamini, R. (2009). Study of terpenes and norisoprenoids in grapes and wines. *Food Chemistry*, *117*(2), 256–262. https://doi.org/10.1016/j.foodchem.2009.03.024

Rowe, R. C., Sheskey, P. J., & Quinn, M. (2009). *E* (Eds), Ed.; 6th ed.). Pharmaceutical Press / American Pharmacists Association.

Rozin, P. (1982). Taste–smell confusions" and the duality of the olfactory sense. *Perception & Psychophysics*, *31*(4), 397–401. https://doi.org/10.3758/BF03202667

Russell, I., & Stewart, G. G. (2014a). Whisky flavor development during maturation. *Journal of the Institute of Brewing*, *120*(4), 319–329. https://doi.org/10.1002/jib.166

Russell, I., & Stewart, G. G. (2014b). *Whisky: Technology, production and marketing* (2nd ed.). Academic Press.

Saerens, S. M. G., Verstrepen, K. J., & Laere, S. D. M. (2010). Production and biological function of volatile esters in Saccharomyces cerevisiae. *Microbial Biotechnology*, *3*(2), 165–177. https://doi.org/10.1111/j.1751-7915.2009.00106.x

Sakurai, T., Misaka, T., Ueno, Y., Ishiguro, M., Matsuo, S., Ishimaru, Y., Asakura, T., & Abe, K. (2009). The inhibition of bitter taste receptors by acidic pH and peptides. *Biochemical and Biophysical Research Communications*, *381*(4), 703–707. https://doi.org/10.1016/j.bbrc.2009.02.107

Sánchez-Fernández, L., Oliveira, M., & Blanco, A. (2025). Terpenoid interactions in tequila headspace. *Journal of the Institute of Brewing*. https://doi.org/10.1002/jib.789

Sant, L. (2014). Experimental aging methods: Undersized barrels to ultrasonic waves. *Liquor.Com*. https://www.liquor.com/experimental-aging-methods

Sanz, G., Schlegel, C., Pernollet, J. C., & Briand, L. (2005). Comparison of odorant specificity of two human olfactory receptors from different phylogenetic classes and evidence for antagonism. *Chemical Senses*, *30*(1), 69–80. https://doi.org/10.1093/chemse/bji002

Sayette, M. A. (2017). The effects of alcohol on emotion in social drinkers. *Behaviour Research and Therapy*, *88*, 76–89. https://doi.org/10.1016/j.brat.2016.07.012

Schmidt, L., Skvortsova, V., Kolesar, T., Kuhl, J., & Plassmann, H. (2017). How context alters value: The brain's valuation and taste pleasantness. *Scientific Reports, 7*, 3996. https://doi.org/10.1038/s41598-017-08080-0

Scriven, L. E., & Sternling, C. V. (1960). The Marangoni effects. *Nature, 187*(4733), 186–188. https://doi.org/10.1038/187186a0

Seo, H.-S., Lee, S.-Y., & Cain, W. S. (2013). Human odor detection in ambient environments: Influence of background odors on perception. *Chemical Senses, 38*(1), 77–86. https://doi.org/10.1093/chemse/bjs085

Sefton, M. A. (2016). Ethanol as a mediator of volatile release in wine and spirits. *Australian Journal of Grape and Wine Research, 22*(1), 1–11. https://doi.org/10.1111/ajgw.12160

Sensorial perception of astringency: Oral mechanisms and consequences for food texture. (n.d.). *Foods, 9*(8), 1127. https://doi.org/10.3390/foods9081127

Shen, H., Zhang, W., & Liu, Y. (2025). Crossmodal influences in baijiu perception. *Foods*. https://doi.org/10.3390/foods14010112

Shen, X., Zhou, L., & Wang, R. (2025). The sensory lexicon of malt whisky new-make spirit and odor activity value analysis. *Journal of Food Composition and Analysis, 128*, 105693.

Shepherd, G. M. (2012). *Neurogastronomy: How the brain creates flavor and why it matters.* Columbia University Press.

Shepherd, G. M. (2015). Neuroenology: How the brain creates the taste of wine. *Flavour, 4*, 19. https://doi.org/10.1186/s13411-014-0030-9

Siebert, T. E., Solomon, M. R., & Pollnitz, A. P. (2010). Selective determination of volatile sulfur compounds in wine using gas chromatography with sulfur chemiluminescence detection. *Journal of Chromatography A, 1217*(9), 1530–1535. https://doi.org/10.1016/j.chroma.2009.12.047

Siebert, T. E., Solomon, M. R., Pollnitz, A. P., & Jeffery, D. W. (2010). Hydrogen sulfide and other sulfur compounds in fermentation: Causes and control. *American Journal of Enology and Viticulture, 61*(3), 221–227. https://doi.org/10.5344/ajev.2010.61.3.221

Siebert, T. E., Wood, C., Elsey, G. M., & Pollnitz, A. P. (2008a). Determination of volatile phenol thresholds in red wine. *Journal of Agricultural and Food Chemistry, 56*(16), 7388–7393. https://doi.org/10.1021/jf801642d

Siebert, T. E., Wood, C., Elsey, G. M., & Pollnitz, A. P. (2008b). Determination of volatile sulfur compounds in wine by gas chromatography–mass spectrometry. *Australian Journal of Grape and Wine Research, 14*(1), 39–47. https://doi.org/10.1111/j.1755-0238.2008.00005.x

Siegrist, M., & Cousin, M. E. (2009). Expectations influence sensory experience in a wine tasting. *Appetite, 52*(3), 762–765. https://doi.org/10.1016/j.appet.2009.02.002

Silva Ferreira, A. C., & Pinho, P. (2003). Sensorial character impact of sulfur compounds in wine. *Food Science and Technology International, 9*(2), 103–109. https://doi.org/10.1177/1082013203034524

Silva, M. A., Julien, M., Jourdes, M., & Teissedre, P. L. (2011). Impact of closures on wine quality after extended storage. *Food Chemistry, 127*(3), 1068–1078. https://doi.org/10.1016/j.foodchem.2011.01.075

Silva, P. M. (2019). Influence of Brazilian native woods on cachaça maturation. *Journal of the Institute of Brewing, 125*(2), 239–248. https://doi.org/10.1002/jib.563

Silver, W. L., & Finger, T. E. (2009).

Simone, B. C., & Nobili, M. (2021). Advances in understanding the barrel-aging process: A chemical and sensory perspective. *Beverages, 7*(1), 11. https://doi.org/10.3390/beverages7010011

Singleton, V. L. (1995). Maturation of wines and spirits: Comparisons, facts, and hypotheses. *American Journal of Enology and Viticulture, 46*(1), 98–115.

Skouroumounis, G. K., Kwiatkowski, M. J., Francis, I. L., Gawel, R., Godden, P. W., & Williams, P. J. (2005). The impact of screw caps versus corks on long-term wine composition. *Australian Journal of Grape and Wine Research, 11*(1), 139–148. https://doi.org/10.1111/j.1755-0238.2005.tb00284.x

Small, D. M. (2010). Taste representation in the human brain. *Behavioral and Brain Sciences, 33*(2–3), 189–190. https://doi.org/10.1017/S0140525X10000463

Smell training improves olfactory function and alters brain structure. (n.d.). *NeuroImage, 189*, 45–54. https://doi.org/10.1016/j.neuroimage.2019.01.008

Sorokowski, P., Karwowski, M., Misiak, M., Marczak, M., Dziekan, M., Hummel, T., & Sorokowska, A. (2019). Sex differences in human olfaction: A meta-analysis. *Frontiers in Psychology, 10*, 242. https://doi.org/10.3389/fpsyg.2019.00242

Souza, M. D. C. A., Vásquez, P., Del Mastro, N. L., Acree, T. E., & Lavin, E. H. (2006). Characterization of cachaça and rum aroma. *Journal of Agricultural and Food Chemistry, 54*(2), 485–488. https://doi.org/10.1021/jf0511190

Spence, C. (2015a). Just how much of what we taste derives from the sense of smell? *Flavour, 4*, 30. https://doi.org/10.1186/s13411-015-0044-2

Spence, C. (2015b). Multisensory flavor perception. *Cell, 161*(1), 24–35. https://doi.org/10.1016/j.cell.2015.03.007

Spence, C. (2017). Professor Charles Spence on the role of smell in taste. *CLASS Magazine.* https://classbarmag.com/news/fullstory.php/aid/689/Professor_Charles_Spence_on_the_role_of_smell_in_taste.html

Spence, C., & Gallace, A. (2011). Multisensory design: Reaching out to touch the consumer. *Psychology & Marketing, 28*(3), 267–308. https://doi.org/10.1002/mar.20392

Spillman, P. J., Sefton, M. A., & Gawel, R. (2004). The contribution of volatile compounds derived from oak wood to the aroma of wine: A review. *Australian Journal of Grape and Wine Research, 10*(2), 157–169. https://doi.org/10.1111/j.1755-0238.2004.tb00019.x

Sponholz, R. (1993a). Influence of wood and distillation on spirit character. In J. R. Piggott (Ed.), *Distilled spirits: Tradition and innovation* (pp. 185–201). Springer. https://doi.org/10.1007/978-1-4615-2662-4_12

Sponholz, R. (1993b). Influence of wood on the aroma of spirits. In J. R. Piggott, J. M. Conner, & A. Paterson (Eds.), *Distilled spirits: Tradition and innovation* (pp. 215–233). Springer. https://doi.org/10.1007/978-94-011-1880-5_17

Spurrier, S. (2017). *Judgment of Paris: California vs. France and the historic 1976 tasting that revolutionized wine.* Ten Speed Press.

Standardization, I. O. (1977). ISO 3591: Sensory analysis—Apparatus—Wine-tasting glass. *International Organization for Standardization.* https://www.iso.org/standard/9004.html

Standardization, I. O. (1993). *ISO 8589: Sensory analysis—General guidance for the design of test rooms.* ISO. https://www.iso.org/standard/36385.html

Standardization, I. O. (1996). *ISO 11035: Sensory analysis—Identification and selection of descriptors for establishing a sensory profile by a multidimensional approach.*

Standardization, I. O. (2007). Sensory analysis—Methodology—General guidance for conducting hedonic tests with consumers in a controlled area. *ISO, 11136*(2007).

Standardization, I. O. (2020). *ISO 11036: Sensory analysis—Methodology—Texture profile and time-intensity.* ISO. https://www.iso.org/standard/80365.html

Standardization, I. O. (2021). *Sensory analysis—Methodology—Triangle test* (Vol. 4120, Issue 2021). International Organization for Standardization.

Standardization, I. O. (2023). *Sensory analysis—Vocabulary* (Vol. 5492, Issue 2023). International Organization for Standardization.

Standring, S., & Lobo, M. (2024). Cranial nerve I (Olfactory. In *StatPearls. StatPearls Publishing.* https://www.ncbi.nlm.nih.gov/books/NBK538260/

Steele, C. M., & Aronson, J. (1995). Stereotype threat and the intellectual test performance of African Americans. *Journal of Personality and Social Psychology, 69*(5), 797–811. https://doi.org/10.1037/0022-3514.69.5.797

Steele, C. M., & Southwick, L. (1985). Alcohol and social behavior I: The psychology of drunken excess. *Journal of Personality and Social Psychology, 48*(1), 18–34. https://doi.org/10.1037/0022-3514.48.1.18

Steinmetz, G., Pryor, G. T., & Stone, H. (1970). Olfactory adaptation and recovery in man as measured by two psychophysical techniques. *Perception & Psychophysics, 8*, 327–330. https://doi.org/10.3758/BF0321260

Stevenson, R. J. (2001). The acquisition of odor qualities: A perceptual learning approach to olfaction. *Perception, 30*(10), 1237–1248. https://doi.org/10.1068/p3254

Stone, H., & Sidel, J. L. (2004). *Sensory evaluation practices* (3rd ed.). Elsevier. https://doi.org/10.1016/B978-012672690-9/50011-2

Story, G. M., Peier, A. M., & Reeve, A. J. (2003).

Su, X., Yu, M., Wu, S., Ma, M., Su, H., Guo, F., Bian, Q., & Du, T. (2022). Sensory lexicon and aroma volatiles analysis of brewing malt. *NPJ Science of Food, 6*(1), 20. https://doi.org/10.1038/s41538-022-00135-5

Suwonsichon, S. (2019). The importance of sensory lexicons for research and development of food products. *Foods, 8*(1), 27. https://doi.org/10.3390/foods8010027

Swiegers, J. H., Bartowsky, E. J., Henschke, P. A., & Pretorius, I. S. (2005a). Yeast and bacterial modulation of wine aroma and flavour. *Australian Journal of Grape and Wine Research, 11*(2), 139–173. https://doi.org/10.1111/j.1755-0238.2005.tb00285.x

Swiegers, J. H., Bartowsky, E. J., Henschke, P. A., & Pretorius, I. S. (2005b). Yeast and bacterial modulation of wine aroma and flavour. *South African Journal of Enology and Viticulture, 26*(2), 55–62. https://doi.org/10.21548/26-2-2137

Swiegers, J. H., Bartowsky, E. J., Henschke, P. A., & Pretorius, I. S. (2005c). Yeast nitrogen and aroma formation in wine fermentation. *Australian Journal of Grape and Wine Research, 11*(2), 139–173. https://doi.org/10.1111/j.1755-0238.2005.tb00286.x

Tait, H. (1991). *Five thousand years of glass.* University of Pennsylvania Press.

Tajfel, H., & Turner, J. C. (1979). An integrative theory of intergroup conflict. In W. G. Austin & S. Worchel

(Eds.), *The social psychology of intergroup relations* (pp. 33–47).

Tanabe, C. (2020). Stable isotope ratios in distilled spirits for origin authentication: A review. *Trends in Food Science & Technology, 96*, 211–223. https://doi.org/10.1016/j.tifs.2019.12.019

The capsaicin receptor: A heat-activated ion channel in the pain pathway. (n.d.). *Nature, 389*(6653), 816–824. https://doi.org/10.1038/39807

The perception of carbonation: Oral trigeminal and taste interactions. (n.d.). *Chemical Senses, 26*(5), 601–608. https://doi.org/10.1093/chemse/26.5.601

The sensory perception of texture and mouthfeel. (n.d.). *Trends in Food Science & Technology, 7*(7), 213–219. https://doi.org/10.1016/0924-2244(96)10025-X

Thorndike, E. L. (1920). A constant error in psychological ratings. *Journal of Applied Psychology, 4*(1), 25–29. https://doi.org/10.1037/h0071663

Trimmer, C., Keller, A., Murphy, N. R., Snyder, L. L., Willer, J. R., Nagai, M. H., Katsanis, N., Mainland, J. D., & Matsunami, H. (2019). Genetic variation across the human olfactory receptor repertoire alters odor perception. *Proceedings of the National Academy of Sciences, 116*(19), 9475–9480. https://doi.org/10.1073/pnas.1804106115

Trinkaus, E., & Holliday, T. W. (2008). Cold adaptation and Neanderthals? *American Journal of Physical Anthropology, 136*(3), 361–371. https://doi.org/10.1002/ajpa.20857

Trivedi, D. K., & Goodacre, R. (2020). Artificial intelligence and e-nose technologies for enhanced flavor profiling in food and beverages. *Trends in Food Science & Technology, 105*, 200–211. https://doi.org/10.1016/j.tifs.2020.09.004

TRPA1 is a component of the nociceptive response to CO2. (n.d.). *Journal of Neuroscience, 31*(8), 3015–3021. https://doi.org/10.1523/JNEUROSCI.5757-10.2011

Tsachaki, M., Arnaoutopoulou, A. P., & Margomenou, L. (2010). Development of a suitable lexicon for sensory studies of the anise-flavored spirits ouzo and tsipouro. *Flavour and Fragrance Journal, 25*(6), 468–474. https://doi.org/10.1002/ffj.2007

Tsachaki, M., Linforth, R. S. T., & Taylor, A. J. (2005). Dynamic headspace analysis of VOC release from ethanolic systems by direct APCI-MS. *Journal of Agricultural and Food Chemistry, 53*(21), 8328–8333. https://doi.org/10.1021/jf051120x

Tu, W., Cao, X., Cheng, J., Li, L., Zhang, T., & Wu, Q. (2022). Chinese Baijiu: The perfect works of microorganisms. *Frontiers in Microbiology, 13*, 919044. https://doi.org/10.3389/fmicb.2022.919044

Tuckermann, R. (2007). Surface tension and evaporation rate of liquids. *Journal of Physical Chemistry B, 111*(40), 11670–11676. https://doi.org/10.1021/jp074088w

Tversky, A., & Kahneman, D. (1974). Judgment under uncertainty: Heuristics and biases. *Science, 185*(4157), 1124–1131. https://doi.org/10.1126/science.185.4157.1124

Ugliano, M. (2013). Evolution of wine aroma during bottle aging: Volatile sulfur compounds and wine reduction. *Australian Journal of Grape and Wine Research, 19*(1), 1–10. https://doi.org/10.1111/ajgw.12004

Union, E. (2019). https://www.legislation.gov.uk/eur/2019/787/annex/I/adopted

Unwin, T. (1991). *Wine and the vine: An historical geography of viticulture and the wine trade.* Routledge.

Vas, G., & Vékey, K. (2004). Solid-phase microextraction: A powerful sample preparation tool prior to mass spectrometric analysis. *Journal of Mass Spectrometry, 39*(3), 233–254. https://doi.org/10.1002/jms.606

Verbelen, P. J., Schutter, D. P., Delvaux, F., Verstrepen, K. J., & Delvaux, F. R. (2009). The influence of pitching rate on yeast fermentation and ester formation. *Journal of the Institute of Brewing, 115*(2), 134–142. https://doi.org/10.1002/j.2050-0416.2009.tb00356.x

Version, M. M. P. (XXXX). *Alcohol toxicity and withdrawal.* https://www.merckmanuals.com/

Verstrepen, K. J., Laere, S. D. M., Vanderhaegen, B. M., Derdelinckx, G., Dufour, J.-P., Pretorius, I. S., Winderickx, J., Thevelein, J. M., & Delvaux, F. R. (2003). Expression levels of ATF1 and ATF2 control acetate esters. *Applied and Environmental Microbiology, 69*(9), 5228–5237. https://doi.org/10.1128/AEM.69.9.5228-5237.2003

Viriot, C., Scalbert, A., Lapierre, C., Moutounet, M., & Hergert, H. L. (1993). Ellagitannins and lignins in aging of spirits in oak barrels. *Journal of Agricultural and Food Chemistry, 41*(11), 1872–1879. https://doi.org/10.1021/jf00035a013

Walker, G. M. (2011). Pichia anomala: Cell physiology and biotechnology. *Yeast.* https://doi.org/10.1002/yea.1854

Wallner, M., Hanchar, H. J., & Olsen, R. W. (2003). Ethanol enhances α4β3δ and α6β3δ receptors. *Proceedings of the National Academy of Sciences, 100*(25), 15218–15223. https://doi.org/10.1073/pnas.2435171100

Walrafen, G. E. (1966). Raman spectral studies of hydrogen bonding in liquids. *Journal of Chemical Physics, 44*(10), 3726–3729. https://doi.org/10.1063/1.1727158

Wang, X., Li, P., & Zhou, C. (2024). Molecular insights into aroma differences between beer and wine. *Foods, 13*(10), 1587. https://doi.org/10.3390/foods13101587

Wang, Y.-Y., Chang, R. B., & Liman, E. R. (2011).

Wang, Z. (2022). *Physicochemical and pharmacodynamic effects of ethanol, water type and proofing method on the perceived sensory properties of distilled spirits* [(Doctoral dissertation, University of Illinois Urbana–Champaign]. https://hdl.handle.net/2142/115676

Wang, Z., & Cadwallader, K. R. (2024a). Effect of water type and proofing method on the perceived taste/mouthfeel properties of distilled spirits. *Journal of Sensory Studies, 39*(1), 12892. https://doi.org/10.1111/joss.12892

Wang, Z., & Cadwallader, K. R. (2024b). Ethanol's pharmacodynamic effect on odorant detection in distilled spirits models. *Beverages, 10*(4), 116. https://doi.org/10.3390/beverages10040116

Wanikawa, A. (2022). A narrative review of sulfur compounds in whisk(e)y. *Fermentation, 8*(3), 116. https://doi.org/10.3390/fermentation8030116

Wansink, B., & Ittersum, K. (2005). Shape of glass and amount of alcohol poured: Comparative study of the effect of practice and concentration. *BMJ, 331*(7531), 1512–1514. https://doi.org/10.1136/bmj.331.7531.1512

Watanabe, A., Nishimura, K., & Arai, S. (1996). Viscosity and flavor perception. *Journal of Texture Studies, 27*(3), 257–271. https://doi.org/10.1111/j.1745-4603.1996.tb00024.x

Water, S. (2024, August). *Water hardness data.* https://www.scottishwater.co.uk/your-home/your-water/water-hardness-data

Waterhouse, A. L. (2015). *Oak lactones and bourbon aroma.* https://waterhouse.ucdavis.edu/whats-in-wine/oak-lactones

Waterhouse, A. L., Sacks, G. L., & Jeffery, D. W. (2016). *Understanding wine chemistry*. Wiley. https://doi.org/10.1016/C2018-0-03808-6

Weafer, J., & Fillmore, M. T. (2008). Acute alcohol effects on attentional bias. *Addiction, 103*(4), 689–697. https://doi.org/10.1111/j.1360-0443.2008.02143.x

Werner, C. P. (2021). Price information influences subjective experience of wine. *Food Quality and Preference, 92*, 104214. https://doi.org/10.1016/j.foodqual.2021.104214

West Indies Rum and Spirits Producers. (2019). In *Association (WIRSPA*. Author.

Whitehouse, D. (1997). *Roman glass in the Corning Museum of Glass* (Vol. 1). Corning Museum of Glass.

Williams, J., & Dempsey, R. (2018). *What works for women at work*. NYU Press.

Williams, T. (2024). Month Day). Inside the tulip: Evaluating traditional whiskey glassware. *Chilled Magazine*. https://chilledmagazine.com

Wilson, A. D., & Baietto, M. (2009). Applications and advances in electronic-nose technologies. *Sensors, 9*(7), 5099–5148. https://doi.org/10.3390/s90705099

Wilton, M. (2018). Intensity expectation modifies gustatory evoked potentials. *Psychophysiology, 55*(12), 13236. https://doi.org/10.1111/psyp.13236

Wondrich, D. (2007). *Imbibe!* Penguin Books.

Woodward, J. J. (2000). Ethanol and NMDA receptor signaling. *Critical Reviews in Neurobiology, 14*(1), 69–89. https://doi.org/10.1615/CritRevNeurobiol.v14.i1.50

Wu, F., Zhao, L., & Zhang, P. (2024). Comparison of aroma compounds in Baijiu using GC-O-MS and OAV/PLS methods. *Foods, 13*(5), 681.

Yoshizaki, Y. (2010). Characteristics of aroma compounds generated by Aspergillus oryzae in rice fermentation. *Journal of Bioscience and Bioengineering, 110*(2), 152–157. https://doi.org/10.1016/j.jbiosc.2010.02.013

Zakhari, S. (2006). Overview: How is alcohol metabolized by the body? *Alcohol Research & Health, 29*(4), 245–254.

Zhang, H., Liu, X., Zhang, L., & Guo, X. (2021). Chemical profiling of head and tail fractions in Baijiu distillation. *Journal of the Institute of Brewing, 127*(2), 197–208. https://doi.org/10.1002/jib.649

Zhang, J. (2022). Key aroma compounds in melon spirits revealed by sensomics and recombination. *Journal of Food Science, 87*(10), 4424–4436. https://doi.org/10.1111/1750-3841.16234

Zheng, X.-W., & Han, B.-Z. (2016). Baijiu, Chinese liquor: History, classification and manufacture. *Journal of Ethnic Foods, 3*(1), 19–25. https://doi.org/10.1016/j.jef.2016.03.001

Zheng, Y. (2024). Identification of key odorants in goji wines via sensomics and recombination. *Food Chemistry, 441*, 137160. https://doi.org/10.1016/j.foodchem.2024.137160

Zucco, G. M., Paolini, M., & Schaal, B. (2011). Olfactory memory: The role of verbalization in odor naming and recognition. *Memory & Cognition, 39*(3), 451–461. https://doi.org/10.3758/s13421-010-0043-5

Glossary

Acuity (sensory); Fineness of sensory perception; ability to discriminate small differences.

Adaptation; Reduced sensitivity to a stimulus after continuous or repeated exposure.

Affective Testing; Sensory testing method focused on preference or liking rather than description.

Aftertaste; Lingering flavor perception after swallowing.

Aggressiveness (spirit); Subjective impression of harsh ethanol or intense flavor attack.

Agrochemical Signature; Composite chemical fingerprint in raw materials resulting from agricultural practices—such as pesticide or fertilizer use, soil amendments, or environmental exposures—that may carry through fermentation and distillation to influence a spirit's minor volatile profile.

Alcohol Acetyltransferase 1 (ATF1); Yeast enzyme catalyzing acetate-ester formation from higher alcohols and acetyl-CoA during fermentation, shaping fruity aroma.

Alcohol Acetyltransferase 2 (ATF2); Related yeast enzyme with overlapping but distinct substrate specificity; relative expression of ATF1 / ATF2 governs ester spectrum.

Alcohol Tolerance Threshold (ATT); Ethanol concentration at which trigeminal irritation or cognitive impairment begins.

Amino Acid Catabolism; Yeast breakdown of amino acids yielding higher alcohols, aldehydes, and acids—precursors of esters and fusel oils.

Amphipathic; Molecule containing both hydrophilic and hydrophobic regions, critical in solubility and membrane interaction.

Anosmia; Loss or absence of smell perception from temporary blockage or receptor damage.

Aroma Lexicon; Structured, standardized vocabulary for describing spirit aromas.

Aromatic Complexity; Layered, multidimensional perception of multiple volatiles.

Artisanal Distilling; Small-scale production emphasizing craft, locality, and heritage.

Astringency; Dry, puckering mouthfeel from tannin binding with salivary proteins.

Attention (sensory focus); Directed cognitive effort toward a specific sensory attribute during evaluation.

Azeotrope; Mixture whose vapor and liquid compositions are identical; cannot be separated by simple distillation (ethanol–water 95.6 %).

Balance; Harmony among aroma, taste, and mouthfeel components.

Barrel Influence; Aggregate impact of oak maturation—compound extraction, oxidation, and evaporation—on spirit character.

Bias (panel evaluation); Systematic deviation in results from expectation, suggestion, or predisposition.

Blind Testing; Evaluation conducted without revealing identity to minimize bias.

Body; Sense of weight or fullness on the palate.

Bouquet; Combined aromatic impression of a spirit; legacy wine term.

Burn; Trigeminal heat sensation from ethanol or pungent volatiles.

Calibration; Alignment of panelists' perceptions using reference standards.

Carryover; Residual sensory impression from one sample affecting the next.

Cask Strength; Spirit bottled at barrel proof, generally > 55 % ABV.

Central Nervous System (CNS); Brain and spinal cord integrating sensory input and response.

Chemesthesis; Chemical somatosensory sensations—burning, cooling, tingling—mediated by trigeminal pathways.

Clean Finish; Aftertaste free of off-notes or residual harshness.

Cognitive Expectation; Perceptual bias caused by prior knowledge, labeling, or assumption.

Complexity; Multidimensionality of perceived aromas and flavors.

Congener; Any compound other than ethanol + water contributing to aroma, flavor, or mouthfeel.

Consensus; Agreement among panelists after structured discussion.

Context Effect; Influence of surrounding samples or conditions on judgment.

Craft Tradition; Cultural and historical small-batch practices defining flavor identity.

Crispness; Sharp, clean sensory impression often tied to acidity or freshness.

Crossmodal Effects; Interaction among senses, e.g., smell influencing taste.

Cultural Symbolism; Meanings and associations of spirits in ritual or identity.

Decarboxylation; Chemical reaction removing a carboxyl group (–COOH) from an organic molecule and releasing CO_2; in fermentation and maturation, converts acids to aldehydes, alcohols, or esters.

Degrees Plato; Brewing scale expressing wort sugar concentration (% by weight) to estimate potential alcohol yield.

Descriptor; Word or phrase characterizing a sensory attribute.

Descriptive Analysis; Sensory method in which trained panelists generate and quantify product attributes.

Detection Threshold; Lowest concentration of a substance that can be perceived.

Diffusion Gradient: Rate and direction of volatile molecule movement from the liquid surface into the **headspace,** governed by partial-pressure differences.

Dilution; Reduction of ethanol concentration, altering volatility and perception balance.

Distillation Cut; Selection of heads, hearts, and tails fractions determining spirit quality.

Dryness; Absence of sweetness or presence of puckering, tannic impression.

Duration; Length of time an aroma or flavor persists after exposure.

Effect Size; Magnitude of a true sensory difference between products; larger values yield higher detection probability.

Ehrlich Pathway; Yeast metabolic route converting amino acids to higher alcohols and flavor volatiles.

Elegance; Balance and refinement in aroma or flavor expression.

Epithelium / Cribriform Plate; Olfactory epithelium housing receptor neurons; axons pass through the cribriform plate to the olfactory bulb.

Ester Biosynthesis; Enzymatic formation of esters from alcohols + acyl-CoA derivatives during fermentation.

Ethanol; Primary psychoactive alcohol in spirits; influences aroma release, mouthfeel, and trigeminal burn.

Ethanol Partition Ratio; Proportion of ethanol vapor concentration at headspace relative to liquid phase at equilibrium; governs sensory intensity.

Evaluation Protocol; Structured procedure for conducting sensory analysis.

Expectation Bias; Influence of prior belief on perception.

Expert Panel: A highly trained group providing authoritative sensory judgments.

Fault; Undesirable sensory attribute signaling contamination, faulty fermentation, or poor distillation.

Feints; Late distillation fraction (tails) containing heavier, often undesirable volatiles; may be recycled or discarded.

Fermentation; Biochemical conversion of sugars by yeast, producing ethanol, CO_2, and flavor compounds.

Finish; Aftertaste and its persistence following swallowing.

Flabby; Lacking structure, liveliness, or balance.

Flavor; Integrated perception of aroma, taste, and trigeminal sensation.

Flavor Wheel; Visual reference organizing sensory descriptors into hierarchical categories.

Foreshots; Initial distillation fraction of low-boiling volatiles (e.g., methanol, acetaldehyde) is mostly discarded.

Free Amino Nitrogen (FAN); Concentration of amino acids + small peptides in wort/wash available to yeast; affects fermentation kinetics and flavor formation.

Free-Choice Profiling; Sensory method permitting panelists to use self-generated descriptors.

Fullness; Richness and completeness of mouthfeel and flavor.

Gas Chromatography–Mass Spectrometry (GC–MS); Analytical technique combining compound separation with mass-based identification of volatiles.

Glycolysis; Central pathway converting glucose to pyruvate, yielding ATP and NADH and precursors for ethanol fermentation.

Grainy; Raw cereal or husk note typical of immature distillate.

Guided Evaluation; Structured training using reference standards and controlled conditions.

Gustatory; Pertaining to taste reception via tongue and palate receptors.

Gustatory Receptor Cell; Specialized cell in taste buds responsive to one of the five primary tastes.

Harshness; Excessive ethanol burn or aggressive flavor impact.

Headspace; Gaseous volume above the liquid in a container or glass where volatiles accumulate before sampling.

Hearts; Central distillation fraction rich in ethanol and desirable flavor compounds; forms the core of the spirit.

Hedonic Scale; Rating scale measuring liking or preference rather than descriptive quality.

Henry's Law Constant; Proportionality constant describing volatile solubility and vapor pressure equilibrium between liquid and gas phases.

Heritage Expression; Spirit tied to a longstanding cultural or geographic identity.

Hydrogen Bond; Weak electrostatic attraction between hydrogen covalently bound to an electronegative atom and another electronegative atom; crucial in ethanol–water structure.

Hydrophilic; Attracted to and soluble in water due to polarity or charge.

Hydrophobic; Repelled by water; non-polar, influencing solubility and partitioning of aroma compounds.

Hypogeusia; Reduced sensitivity to taste stimuli.

Identification Threshold; Concentration at which a stimulus can be recognized and correctly named.

Index of Refraction; Optical property of ethanol–water mixtures affecting light transmission.

Intensity Scaling; Quantitative rating of stimulus magnitude on a defined numerical scale.

Iterative Training; Repeated cycles of evaluation and feedback to improve sensory acuity.

Lactones; Cyclic esters, mainly from oak, imparting coconut, woody, or creamy notes.

Laminar Layer Stability; Condition of undisturbed vapor strata above a liquid where molecular diffusion rather than turbulence governs odorant distribution.

Language of Sensory; Structured vocabulary used to communicate sensory perceptions.

Linalool; Terpene alcohol common in botanicals and oak contributing floral–citrus aroma.

Lock-and-Key Theory; Model describing specific geometric fit between receptor and ligand or enzyme and substrate.

Maillard Reaction; Non-enzymatic browning reaction between amino acids and reducing sugars producing color and flavor compounds during heating of malt or from oak.

Malting; Controlled germination and drying of grains to activate enzymes that convert starch to fermentable sugars.

Maturation; Transformation of a spirit during cask storage including extraction, oxidation, esterification, and evaporation.

Mellow; Smooth integration of ethanol and flavor components yielding soft balance.

Memory (sensory); Stored sensory experience guiding recognition and description.

Microheterogeneity: Small, short-lived clusters of molecules within a liquid—such as ethanol-rich and water-rich micro-domains—that make an otherwise uniform mixture behave unevenly at tiny scales.

Mixture Suppression; Interaction where one stimulus reduces perception of another in aroma mixtures.

Mouthfeel; Tactile and trigeminal sensations perceived in the mouth.

N-Methyl-D-Aspartate Receptor (NMDA); Ionotropic glutamate receptor critical to synaptic plasticity and sensory learning; modulated by ethanol.

Neophyte; Beginner in sensory training or evaluation.

Neurophysiology of Olfaction; Study of neural mechanisms underlying smell perception.

Nociceptor; Sensory receptor responding to noxious chemical, thermal, or mechanical stimuli; mediates pain or burning sensations.

Nose; Orthonasal aroma profile perceived before tasting.

Nosefeel; Trigeminal sensations within the nasal cavity such as burn, prickle, or cooling.

Notch Filtering (sensory); Cognitive suppression of irrelevant sensory input to focus on target attributes.

Nutritional Signature; Distinct chemical imprint left by nutrient composition of raw materials.

Odorant; Volatile chemical compound capable of stimulating olfactory receptors.

Olfaction; Sense of smell; detection of odorants by receptor neurons.

Olfactory Bulb; Brain structure processing input from olfactory receptor neurons.

Olfactory Receptor Neurons (ORNs); Specialized nasal cells that detect odorants and transmit signals to the brain.

Olfactory Threshold; Lowest concentration of an odorant that can be perceived.

Organoleptic; Relating to food or drink qualities as perceived by the senses.

Orthonasal Olfaction; Detection of volatiles through the nostrils before ingestion.

Overproof; Spirit bottled above standard strength, typically above 57 % ABV.

Oxidative Maturation; Slow reaction of ethanol and oxygen-derived radicals in barrel aging producing aldehydes, acids, and esters that round flavor.

Palate; Flavor profile perceived when spirit is in the mouth including gustatory, trigeminal, and retronasal components.

Panel; Group of trained evaluators conducting structured sensory tests.

Panel Calibration; Process of standardizing judgments across panelists to reduce variance.

Panelist Training; Structured instruction improving consistency and accuracy of evaluators.

Partial Pressure; Individual gas pressure in a mixture; governs volatile release and headspace composition.

Perception Bias; Systematic distortion of sensory judgment caused by expectation or context.

Peppery; Spicy, tingling impression often related to ethanol or trigeminal activation.

Phenols; Aromatic compounds contributing smoky, medicinal, or tar-like aromas.

Principal Component Analysis (PCA): a Multivariate statistical method that reduces descriptor data to visualize product differences in sensory space.

Prickliness; Tingling or sharp trigeminal sensation from carbonation or ethanol.

Provenance (cultural); Documented origin or heritage influencing spirit identity and authenticity.

Psychophysics; Quantitative study of relationships between physical stimuli and sensory response.

Pyruvate; Central metabolic intermediate linking sugar metabolism to ethanol or citric-acid-cycle pathways.

Raw Material Provenance (RMP); Geographic and agricultural source information of fermentation substrates.

Recognition Threshold; Concentration at which a stimulus can be identified by name.

Refinement; Impression of polish, balance, and precision in a spirit's sensory profile.

Reliability; Consistency of sensory results across replicates or evaluators.

Retronasal Olfaction; Perception of aromas traveling from oral to nasal cavity during swallowing or exhalation.

Ritual Use; Cultural or ceremonial consumption of spirits tied to tradition or belief.

Roundness; Sense of harmony and completeness in overall flavor impression.

Sample Size; Number of panelists or replicates in a sensory test; affects statistical power and reliability.

Sensory Adaptation Rate; Temporal decline in receptor sensitivity with sustained exposure; varies among volatiles and evaluators.

Sensory Evaluation; Systematic assessment of products using trained human perception under controlled conditions.

Sensory Fatigue; Temporary decline in sensitivity from repeated or intense exposure to stimuli.

Sensory-designed glass (vessel); any vessel engineered to minimize ethanol interference through passive diffusion.

Sensory Memory; Short-term retention of sensory experiences used to compare samples.

Sensory Threshold; Minimum level at which a stimulus can be perceived.

Signal-to-Noise Ratio; The Degree to which a sensory signal stands out from background variability.

Significance Level (α); Probability cutoff for concluding a sensory difference is real (commonly 0.05).

Smoothness; Soft, integrated mouthfeel with minimal ethanol harshness.

Social Influence; Effect of authority or peer context on sensory judgments.

Solubility Coefficient; Quantitative measure of a compound's ability to dissolve in ethanol–water mixtures; key to volatility.

Statistical Power; Probability that a sensory test correctly detects a true difference when one exists.

Structure; Framework of flavors providing balance and coherence.

Sulfur Compounds; Volatile molecules producing aromas like cabbage, onion, or rubber; often faults.

Symbolic Value; Non-sensory meaning of spirits tied to culture or status.

Tannic; Drying, puckering mouthfeel caused by tannins or wood extracts binding salivary proteins.

Taste bud: A Cluster of gustatory cells within papillae on the tongue that detects the five primary tastes.

TCATA; Temporal Check-All-That-Apply, A dynamic sensory profiling technique in which assessors continuously monitor and record all sensory attributes that apply to a sample as they perceive them over a defined time period.

Temporal Dominance of Sensations (TDS); Method tracking which sensation dominates perception over time during tasting.

Terpenes; Plant-derived hydrocarbons imparting citrus, floral, or pine aromas.

Threshold; Lowest stimulus level detectable or recognizable by the senses.

Titratable Acidity; Total acid content measured by titration; indicates potential sensory sharpness.

Training: instruction that develops evaluator skills and consistency through repetition and feedback.

Tradition (spirits); Historical production methods and cultural practices shaping flavor identity.

Trigeminal Sensation; Chemesthetic responses such as burn, prickle, or cooling mediated by trigeminal nerve endings.

Turbulent Mixing; Chaotic airflow disrupting laminar vapor strata in glass headspace; increases ethanol dominance.

Typicity; Degree to which a spirit conforms to its expected regional or category profile.

Type I Error (False Positive); Incorrect conclusion that a sensory difference exists when it does not.

Type II Error (False Negative); Failure to detect a true sensory difference that actually exists.

Type III Error; Correctly detecting a difference but misidentifying its cause or direction.

Umami; Savory or brothy basic taste associated with glutamates.

Utility Glass; Versatile but nonspecialized glassware sacrificing sensory optimization.

Validity; Extent to which a sensory test measures what it is intended to measure.

Variability; Differences in perception among individuals or sessions.

Vapor Stratification; Layered concentration gradient of volatiles above a liquid surface caused by molecular weight differences and diffusion rates.

Viscosity; Physical thickness or perceived weight of a spirit on the palate.

Volatile; Compound that vaporizes easily and contributes to aroma.

Volatile Organic Compounds (VOCs); Organic molecules with sufficient vapor pressure to enter gas phase and be perceived as aroma; includes esters, aldehydes, terpenes, acids, and phenols.

Weber–Fechner Law; Psychophysical relationship showing that perceived intensity grows logarithmically with stimulus magnitude.

Yeast Autolysis; Breakdown of yeast cells after fermentation releasing amino acids, lipids, and enzymes that affect maturation.

Zymurgy; Applied science of fermentation as it pertains to brewing and distilling.

Index

Epilogue—The Future of Measurable Truth

Sensory science remains a relatively young discipline in the context of distilled spirits. Chemistry, physics, and biology have advanced far beyond the poorly descriptive vocabulary and casual traditions that dominate spirits industry practice. Nevertheless, every scientific field begins at the intersection of curiosity and measurement. Once observation becomes quantifiable, truth gains permanence. The same must now occur for the evaluation of spirits.

For centuries, tasting was considered a private act of judgment. Perception was assumed to be too subjective for scientific organization. Distillers worked by intuition, judges by habit, and critics by metaphor. The result has been progress in technology but stagnation in sensory understanding, where opinion continues to outweigh evidence. The future of this discipline depends on breaking that pattern: replacing the vocabulary of persuasion with the vocabulary of mechanism, and replacing the authority of tradition with the evidence of data.

This book has shown that the tools for transition already exist. The sensory pathway, from molecule to neuron to language, can be mapped, measured, and repeated. Glass geometry can be optimized to control ethanol dispersion. Aroma lexicons can be tied to compound families and concentration thresholds. Even panel variability can be bounded by psychophysical principles. When perception is reproducible, learning becomes cumulative; when measurement is transparent, expertise becomes transferable. The next generation of evaluators must inherit a system, not a ceremony, ritual, or the careless, "We have always done it that way."

However, reproducibility is not the enemy of artistry. Understanding why aromas behave as they do gives the distiller greater creative freedom, not less. When a process parameter is known to influence ester distribution, or a maturation condition alters aldehyde oxidation, intent becomes measurable. Craft is elevated when its outcomes can be predicted, verified, and shared. The modern distiller stands not at the edge of tradition, but at the frontier of controllable complexity.

In the decades ahead, scientific literacy will define credibility in the spirits world. The language of evaluation will continue to move from hedonic metaphor to physical reference—expressing not what something seems to be, but what it *is*. Competitions, educators, and producers will increasingly adopt standardized sensory protocols rooted in human physiology and environmental control. Judges will calibrate to chemical anchors instead of brand exemplars. Students will learn sensory methodology before rhetoric. Moreover, consumers, empowered by better information, will recognize that quality experience is not mystical but measurable.

Technology will accelerate this transformation. Portable spectrometry, real-time headspace sensors, and AI-assisted pattern recognition will support evaluators by linking sensory impressions to quantifiable profiles. Yet instruments alone cannot define flavor—only humans can assign value to experience. The essential partnership between instrument and evaluator will remain: machines can detect, but only trained perception can interpret. The mission of sensory science is to make that interpretation reliable.

True progress, however, requires more than new tools. It requires the will to abandon language that flatters bias and obscures evidence. The industry must stop equating harshness with strength, age with quality, or burn with authenticity. These are residues of marketing, not merit. To teach sensory literacy is to teach discernment; the ability to separate chemistry from folklore and perception from assumption. When language reflects mechanism, communication becomes universal.

The ultimate goal is integration. Distillers, scientists, and educators must no longer occupy separate territories. Each discipline contributes to the same continuum: the creation and perception of volatile compounds. A chemist without sensory literacy lacks context; an evaluator without chemical knowledge lacks foundation. The bridge between them is the measurable truth that links molecular causes to human experience and defines the future of this field.

Education is the vehicle. Every tasting session, every judging event, every distillery lab is an opportunity to model precision. Students who learn to describe aromas by functional category—esters, acids, phenols—learn to think scientifically about flavor. In doing so, they not only advance their personal skills but also enhance their professional credibility. The success of this book's mission will be a refined vocabulary, standardized protocols, and sharpened minds to see what was always present but poorly observed.

Scientific literacy deepens enjoyment. Knowing *why* a spirit behaves as it does not strip away wonder—it replaces mystery with mastery. The discovery that ethanol vapor masks sweetness or that glass design alters diffusion does not lessen appreciation; it transforms it into informed respect. The pleasure of drinking well is inseparable from the pleasure of knowing why it is good.

The future of measurable truth in spirits will belong to those who see observation as both art and obligation—to those who understand that precision is not pedantry but respect for reality. Every sensory decision, every glass design, and every evaluation sheet filled out with care contribute to a discipline that may one day rival the rigor of enology or perfumery. When that day arrives, tasting will no longer be ritual—it will be science practiced with artful intention.— *George F. Manska*

About the Author

George F Manska is Chief of Research & Development at Arsilica, Inc., a company dedicated to advancing the sensory science of alcoholic beverages, and co-inventor of the NEAT glass. Drawing on his background in engineering, product design, mathematics, and chemistry, Manska has developed patented glassware and sensory methodologies that are reshaping professional judging standards worldwide. His work unites physical modeling, vapor-phase chemistry, and psychophysical analysis to define how ethanol concentration and aroma distribution influence perception.

Over 15 years as a consultant to major spirits competitions, he has analyzed judges' behavior, judging processes, scoring, ratings, and administrative practices, and has identified areas for improving accuracy and truthfulness, which have provided the basis for this book.

A member of ASTM International Committee E18 on Sensory Evaluation and the Society of Sensory Professionals, Manska contributes to the development of global standards for sensory testing and measurement reliability. He frequently serves as a guest lecturer at local colleges and universities, delivers educational programs, organizes spirits tastings focused on sensory accuracy and evaluation discipline, and consults for numerous competitions.

Manska is an active member of the Advisory Board of the College of Sciences at the University of Nevada, Las Vegas, Society of Sensory Professionals, Wine and Spirits Wholesale Association, and U. S. Bartenders' Guild. He also writes a sensory science column for PR%F Magazine and contributes articles on distilled spirits sensory science to various magazines and periodicals.

His research encompasses the sensory analysis of spirits, beer, and wine, yielding multiple patents for scientifically engineered drinking vessels. Ongoing studies extend this work to the development of wine glasses designed using sensory-science principles, complementing sensory-designed craft beer glasses to complete a continuum of functionally optimized vessels across beverage categories.

Manska's mission is threefold: (1) to connect the art of creating beverages to the current level and tools of sensory science with a bridge of practical application—uniting educators, researchers, and producers around measurable truth in flavor and aroma, (2) to bring industry into the sensory education process through truth in marketing, and (3) to provide the student with a clear, practical understanding of the disciplines needed to excel in the field of spirits' sensory science.

More Quotable Quotes on Distilling

"Gunpowder makes an honest gauge when men do not."—James B. Lockhart, Excise Officer, validating the sailor's proof test (London, 1824)

"Yeast never improvises—it follows enzymatic law. Change the stress, and it will change its esters accordingly."—Paraphrased from K Verstrepen (KU Leuven, 2017 lecture)

"You can't blend away a fault—you can only dilute it to statistical insignificance."—Dr. Don Livermore (Hiram Walker)

"A cask can improve what's good, but it cannot forgive what's bad."— Dr. James Swan

"Flavor architecture begins in yeast, not in oak. Wood only edits the manuscript."—Chris Morris (Woodford Reserve)

"Congeners are chemistry's fingerprints—each one identifies a choice made upstream."— Dr. Matthew Crow (Diageo R&D)

"Whisky making is not imitation but interpretation—fermentation and distillation are the vocabulary."—Masataka Taketsuru (Nikka)

"Consistency is the hardest flavor to make."—Harlan Wheatley (Buffalo Trace)

"Tropical maturation gives speed but not immunity; balance must still be earned."—Ian Chang (Kavalan)

"If you can smell the barrel before the whiskey, you've waited too long."—Jimmy Russel (Wild Turkey)

"Every distillery has its own music; the yeast keeps the rhythm, the stills set the pitch."—Jim Rutledge (Four Roses)

"You can't distill out bad farming."— Dave Pickerell (Maker's Mark, Whistle Pig)

"Fermentation is the heartbeat of whisky; change its rhythm, and the spirit's voice changes forever."—Jim McEwan (Bruichladdich)

"Even in rum, mash defines character—the molasses is the melody, but the mash sets the rhythm." —Joy Spence (Appleton Estate)

"If you can't sleep good at night, you're probably makin' the wrong liquor." — Marvin "Popcorn" Sutton

These quotations are attributed to those who have lived and worked in the world of spirits day after day. Distillation is both art and science. Sensory science assesses the validity of intuition and preserves its wisdom through quantified evidence. Daring to be different and experimenting to discover reveals the true artist and visionary.

www.ingramcontent.com/pod-product-compliance
Lightning Source LLC
Chambersburg PA
CBHW081800200326
41597CB00023B/4097